Ecological Studies

Analysis and Synthesis

Volume 233

Ecological Studies is Springer's premier book series treating all aspects of ecology. These volumes, either authored or edited collections, appear several times each year. They are intended to analyze and synthesize our understanding of natural and managed ecosystems and their constituent organisms and resources at different scales from the biosphere to communities, populations, individual organisms and molecular interactions. Many volumes constitute case studies illustrating and synthesizing ecological principles for an intended audience of scientists, students, environmental managers and policy experts. Recent volumes address biodiversity, global change, landscape ecology, air pollution, ecosystem analysis, microbial ecology, ecophysiology and molecular ecology.

More information about this series at http://www.springer.com/series/86

Madeleine J. H. van Oppen • Janice M. Lough
Editors

Coral Bleaching

Patterns, Processes, Causes and Consequences

Second Edition

Springer

Editors
Madeleine J. H. van Oppen
Australian Institute of Marine Science
Townsville, Queensland, Australia

School of BioSciences
The University of Melbourne
Parkville, Victoria, Australia

Janice M. Lough
Australian Institute of Marine Science
Townsville, Queensland, Australia

ISSN 0070-8356 ISSN 2196-971X (electronic)
Ecological Studies
ISBN 978-3-030-09219-1 ISBN 978-3-319-75393-5 (eBook)
https://doi.org/10.1007/978-3-319-75393-5

Printed on acid-free paper

This Springer imprint is published by the registered company Springer International Publishing AG part of Springer Nature.
The registered company address is: Gewerbestrasse 11, 6330 Cham, Switzerland

Acknowledgements

The editors thank all the contributing authors for their time, effort and patience in the preparation of this second edition. Chapters were reviewed by at least one specialist in the field as well as the editors, and we acknowledge the time and expertise of our reviewers: Barbara Brown, Katharina Fabricius, James Guest, John Hedley, Phil Munday, David Obura, Tom Oliver, Mauricio Rodriguez-Lanetty, Michael Stat, Jaroslaw Stolarski, David Suggett, Ruben van Hooidonk, Rob van Woesik, Bernado Vargas-Angel and Rebecca Vega-Thruba.

The editors acknowledge the support of the Australian Institute of Marine Science and the University of Melbourne. We also thank Andrea Schlitzberger of Springer for her advice and assistance with this volume.

Contents

Chapter 1
Introduction: Coral Bleaching–Patterns, Processes, Causes and Consequences

J. M. Lough and M. J. H. van Oppen

Since the first edition of this book (van Oppen and Lough 2009), the global climate system has continued to change (Stocker et al. 2013; Blunden and Arndt 2017). Global land and sea temperatures (HadCRUT4; https://crudata.uea.ac.uk/cru/data/temperature/; Morice et al. 2012) had warmed by +0.79 °C between 1880 and 2009. Warming through 2016 was +0.90 °C—an additional warming of +0.11 °C. 2017 was the third warmest year in the instrumental record period since 1880 after 2016 and 2015 (https://www.ncdc.noaa.gov/sotc/global/201710). In 2009 the global atmospheric concentration of carbon dioxide, the main greenhouse gas, as measured at Mauna Loa, Hawaii (www.esrl.noaa.gov/gmd/ccgg/trends/data.html) was 387.4 ppm. By 2017 the concentration was 406.5 ppm, an increase of 19 ppm and a 45% increase above pre-industrial levels of 280 ppm (WMO 2017). As the global climate system responds to increasing levels of atmospheric greenhouse gases, the tropical oceans are warming at ~70% of the global average rate (Lough 2012), and this is bad news for tropical coral reef ecosystems.

At the heart of these complex ecosystems is an obligate symbiosis between the coral animal and single-celled photosynthetic algae (*Symbiodinium* spp. aka zooxanthellae) living in the coral tissue. Photosynthetic products provide the coral host with cheap energy. The zooxanthellae also play a role in light-enhanced calcification of scleractinian corals (Barnes and Chalker 1990), allowing the rapid calcification necessary to form reef structures. In return the algae obtain protection and essential

J. M. Lough (✉)
Australian Institute of Marine Science, Townsville MC, QLD, Australia

ARC Centre of Excellence for Coral Reef Studies, James Cook University, Townsville, QLD, Australia
e-mail: j.lough@aims.gov.au

M. J. H. van Oppen
Australian Institute of Marine Science, Townsville MC, QLD, Australia

School of BioSciences, The University of Melbourne, Parkville, Victoria, Australia

M. J. H. van Oppen, J. M. Lough (eds.), *Coral Bleaching*, Ecological Studies 233,
https://doi.org/10.1007/978-3-319-75393-5_1

nutrients (e.g. nitrogen, inorganic carbon; Davies 1984) from their coral host. The photosynthetic pigments within the algae give the corals their deep brown colour.

Coral bleaching is the term used to describe the loss by the coral animal of all or some of their symbiotic algae and photosynthetic pigments—with the result that the white calcium carbonate skeleton becomes visible through the now translucent tissue layer. Coral bleaching is not a new phenomenon. Corals are known to bleach in response to a range of environmental stresses (e.g. low salinity, pollution, unusually high or low water temperatures). In the past, however, such occurrences of bleaching were only observed on small spatial scales in response to localized stresses. What is new since the 1980s (e.g. Glynn 1983), and now clearly related to human-induced global warming, is an increase in frequency and extent of large-scale, mass coral bleaching events where entire reef systems are affected. At the time of the first edition of this book, mass bleaching on nearly every reef system in 1997–1998 had been a wake-up call for reef scientists regarding the sensitivity and vulnerability of these ecosystems to a warming climate (Wilkinson 1998). Since then, regional-scale bleaching events have affected many coral reef ecosystems, such as the Great Barrier Reef, Australia in 2002 (Berkelmans et al. 2004), the Caribbean in 2005 (Eakin et al. 2010) and Western Australian reefs in 2011 (Moore et al. 2012). Significantly, global-scale events assaulting many of the world's reefs as happened in 1997–1998 occurred again in 2010 and from 2014 to 2017 (Eakin et al. 2017; Hughes et al. 2018). Also, some locations such as Hawaii and the Great Barrier Reef experienced consecutive years of bleaching (Eakin et al. 2017)—an occurrence foreshadowed by Hoegh-Guldberg (1999).

The phenomenon of coral bleaching is clearly here to stay and inextricably linked to global climate change due to human activities. The first (1990), second (1995), third (2001) and fourth (Solomon et al. 2007; Parry et al. 2007) Assessment Reports of the Intergovernmental Panel on Climate Change provided mounting evidence of a changing world climate with the IPCC-AR4 (IPCC 2007) concluding that the evidence for a human influence on global climate was considered *very likely*. The most recent report IPCC-AR5 concluded that *the human influence on the climate system is now clear* (IPCC 2013). We have entered a new era for the world and tropical coral reef ecosystems which were identified as *unique and threatened ecosystems* as early as IPCC-AR3 (McCarthy et al. 2001) and reaffirmed in the most recent assessment (IPCC 2014). Human activities since the late eighteenth century have, unwittingly, led to already observable biological responses (coral bleaching) on one of the world's most charismatic ecosystems—coral reefs. This is in addition to the coral reef crisis, where direct local and regional human pressures on coral reef environments (such as overfishing, destructive fishing, decline in water quality due to land-use changes, nutrient and chemical pollution and development on coasts, mining of coral, etc.) have caused declines in the health of many of the world's coral reef ecosystems (e.g. Wilkinson and Buddemeier 1994; Hughes et al. 2003, 2017a, b; Buddemeier et al. 2004; Veron et al. 2009). Human activities, including climate change, have had such profound impacts on planet Earth that a new geological period, the Anthropocene, has been proposed (Zalasiewicz et al. 2017). The suggested start of this new epoch is the mid-twentieth century, when a range of socioeconomic and earth system indicators increased rapidly, known as the *Great Acceleration* (Stefan et al. 2015). The challenge is to enable coral reefs to survive this new environment, though recognizing that reefs of the future are unlikely to be the same as those of the past (Hughes et al. 2017b).

The Paris Agreement (UNFCC 2015) has provided some cause for optimism, albeit over 25 years after the majority of countries agreed, under Article 2 of the 1992 UN Framework Convention on Climate Change (http://www.globelaw.com/Climate/fcc.htm), to *stabilization of greenhouse gas concentrations in the atmosphere at a level that would prevent dangerous anthropogenic interference with the climate system. Such a level should be achieved within a time frame sufficient to allow ecosystems to adapt naturally to climate change*. The historic Paris Agreement (currently signed by 195 members of the United Nations Framework Convention on Climate Change, unfccc.int/paris_agreement/items/9444.php) committed the world's countries to keeping the increase in global average temperatures above pre-industrial levels to *well below 2 °C* whilst *pursuing efforts to limit the temperature increase to 1.5 °C* (UNFCC 2015). It should be recognized, however, that these limits include the global warming of nearly 1.0 °C that has already occurred since pre-industrial times. Even the 1.5 °C target is likely to result in continued bleaching impacts on the majority of the world's coral reefs (Frieler et al. 2012; Gattuso et al. 2015). Time is, however, rapidly running out for constraining the magnitude of global warming to the Paris targets of 2.0 °C or anything close to 1.5 °C given current rates of CO_2 emissions (Jackson et al. 2017), though rapid decarbonization of the world is still considered possible (Figueres et al. 2017; Millar et al. 2017). This will, however, require *accelerated short-term action and enhanced long-term ambition if the goals of the Paris Agreement are to remain achievable—and that practical and cost-effective options are available to make this possible* (UNEP 2017).

Tropical coral reefs are the most biologically diverse of marine ecosystems *equalling in beauty and excelling in grandeur the most favourite parterre of the curious florist* (Matthew Flinders, October 1802). They are complex ecosystems at all levels from their geological history, growth and structure, biological adaptation, evolution and biogeography, community structure, organisms and ecosystem metabolism and physical regimes. Coral reefs lay down enormous amounts of calcium carbonate to form massive reef structures that are able to withstand the forces of erosion and create diverse habitats for many organisms. Despite their relatively small area (estimated at only 0.1–0.5% of the ocean floor), coral reefs contain about one-third of the world's marine fish and reef fish account for ~10% of fish consumed by humans. Tens of millions of people in over 100 countries with coral reefs along their coastline depend on the economic and social goods and services provided by these rich ecosystems (Moberg and Folke 1999), valued at US\$375 $\times$ 10^{12}/year (Pandolfi et al. 2003).

Warming ocean temperatures in the vicinity of coral reefs are already having observable consequences for coral reef ecosystems. Other aspects of projected climate change will also impact coral reefs. Although in some cases less certain (see Hoegh-Guldberg et al. 2007; Lough 2008; Stocker et al. 2013) these include maybe fewer but more intense tropical cyclones which are a source of localized physical destruction on reefs (e.g. De'ath et al. 2012); changes in regional rainfall and river flow regimes with likely more extreme rainfall events that could affect the periodic extent of freshwater onto reefs (e.g. Lough et al. 2015); gradual rise in sea level that will affect light penetration and also the availability of suitable areas for corals to live; changes in large-scale and regional atmospheric (e.g. El Niño–Southern Oscillation events, Cai et al. 2015) and ocean circulation patterns that will affect connectivity between reefs; and changes in ocean chemistry (ocean acidification) due to about one-third of the excess atmospheric CO_2 being absorbed by the oceans, which can have substantial direct and indirect impacts on corals and reef-associated organisms (e.g. Fabricius et al. 2011). These rapid climate changes are occurring against a backdrop of near-worldwide reef degradation due to local human activities (Hughes et al. 2003; Kleypas and Eakin 2007). In an ideal world, these localized sources of stress to coral reefs should be minimized to enhance the resilience of these remarkable ecosystems to global climate change (Hughes et al. 2017b).

In this book we bring together available scientific information on coral bleaching at different space and time scales from the deep geological record through to future projections and from the cellular to ecosystem levels. The focus is on the many facets of the coral bleaching phenomenon, and this 2nd edition, with over 50% new references (2009–2017), illustrates how much more we have learned about coral bleaching since the original publication. All but one of the original chapters have been updated and three new chapters added. The geological history and evolution of the critical coral–algal symbiosis at the heart of coral reefs is introduced by Stanley and van de Schootbrugge (Chap. 2). Oliver et al. (Chap. 3) assess the quality of observations of coral bleaching events in space and time and how the reliability of such observations (in the absence of globally widespread and standardized observations) can compromise our ability to determine significant changes in the frequency and occurrence of coral bleaching events. The observational record of the physical environment of coral reefs (particularly sea surface temperatures, SST) is much better than the biological record of coral bleaching events. Eakin et al. (Chap. 4) demonstrate how tropical SST are warming, the links between unusual warming and ENSO events and the now sophisticated remote-sensing products that allow identification of oceanic 'hotspots' and conditions conducive to coral bleaching in near real time. The various tools available for detecting and observing coral bleaching are discussed by Cantin and Spalding (Chap. 5). These range from remote sensing to the detail necessary in the field extending from whole reefs to individual colonies and, most importantly, the necessity for follow-on surveys to determine the consequences of a coral bleaching event. The possible role that the now recognized diversity of algal symbionts play in conferring thermal resilience on corals is considered by Quigley et al. (Chap. 6). This also highlights the developing application of genetic analyses to determine algal symbiont diversity and their spatial patterns. A new chapter by Morrow et al. (Chap. 7) explores the role in coral bleaching of the many prokaryote symbionts which we now know are an important component of the 'coral microbiome'. Another new chapter (Chap. 8) by Oakley and Davy explores what we know about the molecular and cellular pathways that underpin the bleaching response of the coral host. Having undergone several bleaching events, is it possible that corals can increase their thermal tolerance? This is addressed by Berkelmans (Chap. 9) who also considers the relationship between thermal bleaching thresholds and the threshold that draws the line between coral recovery and mortality. McClanahan et al. (Chap. 8) consider the range of consequences of coral bleaching events for corals. They tease out the observed range of responses that vary between taxa and also through longer-term effects on reproduction, growth and the incidence of disease, etc. The complex structure of tropical coral reefs, built as a result of the coral–algal symbiosis, provides a habitat for many other motile reef organisms. Pratchett et al. (Chap. 10) assess the consequences of coral bleaching events for the most-studied motile reef organisms, fishes, and how these effects operate on both short and long time scales. Warming SST is just one aspect of human interference in the global climate system affecting coral reefs. An additional global-scale problem is acidification of the oceans as they absorb about 30% of the additional CO_2 that human activities have injected into the atmosphere—without this oceanic (and terrestrial) sink of CO_2, the amount of global warming to date would have been greater. In a new chapter, Albright (Chap. 12) reviews how ocean acidification and nutrients can influence bleaching responses and their interactions with rising water temperatures. Predicting what might happen to coral reefs in the future depends on understanding coral reef processes and reliably estimating how coral reef climates may change as global climate continues to warm. Donner et al. (Chap. 13) discuss how well current large-scale climate models can provide such information and the possible range of future climates for coral reefs. The findings of the various chapters are synthesized in Chap. 14, which also considers novel interventions that may assist tropical coral reef transition through the Anthropocene.

References

Barnes DJ, Chalker BE (1990) Calcification and photosynthesis in reef-building corals and algae. In: Dubinsky Z (ed) Coral reefs. Elsevier, Amsterdam, pp 109–131

Berkelmans R, De'ath G, Kininmonth S, Skirving W (2004) A comparison of the 1998 and 2002 coral bleaching events on the Great Barrier Reef: spatial correlation, patterns and predictions. Coral Reefs 23:74–83. https://doi.org/10.1007/s00338-003-0353-y

Blunden J, Arndt DS (eds) (2017) State of the climate in 2016. Bull Am Meteorol Soc 98:Si–S277. https://doi.org/10.1175/2017BAMSStateoftheClimate.1

Buddemeier RW, Kleypas JA, Aronson RB (2004) Coral reefs and global climate change. Potential contributions of climate change to stresses on coral reef ecosystem. Pew Center on Global Climate Change, Arlington

Cai W, Santoso A, Wang G, Yeh S-W, An S-I, Cobb KM et al (2015) ENSO and greenhouse warming. Nat Clim Chang 5:849–859. https://doi.org/10.1038/nclimate2743

Davies PS (1984) The role of zooxanthellae in the nutritional energy requirements of *Pocillopora eydouxi*. Coral Reefs 2:181–186. https://doi.org/10.1007/BF00263571

De'ath G, Fabricius KE, Sweatman H, Puotinen M (2012) The 27-year decline of coral cover on the Great Barrier Reef and its causes. Proc Natl Acad Sci U S A 109:17995–17999. https://doi.org/10.1073/pnas.1208909109

Eakin CM, Morgan JA, Heron SF, Smith TB, Liu G, Alvarez-Filip L et al (2010) Caribbean corals in crisis: record thermal stress, bleaching, and mortality in 2005. PLoS One 5. https://doi.org/10.1371/journal.pone.0013969

Eakin CM, Liu G, Gomez AM, De La Cour JL, Heron SF, Skirving WJ et al (2017) Ding, dong, the witch is dead (?) – three years of global coral bleaching 2014-2017. Reef Encounter 32:31–36

Fabricius KE, Langdon C, Uthicke S, Humphrey C, Noonan S, De'ath G et al (2011) Losers and winners in coral reefs acclimatized to elevated carbon dioxide concentrations. Nat Clim Chang 1:165–169. https://doi.org/10.1038/nclimate1122

Figueres C, Schnellnhuber HJ, Whiteman G, Rockström J, Hobley A, Rahmstorf S (2017) Three years to safeguard our climate. Nature 546:593–595. https://doi.org/10.1038/546593a

Frieler K, Meinshausen M, Golly A, Mengel M, Lebek K, Donner SD et al (2012) Limiting global warming to 2°C is unlikely to save most coral reefs. Nat Clim Chang 3:165. https://doi.org/10.1038/nclimate1674

Gattuso J-P, Magnan A, Billé R, Cheung WWL, Howes EL, Joos F et al (2015) Contrasting futures for ocean and society from different anthropogenic CO_2 emissions scenarios. Science 349: aac4722. https://doi.org/10.1126/science.aac4722

Glynn PW (1983) Extensive "bleaching" and death of reef corals on the Pacific coast of Panama. Environ Conserv 10:149–154. https://doi.org/10.1017/s0376892900012248

Hoegh-Guldberg O (1999) Climate change, coral bleaching and the future of the world's coral reefs. Mar Freshw Res 50:839–866. https://doi.org/10.1071/MF99078

Hoegh-Guldberg O, Mumby PJ, Hooten AJ, Steneck RS, Greenfield P, Gomez E et al (2007) Coral reefs under rapid climate change and ocean acidification. Science 318:1737–1742. https://doi.org/10.1126/science.1152509

Hughes TP, Baird AH, Bellwood DR, Card M, Connolly SE, Folke C et al (2003) Climate change, human impacts and the resilience of coral reefs. Science 301:929–933. https://doi.org/10.1126/science.1085046

Hughes TP, Kerry JT, Álvarez-Noriega M, Álvarez-Romero JG, Anderson KD, Baird AH et al (2017a) Global warming and recurrent mass bleaching of corals. Nature 543:373–377. https://doi.org/10.1038/nature21707

Hughes TP, Barnes ML, Bellwood DR, Cinner JE, Cumming GS, Jackson JBC et al (2017b) Coral reefs in the Anthropocene. Nature 546:82–90. https://doi.org/10.1038/nature22901

Hughes TP, Anderson KD, Connolly SR, Heron SF, Kerry JT, Lough JM et al (2018) Spatial and temporal patterns of mass bleaching of corals in the Anthropocene. Science 359:80–83. https://doi.org/10.1126/science.aan8048

IPCC (2007) Summary for Policymakers. In: Solomon SD, Qin D, Manning M, Chen Z, Marquis M, Averyt KB et al (eds) Climate change 2007: the physical science basis. Contribution of Working Group 1 to the fourth assessment report of the Intergovernmental Panel on Climate Change. Cambridge University Press, Cambridge

IPCC (2013) Summary for policymakers. In: Stocker TF, Qin D, Plattner G-K, Tignor M, Allen SK, Boschung J et al (eds) Climate change 2013: the physical science basis, Contribution of working group I to the fifth assessment report of the intergovernmental panel on climate change. Cambridge University Press, Cambridge

IPCC (2014) Summary for Policymakers. In: Field CB, Barros VR, Dokken DJ, Mach KJ, Mastrandrea MD, Bilir TE et al (eds) Climate change 2014: impacts, adaptation, and vulnerability. Part A: global and sectoral aspects. Contribution of Working Group II to the fifth assessment report of the Intergovernmental Panel on Climate Change. Cambridge University Press, Cambridge

Jackson RB, Le Quéré C, Andrew RM, Canadell JG, Peters GP, Roy J et al (2017) Warning signs for stabilizing global CO_2 emissions. Environ Res Lett 12(11):110202. https://doi.org/10.1088/1748-9326/aa9662

Kleypas JA, Eakin CM (2007) Scientists' perceptions of threats to coral reefs: results of a survey of coral reef researchers. Bull Mar Sci 80:419–436

Lough JM (2008) A changing climate for coral reefs. J Environ Monit 10:21–29. https://doi.org/10.1039/b714627m

Lough JM (2012) Small change, big difference: sea surface temperature distributions for tropical coral reef ecosystems, 1950-2011. J Geophys Res 117. https://doi.org/10.1029/2012JC008199

Lough JM, Lewis SE, Cantin NE (2015) Freshwater impacts in the central Great Barrier Reef: 1648-2011. Coral Reefs 34:739–751. https://doi.org/10.1007/s00338-015-1297-8

McCarthy JJ, Canziani OF, Leary NA, Dokken DJ, White KS (2001) Climate change 2001: impacts, adaptation, and vulnerability. Contribution of Working Group II to the Third Assessment Report of the Intergovernmental Panel on Climate Change. Cambridge University Press, Cambridge

Millar RJ, Fuglestvedt JS, Friedlingstein P, Rogelj J, Grubb MJ, Matthews HD et al (2017) Emission budgets and pathways consistent with limiting warming to 1.5°C. Nat Geosci 10:741–747. https://doi.org/10.1038/ngeo3031

Moberg F, Folke C (1999) Ecological goods and services of coral reef ecosystems. Ecol Econ 29:215–233. https://doi.org/10.1016/S0921-8009(99)00009-9

Moore JAY, Bellchambers LM, Depczynski MR, Evans RD, Evans SN, Field SN et al (2012) Unprecedented mass bleaching and loss of coral across 12° of latitude in 2010-11. PLoS One 7: e51807. https://doi.org/10.1371/journal.pone.0051807

Morice CP, Kennedy JJ, Rayner NA, Jones PD (2012) Quantifying uncertainties in global and regional temperature change using an ensemble of observational estimates: the HadCRUT4 dataset. J Geophys Res 117. https://doi.org/10.1029/2011JD017187

Pandolfi JM, Bradbury RH, Sala E, Hughes TP, Bjorndal KA, Cooke RG et al (2003) Global trajectories of the long-term decline of coral reef ecosystems. Science 301:955–958. https://doi.org/10.1126/science.1085706

Parry ML, Canziani OF, Palutikof JP, van der Linden PJ, Hanson CE (eds) (2007) Climate change 2007: impacts, adaptation and vulnerability. Contribution of Working Group II to the Fourth Assessment Report of the Intergovernmental Panel on Climate Change. Cambridge University Press, Cambridge

Solomon S, Qin D, Manning M, Chen Z, Marquis M, Averyt KB et al (eds) (2007) Climate change 2007: the physical science basis. Contribution of Working Group I to the Fourth Assessment Report of the Intergovernmental Panel on Climate Change. Cambridge University Press, Cambridge

Stefan W, Broadgate W, Deutsch L, Gaffney O, Ludwig C (2015) The trajectory of the Anthropocene: the great acceleration. Anthropol Rev 2:81–98. https://doi.org/10.1177/2053019614564785

Stocker TF, Qin D, Plattner G-K, Alexander LV, Allen SK, Bindoff NL et al (2013) Technical summary. In: Stocker TF, Qin D, Plattner G-K, Tignor M, Allen SK, Boschung J et al (eds) Climate change 2013: the physical science basis. Contribution of Working Group I to the Fifth Assessment Report of the Intergovernmental Panel on Climate Change. Cambridge University Press, Cambridge

UNEP (2017) The emissions gap report 2017. United Nations Environment Programme (UNEP), Nairobi

UNFCC (2015) The paris agreement. The conference of the parties. unfccc.int/resource/docs/2015/cop21/eng/l09r01.pdf

van Oppen MJH, Lough JM (2009) Coral bleaching – patterns, processes, causes and consequences. Springer Ecol Stud 205:178

Veron JEN, Hoegh-Guldberg O, Lenton TM, Lough JM, Obura DO, Pearce-Kelly P et al (2009) The coral reef crisis: the critical importance of <350 ppm CO_2. Mar Pollut Bull 58:1428–1436. https://doi.org/10.1016/j.marpolbul.2009.09.009

Wilkinson CR (ed) (1998) Status of coral reefs of the world: 1998. Global coral reef monitoring network. Australian Institute of Marine Science, Townsville

Wilkinson CR, Buddemeier RW (1994) Global climate change and coral reefs: implications for people and reefs. (Report of the UNEP-IOC-ASPEI-IUCN global task team on the implications of climate change on coral reefs) International Union for the Conservation of Nature and Natural Resources, Gland

WMO (2017) WMO greenhouse gas bulletin No. 13. public.wmo.int/en/resources/library/wmo-greenhouse-gas-bulletin

Zalasiewicz J, Waters CN, Summerhayes CP, Wolfe AP, Barnosky AD, Cearreta A et al (2017) The Working Group on the Anthropocene: summary of evidence and interim recommendations. Anthropocene 19:55–60. https://doi.org/10.1016/j.ancene.2017.09.001

Chapter 2
The Evolution of the Coral–Algal Symbiosis and Coral Bleaching in the Geologic Past

George Stanley and B. van de Schootbrugge

2.1 Introduction

Scleractinians extract calcium (Ca^{2+}) and carbonate ($CO_3{}^{2-}$) from seawater to construct aragonite skeletons, and they have come to dominate well-lit, shallow-water coral reefs, evolving into a diverse array of growth forms with complicated corallum morphologies. Dinoflagellate algae (zooxanthellae) invade the tissues of corals and a wide variety of other calcified and non-calcified marine invertebrates. These symbionts have developed abilities to avoid the host's immune system to develop close mutualistic associations with their hosts. Scleractinian reef corals offer a spectacular example of zooxanthellae symbiosis. Azooxanthellate corals without this symbiosis may be found in shallow-water reefs, but given adequate nutrients, they are best developed in cold- and deep-water settings. Most shallow-water reef corals have a successful partnership with these endosymbiotic dinoflagellates belonging to the genus *Symbiodinium*. Prospering in warm, well-lit, tropical to subtropical settings, the symbiosis is astonishingly successful, resulting in huge colonies and massive framework. Growth rates of corals on reefs far exceed the ability of physical erosion and boring organisms to break them down. Thanks to rapid reef growth conferred by the symbiotic relationship, they create spectacular ecosystems with calcified and non-calcified algae, other invertebrates, and fish. The symbiosis also provides efficient unparalleled nutritional advantages, enabling reefs to flourish in low-nutrient, oligotrophic waters of the tropics (Hallock 1997, 2001).

G. Stanley (✉)
Paleontology Center, University of Montana, Missoula, MT, USA
e-mail: george.stanley@umontana.edu

B. van de Schootbrugge
Marine Palynology and Paleoceanography Group, Department of Earth Sciences, Utrecht University, Utrecht, Netherlands

M. J. H. van Oppen, J. M. Lough (eds.), *Coral Bleaching*, Ecological Studies 233,
https://doi.org/10.1007/978-3-319-75393-5_2

The capacity of so many living shallow-water corals for rapid and sustained skeletal growth is linked to light-enhanced calcification (Goreau and Goreau 1959). The physiochemical processes and pathways by which the symbionts increase calcification have received intensive study (Marshall 1996; Goreau et al. 1996; Gattuso et al. 1999). Although precise mechanisms by which calcification occurs are not fully resolved, photosymbiosis is closely linked to calcification, and the topic has great relevance to global warming and coral bleaching. There seems little doubt that photosymbiosis is and was the driving force behind both recent and ancient reef building (Stanley and Lipps 2011). Although we have a rich fossil record of scleractinian corals beginning some 240 Ma (million years ago), we have precious little information about the nature and evolution of their symbionts. This chapter summarizes from a geological perspective the early evolution of scleractinian corals, their photosymbionts, and their relationship to mass extinction and coral bleaching.

The dinoflagellate *Symbiodinium*, partnering with corals today, likely evolved from a single adaptive radiation (Wood 1999; Pochon et al. 2006), and molecular studies confirm that this symbiont forms a monophyletic group. However, today it is a genetically diverse complex, inhabiting corals as well as foraminifers, sponges, mollusks, and other unrelated taxa (Stat et al. 2006). Early molecular studies of *Symbiodinium microadriaticum* in corals and other hosts revealed a number of distinct clades (Blank and Trench 1985; Rowan 1998; Rowan and Powers 1991). With continued molecular study, many clades and subclades are known (LaJeunesse 2002; LaJeunesse et al. 2010; van Oppen et al. 2009), and with new investigations, their numbers are increasing (Chap. 6). Regarded as "holosymbionts," different clades inhabit different corals, and many have distinct preferences for depth, temperature, and light intensity. The composition of these clades may change through time, and even clade competition may be at play. Some associations of symbionts appear to confer resistance to corals, and these discoveries challenge the simplistic assumption of single symbiont–host coevolution. Reshuffling of algal symbionts in response to global warming and bleaching is hypothesized to have produced innovative adaptive responses to bleaching (Fautin and Buddemeier 2004). This hypothesis has not gone without challenge (Goulet 2006), but other studies render support for reshuffling (Mieog et al. 2007). The mechanism of shuffling among coral symbionts is complex, but the adaptive significance for corals is high. Important aspects appear to be the present ecological environment, the severity of bleaching, and the recovery environment (Cunning et al. 2015). Fertile research in this direction has great potential for better understanding coral bleaching (Chap. 6).

Coral bleaching, global climate change, and ocean acidification portend a bleak future, but the fossil record has many lessons to teach us. Mass extinctions have toppled whole marine ecosystems and virtually decimated reefs of the past. Ancient ecosystem collapse and recovery have relevance to photosymbiosis and the increasingly severe episodes of coral bleaching today. Reef building in the fossil record was not limited to corals but included a variety of other ancient calcifying metazoans which successfully exploited the photosymbiotic way of life (Lipps and Stanley 2016a). Photosymbiosis explains the success of coral reefs today, and it was a pervasive and unifying theme among reefs extending back 430 million years to

Silurian time when extinct tabulate corals and stony sponges built reefs (Copper 1989).

In addition to dinoflagellates, a variety of other symbionts (diatoms, nitrogen-fixing cyanobacteria, etc.) partner with calcifying organism today, but dinoflagellates bestow the most benefits and likely evolved in association with corals of the Mesozoic (Stanley and Lipps 2011). When did the coral–symbiont relationship evolve among the ancestors of modern corals, and how did it change through time, especially during mass extinctions? Wells (1956) assumed the original condition for the first corals of the Triassic was zooxanthellate. Stanley (1981) proposed an alternative idea that the earliest scleractinians were initially azooxanthellate but, later in the Triassic, coevolved a symbiosis with zooxanthellate symbionts. The timing of this event was based, in part, on an absence of reef building in Middle Triassic corals and the adaptive radiation among Late Triassic corals. Massive, platy corals of the Middle Triassic did not build reefs yet show evidence of being photosymbiotic (Stanley and Helmle 2010). The Late Triassic (Norian–Rhaetian) was a time of major reef building by corals, and carbonate rocks show volumetric increases in reef complexes in a warm, shallow Tethys seaway extending from Western Europe across what is now Eurasia. It was questioned whether Triassic and Jurassic corals of this time were ecologically comparable to living zooxanthellate counterparts (Flügel 2002; Stanton 2006). Clearly reef corals that emerged later in the Cenozoic (Rosen 2000) were more direct ancestors of modern corals and were efficient at constructing reefs (Perrin 2002).

2.2 Detecting Photosymbiosis in the Fossil Record

There is a problem in identifying photosymbiosis among fossils which do not preserve their algal symbionts. Paleobiologists must assess their former presence by a variety of indirect methods (Stanley and Lipps 2011). For corals such criteria include colony size, indicative of massive hypercalcifiers, shape (the solar panel effect), corallite size and level of integration, as well as the presence of the edge zone (Coates and Jackson 1987; Rosen 2000; Stanley 2003; Lipps and Stanley 2016a). Since scleractinian corals and other rapidly growing hypercalcifiers produce discrete, periodic (annual) skeletal growth bands, their measurement in the ancient organisms provides rates of growth. Comparisons with modern counterparts can be used to indicate photosymbiotic growth rates (Copper 2002; Stanley and Helmle 2010). Microfeatures in the skeletons of corals also can be useful. Stolarski (2003) suggested that the regularity of mineral/organic phase alternations in thickening deposits among nanostructural aspects of the skeleton might distinguish zooxanthellate species from azooxanthellate counterparts. It is well known that colony shape, corallite size, and corallite complexity correlate with living photosymbiotic corals as well as some other hypercalcifying organisms. This is true of the platy growth habit to maximize light capture in corals (Rosen et al. 2000). Thin tissue syndrome in the coral skeleton was discussed as indicative of photosymbiosis (Wood 1999). Cowen (1988) inferred by such indirect methods that many extinct reef organisms, extending back over 500 million years, were photosymbiotic. These included large calcified sponges (stromatoporoids) and Cretaceous rudistid bivalves. These conclusions led to confirmation of the photosymbiotic hypothesis and that photosymbiosis evolved very early in Paleozoic time, opening new metabolic pathways as well as enhancing calcification rates.

A different approach to the detection of ancient photosymbiosis lies in geochemistry and stable isotopes. Stable isotopes ^{13}C and ^{18}O fractionated in the skeletons of living corals were detected and used to differentiate azooxanthellate from zooxanthellate species (Swart 1983). The photosymbiosis hypothesis was successfully tested and confirmed with stable isotopes in Late Triassic corals (Stanley and Swart 1995). Similar stable isotope signals were detected in Paleozoic corals (Zapalski 2014), but diagenetic alteration of these signals remains a major problem for such ancient samples. In other geochemical approaches, an analysis of microstructure in Late Triassic corals using $^{13}C/^{12}C$ and $^{18}O/^{16}O$, as well as intracrystalline $^{15}N/^{14}N$ (Frankowiak et al. 2016), confirmed the prevalence of photosymbiosis in the ancient past, supporting also the conclusion reached by Muscatine et al. (2005) based on the isotopic composition of the organic matrix in the skeleton.

2.3 Mesozoic Reef History and Coral Evolution

Following the world's greatest mass extinction at the end of the Permian (Erwin 2015), there was a lengthy Early Triassic interval marked by an absence of metazoan reefs and general suppression of carbonate production, a notable exception coming from South China (Lehrmann 1999). The Early Triassic was thus a whole geologic epoch, 8–10 million years in duration, devoid of corals and significant metazoan reefs. The Middle Triassic appearance of modern corals occurred in the Tethys Sea, coinciding with an interval of warm climate and expanded carbonate shelves (Fig. 2.1). Shallow-water calcified organisms of the Middle Triassic, including corals, were part of a delayed recovery following the largest mass extinction at the end of the Permian (Chen and Benton 2012). The first reef-like features were constructed by sponges, bryozoans, calcified algae, and non-colonial invertebrates. Scleractinians appeared around the world in the Middle Triassic interval (Fig. 2.1) and were unrelated to ancient orders of Paleozoic corals, all of which perished in the end-Permian extinction. Scleractinians are thus separated from the last Paleozoic corals by a considerable time gap. They also differ in composition (Paleozoic corals secreted calcite rather than aragonite), with a different symmetry, and different patterns of septal insertion. Morphologic differences and the temporal separation of scleractinians from Paleozoic corals led to the idea of existence as soft-bodied, anemone-like forms that left no fossil record during the Early Triassic interval. This was the "naked coral" hypothesis (Stanley and Fautin 2001). It was supported by

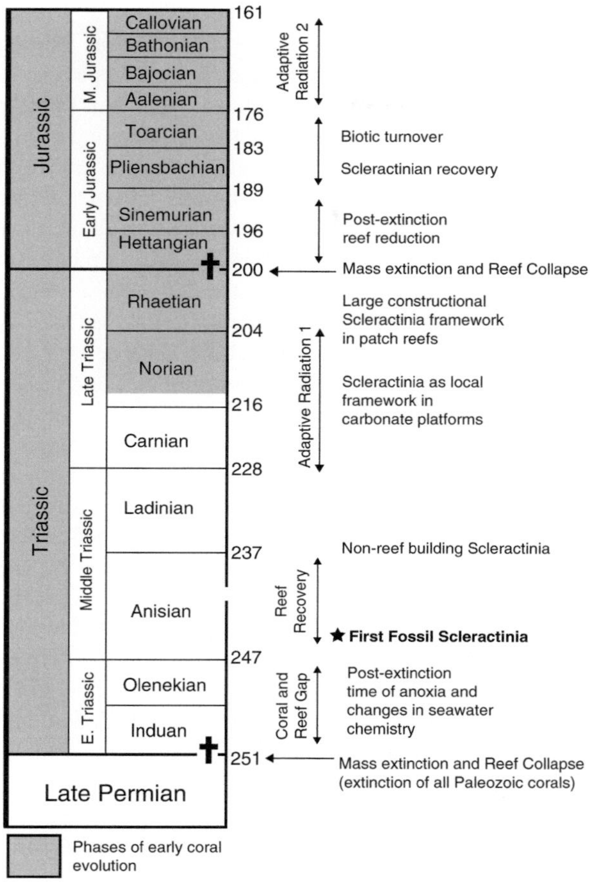

Fig. 2.1 Signposts and highlights in the evolution of scleractinian corals during the early Mesozoic. The smallest subdivisions of the time column are stages discussed in the text. Geologic time in millions of years (*Ma*) is given to the right of the column

molecular analyses (Medina et al. 2006) and by decalcification experiments in living corals (Fine and Tchernov 2007; Stanley 2007).

Although diverse and complex in corallum morphologies, the first calcified scleractinians of the Middle Triassic interval might have been photosymbiotic, but surprisingly they did not build reefs. These appear abruptly worldwide and with calcareous algae, foraminifers, bryozoans, and non-colonial invertebrates contributing to reef mounds and carbonate buildups. The Late Triassic witnessed sea-level rise, climatic warming, the emergence of large platform complexes, and an adaptive radiation among corals (Fig. 2.1). The emergence of scleractinian corals and their potential as framework builders of reefs have been discussed extensively (Flügel 2002; Stanley 2003). Prior to the Late Triassic, scleractinians were merely reef dwellers and did not participate as the primary constructors of reefs.

Following a Late Triassic (Carnian–Norian) turnover and a smaller-scale extinction, the early ancestors of modern corals experienced major changes in taxonomy and dominance (Stanley 1988; Roniewicz 2011). These took place within the great expanses of the shallow-water Tethys seaway, now represented by deposits in mountain ranges widespread through Central Europe across Eurasia. During a subsequent Norian–Rhaetian "reef bloom," large colonial corals increased in importance and evolved more complex structure. This correlates with the start of the long prevalence of platy coral growth, likely indicating a photosymbiotic response (Rosen et al. 2000). Coral-dominated reefs of this age are well known from reef complexes of the Tethys (Flügel 2002). During this Triassic reef bloom, the latitudinal range of reefs expanded. Photosymbiosis was assessed by Kiessling et al. (2009) to have been a "key driver" of Triassic coral evolution and reef expansion.

The major mass extinction of the end-Triassic (Fig. 2.1) affected not only corals of the shallow Tethys but the Americas as well (Hodges and Stanley 2015). The cause of this end-Triassic mass extinction, one of the "big five" of the Phanerozoic, has been variously related to the eruption of flood basalts and release of aerosols and greenhouse gases and carbon dioxide (CO_2) and a sudden release of methane hydrates (Tanner et al. 2004). This triggered major perturbations of the marine environment as well as severe ocean acidification. Although an Early Jurassic coral recovery began soon after, the extinction diversity remained low. Compared to other calcified biotas after the extinction, reef-building corals and sponges experienced proportionately greater losses (corals, 96.1%; sponges 91.4%), possibly related to their reduced physiological control of calcification and changes in aragonite saturation in seawater (Hautmann et al. 2008).

The first two stages of the succeeding Early Jurassic record a time of global reef reduction followed by recovery. They represent an interval of some 4–5 $\times$ 10^6 years when a vast number of Triassic coral species died out and reefs collapsed, save for a few in isolated locations in the Tethys (Kiessling et al. 2009; Gretz et al. 2013). In the following stages of the Lower Jurassic, despite an anoxic event at the end of the Early Jurassic (Toarcian), there is evidence of recovery and biotic turnover leading to a renewal of coral and reef diversity during the Middle to Late Jurassic (Lathuilière and Marchal 2009). During the Middle and Late Jurassic, coral, sponge, and microbial reefs are known, but it was during this period that corals again became principal builders of shallow reefs, while siliceous sponges and microbial deposits predominated in deeper water settings (Leinfelder 2001). Some deeper water coral biostromes show evidence of an ecology quite different from the nutrient-limited model of coral reefs (Insalaco 1996). The end of the Jurassic was marked by a small-scale biotic extinction, but it was only slightly felt among reef communities.

During much of Cretaceous time, tropical shallow-water reefs became occupied by rudistid bivalves (Johnson 2002). Rudists were gregarious bivalves exploiting bizarre adaptive morphologies. They were ecologically successful and by the Late Cretaceous were remarkably diverse, forming reefs or large-scale buildups. Rudists mimicked colonial corals in their shapes, close packing, and interlocking margins, and many taxa show evidence of photosymbiosis. For nearly 30 million years of the Early Cretaceous, these uniquely reef-adapted bivalves coexisted alongside corals, sponges, and other organisms. From the Middle to Late Cretaceous time, they became more predominant in reef construction, inhabiting carbonate platforms of the warm tropical Tethys. This rudistid dominance coincided with global warming during a super greenhouse interval of high sea surface temperature (Johnson et al. 2001). Interestingly, throughout the latest Cretaceous interval of maximum rudistid development, corals were present and actually increased in diversity. Although corals lost their former dominance on the reef and their ability for reef construction, they nevertheless remained diverse and continued to prosper, especially in deeper, downslope settings.

The highly modified rudists show good evidence for photosymbiosis (Lipps and Stanley 2016a, b). Photosymbiosis likely was widespread at this time, and the presence of symbionts was even detected with stable isotopes in Late Cretaceous planktonic foraminifers (Houston and Huber 1998). The end of the Cretaceous at the Cretaceous/Paleogene boundary (K/Pg mass extinction) clearly led to severe extinctions of marine life, including the demise of all rudistids. This occurred during a warm greenhouse supercycle. Corals suffered great extinction but survived the devastating K/Pg mass extinction while rudistids did not. Relative to azooxanthellate taxa, Cretaceous zooxanthellate corals were most severely affected, but both zooxanthellate and azooxanthellate corals died out with 33% of all families and 70% of species removed (Veron 2008).

There was a decidedly latitudinal effect to coral extinction with better survival of azooxanthellate taxa at cooler, higher latitudes (Kiessling and Baron-Szabo 2004). In addition to the well-known meteorite impact at the K/Pg boundary, extinction also was induced by mass volcanism underway during K/Pg time. These events produced surface ocean acidification (D'Hondt et al. 1994; Hautmann et al. 2008), but the full impact on corals is not known in detail. Many causes have been posited for this extinction, but cogent arguments were made for ocean acidification as cause for the collapse of Cretaceous and other reef ecosystems in geologic time (Veron 2008; Kiessling and Simpson 2011).

The recovery after the K/Pg mass extinction has been studied intensively, but not many investigations have focused on corals. Danian corals found in the immediately succeeding Paleogene interval were all azooxanthellate species and some built deep-water mounds in the aphotic zone (Bernecker and Weidlich 1990). Data shows that preferentially more azooxanthellate than zooxanthellate taxa survived to mark the start of the Cenozoic (Kiessling and Baron-Szabo 2004). Scleractinians diversified soon after the extinction, but reefs took considerably longer to return.

Not long after the K/Pg event, the late Paleocene to early Eocene is well known as a time of great global warming. Within this time a marine perturbation called the Paleocene–Eocene Thermal Maximum (PETM) occurred. It has been cited as an analogy of what could happen in the near future for today's marine ecosystem. The PETM was an unprecedented 100,000-year interval of warming brought about by sudden carbon release. Surface seawater temperature during that time was calculated to have risen 5–6 °C (Wright and Schaller 2013). Deep ocean water heated and sea surface temperatures are estimated to have reached 38–40 °C. Also ocean acidification was pervasive. Coral diversity responded to the initial warming by bleaching and subsequently retreating to cooler, northerly latitudes, and during the PETM, coral reefs vanished altogether (Scheibner and Speijer 2008). Following this event, corals and reefs slowly recovered. The remaining Cenozoic record of corals and reefs reveals the persistence of coral reef framework, starting in the Middle Eocene with an Oligocene diversity plateau after which more coral extinctions occurred (Perrin 2002). The Neogene offers cases of reef collapse and expansion. At the end of the Neogene corals survived a number of ecologic shocks such as the rising of the Central American Seaway, now separating taxa of the Pacific and Caribbean realms and the major Ice Age glaciation events of the Pleistocene.

Reef gaps are clearly following most if not all mass extinctions. They ranged from a few to as much as 8 million years in duration. These gaps have been equated with the breakdown of photosymbiosis and used to explain the repetitive pattern of reef collapse and recovery throughout much of geologic history (Talent 1988). Some believe we are headed for another major mass extinction of similar magnitude and with a similar reef collapse (Payne et al. 2016).

2.4 Late Triassic Dinoflagellate Symbionts?

Although the dinoflagellate *Symbiodinium* is the dominant photosymbiont today, other groups such as diatoms, chlorophytes, and cyanobacteria are photosynthetic and are known to inhabit living foraminifers, sponges, and bivalves (Wood 1999). In terms of greatest metabolism and calcification benefits, dinoflagellates today clearly are the most efficient, and their presence seems evident during Triassic times (Stanley and Swart 1995; Lipps and Stanley 2016a; Tornabene et al. 2017). Mesozoic dinoflagellate symbionts may not have belonged to the same taxon as modern *Symbiodinium*, which according to molecular analyses of living corals most likely evolved over 60 million years ago during the Paleocene epoch (Pochon et al. 2006). Interestingly, the radiation of scleractinian reef corals during the Late Triassic, Carnian–Norian interval (230–210 Ma) coincided with a comparable radiation of modern dinoflagellates (MacRae et al. 1996). In the following we briefly discuss the evidence for a possible evolutionary relationship between Triassic dinoflagellates and corals.

In a seminal paper on the recognition of animal–algal symbiosis in the fossil record, Cowen (1983) was adamant that: “Direct evidence of symbiosis (discovery of the symbionts themselves) is highly unlikely in the fossil record. No symbionts are embedded in hard tissues of [the] host.” *Symbiodinium* can exist in both a vegetative encysted form in a host and as a free-living zoospore. A possible, and surely not the only, exception may be observed in the Late Triassic to Early Jurassic interval, where we find fossilized dinoflagellate cysts that belong to the family Suessiaceae, order Suessiales (Fig. 2.2). The evolution of the family Suessiaceae closely relates to the evolution of corals. This is based on similarities of their evolutionary patterns and geographical palaeodistributions. Fossil dinoflagellate cysts are distinguished and classified based on the number and arrangement of para-plates on the cyst wall that are known to reflect the plate tabulation on the cell wall in the motile stage (Fensome et al. 1999). Apart from being practically the oldest unequivocal dinoflagellate body fossils, members of the Suessiaceae share a

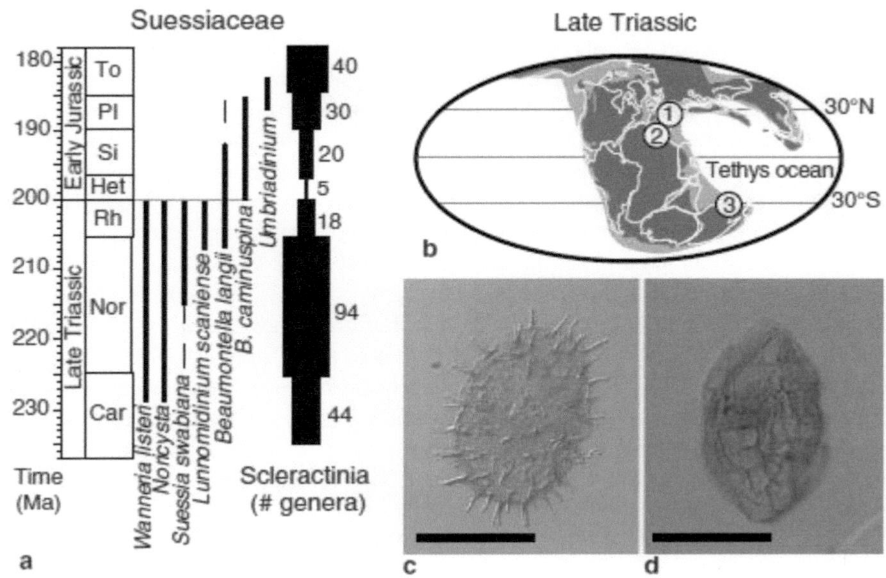

Fig. 2.2 (**a**) Late Triassic to Early Jurassic timescale with range chart showing most important Suessiales and scleractinian coral diversity. Coral data from Stanley and Beauvais (1994) and Flügel (2002). Stages: *CAR* Carnian, *NOR* Norian, *RH* Rhaetian, *HET* Hettangian, *SI* Sinemurian, *PLI* Pliensbachian, *TOA* Toarcian. (**b**) Late Triassic paleogeography showing the one-world continent of Pangea (modified after Scotese Paleomap Project 2000; www.scotese.com/earth.htm). Map shows reef sites with suessiacean dinoflagellate cysts: *1* Austria, *2* Libya, *3* NW Australia. Note the co-occurrence of reefs and Suessiaceae dinoflagellates between 30 °N and 30 °S and along the edges of the Tethys Ocean. (**c**) *Beaumontella langii*. *Beaumontella* was the only genus that survived the end-Triassic mass-extinction event. This dinoflagellate cyst was obtained from the earliest Jurassic in the Mingolsheim core, SW Germany (van de Schootbrugge, unpublished data). *Bar* 20 μm. (**d**) *Suessia swabiana*. This specimen is from the latest Triassic in the FFC-1 core from southern Sweden. Photograph from Lindström and Erlström (2006), reprinted with permission of the authors and Elsevier Science. *Bar* 20 μm

unique para-tabulation of seven to ten latitudinal and two cingular para-plate series (Bucefalo Palliani and Riding 2000), which is strikingly similar to plate tabulation patterns observed in present-day *Symbiodinium*. Based on this similarity, it has been suggested that Triassic Suessiaceae were symbionts of the earliest scleractinian corals (Bucefalo Palliani and Riding 2000).

Although it is tempting to attribute the proliferation of scleractinian corals during the Late Triassic to coevolution with dinoflagellate symbionts, many questions remain. Cowen (1983) noted that: "Symbionts escape or decay or are ingested on the death of the host and would in any case be impossible to distinguish from free-living species even if preserved." The fossil record of Triassic suessiacean dinoflagellate species is reconstructed from the encysted life-cycle stage that must have developed independently from their coral hosts, during a free-living stage in the water column. Extant *Symbiodinium* is known to survive in the water column, where it does “reinfect” the embryos of coral species that do not transmit their symbionts vertically (Smith and Douglas 1987). However, in contrast to Triassic suessiacean dinoflagellates, *Symbiodinium* does not produce cysts that would allow it to become fossilized. Even though it will be virtually impossible to tie individual Triassic Suessiaceae to single Triassic coral species, there is much indirect evidence to suggest a link existed between Suessiaceae and early modern corals of the Triassic. Paleogeography may also hold a key (Fig. 2.2). A good match exists between the paleobiogeographic distribution of Late Triassic reefs and suessiacean dinoflagellates (Bucefalo Palliani and Riding 2000). For most of the Late Triassic, coral reefs were bimodally distributed across the equator from 40 °N to 40 °S (Kiessling 2001). This reef distribution is mimicked by the distribution of suessiacean cysts (Fig. 2.2), which are most abundant from Austria (Morbey and Dunay 1978) to Australia (Backhouse et al. 2002). Nearly all reported occurrences of Suessiaceae cysts are from sediments interbedded with reef limestones. The oldest suessiacean genus *Noricysta* has been found in shales interbedded with Upper Carnian reef limestones in the Swiss Alps (Hochuli and Frank 2000). The best example, however, has been described from Ocean Drilling Program cores (ODP Leg 122) obtained from the Wombat Plateau, off the northwestern coast of Australia (Fig. 2.2). There, *Suessia* and *Wanneria* are the dominant dinoflagellate species occurring in massive *Astraeomorpha–Retiophyllia* reef buildups (Brenner 1992; Stanley 1994).

The genus *Beaumontella* (Fig. 2.2) occurs abundantly in the uppermost Triassic of Northwest Europe and is the only suessiacean genus that survived the end-Triassic mass-extinction event (200 Ma). The Triassic–Jurassic boundary mass-extinction event wiped out more than 95% of all scleractinian coral species (Flügel and Senowbari-Daryan 2001; Flügel 2002), and coral reef buildups are extremely rare during the beginning of the Jurassic (Stanley 2001, 2003). *Beaumontella* also occurs abundantly in Lower Jurassic shallow marine sediments (van de Schootbrugge et al. 2007), and it may thus be hypothesized that *Beaumontella* thrived in response to severe environmental stress. Interestingly, the highly spinose cysts of *Beaumontella* resemble *Polarella glacialis*, the only extant suessialean dinoflagellate known to build cysts (Montresor et al. 1999).

The last suessiacean cyst genus, *Umbriadinium*, is known from the Early Jurassic in the latest Pliensbachian (185 Ma) sediments in Italy (Bucefalo Palliani and Riding 1997). By this time scleractinian corals which survived the end-Triassic extinction had started a renewed radiation (Fig 2.1). Suessiacean dinoflagellate cysts have not been identified from the earliest Jurassic (Bucefalo Palliani and Riding 2003), leaving a lengthy gap in our knowledge of the group. One explanation for the disappearance of the Suessiaceae from the fossil record is that they lost the ability to produce fossilizable cysts or stopped making cysts altogether. An alternative, more easily testable, hypothesis is that scleractinian corals during the Jurassic and Cretaceous did not rely on symbiosis with dinoflagellates (i.e., *Symbiodinium*), and this was reinvented during the Cenozoic. There are independent lines of evidence to suggest that Late Triassic photosymbiosis between corals and dinoflagellates evolved in response to oligotrophic conditions (Riedel 1991). Middle Jurassic scleractinian corals that appeared after the Toarcian bottleneck are markedly different in taxonomic composition from those of the Late Triassic. According to Leinfelder (2001), Jurassic reef corals in the central Tethys lived mostly in mesotrophic waters of high siliciclastic influx, and, as previously mentioned, it was speculated that their zooxanthellate symbiosis was not as "effective" for metabolism and calcification as in modern reef corals. Molecular phylogenetic analyses indicate, as previously mentioned, that *Symbiodinium* taxa evolved after the end-Cretaceous (Shaked and de Vargas 2006) and that they diversified rapidly during the Miocene (LaJeunesse 2005). At this time the modern ecologic structure of coral reefs took shape (Perrin 2002).

2.5 Geological Perspectives on Photosymbiosis and Coral Bleaching

Present-day coral reefs exist within narrow limits of temperature, light, and seawater aragonite saturation states. In the broader, longer-term perspective of geologic time, we are nearing the end of an icehouse supercycle with an ocean favoring the precipitation of aragonite. Current climate change induced by rising levels of CO_2 may be prematurely moving us into greenhouse conditions and is subjecting photosymbiotic corals and living reefs to major stresses (Hoegh-Guldberg 2005). It has been known that the atmospheric partial pressure of CO_2 relates to the calcification process. Woodridge (2017) proposed that zooxanthellae density and atmospheric partial pressure of CO_2 (pCO_2) are critical to preventing breakdown of the coral–zooxanthellae relationship. Radical changes in the carbon cycle affecting the optimal threshold (pCO_2 260 ppmv) could adversely affect corals and trigger extinctions. This idea should be further explored by experiments with modern corals and tested against the fossil record of corals and reefs.

Thermal stress in recent times, caused by rising levels of CO_2, has triggered mass bleaching events among living photosymbiotic reef corals (Chap. 3). In addition to warming, rising CO_2 levels are resulting in ocean acidification known as carbon dioxide's "evil twin," and with continued projected increases in greenhouse gases, the prediction for reefs is dire (Kleypas et al. 2001). It was calculated, for example, that in the immediate future, one-third of reef corals face imminent extinction (Carpenter et al. 2008). Global bleaching events signal a breakdown of symbiosis. Long-term repercussions for reefs, their capacity for adaptation, and possible speed of recovery into the future can only be estimated. The rich fossil record of coral reefs is, therefore, a valuable key to better understand and predict possible future changes.

Modern tropical coral reefs are considered both fragile and robust (Lipps and Stanley 2016b), but the geologic record shows repetitive examples of global reef collapse followed by millions of years of reef eclipse (Fig. 2.3). Phanerozoic patterns of collapse are followed by extended harsh conditions inimical to reefs. During this time algal/bacteria associations surged, and metazoan reef mounds and carbonate sediments were greatly reduced. This is succeeded by reef recoveries, eventually

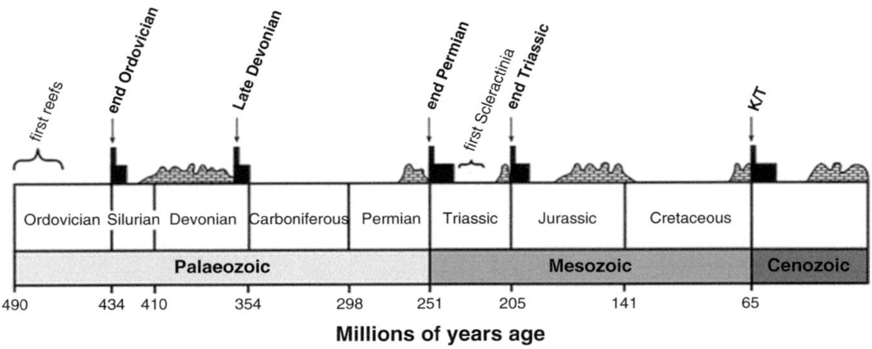

Fig. 2.3 The five big mass extinctions in geologic time and the occurrence and relative response of carbonate sediments and reefs (brick symbols). Black intervals are post-extinction reef eclipses lasting for millions of years. Collapse of photosymbiosis and bleaching are postulated to have occurred during these and other intervals (diagram courtesy of J E N Veron)

leading to new reef ecosystems. The geologically "sudden" response of many reef ecosystems to mass extinction and the extended post-extinction reef eclipse following collapse have direct relevance to current problems of coral bleaching. While difficult to detect bleaching in the fossil record, crises recorded when reef ecosystems collapse certainly would have terminated photosymbiosis, and global mass bleaching must have been a consequence. This appears to have been the case for corals during several mass extinctions: the end-Triassic when 96% of coral species became extinct (Hautmann et al. 2008; Hodges and Stanley 2015), when most or all of which were judged to have been photosymbiotic; during the Cretaceous–Paleogene event when an estimated 45% of coral species were extinguished (Kiessling and Baron-Szabo 2004); and during lengthy warm intervals of the Cenozoic. During Cenozoic warming brought on by rise in greenhouse gases, some photosymbiotic corals survived by migrating to higher paleolatitudes, and during the sudden event of the PETM, reefs disappeared altogether. During the highest levels of extinction and ocean acidification, such as the end-Permian and end-Triassic, coral skeletons may have survived in naked form to calcify much later when sea chemistry improved (Stanley 2011; Kvitt et al. 2015). Another theory is that corals survived by moving offshore, becoming solitary and non-photosymbiotic in deeper water refuges. Molecular findings support this idea, suggesting that loss of symbiosis occurred multiple times with repeated loss of coloniality (Barbeitos et al. 2010). The conclusions of Kitahara et al. (2010) and Stolarski et al. (2011), that modern shallow-water corals had multiple independent origins from deep-water (azooxanthellate and solitary) ancestors, would support this.

Although paleoecological changes within ancient reefs cannot be resolved as precisely as for their Holocene counterparts, stresses associated with global mass extinction on ancient reefs most certainly caused bleaching and major disruption of photosymbiosis. Several workers were quick to equate the sudden collapse of ancient reef ecosystems following mass extinctions to the breakdown of symbiosis

(Talent 1988; Copper 1989; Rosen and Turnšek 1989; Stanley 1992; Perrin 2002; Stanley and Lipps 2011).

Compared with their Mesozoic ancestors, late Cenozoic reef-building corals are enigmatic in terms of their apparent fragility. More geologically recent photosymbiotic corals in comparison with their more ancient ancestors reveal resilience in the face of major climate change and other upheavals. A solution to this conundrum could be the discovery that clades of *Symbiodinium* have adaptations to different temperature regimes (Little et al. 2004). Shuffling or switching of stress-tolerant symbionts, in concert with physiological resistance of the host, could account for the amazing adaptive potential (LaJeunesse et al. 2010; Chap. 6). This mechanism could be a driving force explaining the resilience and evolutionary success of Cenozoic to Holocene corals (Stanley 2006). The conclusions presented above, regarding the lower efficiency of Mesozoic corals for reef building relative to younger corals (Leinfelder 2001; Flügel 2002), might be explained by a lack of genetic diversity among their symbionts, resulting in limitations to adapt to change and, with their hosts, to build reefs. Perhaps these ancient symbionts had not yet evolved the modern adaptive repertoire with coral hosts. Finally, an alternate hypothesis is that after the devastating end-Triassic extinction, the coral–dinoflagellate symbiosis was not reinvented until the start of the Cenozoic when modern reefs took shape.

The fossil record chronicles the rise, fall, and recovery of reefs. It is a sobering record because of the longevity of post-extinction global reef gaps and the length of time before reef recovery. Intervals when reefs were essentially absent ranged from a few to more than 10 million years in duration (Fig. 2.3). The length of time for recovery has implications for the current environmental crisis. Put into perspectives of the current biotic marine crisis, in which humans are involved, the implications are bleak for the future evolution of coral reefs (Myers and Knoll 2001; Carpenter et al. 2008). Although evolution is not predictable, meaningful estimates on diversity loss and rates of recovery following mass extinction should come from the fossil record. A study of the role of zooxanthellate photosymbiosis in the geologic past provides new insight into both success and failure of living coral reefs. The integration of biology and the fossil record, especially ecology, molecular biology, and the life history of corals and symbionts, offers potential to better understand the current coral reef problems, including the bleaching phenomenon.

References

Backhouse J, Balme BE, Helby R, Marshall N, Morgan R (2002) Palynological zonation and correlation of the latest Triassic, Northern Carnarvon Basin. In: Keep M, Moss SJ (eds) The sedimentary basins of Western Australia. Proceedings of Petroleum Exploration Society of Australia Symposium, pp 179–201

Barbeitos MS, Romano SL, Lasker HR (2010) Repeated loss of coloniality and symbiosis in scleractinian corals. Proc Natl Acad Sci U S A 107:11877–11882. https://doi.org/10.1073/pnas.0914380107

Bernecker M, Weidlich O (1990) The Danian (Paleocene) coral limestone of Fakse, Denmark: a model for ancient aphotic, azooxanthellate coral mounds. Facies 22:103–138. https://doi.org/10.1007/BF02536947

Blank RJ, Trench RK (1985) Speciation and symbiotic dinoflagellates. Science 229:656–659. https://doi.org/10.1126/science.229.4714.656

Brenner W (1992) First results of late Triassic palynology of the Wombat Plateau, northwestern Australia. Proc Ocean Drill Progr Sci Res 122:413–426

Bucefalo Palliani R, Riding JB (1997) *Umbriadinium mediterraneense* gen. et sp. nov. and *Valvaeodinium hirsutum* sp. nov.: two dinoflagellate cysts from the lower Jurassic of the Tethyan realm. Palynology 21:197–206. https://doi.org/10.1080/01916122.1997.9989494

Bucefalo Palliani R, Riding JB (2000) Subdivision of the dinoflagellate cyst family Suessiaceae and discussion of its evolution. J Micropalaeontol 19:133–137. https://doi.org/10.1144/jm.19.2.133

Bucefalo Palliani R, Riding JB (2003) *Umbriadinium* and *Polarella*: an example of selectivity in the dinoflagellate fossil record. Grana 42:108–111. https://doi.org/10.1080/00173130310012495

Carpenter KE, Abrar M, Aeby G, Aronson RB, Banks S, Chiriboga A et al (2008) One-third of reef-building corals face elevated extinction risk from climate change and local impacts. Science 321:560–563. https://doi.org/10.1126/science.1159196

Chen Z-Q, Benton MJ (2012) The timing and pattern of biotic recovery following the end-Permian mass extinction. Nat Geosci 5:375–383. https://doi.org/10.1038/ngeo1475

Coates AG, Jackson JBC (1987) Clonal growth, algal symbiosis and reef formation by corals. Paleobiology 13:363–378. https://doi.org/10.1017/S0094837300008988

Copper P (1989) Enigmas in Phanerozoic reef development. Mem Assoc Australas Paleontol 8:371–385

Copper P (2002) Silurian and Devonian reefs: 80 million years of global greenhouse between two ice ages. In: Flügel E, Kiessling W, Golonka J (eds) Phanerozoic reef patterns. Society of Economic Paleontologists and Mineralogists, Tulsa, pp 181–238

Cowen R (1983) Algal symbiosis and its recognition in the fossil record. In: Tevesz MJS, McCall PL (eds) Biotic interactions in recent and fossil benthic communities. Plenum, New York, pp 431–478

Cowen R (1988) The role of algal symbiosis in reefs through time. Palaios 3:221–227. https://doi.org/10.2307/3514532

Cunning R, Silverstein RN, Baker AC (2015) Investigating the causes and consequences of symbiont shuffling in a multi-partner reef coral symbiosis under environmental change. Proc R Soc B 282:20141725. https://doi.org/10.1098/rspb.2014.1725

D'Hondt S, Pilson MEQ, Sigurdsson H, Hanson AK, Carey S (1994) Surface-water acidification and extinction at the Cretaceous-Tertiary boundary. Geology 22:983–986. https://doi.org/10.1130/0091-7613(1994)022<0983:SWAAEA>2.3.CO;2

Erwin D (2015) Extinction: how life on earth nearly ended 250 million years ago. Princeton University Press, Princeton

Fautin DG, Buddemeier RW (2004) Adaptive bleaching: a general phenomenon. Hydrobiologia 530/531:459–467. https://doi.org/10.1007/s10750-004-2642-z

Fensome RA, Saldarriaga JF, Taylor FJR (1999) Dinoflagellate phylogeny revisited: reconciling morphological and molecular based phylogenies. Grana 38:66–80. https://doi.org/10.1080/00173139908559216

Fine M, Tchernov D (2007) Scleractinian corals survive and recover from decalcification. Science 317:1811. https://doi.org/10.1126/science.1137094

Flügel E (2002) Triassic reef patterns. In: Flügel E, Kiessling W, Golonka J (eds) Phanerozoic reef patterns. Soc Econ Paleontol Mineral, Tulsa, pp 391–463

Flügel E, Senowbari-Daryan B (2001) Triassic reefs of the Tethys. In: Stanley GD Jr (ed) The history and sedimentology of ancient reef systems. Kluwer/Plenum, New York, pp 217–249

Frankowiak K, Wang XT, Sigman DM, Gothmann AM, Kitahara MV, Mazur M et al (2016) Photosymbiosis and the expansion of shallow-water corals. Sci Adv 2:e1601122. https://doi.org/10.1126/sciadv.1601122

Gattuso J-P, Allemand D, Frankignoulle M (1999) Photosynthesis and calcification at cellular, organismal and community levels in coral reefs: a review on interactions and control by carbonate chemistry. Am Zool 39:160–183. https://doi.org/10.1093/icb/39.1.160

Goreau TJ, Goreau NI (1959) The physiology of skeleton formation in corals. II. Calcium deposition by hermatypic corals under various conditions in the reef. Biol Bull 117:239–250. https://doi.org/10.2307/1538903

Goreau TJ, Goreau NI, Trench RK, Hayes RL (1996) Calcification rates in corals. Science 274:117–118. https://doi.org/10.1126/science.274.5284.117

Goulet TL (2006) Most corals may not change their symbionts. Mar Ecol Prog Ser 321:1–7. https://doi.org/10.3354/meps321001

Gretz M, Lathuilière B, Martini R, Bartolini A (2013) The Hettangian corals of the Isle of Skye (Scotland): an opportunity to better understand the palaeoenvironmental conditions during the aftermath of the Triassic–Jurassic boundary crisis. Palaeogeogr Palaeoclimatol Palaeoecol 376:132–148. https://doi.org/10.1016/j.palaeo.2013.02.029

Hallock P (1997) Reefs and reef limestone in earth history. In: Birkeland C (ed) Life and death of coral reefs. Chapman and Hall, New York, pp 13–42

Hallock P (2001) Coral reefs, carbonate sediments, nutrients, and global change. In: Stanley GD Jr (ed) The history and sedimentology of ancient reef systems. Kluwer/Plenum, New York, pp 388–427

Hautmann M, Benton MJ, Tomaový́ch A (2008) Catastrophic ocean acidification at the Triassic-Jurassic boundary. Neues Jahrbuch für Geologie und Palaontologie. Abhandlungen 249:119–127. https://doi.org/10.1127/0077-7749/2008/0249-0119

Hochuli PA, Frank SM (2000) Palynology (dinoflagellate cysts, spore-pollen) and stratigraphy of the Lower Carnian Raibl Group in the eastern Swiss Alps. Eclogae Geol Helv 93:429–443

Hodges MS, Stanley GD Jr (2015) North American coral recovery after the end-Triassic mass extinction, New York Canyon, Nevada. GSA Today 25:4–7. https://doi.org/10.1130/GSATG249A.1

Hoegh-Guldberg O (2005) Low coral cover in a high CO_2 world. J Geophys Res 110:1–11. https://doi.org/10.1029/2004JC002528

Houston RM, Huber BT (1998) Evidence of photosymbiosis in fossil taxa? Ontogenetic stable isotope trends in some Late Cretaceous planktonic foraminifera. Mar Micropaleontol 34:29–46. https://doi.org/10.1016/S0377-8398(97)00038-8

Insalaco E (1996) Upper Jurassic microsolenid biostromes of northern and central Europe: facies and depositional environment. Palaeogeogr Palaeoclimatol Palaeoecol 121:169–194. https://doi.org/10.1016/0031-0182(95)00072-0

Johnson CC (2002) The rise and fall of rudist reefs. Am Sci 90:148–153. https://doi.org/10.1511/2002.2.148

Johnson CC, Sanders D, Kauffman EG, Hay WW (2001) Patterns and processes influencing upper Cretaceous reefs. In: Flügel E, Kiessling W, Golonka J (eds) Phanerozoic reef patterns. Soc Econ Paleontol Mineral, Tulsa, pp 549–585

Kiessling W (2001) Paleoclimatic significance of Phanerozoic reefs. Geology 29:751–754. https://doi.org/10.1130/0091-7613(2001)029<0751:PSOPR>2.0.CO;2

Kiessling W, Baron-Szabo RC (2004) Extinction and recovery patterns of scleractinian corals at the Cretaceous–Tertiary boundary. Palaeogeogr Palaeoclimatol Palaeoecol 214:195–223. https://doi.org/10.1016/j.palaeo.2004.05.025

Kiessling W, Simpson C (2011) On the potential for ocean acidification to be a general cause of ancient reef crises. Glob Chang Biol 17:56–67. https://doi.org/10.1111/j.1365-2486.2010.02204.x

Kiessling W, Roniewicz E, Villier L, Léonide P, Struck U (2009) An early Hettangian coral reef in southern France: implications for the end-Triassic reef crisis. Palaios 24:657–671. https://doi.org/10.2110/palo.2009.p09-030r

Kitahara MV, Cairns SD, Stolarski J, Blair D, Miller DJ (2010) A comprehensive phylogenetic analysis of the Scleractinia (Cnidaria, Anthozoa) based on mitochondrial CO1 sequence data. PLoS One 5:e11490. https://doi.org/10.1371/journal.pone.0011490

Kleypas JA, Buddemeier RW, Gattuso JP (2001) The future of coral reefs in an age of global change. Int J Earth Sci 90:426–437. https://doi.org/10.1007/s005310000125

Kvitt H, Kramarsky-Winter E, Maor-Landaw K, Zandbank K, Kushmaro A, Rosenfeld H et al (2015) Breakdown of coral colonial form under reduced pH conditions is initiated in polyps and mediated through apoptosis. Proc Natl Acad Sci U S A 112:2082–2086. https://doi.org/10.1073/pnas.1419621112

LaJeunesse TC (2002) Diversity and community structure of symbiotic dinoflagellates from Caribbean coral reefs. Mar Biol 141:387–400. https://doi.org/10.1007/s00227-002-0829-2

LaJeunesse TC (2005) "Species" radiations of symbiotic dinoflagellates in the Atlantic and Indo-Pacific since the Miocene–Pliocene transition. Mol Biol Evol 22:570–581. https://doi.org/10.1093/molbev/msi042

LaJeunesse TC, Pettay DT, Sampayo EM, Phongsuwan N, Brown B, Obura DO et al (2010) Long-standing environmental conditions, geographic isolation and host–symbiont specificity influence the relative ecological dominance and genetic diversification of coral endosymbionts in the genus *Symbiodinium*. J Biogeogr 37:785–800. https://doi.org/10.1111/j.1365-2699.2010.02273.x

Lathuilière B, Marchal D (2009) Extinction, survival and recovery of corals from the Triassic to Middle Jurassic time. Terra Nova 21:57–66. https://doi.org/10.1111/j.1365-3121.2008.00856.x

Lehrmann DJ (1999) Early Triassic calcimicrobial mounds and biostromes of the Nanpanjiang Basin, south China. Geology 27:359–362. https://doi.org/10.1130/0091-7613(1999)027<0359:ETCMAB>2.3.CO;2

Leinfelder RR (2001) Jurassic reef ecosystems. In: Stanley GD Jr (ed) The history and sedimentology of ancient reef systems. Kluwer/Plenum, New York, pp 251–309

Lindström S, Erlström M (2006) The late Rhaetian transgression in southern Sweden: regional (and global) recognition and relation to the Triassic–Jurassic boundary. Palaeogeogr Palaeoclimatol Palaeoecol 241:339–372

Lipps JH, Stanley GD Jr (2016a) Photosymbiosis in past and present reefs. In: Hubbard DK, Rogers CS, Lipps JH, Stanley GD Jr (eds) Coral reefs at the crossroads, Coral reefs of the world, vol 6. Springer, Dordrecht, pp 47–68

Lipps JH, Stanley GD Jr (2016b) Reefs through time: An evolutionary view. In: Hubbard DK, Rogers CS, Lipps JH, Stanley GD Jr (eds) Coral reefs at the crossroads, Coral reefs of the world, vol 6. Springer, Dordrecht, pp 175–196

Little AF, van Oppen MJH, Willis BL (2004) Flexibility in algal endosymbioses shapes growth in reef corals. Science 304:1492–1494. https://doi.org/10.1126/science.1095733

MacRae RA, Fensome RA, Williams GL (1996) Fossil dinoflagellate diversity, originations, and extinctions and their significance. Can J Bot 74:1687–1694. https://doi.org/10.1139/b96-205

Marshall AT (1996) Calcification in hermatypic and ahermatypic corals. Science 271:637–639. https://doi.org/10.1126/science.271.5249.637

Medina M, Collins AG, Takaoka TL, Kuehl JV, Boore JL (2006) Naked corals: skeleton loss in Scleractinia. Proc Natl Acad Sci U S A 103:9096–9100. https://doi.org/10.1073/pnas.0602444103

Mieog JC, van Oppen MJH, Cantin NE, Stam WT, Olsen JL (2007) Real-time PCR reveals a high incidence of *Symbiodinium* clade D at low levels in four scleractinian corals across the Great Barrier Reef: implications for symbiont shuffling. Coral Reefs 36:449–457. https://doi.org/10.1007/s00338-007-0244-8

Montresor M, Procaccini G, Stoeker DK (1999) *Polarella glacialis*, gen. nov., sp. nov. (Dinophyceae): Suessiaceae are still alive! J Phycol 35:186–197. https://doi.org/10.1046/j.1529-8817.1999.3510186.x

Morbey SJ, Dunay RE (1978) Early Jurassic to late Triassic dinoflagellate cysts and miospores. In: Thusu B (ed) Distribution of biostratigraphically diagnostic dinoflagellate cysts and miospores from the Northwest European continental shelf and adjacent areas, vol 100. Continental Shelf Institute, London, pp 47–59

Muscatine L, Goiran C, Land L, Jaubert J, Cuif JP, Allemand D (2005) Stable isotopes ($\delta^{13}C$ and $\delta^{15}N$) of organic matrix from coral skeleton. Proc Natl Acad Sci U S A 102:1525–1530. https://doi.org/10.1073/pnas.0408921102

Myers N, Knoll AH (2001) The biotic crisis and the future of evolution. Proc Natl Acad Sci U S A 98:5389–5392. https://doi.org/10.1073/pnas.091092498

Payne JL, Bush AM, Heim NA, Knope ML, McCauley DJ (2016) Ecological selectivity of the emerging mass extinction in the oceans. Science 353:1284–1286. https://doi.org/10.1026/science.aaf2416

Perrin C (2002) Tertiary: the emergence of modern reef ecosystems. In: Flügel E, Kiessling W, Golonka J (eds) Phanerozoic reef patterns. Soc Econ Paleontol Mineral, Tulsa, pp 587–621

Pochon X, Montoya-Burgos J, Stadelmann B, Pawlowski J (2006) Molecular phylogeny, evolutionary rates, and divergence timing of the symbiotic dinoflagellate genus *Symbiodinium*. Mol Phylogenet Evol 38:20–30. https://doi.org/10.1016/j.ympev.2005.04.028

Riedel P (1991) Korallen in der Trias der Tethys: Stratigraphische Reichweiten, Diversitätsmuster, Entwicklungstrends und Bedeutung als Rifforganismen. Mitt Geol Bergbaustud Oesterr 37:97–118

Roniewicz E (2011) Early Norian (Triassic) corals from the Northern Calcareous Alps, Austria, and the intra−Norian faunal turnover. Acta Palaeontol Pol 56:401–428. https://doi.org/10.4202/app.2009.0092

Rosen BR (2000) Algal symbiosis and the collapse and recovery of reef communities: Lazarus corals across the K-T boundary. In: Culver SJ, Rawson PF (eds) Biotic response to global change, vol 12. Cambridge University Press, Cambridge, pp 164–180

Rosen BR, Turnšek D (1989) Extinction patterns and biogeography of scleractinian corals across the Cretaceous/Tertiary boundary. Mem Assoc Australas Palaeontol 8:355–370

Rosen BR, Aillud GS, Bosellini FR, Clack NJ, Insalaco E, Valldeperas FX, Wilson MEJ et al. (2000) Platy coral assemblages: 200 million years of functional stability in response to the limiting effects of light and turbidity. Proceedings of the 9th International Coral Reef Symposium, vol 1, pp 255–264

Rowan R (1998) Diversity and ecology of zooxanthellae on coral reefs. J Phycol 34:407–417. https://doi.org/10.1046/j.1529-8817.1998.340407.x

Rowan R, Powers DA (1991) A molecular genetic classification of zooxanthellae and the evolution of animal-algal symbioses. Science 251:1348–1351. https://doi.org/10.1126/science.251.4999.1348

Scheibner C, Speijer RP (2008) Decline of coral reefs during late Paleocene to early Eocene global warming. eEarth 3:19–27. www.electronic-earth.net/3/19/2008/

Shaked Y, de Vargas C (2006) Pelagic photosymbiosis: rDNA assessment of diversity and evolution of dinoflagellate symbionts and planktonic foraminiferal hosts. Mar Ecol Prog Ser 325:59–71. https://doi.org/10.3354/meps325059

Smith DC, Douglas AE (1987) The biology of symbiosis. Edward Arnold, London

Stanley GD Jr (1981) The early history of scleractinian corals and its geologic consequences. Geology 9:507–511. https://doi.org/10.1130/0091-7613(1981)9<507:EHOSCA>2.0.CO;2

Stanley GD Jr (1988) The history of early Mesozoic reef communities: a three-step process. Palaios 3:170–183. https://doi.org/10.2307/3514528

Stanley GD Jr (1992) Tropical reef ecosystems. In: Nierenberg WA (ed) Encyclopedia of Earth system science, vol 4. Academic, New York, pp 375–388

Stanley GD Jr (2001) Introduction to reef ecosystems and their evolution. In: Stanley GD Jr (ed) The history and sedimentology of ancient reef systems. Kluwer/Plenum, New York, pp 1–39

Stanley GD Jr (2003) The evolution of modern corals and their early history. Earth Sci Rev 60:195–225. https://doi.org/10.1016/S0012-8252(02)00104-6

Stanley GD Jr (2006) Photosymbiosis and the evolution of modern coral reefs. Science 312:857–858. https://doi.org/10.1126/science.1123701

Stanley GD Jr (2007) Ocean acidification and scleractinian corals. Science 317:1032–1033. https://doi.org/10.1126/science.317.5841.1032c
Stanley GD Jr (2011) The naked "Lazarus effect" and the recovery of corals after the end-Permian mass extinction. Kölner Forum für Geologie und Paläontologie 19:64–165
Stanley GD Jr, Beauvais L (1994) Corals from an early Jurassic reef in British Columbia: refuge on an oceanic island reef. Lethaia 27:35–47. https://doi.org/10.1111/j.1502-3931.1994.tb01553.x
Stanley GD Jr, Fautin DG (2001) The origins of modern corals. Science 291:1913–1914. https://doi.org/10.1126/science.1056632
Stanley GD Jr, Helmle KB (2010) Middle Triassic coral growth bands and their implication for photosymbiosis. Palaios 25:754–763. https://doi.org/10.2110/palo.2010.p10-039r
Stanley GD Jr, Lipps JH (2011) Photosymbiosis: the driving force for reef success and failure. Paleontol Soc Paper 17:33–59. https://doi.org/10.1017/S1089332600002436
Stanley GD Jr, Swart PK (1995) Evolution of the coral–zooxanthellae symbiosis during the Triassic: a geochemical approach. Paleobiology 21:179–199. https://doi.org/10.1017/S0094837300013191
Stanley GD Jr (1994) Upper Triassic spongiomorph and coral association dredged off the north western Australian shelf. AGSO J Aust Geol Geophys 15:127–133
Stanton RJ Jr (2006) Nutrient models for the development and location of ancient reefs. Geo Alps 3:191–206
Stat M, Carter D, Hoegh-Guldberg O (2006) The evolutionary history of *Symbiodinium* and scleractinian hosts – symbiosis, diversity and the effect of climate change. Perspect Plant Ecol Evol Syst 8:23–43. https://doi.org/10.1016/j.ppees.2006.04.001
Stolarski J (2003) Three-dimensional micro- and nanostructural characteristics of the scleractinian coral skeleton: a biocalcification proxy. Acta Palaeontol Pol 48:497–530
Stolarski J, Kitahara VM, Miller D, Cairns SD, Mazur M, Meibom A (2011) The ancient evolutionary origins of Scleractinia revealed by azooxanthellate corals. BMC Evol Biol 11:316. https://doi.org/10.1186/1471-2148-11-316
Swart PK (1983) Carbon and oxygen isotope fractionation in scleractinian corals: a review. Earth Sci Rev 19:51–80. https://doi.org/10.1016/0012-8252(83)90076-4
Talent JA (1988) Organic reef-building: episodes of extinction and symbiosis? Senckenbergiana Lethaea 69:315–368
Tanner LH, Lucas SG, Chapman MG (2004) Assessing the record and causes of late Triassic extinctions. Earth Sci Rev 65:103–139. https://doi.org/10.1016/S0012-8252(03)00082-5
Tornabene C, Martindale RC, Wang XT, Schaller MF (2017) Detecting photosymbiosis in fossil scleractinian corals. Sci Rep 7. https://doi.org/10.1038/s41598-017-09008-4
van de Schootbrugge B, Tremolada F, Rosenthal Y, Bailey TR, Feist-Burkhardt S, Brinkhuis H et al (2007) End-Triassic calcification crisis and blooms of organic-walled disaster species. Palaeogeogr Palaeoclimatol Palaeoecol 244:126–141. https://doi.org/10.1016/j.palaeo.2006.06.026
van Oppen MJ, Baker AC, Coffroth MA, Willis BL (2009) Bleaching resistance and the role of algal endosymbionts. Ecol Stud 205:83–102. https://doi.org/10.1007/978-3-540-6975-6_6
Veron JEN (2008) Mass extinctions and ocean acidification: biological constraints on geological dilemmas. Coral Reefs 27:459–472. https://doi.org/10.1007/s00338-008-0381-8
Wells JW (1956) Scleractinia. In: Moore RC (ed) Treatise on invertebrate paleontology. University Kansas Press, Lawrence, pp 328–444. pt. F
Wood R (1999) Reef evolution. Oxford University Press, New York
Woodridge SA (2017) Instability and breakdown of the coral–algae symbiosis upon exceedence of the interglacial pCO_2 threshold (>260 ppmv): the "missing" Earth-System feedback mechanism. Coral Reefs 36:1025–1037. https://doi.org/10.1007/s00338-017-1594-5
Wright JD, Schaller MF (2013) Evidence for a rapid release of carbon at the Paleocene-Eocene thermal maximum. Proc Natl Acad Sci U S A 110:15908–15913. https://doi.org/10.1073/pnas.130918810
Zapalski JK (2014) Evidence of photosymbiosis in Palaeozoic tabulate corals. Proc R Soc B 281:20132663. https://doi.org/10.1098/rspb.2013.2663

Chapter 3
Coral Bleaching in Space and Time

J. K. Oliver, R. Berkelmans, and C. M. Eakin

3.1 Introduction

Coral reefs face a range of serious anthropogenic threats that may significantly alter their ecological composition and reduce their capacity to deliver essential ecosystem services. Human influences such as destructive and overfishing, terrestrial runoff, pollution, and uncontrolled coastal development have a direct and immediately apparent impact on reefs. However, the impacts of human-induced climate change are a more pernicious and intractable problem to resolve, and this requires concerted social action at a global scale and over many generations (Hoegh-Guldberg 1999; Gattuso et al. 2015; Hughes et al. 2003, 2017; Spalding and Brown 2015). Large-scale bleaching of reef corals, resulting in mass mortality, is now a critical global threat to coral reefs and is clearly attributable to thermal stress (Baker et al. 2008; Heron et al. 2016) with excess light playing a key additional role (Brown 1997; Fitt et al. 2001). Thermal stress on coral reefs has clearly increased over the past century (Heron et al. 2016; Chap. 4). As global temperatures continue to rise, due to anthropogenic greenhouse gas emissions, the threat to coral reefs is increasing significantly.

Predictions based on climate models and thermal tolerance of corals suggest regular widespread catastrophic bleaching within the next 15–25 years (Hoegh-Guldberg 1999; Donner et al. 2005; Logan et al. 2014; van Hooidonk et al. 2016; Chap. 13). However, climate models deal with large-scale atmospheric and oceanic processes, which in themselves are highly complex with many parameters and feedback loops that are difficult to quantify. There is additional uncertainty in

J. K. Oliver (✉) · R. Berkelmans
Australian Institute of Marine Science, Townsville MC, QLD, Australia

C. M. Eakin
U.S. National Oceanic and Atmospheric Administration, Coral Reef Watch, College Park, MD, USA

M. J. H. van Oppen, J. M. Lough (eds.), *Coral Bleaching*, Ecological Studies 233,
https://doi.org/10.1007/978-3-319-75393-5_3

foreseeing human trajectories of resource use and change. Predictions of the impacts of climate change are thus uncertain even over large (ocean basin) scales. Even satellite-based observations, while improving both in spatial and temporal resolution and in accuracy of prediction, only measure the heat stress to which corals are exposed, not the response of the corals (Heron et al. 2016). At the scale of coral reefs, this uncertainty is compounded further by the physical and biological complexity of coral reef environments. Physical complexities include currents, tides, bathymetry, depth, water quality, 3D habitat structure, and weather, all of which can affect the temperature and light environment and hence the susceptibility of corals to bleaching. Biological complexities include the variable responses of coral species, the interaction of corals with their diverse and differentially susceptible microbial symbionts, interactions with pathogens, acclimatisation, and adaptation processes (Chap. 9). Coral mortality and reef recovery depend on numerous local factors, human use, and conservation status. It follows then that the degree of destruction and the permanence of the impacts are uncertain over large spatial scales and that the extent and severity of coral bleaching actually observed may not be as simple as predicted from climate models. These uncertainties increase as scales become smaller. Many of these complexities may never be fully understood, and in any case each bleaching event is a unique case with its own suite of causal factors. However, much can be learnt from studying spatial and temporal patterns in bleaching records. An analysis of past records is, therefore, useful in identifying large-scale spatial and temporal patterns in coral bleaching and identifying key data gaps and data deficiencies which can be addressed in the future.

In this chapter, we investigate the spatial and temporal patterns of coral bleaching that can be detected in the global database of bleaching records published by Donner et al. (2017) and supplemented by additional records from the recent literature. In particular we address the following questions:

1. Can discrete global bleaching events be identified from the records of bleaching?
2. How many global events have occurred in the past three decades?
3. Are major bleaching events increasing in frequency and intensity?
4. Is background (low-level) bleaching increasing in frequency and extent?
5. Are there any clear spatial patterns of coral bleaching at global, regional, and subregional scales?
6. Do the observed periods of significant global bleaching correspond with the periodic occurrence of El Niño-Southern Oscillation (ENSO) events and, if so, where?

3.1.1 Early Bleaching Records (Pre-1982)

There is a total of 26 records of coral bleaching before the first well-documented global-scale coral bleaching event of 1982–1983. The earliest confirmed record of reef-wide bleaching due to thermal stress is probably that of Yonge and Nicholls

(1931). They mention that, during a period of high summertime temperatures at Low Isles (Great Barrier Reef; GBR) in 1929, many corals died, and several corals (particularly *Goniastrea* spp., *Favia* spp.) were observed to have lost their zooxanthellae and turned white. Some weeks later these corals were observed to be recovering their colouration, and histological inspection revealed that they had lost and then started to recover their zooxanthellae populations.

Shinn (1966) notes that *Acropora cervicornis* that had been transplanted to an inshore site in the Florida Keys bleached on their upper surfaces during periods of maximum summertime temperatures. This was not, however, a normal habitat for these corals, and no observations of bleaching in normal populations were recorded. Goreau (1964) is probably the first person to publish a specific report on mass bleaching of corals in the reefs around Port Royal, Jamaica, during the aftermath of Hurricane Flora in 1963. He concluded, however, that the main cause of this bleaching was low salinity following heavy rains and floodwaters, rather than the high temperatures associated with current mass bleaching events. It is also possible that the report by Mayer (1914), in which he refers to corals not exposed to the air being "injured" after exposure to several hot calm days, represents a bleaching event. There is no mention, however, in his paper of loss of colour or bleaching.

Coffroth et al. (1990), Williams and Bunkley-Williams (1990), Glynn (1991, 1993), and Goreau and Hayes (1994) have compiled records of much earlier mass coral mortality, dating back to 1876. These earlier reports provide details for specific sites in terms of the pattern and extent of mortality and the probable causes. Factors other than high temperatures (e.g. aerial exposure, freshwater, "dark water") are most often used to explain the mortality, and in these cases there is no firm evidence to suggest that the cause of the mortality was high temperatures and that corals had bleached prior to dying.

3.1.2 Bleaching Records from 1982 Onwards

It was not until the global bleaching event of 1982–1983, first documented by Glynn (1983, 1984) and Coffroth et al. (1990), that widespread bleaching and mortality were recognised as a major phenomenon that could impact coral reef status and health at regional and global scales. The interest generated by Glynn's early papers on bleaching in the Eastern Pacific led to anecdotal reports from a wide range of sites across the world in subsequent years. These early reports were compiled by Brown (1987), Glynn (1990, 1991, 1993), Williams and Bunkley-Williams (1990), and Goreau and Hayes (1994). More recent summaries of coral bleaching records were published by a number of authors (Wilkinson 1998, 2000, 2002; Wellington and Glynn 2007).

In the late 1990s, the World Conservation Monitoring Centre and the WorldFish Center (then called ICLARM) both developed databases that compiled published and unpublished records of coral bleaching from throughout the world. In 2001 these

Table 3.1 Coral bleaching categories used in ReefBase and Donner et al. (2017) and adopted here

Code	Category	Description	Notes
0	No bleaching	No bleaching observed	For quantitative surveys, no bleaching is recorded if the percentage of live coral cover bleached is less than 1%
−1	Bleaching (unknown severity)	Bleaching recorded	Bleaching observed but no information on severity. For the purposes of analysis, this category is converted to 2—moderate bleaching
1	Mild bleaching	1–10% of coral cover bleached	If no estimate of % of bleached corals or coral cover is provided, then terms such as "light", "mild", "scattered", and "occasional" are used to identify this category
2	Moderate bleaching	11–50% of coral cover bleached	If no estimate of % of bleached corals or coral cover is provided, then terms such as "significant", "common", "frequent", and "moderate" are used to identify this category
3	Severe bleaching	More than 50% of coral cover bleached	If no estimate of % of bleached corals or coral cover is provided, then terms such as "heavy", "abundant", and "severe" are used to identify this category

datasets were combined and updated into a single database within ReefBase. More recently, Donner et al. (2017) have comprehensively revised the ReefBase bleaching database, nearly doubling the number of observations and including records up until 2010. This new database is available online and, with some revisions, is used as the basis for the present analysis.[1]

Many available bleaching records consist of descriptive accounts of the location of bleaching and, with varying degrees of detail, an assessment of the extent and severity of bleaching. Water depth and coral species affected are often not recorded. In a growing number of cases, formal surveys using quantitative or semi-quantitative measurements provide estimates of the percentage of coral that bleached.

The minimum information in each bleaching record in the current database is the date of observation, location, bleaching severity, and source of the information. Bleaching severity is a categorical variable. Table 3.1 shows the different categories, their descriptions, and notes on how verbal reports have been translated into one of the categories.

For the global analysis conducted here, a total of 9005 records in the bleaching database were included, up to 2010. Only a small number of records (<1%) did not include information on bleaching severity. Most of these were reports for early bleaching events prior to the major bleaching event of 1997–1998. For the purpose of the analysis, these records were reclassified as "moderate bleaching" since it is most likely that these early reports related to bleaching of more than 10% of coral cover. Very few (<3%) reports of mild bleaching were recorded in the database prior to 1997.

[1] simondonner.com/bleachingdatabase/

While the Donner et al. (2017) bleaching database contains the most comprehensive archive of coral bleaching records and all records are referenced either to a publication or a formal source, analysis of these records is hampered by the presence of multiple records at a single location. Donner et al. (2017) addressed this issue by grouping all records into 0.04° grid cells. For our analysis at a global scale, we chose a 1° grid scale to concentrate the bleaching phenomena at a subregional scale and to minimise the distorting influence of highly concentrated monitoring at some locations, especially the Great Barrier Reef (GBR), Australia. A further issue is that the number of reports received can vary both as a function of the severity and extent of bleaching and as a result of increased numbers of observers with an interest in bleaching. This "reporting effect" represents an important potential sampling bias that cannot be discounted or completely eliminated. While there are no data on the extent of this reporting effect, it is likely that the total number of potential observers of bleaching and the proportion of these who would be motivated to report on bleaching have increased over the past two decades as field-based coral studies have increased, media attention on the destruction of coral reefs from bleaching and climate change has grown, and easier systems for online reporting have been provided. Despite these sources of bias, we believe that a cautious and conservative analysis of the bleaching records can yield important insights into the extent of coral bleaching in space and time and the relationship between bleaching and climate variability and change at regional and global scales.

For our detection of bleaching at a regional scale through time, 1098 records were added to the Donner et al. (2017) database from the following sources: (1) Reef Check,[2] new ReefBase data,[3] and additional records obtained from the literature. These records (from 2011 to 2016) are not a comprehensive list for all locations, but the data were sufficiently detailed to allow a more comprehensive determination of the presence/absence of bleaching in each of seven regions[4] up to 2016:

1. Caribbean
2. Eastern Pacific
3. Central and western Pacific
4. Southeast Asia
5. Eastern Indian Ocean
6. Gulf Region
7. Central and Western Indian Ocean

[2]Online data from data.reefcheck.us

[3]www.reefbase.org

[4]These correspond to the regions in Burke et al. (2011) but divide the Pacific into east (east of longitude 120 °E) and central west (including eastern Australia) and the Indian Ocean into east (east of the Andaman's & Christmas I. including Western Australia) and west. The Australian region is not used.

Since this extended dataset (up to 2016) included observations based only on a named location rather than geographic coordinates, the records were grouped by named location rather than 1° grid cell.

3.2 Global Patterns of Coral Bleaching

3.2.1 Temporal Patterns

3.2.1.1 Location Records

This analysis was restricted to the period 1979–2010 due to the lack of comprehensive quantitative records for later years. Two very clear patterns emerge when examining all levels of bleaching severity (Fig. 3.1). First, there is a clear increase in the number of grid cells for which bleaching has been reported from 1998 to 2010 compared with prior years. Second, several clear peaks show the episodic nature of bleaching through time.

The trend of increasing bleaching occurrence is driven to some extent by mild bleaching records, but all years since 1991 have reports of bleaching in all severity categories. From 1998 there has been a minimum of 36 grids with bleaching (median 64), while prior to 1998 the minimum is 2 (median 12). While this increase in bleaching could be due to an increase in chronic stress to corals or an increase in small outbreaks at different locations, it is also possible that the increased awareness of coral bleaching is due to increased reporting of small amounts of bleaching that largely went unreported before then. Also apparent (Fig. 3.1) is the increase in the number of reports of zero bleaching since 1998. This has been driven by a significant increase in the number of formal monitoring programmes, such as Atlantic and Gulf Rapid Reef Assessment (AGRRA) and Reef Check, and systematic bleaching surveys that formally report the absence as well as the presence of bleaching.

Peaks of varying magnitude can be clearly identified (Fig. 3.1): 1998 is by far the dominant feature, with 2002, and 2005 also clearly distinguishable, but there is some suggestion of peaks in 1983, 1987, and 2010. In 1998 and 2002, a very large number of sites were surveyed on the GBR, using aerial survey techniques, while in 2005 additional survey effort to document the Caribbean bleaching event also occurred, and this has contributed to the disproportionately large peaks for these years. If the GBR and Caribbean records are eliminated from the graph, then the 2002 and 2005 peaks nearly disappear (Fig. 3.1b). The early peaks in 1983 and 1987 are less distinct, and 2010 remains as a fairly discrete peak. In summary, the data on frequency of bleaching reports per 1° grid cell indicate only major peaks (1998 and 2010). While the above analysis was restricted to the period 1979–2010, in the next section, we extend the analysis period up to 2016 by considering only the presence or absence at the regional level rather than actual counts of bleaching reports.

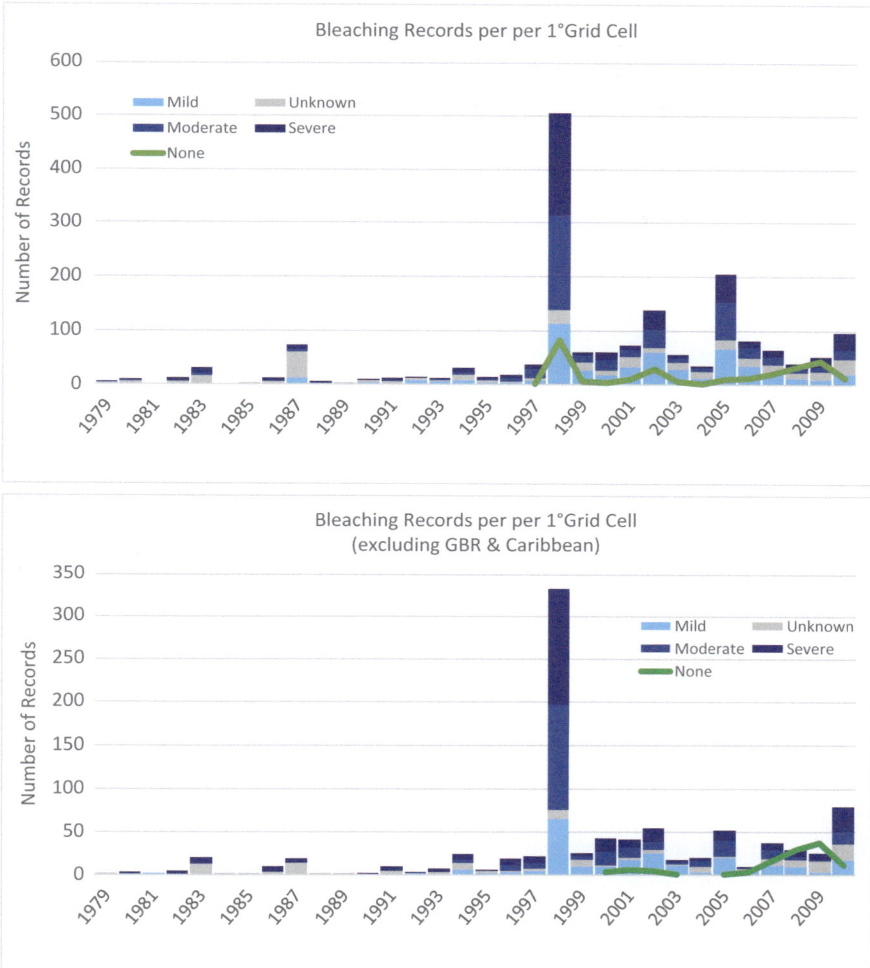

Fig. 3.1 (**a**) Yearly coral bleaching records by 1° grid cell. (**b**) The same graph with records from the GBR and Caribbean removed

3.2.1.2 Identification of Global Bleaching Events

The detection of peaks in the frequency of bleaching over time described above does not necessarily demonstrate global extent. One way of dealing with this problem is to further group records into geographic regions and test for bleaching presence in each region. Recent publications from the US National Oceanic and Atmospheric Administration's (NOAA) Coral Reef Watch (Eakin et al. 2016, 2017; Heron et al. 2016) suggested that there had been two global bleaching events (prior to the 2014–2017 event) based on the occurrence of widespread bleaching in all three major oceans

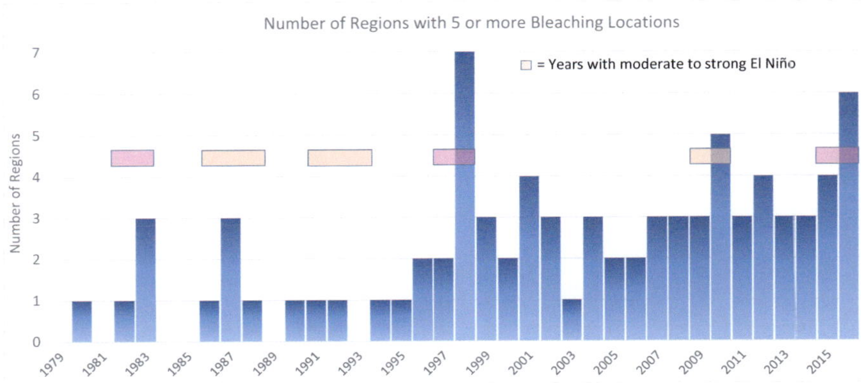

Fig. 3.2 Number of regions reporting bleaching at a moderate to severe level in at least five locations per region. Coloured horizontal bars indicate years with moderate (orange) and major (pink) El Niño events

(Atlantic, Pacific, Indian) in 1998 and 2010. This criterion is further enhanced here by examining bleaching at a more detailed geographic scale across seven regions and specifying that a region must have at least five locations with moderate or severe bleaching levels before bleaching is considered to be "present" and contribute to a global event over the period 1979–2016 (Fig. 3.2).

The two dominant peaks over the period 1979–2016 (Fig. 3.2) correspond to the two major El Niño events in 1997–1998 and 2015–2016 affecting all or all but one of the regions. These two events have also been associated with major, widespread coral mortality (Wilkinson 1998; Hughes et al. 2017). A smaller peak (five of seven regions affected) in 2010 is also clearly distinguishable. Unfortunately, a decline in global coral reef monitoring and reporting programmes in 2009–2010 may have contributed to the lower level of reports during 2010. Interestingly, 1983 and 1987, while exhibiting bleaching in only three regions, still stand out compared to adjacent years with much lower bleaching occurrence. It is plausible that lack of reporting effort masked more significant bleaching in 1983 and 1987. While only three regions had five or more bleaching events in these years, five regions reported bleaching in two or more locations in 1983 and 1987, suggesting that these years could also be considered to have been global events of at least moderate severity.

A clear feature is the increase in the global extent of bleaching from 1997 onwards (Fig. 3.2). During these two recent decades, only 1 year showed bleaching restricted to just 1 region, while 16 years (80%) had bleaching in at least 3 regions, and in 13 of these, some bleaching occurred in all 3 ocean basins. This may indicate the development of chronic albeit scattered bleaching of the world's reefs since 2007.

Overall, the evidence from the analysis of bleaching frequency at the 1° grid scale and regional presence/absence at the global scale suggest that two major and one moderate global bleaching events have occurred over the last two decades. Prior to that, low reporting effort may have masked major global events, but the relative frequencies of bleaching at grid and regional levels suggest that there may have been two widespread bleaching events during the two decades leading up to 1997. There is good evidence that major regional bleaching events occurred in the GBR and the Caribbean in 2002 and 2005, respectively, and this is backed up by separate detailed studies (Berkelmans et al. 2004; Eakin et al. 2010).

3.2.2 Spatial Patterns in Bleaching Reports

Temporal analysis at the grid and regional level indicates that there were three significant bleaching events between 1997 and 2016 and possibly two others prior to that. Detailed and comprehensive spatial records for the 2014–2017 event are still being assembled (Eakin et al. 2017), but the prior events are mapped in Fig. 3.3. The 1998 map clearly shows the high frequency of moderate to severe bleaching in most regions[5] and this year stands out as the most severe and extensive event up to 2010. The 2010 event also shows multiple occurrences of major bleaching in most regions, although there is an almost complete lack of bleaching on the GBR. In 1987 bleaching was mostly reported from the Caribbean and the GBR but still included some records in other regions and all oceans. In 1982–1983 multiple bleaching records predominate in the Caribbean, the Eastern Pacific, and the GBR (in 1982), but a few records (<5) occur in the Indian Ocean and Southeast Asia.

3.2.3 Trends in Bleaching Severity

There are several ways of addressing the question of whether bleaching severity is increasing over time. The first would be to examine only significant bleaching events to determine if the proportion of severe bleaching records increases in more recent events. However, the data for the 2014–2017 bleaching event are still not fully compiled, and earlier bleaching events (prior to 1998) were not well documented, making this type of analysis unfeasible. Another approach is to examine the frequency of severe bleaching across all years since 1979 (with the exception of 2014–2017) and look for increases in the frequency of severe records and a corresponding decrease in mild records (Fig. 3.4). While both severe and mild bleaching show a small positive correlation with year and frequency ($r = 0.32$ and 0.26, respectively), neither of these is significant ($p > 0.05$). These two weak positive trends may be explained by the fact that the proportion of records with unknown severity has significantly decreased ($r = 0.57, p < 0.01$) during this period.

[5]While the Andaman and Nicobar Islands appear to be severely affected, subsequent reports of impacts in the region by Rajasuriya et al. (2004) suggest that these two areas escaped major mortality and that the reports of bleaching during the event may have been an overestimate.

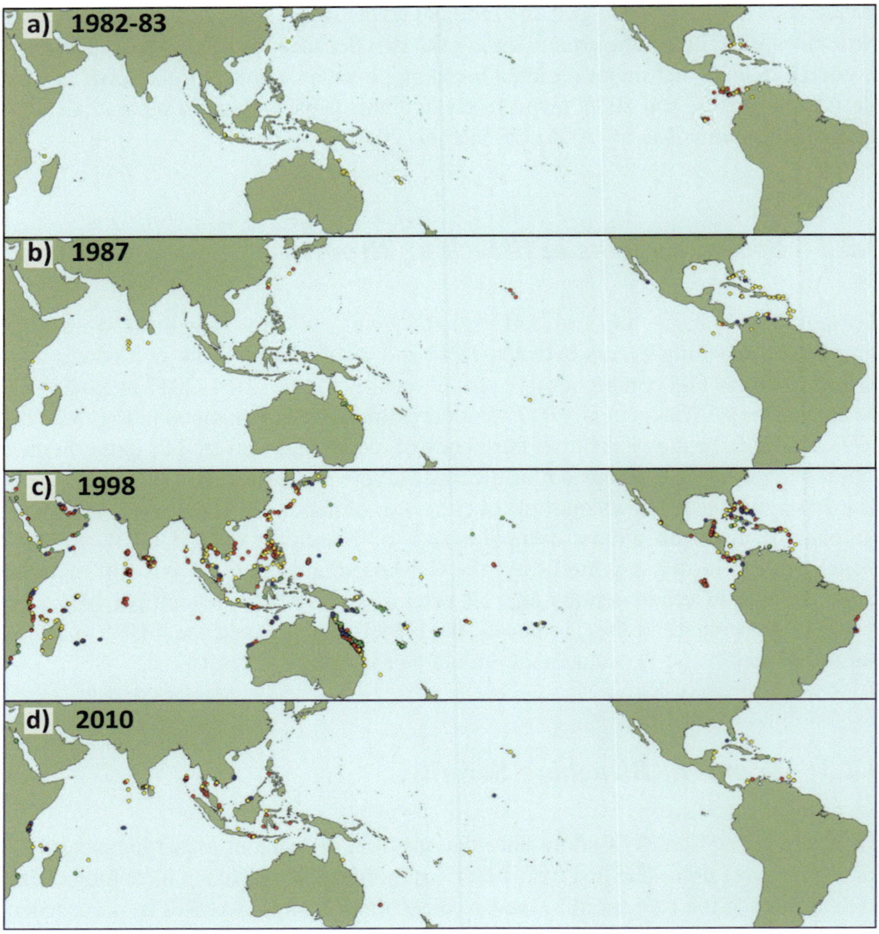

Fig. 3.3 Distribution and intensity of bleaching from records in the Donner et al. (2017) database for the global bleaching years: (**a**) 1982–1983, (**b**) 1987, (**c**) 1998, and (**d**) 2010. *Red dots* severe bleaching, *yellow dots* moderate bleaching, *blue dots* mild bleaching, *green dots* no bleaching

It can be concluded that there is little, if any, evidence that bleaching is becoming more severe over time, at least up until 2010. The recent bleaching in 2014–2017 is widely referred to as the worst ever, and further analysis of this event may change our conclusions. Finally, the severity of a bleaching event might also be determined by its ecological impact. Unfortunately, the database does not contain enough information on ecological impacts, such as mortality, to allow this analysis. In general, given the low number of global bleaching events so far, analysis of trends in their severity is not realistic from bleaching reports alone.

Fig. 3.4 Proportion of bleaching records classified as mild (blue circles), severe (red circles), and unknown severity (grey circles). Linear trends are also shown

3.3 Great Barrier Reef

3.3.1 Time Series

The GBR is the largest contiguous reef system in the world, and much of it has been intensively and continuously monitored for over 30 years. As such it represents a useful case study for a more detailed analysis of spatial and temporal patterns of bleaching than is possible at a global scale. Analysis of GBR records is based on four data sources: Donner et al. (2017), GBRMPA (2017), Reef Check (2017), and COECRS (2017a, b). Because some data did not contain specific coordinates, the data were grouped by named location, and an additional bleaching category of moderate/severe was included to enable the summaries of aerial surveys in 2016 to be included (COECRS 2017a,b). Over the past 25 years, there were three major peaks in bleaching on the GBR, corresponding to the bleaching events of 1998, 2002, and 2016 (Fig. 3.5). These peaks are disproportionately high compared to other years due to the intensive surveys conducted by Berkelmans and Oliver (1999), Berkelmans et al. (2004), and COECRS (2017a, b; Hughes et al. 2017). Although no quantitative data are available at the time of writing, severe bleaching also occurred in 2017 (Fig. 3.7). While the number of records is very low in the early years, there appear to be periods when significant bleaching was observed as early as 1980. Anecdotal reports suggest that there may also have been a bleaching event sometime

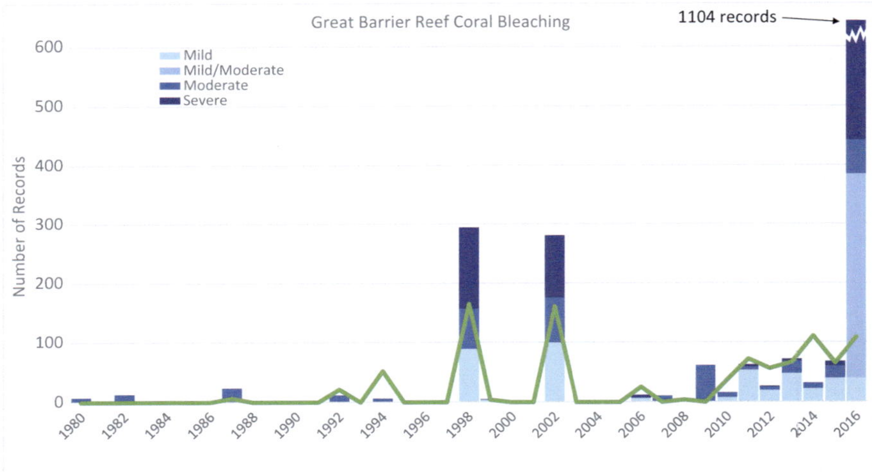

Fig. 3.5 Number of records of coral bleaching on the Great Barrier Reef

in the 1970s, but the year, extent, and intensity are unknown (Oliver 1985).[6] A further feature is that small bleaching events of moderate to severe intensity occur almost every year since 2006 (Fig. 3.5). This parallels a similar pattern already noted in the global time series (starting from the mid-1990s; see Fig. 3.1).

3.3.2 Spatial Patterns

The spatial distribution of bleaching at named sites on the GBR is shown for years of significant bleaching in Figs. 3.6 and 3.7. Overall there is a pattern of more severe bleaching in inner shelf regions, although in the most severe events of 2016–2017 bleaching extended across the entire shelf in some areas. Bleaching was recorded across more than two thirds of the length of the GBR in all bleaching years, but for years when comprehensive aerial surveys were conducted (1998, 2002, 2016, 2017), different latitudinal regions were affected in each year. The northern GBR has only shown severe bleaching during the 2016 event, and the far southern offshore reefs have only rarely recorded bleaching, and this has never been widespread. The specific spatial patterns for each of these years have been demonstrated to be highly correlated to elevated sea surface temperatures (SST, Berkelmans et al. 2004; Hughes et al. 2017).

[6]Oliver (1985) also states that "bleaching at Magnetic Island was not nearly as extensive in 1983 compared with 1982". This indirect reference to a bleaching event in 1983 is in fact a typographical error. It should have been a reference to the 1980 bleaching event.

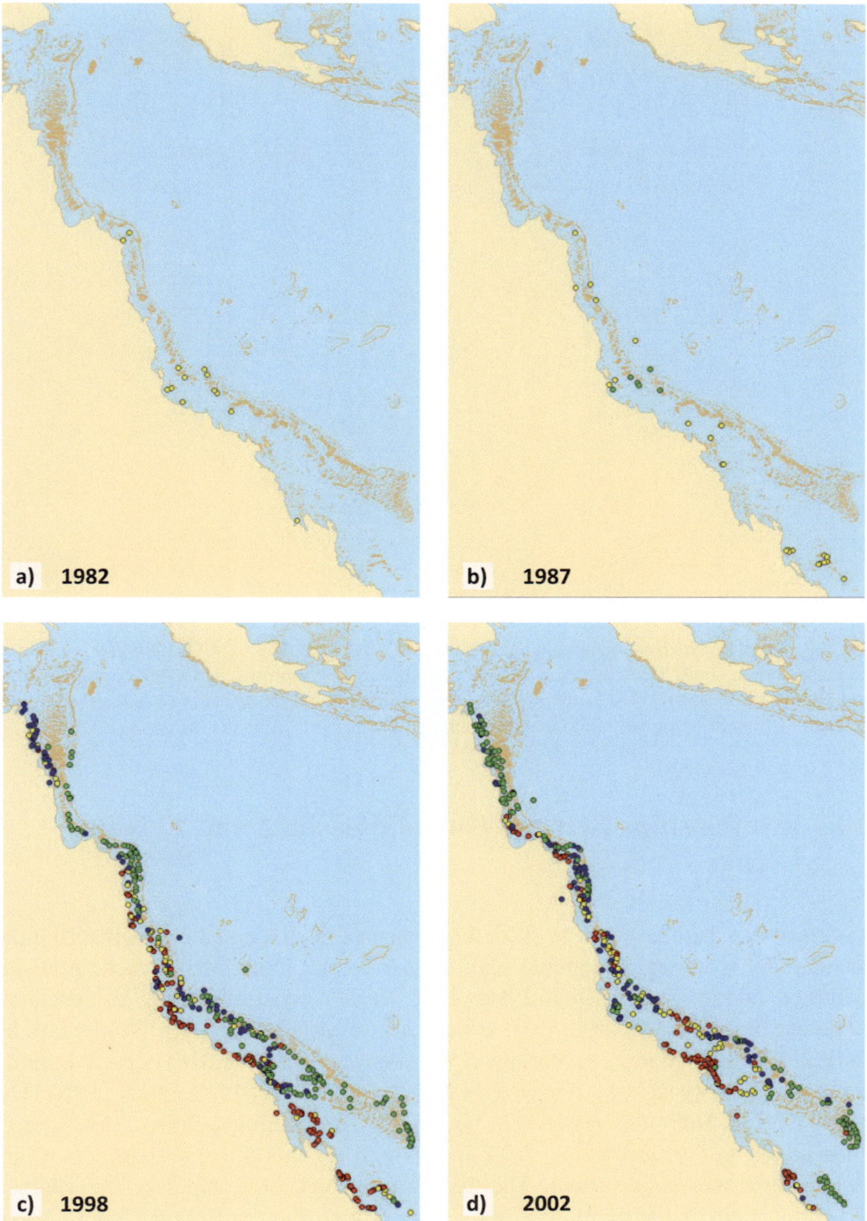

Fig. 3.6 Great Barrier Reef bleaching records. The maps for 1998 and 2002 include aerial survey data. Colours as in Fig. 3.4

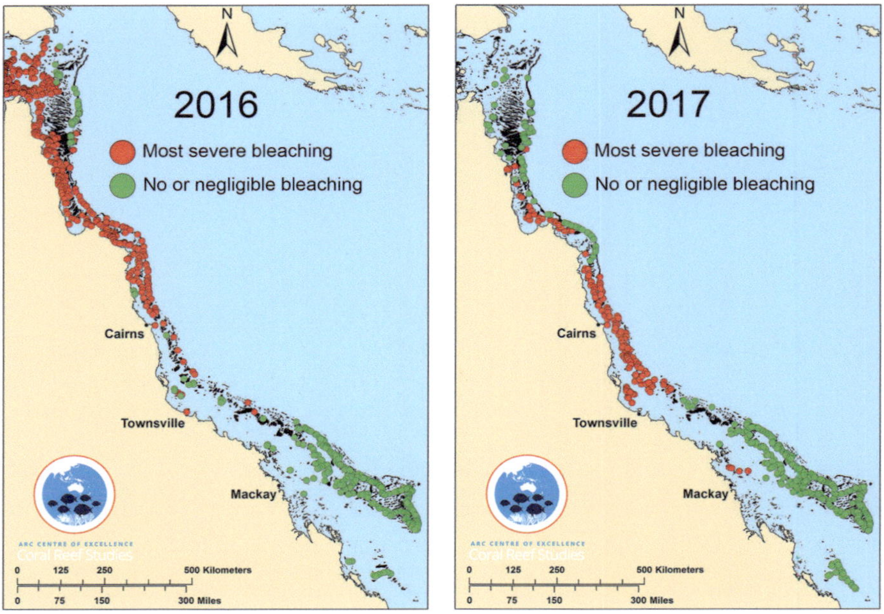

Fig. 3.7 Great Barrier Reef bleaching records for 2016 and 2017 (source: COECRS 2017b). Not all data are shown, only reefs at either end of the bleaching spectrum: red circles indicate reefs undergoing most severe bleaching (60% or more of visible corals bleaching), and green circles indicate reefs with no or only mild or no bleaching (10% or less of corals bleaching)

3.4 Relationships Between Global Bleaching and El Niño Events

It is clear that the link between moderate to severe bleaching and powerful El Niño events is fairly strong at a global scale (Fig. 3.8). All five global bleaching events identified above (1983, 1987, 1998, 2010, 2016) occurred during or just after moderate or major El Niño years. Based on a classification of El Niño years by Null (2017), the two most severe bleaching events (1998 and 2016) occurred during the second year (year +1) of strong El Niño events (1997–1998, 2015–2016). The 1982–1983 El Niño also resulted in widespread, and perhaps global, bleaching as summarised in Coffroth et al. (1990), although records are too sparse to firmly classify this as a major event. The 2010 global bleaching and less well-defined bleaching event of 1987 both occurred during or just after moderate El Niños. Years in which regional bleaching occurred (2002 in the GBR and 2005 in the Caribbean) also coincided with moderate or weak El Niños. Additionally, some regions, particularly the western Pacific, and also southern Western Australia have bleached during the first year of La Niña events, although the quick transition from the 1997–1998 El Niño to the 1998–1999 La Niña can make such distinctions challenging.

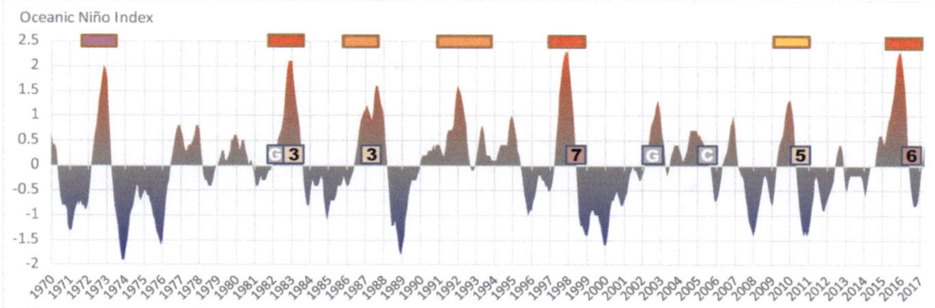

Fig. 3.8 Global bleaching events and the Oceanic Niño Index (ONI, NOAA 2017). The *shaded bars* at the top indicate moderate (orange), strong (pink), and very strong (red) El Niño events [categories from Null (2017)]. Solid boxes indicated years of near-global or global bleaching with numbers of regions affected or areas of regional bleaching in the Caribbean (C) or Great Barrier Reef (G)

3.5 Discussion

The spatial distribution and frequency of bleaching over time are both continuous variables which allow bleaching events to vary from singular minor events to ones that are spread densely across the globe over extended periods. This makes the identification of global bleaching events a somewhat subjective issue, even though there is value, from both an ecological and environmental management perspective, in distinguishing discrete severe events that have had major impacts on reef systems.

The identification of global bleaching events is predicated on a clear definition of the term, but this has not been explicitly addressed in the literature. An implicit definition in many treatments of global bleaching is that it occurs with moderate to high impact in many countries throughout the world, but this leaves questions regarding the required frequency, severity, extent, and uniformity unanswered. A more specific implicit definition, which refers to widespread bleaching in all three ocean basins, was used by NOAA's Coral Reef Watch (Eakin et al. 2016, 2017; Heron et al. 2016). They identified two global events in 1998 and 2010. Eakin et al. (2016) suggested an earlier global event may have occurred in 1983 based on work by Coffroth et al. (1990). Here we present bleaching frequency across 1° grid cells and regional presence/absence data for seven regions and visually identified peaks in the frequency distribution as a first approximation of discrete major bleaching years. This analysis suggests that over the last two decades, during which reporting efforts at a global scale have been high, 1998 and 2016 can be classified as severe global bleaching events and that 2010 is more usefully classified as a moderate global event. Prior to 1998 reporting efforts were lower, but there are discrete peaks suggesting that 1983 and 1987 experienced global-scale bleaching of at least moderate severity (Table 3.2).

Heron et al. (2016) have carried out a similar type of analysis using satellite SST data to plot the incidence of bleaching-level thermal stress (≥4 °C-weeks) over time

Table 3.2 Classification of global bleaching events since 1979

Year	Global event	Comments
2014–2017	Severe	• Data still being compiled • Very high frequency of bleaching reports at 1° grid and regional scale • Bleaching reports extend from 2014 to 2017 with peak in 2016 • Major mortality impacts already reported in some areas
2010	Moderate	• Lower frequency of bleaching at 1° grid and regional scale compared to 2016 and 1998 • Mortality impacts also reported
1998	Severe	• Highest frequency of bleaching at 1° grid and regional scale • Extensive well-documented mortality
1987	Possible moderate?	• Discrete peak in bleaching frequency at 1° grid and regional scale • Absolute frequencies lower than later years that have not been classified as bleaching, but this could be due to lower reporting effort • Heat stress reached critical levels in a small percentage of global reefs
1983	Possible moderate?	• Discrete peak in bleaching frequency at 1° grid and regional scale • Absolute frequencies lower than other years not classified as bleaching, but this could be due to lower reporting effort • Bleaching on GBR occurred in 1982—well-documented mortality • Heat stress may have reached critical levels in a small percentage of global reefs

for different regions during 1985–2012. At the global scale, their analysis shows peaks in bleaching-level stress, corresponding to the 1998 and 2010 events described in Table 3.2. There is also an indistinct peak around 1987–1988 which approximately matches the possible moderate event for 1987 (Table 3.2). Newer analyses by NOAA (Eakin, pers. obs.) find that the extent of heat stress in 1983 and 1987 may have been less than half that seen in 2010. It is possible that other stressors such as light or water motion played a role in bleaching during these periods, but given the doubt surrounding the adequacy of the reporting effort on coral bleaching, and the importance of thermal stress in the development of mass bleaching events, the lack of a major bleaching-level stress peak for 1987 in Heron et al. (2016), comparable to 2010 or 1998, suggests that if there was a global bleaching event in 1987, it was not a major one. The same may also be true for 1983.

Donner et al. (2017) have recently conducted a comprehensive analysis of bleaching records based on a major revision of the ReefBase database. Their analysis of bleaching records over time differs from ours in grouping records to a much finer 0.4° grid and by excluding records with no bleaching or unknown bleaching severity. In general, our analysis agrees with Donner et al. (2017); however in our analysis the peaks for years 2000, 2007, and 2009 are much lower, probably due to less extensive spatial coverage of bleaching in those years.

The proximate cause of virtually all mass bleaching events is widely acknowledged to be thermal stress, with light, water motion, and ocean circulation acting as important modifiers (Brown 1997; Wilkinson 1998; Glynn 1993; Hoegh-Guldberg 1999; Mumby et al. 2001; Hughes et al. 2003, 2017). These reviews and others

(Glynn 1984, 2000, 2002; Wellington and Glynn 2007) have stressed the role of El Niño as one of the higher-level causes, especially for major events in 1983, 1998, and now 2016. The comparison of major El Niño events with the five global and possible global events identified here (Figs. 3.2 and 3.8) indicates that each one was associated with one of the eight most severe El Niño events in the last half century (NOAA 2017). Even the two significant but more regionally restricted bleaching events in the GBR and the Caribbean (2002 and 2005, respectively) were associated with weak or moderate El Niño events. The only exception to the close relationship between moderate to strong El Niño years and global bleaching is the extended moderate El Niño of 1991–1993, during which no major bleaching occurred. Not all teleconnections between the core canonical events of El Niño and thermal anomalies on coral reef areas have been clearly identified. Huppert and Stone (1998) suggest that additional stochastic triggering events may explain records of bleaching in non-El Niño years. The role of the Indian Ocean Dipole and the Madden-Julian Oscillation as modulators of ocean basin-scale heat stress has also been noted by Heron et al. (2016) and Zhang et al. (2017).

On the GBR, the records show a less direct correlation between El Niño strength (positive ONI, NOAA 2017) and bleaching events, although the proximity of major bleaching to these deviations is very suggestive. The ONI may, therefore, be a poor direct indicator of the impacts of ENSO events at a local or subregional scale even though it is possible that many of the GBR bleaching events are indirectly caused by ENSO-related climate anomalies. The reversals of normal ocean current directions, particularly in the equatorial Pacific during ENSO events, have a profound but indirect influence on local weather conditions. The nature and timing of the effects may be highly variable in different locations and from one event to another. For example, the western Pacific is generally cooler during an El Niño event which in theory suggests a low chance of bleaching (Chap. 4). In contrast, the western Pacific is generally warmer during strong La Niña years, increasing the chance of bleaching. However, El Niño conditions in Australia generally bring drought and long periods of cloud-free, doldrum-like conditions. When this coincides with the austral summer, local heating of shallow inshore waters occurs. This would explain the observed inshore-offshore bleaching pattern during major bleaching events. El Niño weather patterns operate at regional scales and affect reef provinces only when they coincide with the regional summer. The GBR probably avoided a major bleaching event in 2005 only because ocean current anomalies returned to normal just before the austral summer.

At the global scale, identification of detailed spatial bleaching patterns is hampered by the lack of detailed records for the most recent events, which have not yet been published, and probable under-reporting of events prior to 1998. In general, records are more likely to be complete in areas where there are concentrations of research activity and programmes of regular monitoring, especially in the GBR and parts of the Caribbean. In this respect 2010 stands out. Despite the presence of bleaching in five regions, very little bleaching occurred in the central and western Pacific and almost none on the GBR (Fig. 3.3). In the analysis of bleaching-level thermal stress carried out by Heron et al. (2016), the Pacific (excluding Australia) is the only region that does not show a peak for 2010. However, the Australian region (mostly comprising the GBR) has a peak in thermal stress for 2010 that is equal to that of the major bleaching year of 1998. The GBR may have been subject to localised influences independent of SST that prevented significant bleaching in 2010, although the lack of major tropical cyclones at this time rules out cyclonic cooling (Carrigan and Puotinen 2014) as a mitigating factor. Outside the GBR and much further south, however, Lord Howe Island reported severe bleaching in 2010 (Harrison et al. 2011).

Spatial bleaching patterns within the GBR (Figs. 3.6 and 3.7) show distinctive longitudinal and cross-shelf patterns that vary between years. In general, more frequent and severe bleaching is seen on inshore reefs and on reefs in the central GBR. The primary driver for these patterns is SST (Berkemans et al. 2004, Hughes et al. 2017). A number of factors, singly or in combination, may contribute to the observed effect. First, inshore shallow waters have a smaller volume and hence a reduced thermal capacity compared with deeper offshore waters. Second, inshore waters may have a reduced ability to mix with cooler deeper water simply because the waters are warm for the full depth of the water column. Third, inshore corals are generally more darkly pigmented compared with their offshore relatives due to reduced light availability (higher turbidity) and higher nutrient loadings. The higher pigment density has been shown to increase solar absorption and raise the effective temperature experienced by corals by up to 1.5 °C, exacerbating the bleaching risk for inshore coral communities (Fabricius 2006). More recently Wiedenmann et al. (2013), Wooldridge (2016), and Pogoreutz et al. (2017) have suggested that increased nutrients can lead to an imbalance between the metabolic activities and zooxanthellae and the coral host, leading to bleaching. Since anthropogenic impacts are also concentrated inshore, the clear management imperative to mitigate against additional stresses is doubly important in inshore environments. Good land management practices (river catchment, coastal development) and prudent fisheries management, especially for herbivorous fisheries, are particularly important priorities for management in these areas (Salm and Coles 2001; Marshall and Shuttenberg 2006). However, at times of severe heat stress, such management may be insufficient to protect corals (Hughes et al. 2017).

Apart from the strong inshore-offshore effect, bleaching in the most recent severe bleaching event on the GBR was patchy over scales of tens of kilometres, reflecting patterns in local weather and oceanography. Specific oceanographic features have also been shown to explain some of the detailed patterns of bleaching. The unusual bleaching of northern GBR reefs in 2016, for instance, has been explained by Wolanski et al. (2017) by a combination of a shutdown in the North Queensland Coastal Current, sea level-mediated transport of warm water from the Gulf of Carpentaria into the Torres Strait, and local solar heating. The lower levels of bleaching in the central and southern GBR compared to previous sever events can also be explained in terms of local weather events: ex-Tropical Cyclone Winston brought increased cloudiness and cooler temperatures to the region (Hughes et al. 2017). These local to subregional processes make prediction and scenario modelling particularly challenging at local scales and highlight the need for a detailed understanding of the oceanographic environment and appropriate tools to draw inferences from diverse data sets (Wooldridge and Done 2004; Skirving et al. 2006; Wooldridge et al. 2006). A positive implication from this patchiness is that there is likely to be a proportion of reefs that will not bleach in successive events, giving some affected reefs longer to recover. Unfortunately, climate change may disrupt some circulation patterns eliminating past "protection" in future years.

Our results clearly show an increase in the number of bleaching reports between 1983 and 2017, with a major increase in the past decade after the 1998 event. In addition to three major global events, there has been a low level of bleaching in all years and in multiple regions (Figs. 3.1 and 3.2). While this may be partly due to an increase in the level of reporting, the level of scientific research and monitoring, together with the dedicated efforts of key institutions to record all bleaching events since 1983, makes it likely that this is a real trend. This confirms the analyses of Heron et al. (2016) and Donner et al. (2017) who found, respectively, a threefold increase in the frequency of bleaching-level stress and an eightfold increase in the number of reefs with a greater than 50% probability of bleaching. When we look only at major peaks in bleaching records, then at a global level we can clearly differentiate three to five global events. This is too small a number to quantitatively determine whether the frequency of severe events is increasing. However, the two lesser events preceded 1990, and the more severe events occurred in 1998, 2010, and 2014–2017. While there is insufficient evidence in the global database of bleaching records to statistically support or refute the hypothesis that major bleaching events are increasing in frequency, there is growing evidence that low-level background bleaching has increased to the point where most regions and ocean basins are reporting some level of bleaching every year. This is most likely linked to the rise in ocean temperatures.

A separate, but related, issue is whether the intensity of bleaching is increasing. As discussed above, our results (Fig. 3.4) do not support this notion based on the relative frequency of the severe bleaching category in the database. This finding contrasts with data that show clear increases in both the frequency and intensity of bleaching-level thermal stress (Heron et al. 2016; Chap. 4). It is possible that a real increase in severity is not discernable from an analysis of bleaching categories but would require more information on subsequent mortality impacts. If the lack of a trend is real, one potential reason is that the corals that survive severe events, such as 1998, are more capable of surviving subsequent thermal stress. Studies on this hypothesis have reported mixed results so far (Carilli et al. 2012; Guest et al. 2012; Pratchett et al. 2013; Hughes et al. 2017). A corollary is that severe bleaching events reduce diversity, removing the more thermally sensitive corals.

The revised global database of Donner et al. (2017) represents an important resource for documenting and understanding the impacts of coral bleaching. Its utility could be greatly increased if monitoring and reporting effort could be standardised. This would enable smaller-scale patterns of milder bleaching events to be reliably detected, thus providing early verification of predicted increases in bleaching frequency due to climatic temperature increases. It would also provide much needed ground truth data for the bleaching HotSpots detected by satellite data

(Chap. 4). Two key improvements in bleaching reporting are standardisation in the measurement of bleaching intensity and standardisation in the number, location, and timing of bleaching surveys. The first issue was addressed by Oliver et al. (2004) and Marshall and Schuttenberg (2006). The adoption and use of a standard reporting protocol would greatly increase the reliability of bleaching records. The second issue will require a much greater level of coordinated effort by coral reef scientists and reef users. Existing global networks such as Reef Check and the currently semi-operative Global Coral Reef Monitoring Network, together with major regional monitoring programmes such as the AGRRA and National Coral Reef Monitoring Program (NCRMP) and the programmes in Australia by the Australian Institute of Marine Science (AIMS) and the Great Barrier Reef Marine Park Authority (GBRMPA), are well placed to promote the regular reporting of bleaching conditions (both presence and absence) from standard locations; and we hope that these can encourage greater monitoring and reporting as a priority.

Overall, we conclude that the existing observations of coral bleaching enable us to confirm that severe bleaching events occurred at a global level on three recent occasions and possibly on two other occasions prior to 1998. Also, we have demonstrated that each of these events occurred in close temporal proximity to a strong El Niño event. While no clear increase in the frequency or intensity of major global bleaching events is so far discernable using bleaching records alone, the observed and predicted increase in ocean temperatures to which El Niños add extra warming throughout much of the tropical oceans (Chap. 4) has been predicted to dramatically increase the frequency and severity of bleaching events. In addition, the frequency and extent of annual bleaching records have clearly increased over the period 1979–2016 to the point where bleaching is now reported at various sites around the world every year. If trends in global ocean warming continue, reefs may be faced with a combination of both chronic bleaching and more frequent and highly destructive events.

Acknowledgements C.M.E. thanks NOAA, including the Coral Reef Watch and the Coral Reef Conservation Program, for work that contributed to this manuscript. The scientific results and conclusions, as well as any views or opinions expressed herein, are those of the author(s) and do not necessarily reflect the views of NOAA or the Department of Commerce. The Great Barrier Reef Marine Park Authority is thanked for providing data on bleaching for the Great Barrier Reef from its Eye on the Reef Program. Barbara Brown, Janice Lough, and Madeleine van Oppen provided very helpful and thoughtful comments on the manuscript.

References

Baker AC, Glynn PW, Riegl B (2008) Climate change and coral reef bleaching: an ecological assessment of long-term impacts, recovery trends and future outlook. Estuar Coast Shelf Sci 80:435–471. https://doi.org/10.1016/j.ecss.2008.09.003

Berkelmans R, Oliver J (1999) Large-scale bleaching of corals on the Great Barrier Reef. Coral Reefs 18:55–60. https://doi.org/10.1007/s003380050154

Berkelmans R, De'ath G, Kininmonth S, Skirving W (2004) A comparison of the 1998 and 2002 coral bleaching events on the Great Barrier Reef: spatial correlation, patterns, and predictions. Coral Reefs 23:74–84. https://doi.org/10.1007/s00338-003-0353-y

Brown BE (1987) Worldwide death of corals – natural cyclical events or man-made pollution? Mar Pollut Bull 18:9–13. https://doi.org/10.1016/0025-326X(87)90649-7

Brown BE (1997) Coral bleaching: causes and consequences. Coral Reefs 16:S129–S138. https://doi.org/10.1007/s003380050249

Burke L, Reytar K, Spalding M, Perry A (2011) Reefs at risk revisited. World Resources Institute, Washington, DC

Carilli J, Donner SD, Hartmann AC (2012) Historical temperature variability affects coral response to heat stress. PLoS One 7:e34418. https://doi.org/10.1371/journal.pone.0034418

Carrigan AD, Puotinen M (2014) Tropical cyclone cooling combats region-wide coral bleaching. Glob Chang Biol 20:1604–1613. https://doi.org/10.1111/gcb.12541

COECRS (2017a) Map of the Great Barrier Reef showing results of aerial surveys for 911 reefs. ARC Centre of Excellence for Coral Reef Studies. https://www.coralcoe.org.au/media-releases/only-7-of-the-great-barrier-reef-has-avoided-coral-bleaching

COECRS (2017b) Composite map showing surveyed coral reefs in 2016 (left panel) and 2017 (right panel). ARC Centre of Excellence for Coral Reef Studies. https://www.coralcoe.org.au/media-releases/two-thirds-of-great-barrier-reef-hit-by-back-to-back-mass-coral-bleaching

Coffroth MA, Lasker HA, Oliver JK (1990) Coral mortality outside of the eastern Pacific during 1982–1983: relationship to El Niño. In: Glynn PW (ed) Global ecological consequences of the 1982–83 El Niño–Southern Oscillation. Elsevier, New York, pp 141–182

Donner SD, Skirving WJ, Little CM, Oppenheimer M, Hoegh-Guldberg O (2005) Global assessment of coral bleaching and required rates of adaptation under climate change. Glob Chang Biol 11:2251–2265. https://doi.org/10.1111/j.1365-2486.2005.01073.x

Donner SD, Rickbeil GJM, Heron SF (2017) A new, high-resolution global mass coral bleaching database. PLoS One 12:e0175490. https://doi.org/10.1371/journal.pone.0175490

Eakin CM, Morgan JA, Heron SF, Smith TB, Liu G, Alvarez-Filip L et al (2010) Caribbean corals in crisis: record thermal stress, bleaching, and mortality in 2005. PLoS One 5:e13969. https://doi.org/10.1371/journal.pone.0013969

Eakin CM, Liu G, Gomez AM, De La Cour JL, Heron SF, Skirving WJ et al (2016) Global coral bleaching 2014-2017. Reef Encount 31:20–26

Eakin CM, Liu G, Gomez AM, De La Cour JL, Heron SF, Skirving WJ et al (2017) Ding, dong, the witch is dead (?) – three years of global coral bleaching 2014-2017. Reef Encount 32:31–36

Fabricius KE (2006) Effects of irradiance, flow, and colony pigmentation on the temperature microenvironment around corals: implications for coral bleaching? Limnol Oceanogr 51:30–37. https://doi.org/10.4319/lo.2006.51.1.0030

Fitt WK, Brown BE, Warner ME, Dunne RP (2001) Coral bleaching: interpretation of thermal tolerance limits in tropical corals. Coral Reefs 20:51–65. https://doi.org/10.1007/s003380100146

Gattuso J-P, Magnan A, Billé R, Cheung WWL, Howes EL, Joos F et al (2015) Contrasting futures for ocean and society from different anthropogenic CO_2 emissions scenarios. Science 349:6243. https://doi.org/10.1126/science.aac4722

GBRMPA (2017) Reef Health Impact Surveys (RHIS) data on coral bleaching supplied courtesy of the Great Barrier Reef Marine Park Authority Eye on the Reef Program. Email: eyeonthereef@gbrmap.gov.au

Glynn PW (1983) Extensive "bleaching" and death of reef corals on the Pacific coast of Panama. Environ Conserv 10:149–154. https://doi.org/10.1017/s0376892900012248

Glynn PW (1984) Widespread coral mortality and the 1982–83 El Niño warming event. Environ Conserv 11:133–146. https://doi.org/10.1017/S0376892900013825

Glynn PW (1990) Coral mortality and disturbances to coral reefs in the tropical eastern Pacific. In: Glynn PW (ed) Global ecological consequences of the 1982–83 El Niño–Southern Oscillation. Elsevier, New York, pp 55–126

Glynn PW (1991) Coral reef bleaching in the 1980s and possible connections with global warming. Trends Ecol Evol 6:175–179. https://doi.org/10.1016/0169-5347(91)90208-F

Glynn PW (1993) Coral reef bleaching: ecological perspectives. Coral Reefs 12:1–17. https://doi.org/10.1007/BF00303779

Glynn PW (2000) El Niño–Southern Oscillation mass mortalities of reef corals: a model of high temperature marine extinctions. In: Insalaco E, Skelton PW, Palmer TJ (eds) Carbonate platform systems: components and interactions. Geological Society, London, Special Publications, vol 178, pp 117–133

Glynn PW (2002) Effects of the 1997–98 El Niño–Southern Oscillation on eastern Pacific corals and coral reefs: an overview. In: Moosa MK, Soemodihardjo S, Soegiarto A, Romimohtarto K, Nontji A, Soekarno et al. (eds) Proceedings of the international 9th coral reef symposium, Bali 9-2, pp 1169–1174

Goreau TF (1964) Mass expulsion of zooxanthellae from Jamaican coral reef communities after Hurricane Flora. Science 145:383–386. https://doi.org/10.1126/science.145.3630.383

Goreau TJ, Hayes RL (1994) Coral bleaching and ocean hot spots. Ambio 23:176–180

Guest JR, Baird AH, Maynard JA, Muttaqin E, Edwards AJ, Campbell SJ et al (2012) Contrasting patterns of coral bleaching susceptibility in 2010 suggest an adaptive response to thermal stress. PLoS One 7:e33353. https://doi.org/10.1371/journal.pone.0033353

Harrison PL, Dalton SJ, Carroll AG (2011) Extensive coral bleaching on the world's southernmost coral reef at Lord Howe Island, Australia. Coral Reefs 30:775. https://doi.org/10.1007/s00338-011-0778-7

Heron SF, Johnston L, Liu G, Geiger EF, Maynard JA, De La Cour JL et al (2016) Validation of reef-scale thermal stress satellite products for coral bleaching monitoring. Remote Sens 8:59. https://doi.org/10.3390/rs8010059

Hoegh-Guldberg O (1999) Climate change, coral bleaching and the future of the world's coral reefs. Mar Freshw Res 50:839–866. https://doi.org/10.1071/MF99078

Hughes TP, Baird AH, Bellwood DR, Card M, Connolly SR, Folke C et al (2003) Climate change, human impacts, and the resilience of coral reefs. Science 301:929–933. https://doi.org/10.1126/science.1085046

Hughes TP, Kerry J, Álvarez-Noriega M, Álvarez-Romero JG, Anderson KD, Baird AH et al (2017) Global warming and recurrent mass bleaching of corals. Nature 543:373–378. https://doi.org/10.1038/nature21707

Huppert A, Stone L (1998) Chaos in the Pacific's coral bleaching cycle. Am Nat 152:447–459. https://doi.org/10.1086/286181

Logan CA, Dunne JP, Eakin CM, Donner SD (2014) Incorporating adaptive responses into future projections of coral bleaching. Glob Chang Biol 20:125–139. https://doi.org/10.1111/gcb.12390

Marshall P, Shuttenberg H (2006) A reef managers guide to coral bleaching. Great Barrier Reef Marine Park Authority, Townsville

Mayer AG (1914) The effects of temperature on tropical marine animals. Car Inst Wash Pap Tort Lab 6:1–24

Mumby PJ, Chisolm JRM, Edwards AJ, Andrefouet S, Jaubert J (2001) Cloudy weather may have saved Society Island reef corals during the 1998 ENSO event. Mar Ecol Prog Ser 222:209–216. https://doi.org/10.3354/meps222209

NOAA (2017) Online ONI data: http://www.cpc.ncep.noaa.gov/products/analysis_monitoring/ensostuff/ensoyears.shtml. Accessed June 2017

Null J (2017) El Niño and La Niña Years and Intensities, based on Oceanic Niño Index (ONI). http://ggweather.com/enso/oni.htm. Updated June 2017, Accessed July 2017

Oliver J (1985) Recurrent seasonal bleaching and mortality of corals on the Great Barrier Reef. In: Gabrie C, Salvat B (eds) Proceedings of the international coral reef congress 5-4, pp 201–212

Oliver J, Marshall P, Setiasih N, Hansen L (2004) A global protocol for assessment and monitoring of coral bleaching. WorldFish Center/WWF, Penang/Indonesia

Pogoreutz C, Rädecker N, Cárdenas A, Gärdes A, Voolstra CR, Wild C (2017) Sugar enrichment provides evidence for a role of nitrogen fixation in coral bleaching. Glob Chang Biol 23:3838–3848. https://doi.org/10.1111/gcb.13695

Pratchett MS, McCowan D, Maynard JA, Heron SF (2013) Changes in bleaching susceptibility among corals subject to ocean warming and recurrent bleaching in Moorea, French Polynesia. PLoS One 8:e70443. https://doi.org/10.1371/journal.pone.0070443

Rajasuriya A, Zahir H, Venkataraman K, Islam Z, Tamelander J (2004) Status of coral reefs in South Asia: Bangladesh, Chagos, India, Maldives and Sri Lanka. In: Wilkinson C (ed) Status of coral reefs of the world: 2004, vol 1. Australian Institute of Marine Science, Townsville, pp 213–234

Reef Check (2017) Online database of ReefCheck global surveys. http://data.reefcheck.us/. Accessed in May 2017

Salm RV, Coles SL (eds) (2001) Coral bleaching and marine protected areas. Proceedings of the workshop on mitigating coral bleaching impact through MPA design, Bishop Museum, Honolulu, Hawaii, 29–31 May 2001. Asia Pacific coastal marine program report 0102. The Nature Conservancy, Honolulu

Shinn EA (1966) Coral growth-rate: an environmental indicator. J Paleontol 40:233–240

Skirving W, Heron M, Heron S (2006) The hydrodynamics of a bleaching event: implications for management and monitoring. In: Phinney JT, Hoegh-Guldberg O, Kleypas J, Skirving W, Strong A (eds) Coral reefs and climate change science and management, Coastal and estuarine studies, vol 61. American Geophysical Union, Washington, DC, pp 145–161

Spalding MD, Brown BE (2015) Warm-water coral reefs and climate change. Science 350:769–771. https://doi.org/10.1126/science.aad0349

van Hooidonk R, Maynard J, Tamelander J, Gove J, Ahmadia G, Raymundo L et al (2016) Local-scale projections of coral reef futures and implications of the Paris agreement. Sci Rep 6:39666. https://doi.org/10.1038/srep39666

Wellington GM, Glynn PW (2007) Responses of coral reefs to El Niño–Southern Oscillation sea-warming events. In: Aronson RB (ed) Geological approaches to coral reef ecology. Springer, New York, pp 342–385

Wiedenmann J, D'Angelo C, Smith EG, Hunt AN, Legiret F-E, Postle AD et al (2013) Nutrient enrichment can increase the susceptibility of reef corals to bleaching. Nat Clim Chang 3:160–164. https://doi.org/10.1038/nclimate1661

Wilkinson CR (1998) Status of coral reefs of the world: 1998. Global coral reef monitoring network. Australian Institute of Marine Science, Townsville

Wilkinson CR (2000) Status of coral reefs of the world: 2000. Global coral reef monitoring network. Australian Institute of Marine Science, Townsville

Wilkinson CR (2002) Status of coral reefs of the world: 2002. Global coral reef monitoring network. Australian Institute of Marine Science, Townsville

Williams EH, Bunkley-Williams L (1990) The worldwide coral bleaching cycle and related sources of coral mortality. Atoll Res Bull 335:1–71. https://doi.org/10.5479/si00775630.335.1

Wolanski E, Andutta F, Deleersnijder E, Li Y, Thomas CJ (2017) The Gulf of Carpentaria heated Torres Strait and the Northern Great Barrier Reef during the 2016 mass coral bleaching event. Estuar Coast Shelf Sci 194:172–181. https://doi.org/10.1016/j.eccs.2017.06.018

Wooldridge S (2016) Excess seawater nutrients, enlarged symbiont densities and bleaching sensitive reef locations: 1. Identifying thresholds of concern for the Great Barrier Reef, Australia. Mar Pollut Bull. https://doi.org/10.1016/j.marpolbul.2016.04.054

Wooldridge S, Done T (2004) Learning to predict large-scale coral bleaching from past events: a Bayesian approach using remotely sensed data, in situ data, and environmental proxies. Coral Reefs 23:96–108. https://doi.org/10.1007/s00338-003-0361-y

Wooldridge S, Brodie J, Furnas M (2006) Exposure of inner-shelf reefs to nutrient enriched runoff entering the Great Barrier Reef lagoon: post-European changes and the design of water quality targets. Mar Pollut Bull 52:1467–1479. https://doi.org/10.1016/j.marpolbul.2006.05.009

Yonge CM, Nicholls AG (1931) The structure, distribution and physiology of the zooxanthellae. (Studies on the physiology of corals IV). Sci Rep Great Barrier Reef Exped 1928–29 1:135–176

Zhang N, Feng M, Hendon HH, Hobday AJ, Zinke J (2017) Opposite polarities of ENSO drive distinct patterns of coral bleaching potentials in the southeast Indian Ocean. Sci Rep 7:2443. https://doi.org/10.1038/s41598-017-02688-y

Chapter 4
Climate Variability and Change: Monitoring Data and Evidence for Increased Coral Bleaching Stress

C. M. Eakin, J. M. Lough, S. F. Heron, and G. Liu

4.1 Introduction

Coral reefs live within a fairly narrow envelope of environmental conditions constrained by water temperatures, light, salinity, nutrients, bathymetry, and the aragonite saturation state of seawater (Buddemeier and Kinzie 1976; Kleypas et al. 1999; Hoegh-Guldberg 2005). Their natural environment, at the interface of land, sea, and the atmosphere, can vary quickly and potentially be stressful. Reef organisms have, over millions of years, evolved strategies to cope with occasional environmental disturbances (such as tropical cyclones). Given sufficient time between disturbances, damage or destruction would normally be followed by recovery and regrowth (Buddemeier et al. 2004). As documented in numerous scientific studies and reports, the world's coral reefs are "in crisis" as a result of direct local- and regional-scale human impacts on their environment. These impacts include overfishing; destructive fishing practices; changed land use that increases sediment, nutrient, and pollutant flows into reef waters; and poorly designed coastal development. Such local impacts were, in the past, the primary ecosystem degradation in

C. M. Eakin (✉) · S. F. Heron
U.S. National Oceanic and Atmospheric Administration, Coral Reef Watch, College Park, MD, USA
e-mail: mark.eakin@noaa.gov

J. M. Lough
Australian Institute of Marine Science, Townsville MC, QLD, Australia

ARC Centre of Excellence for Coral Reef Studies, James Cook University, Townsville, QLD, Australia

G. Liu
U.S. National Oceanic and Atmospheric Administration, Coral Reef Watch, College Park, MD, USA

Global Science and Technology Inc., Greenbelt, MD, USA

M. J. H. van Oppen, J. M. Lough (eds.), *Coral Bleaching*, Ecological Studies 233, https://doi.org/10.1007/978-3-319-75393-5_4

those tropical countries whose increasing populations are heavily dependent on coral reefs yet have insufficient resources to develop appropriate, sustainable management practices (Wilkinson 2004). Now, the greatest threat to coral reefs are destructive global-scale stresses due to the accumulation of anthropogenic greenhouse gases in the atmosphere that are rapidly changing coral reefs' environmental envelopes through both ocean acidification and increased heat stress due to climate change (Hoegh-Guldberg et al. 2007; Heron et al. 2016a).

There are several aspects of global climate change that are already impacting the environments of coral reef ecosystems (Chap. 1). The most immediate is rising sea surface temperature (SST) that is correlated with an increased frequency of mass coral bleaching reports since the early 1980s (Glynn 1993; Chap. 3). Pioneering studies in the 1970s demonstrated just how close (within 1–2 °C) reef-building corals usually live to their upper thermal tolerance limits and further how subtle elevation in temperature often led to bleaching (Coles et al. 1976; Jokiel and Coles 1977; Glynn and D'Croz 1990). These studies and others have identified that temperature thresholds at which corals bleach vary with the ambient water temperatures on each reef, such that corals have adapted to their local environmental conditions over long timescales (Chap. 9).

Alarming first reports of mass coral bleaching events were not immediately linked with unusually warm global SSTs until a connection eventually was made with El Niño warming (Glynn 1983, 1984; Coffroth et al. 1990). This was largely due to limited availability of reliable, long-term records of SSTs and other environmental variables in the vicinity of coral reefs. Gradually, as more mass bleaching events occurred and observations improved, the link was made with unusually warm SSTs (Brown 1987; Glynn 1990, 1991, 1996). Unfortunately, the reporting of bleaching events is still incomplete in many parts of the world, and the ability to tease out causal relationships between bleaching and environmental conditions is confounded by two simultaneous factors: (1) the rising heat stress and (2) the recent increase in monitoring and reporting (Chap. 3). Nevertheless, the evidence linking mass bleaching of coral reefs to global climate change due to increased anthropogenic greenhouse gas emissions (Smith and Buddemeier 1992; Brown 1997), which was considered unconvincing in the early 1990s (Glynn 1993), is now considered incontrovertible (Hughes et al. 2003, 2017a, b; Heron et al. 2016a).

This chapter focuses on the changing physical environment of coral reef ecosystems and especially SSTs that are responsible for most mass coral bleaching events. We use long-term climatic datasets to document SST changes near coral reefs over the past 146 years (future scenarios are considered in Chap. 13). SST conditions conducive to coral bleaching are considered in the context of observed global-scale ocean warming and climatic variability [e.g., El Niño-Southern Oscillation (ENSO)]. Since the 1980s, satellite-based observations of the oceans have dramatically increased our capability to observe ocean variations globally and synoptically and provide the basis for identifying bleaching conditions in near-real-time. We describe the application of satellite-based SST and other products to detect and monitor environmental conditions related to coral bleaching events.

4.2 Data for Understanding Heat Stress and Bleaching Patterns Worldwide

To date, no dataset provides continuous coverage spanning modern satellite and instrumental observations back through multiple decades or centuries. However, good data do exist at various temporal and spatial scales. Two general types of primary datasets are available: (1) century-length reconstructions generated from available instrumental observations of global SST and (2) modern (post-1980), near-real-time satellite observations and reanalyses of these records. Because both of these data are calibrated from similar instrumental datasets for recent years, both are considered to accurately represent large-scale patterns of thermal conditions that influence coral reefs. The datasets used for this study are described below.

4.2.1 Century-Length Global SST Reconstructions from Instrumental and Paleoclimatic Data

Unfortunately, continuous observations of physical parameters have been taken at only a limited number of reef sites and for less than 30 years. Observations from other parts of the global ocean are not much more complete and largely consist of open ocean measurements. The need for long records of SSTs has driven the development of new local to global ocean observing systems, satellite observations, and algorithms to reconstruct past SSTs from instrumental data that are heterogeneous in space and time. Two organizations have developed such reanalyses that are widely used. The Hadley Centre Sea Ice and Sea Surface Temperature Version 1 (HadISST1) dataset was developed at the Hadley Centre of the UK Meteorological Office (available at http://www.metoffice.gov.uk/hadobs/hadisst). It provides monthly, globally replete fields of SST and sea ice concentration on a one-degree latitude-longitude grid from 1870 to present (Rayner et al. 2003). Two separate reanalyses were developed at the National Centers for Environmental Information of the US National Oceanic and Atmospheric Administration (NOAA), the Optimum Interpolation Sea Surface Temperature analysis (OISST) dataset providing ¼-degree daily, globally complete fields of SST from 1981 to present and the Extended Reconstructed Sea Surface Temperature (ERSST) dataset providing two-degree monthly, globally complete fields of SST from 1854 to present (both available at https://www.ncdc.noaa.gov) (Reynolds and Smith 1994; Reynolds et al. 2002; Smith and Reynolds 2003, 2004). While the methods used in developing these datasets are similar, users should examine each one to select the best dataset for their application (e.g., Huang et al. 2017).

Paleoclimatic data also extend our understanding of climate patterns into the past before routine instrumental measurements (Jones and Mann 2004; Jones et al. 2009). Certain annually banded massive coral skeletons contain a rich archive of past climatic and environmental conditions in coral reef environments (Lough 2010;

McGregor 2011), which can be extracted, for example, using the ratios of stable isotopes ($\delta^{18}O/^{16}O$) or trace metals (Sr/Ca) in annual growth bands (Felis and Patzold 2003; Gagan and Abram 2011). Although there are problems with the sparse array of available data from long coral cores (Evans et al. 2002) and, until recently, a lack of replication (Lough 2004), it is now possible to reliably reconstruct regional tropical SST for most of the past 400 years (Tierney et al. 2015). These reconstructions show that tropical SST was cooling prior to the mid-1800s but has significantly warmed in the Indian, western Pacific, and Atlantic Oceans since 1850 by ~0.04 °C per decade. Coral cores have also revealed disruptions in growth and slowing of growth rates in response to coral bleaching events (Carilli et al. 2009; Cantin and Lough 2014; Manzello et al. 2015; Barkley and Cohen 2016) and that these bleaching-growth anomalies appear to be a recent phenomenon (Carilli et al. 2010). There is also evidence from several reef locations for a recent slowing in the growth rates of massive corals (e.g., De'ath et al. 2009, 2013; D'Olivo et al. 2013; Manzello 2010; Tanzil et al. 2009, 2013)—the primary driver of these declines appears to be the local rate of ocean warming (Cooper et al. 2012). This results in setbacks to growth due to heat stress and a general decline in growth due to water temperatures persistently exceeding optimum values (Lough and Cantin 2014).

4.2.2 Satellite Observations of SST and Heat Stress

NOAA Coral Reef Watch's (CRW) original 50-km ("heritage") products used only data from polar-orbiting satellites, which provide near-real-time observations across the globe. Global 0.5° (approx. 50 km) nighttime SSTs and SST-based products were delivered at a twice-weekly time step. As higher spatial resolution was the product improvement most requested by users, CRW developed new daily products at global 0.05° (approx. 5 km) resolution, again using only nighttime SSTs. This increase in both spatial and temporal resolution was accomplished by using both polar-orbiting and geostationary satellite sensors for SST (NOAA's Geo-Polar Blended SST Analysis, Maturi et al. 2017, Fig. 4.1a). Geostationary satellites provide more frequent sampling and better measures of variability and patterns throughout the day, but with coverage limited to part of one hemisphere. International data sharing agreements have provided access to geostationary data for most coral reef areas and hopefully will soon cover all coral reef areas globally.

CRW's current product suite includes 5-km SST, SST anomaly, coral bleaching HotSpot, coral bleaching Degree Heating Week (DHW), and Bleaching Alert Area updated daily (Fig. 4.1; Liu et al. 2014, 2017). These satellite data products are available online in graphical formats and as data files for the period from 2013 to the present (CRW 2017; http://coralreefwatch.noaa.gov). Animations of all product charts over the most recent 30 and 60 days are also available. Other products, such as the Regional Virtual Stations, SST time series, and automated Satellite Bleaching Alert emails provide needed information to coral reef resource managers and scientists (Heron et al. 2016b). The current suite of near-real-time products were

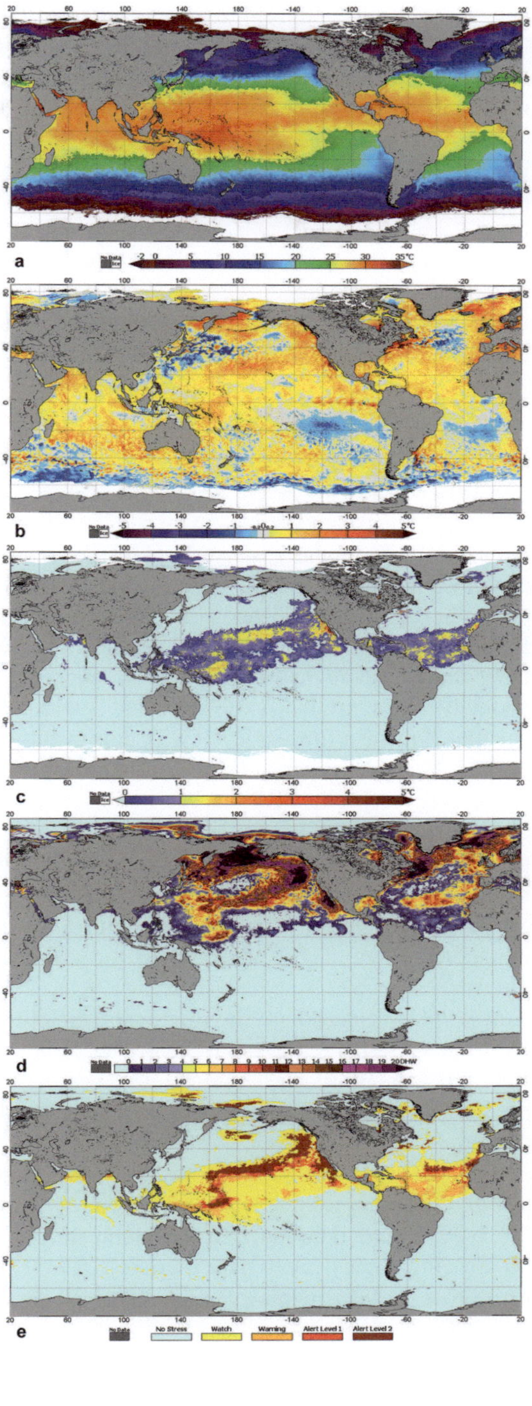

Fig. 4.1 NOAA Coral Reef Watch near-real-time satellite global 5-km nighttime product suite for 17 October 2014: (**a**) sea surface temperature (SST), (**b**) SST anomaly, (**c**) coral bleaching HotSpot, (**d**) coral bleaching Degree Heating Week (DHW), and (**e**) Bleaching Alert Area

developed based on algorithms derived from earlier monthly analyses of satellite and in situ SST data (Montgomery and Strong 1994; Gleeson and Strong 1995; Strong et al. 1997; Goreau et al. 2000) and are direct descendants of the 50-km products CRW has produced for two decades (Liu et al. 2006).

The near-real-time geo-polar blended SST dataset, a geo-polar blended SST reanalysis from earlier satellite data, and other high-resolution satellite-based SST records have been combined to provide a single SST dataset, Coral Reef Watch's CoralTemp, that spans 1985–present (Skirving et al. 2018). Based on CoralTemp, a full suite of 1985–present Coral Reef Watch daily 5-km coral bleaching heat stress product suite, including all of the 5-km products described above (see Sect. 4.2.2.5), will come online in 2018. CRW continues to evaluate and improve products with respect to both input datasets and product algorithms.

4.2.2.1 SST Anomaly

The SST anomaly (Fig. 4.1b) is the positive, negative, and zero departure of temperature from the expected value (i.e., climatology or long-term mean) for each location at the specific time of year, identifying where temperature is warmer or cooler than usual. SST anomaly has been instrumental in visualizing the effects of ENSO across the Pacific and with applications beyond coral reefs including ocean front analyses and whale migrations. To compute the SST anomaly, as well as other anomaly-based products, the expected value for each day at each data pixel is calculated using a time series of SST data spanning 1985–2012. For the first release of the 5-km product suite, the climatology was derived from the Pathfinder record of Advanced Very High Resolution Radiometer (AVHRR) nighttime SSTs (Heron et al. 2015). However, spatially and temporally variant errors were found in the products derived using the Pathfinder-based climatology, especially in regions with significant and persistent seasonal cloud cover (i.e., monsoonal regions). This was especially apparent during the 2014–2017 global coral bleaching event. In 2017, a new climatology (CRW version 3) was derived from the Operational SST and Sea Ice Analysis (OSTIA) reanalysis from 1985 to 2002 (Roberts-Jones et al. 2012) and a new NOAA National Environmental Satellite and Data Information Service (NOAA/NESDIS) reanalysis of geo-polar blended SSTs from 2002 to 2012 (Maturi et al. 2017; A. Harris pers. comm.). This new version 3 climatology was used for the production of the SST anomaly and all subsequent CRW products provided in this chapter.

For all of the anomaly-based 5-km products, the 12 monthly mean (MM) climatologies for each pixel (Jan, Feb, etc.) were based on the period 1985–2012 and de-trended to the temperature at the midpoint of the heritage 50-km climatology previously used by CRW (Heron et al. 2015; Liu et al. 2017). This process of de-trending provided datasets that include the full variability seen in the 1985–2012 record but retain the point in time to be consistent with CRW's original 50-km products. Thus, despite the obviously rising temperatures during

1985–2012, these datasets provide bleaching predictions that do not assume any acclimation or adaptation to higher temperatures on the part of the corals since 1985.

4.2.2.2 HotSpot

The coral bleaching HotSpot (Fig. 4.1c) is the positive anomaly of temperatures that exceed the maximum monthly mean (MMM) climatology for each 5-km pixel, thus identifying regions that are currently exposed to heat stress. For each pixel, the MMM climatology is the warmest among the 12 MM climatologies described in the previous section, representing the expected summertime maximum. CRW originally released the 50-km satellite coral bleaching HotSpot product in 1997 based on the "ocean hot spots" concept introduced by Goreau and Hayes (1994) from analyses by Atwood et al. (1992) and experiments by Glynn and D'Croz (1990). The 50-km HotSpot was produced experimentally over 1997–2002 and became CRW's first operational product in September 2002. This was replaced by the initial 5-km version in 2013 (Liu et al. 2014) and subsequent 5-km version 3 in 2017 (Liu et al. 2017).

4.2.2.3 Degree Heating Week

While the HotSpot product is extremely useful, it only provides an instantaneous measure of heat stress. Corals respond to the cumulative heat stress to which they are exposed. The NOAA Degree Heating Week (DHW) index is a highly reliable predictor of impacts on corals from sustained heat stress. By accumulating the positive anomalies of SST above the MMM, CRW's DHW index provides a measure of the cumulative heat stress that corals experience (Fig. 4.1d). The 50-km DHW product was produced experimentally starting in 2000 and became operational in September 2003. Like the HotSpot, the original DHW was replaced by the initial 5-km version in 2013 (Liu et al. 2014) and subsequent 5-km version 3 in 2017 (Liu et al. 2017). Following Glynn and D'Croz (1990) and Atwood et al. (1992), the threshold for bleaching was established to be 1 °C above the expected summer maximum temperature; and the DHW is calculated by summing all HotSpot values $\geq$1 °C in each pixel over a 12-week period (Liu et al. 2003; Skirving et al. 2006a). In most cases, HotSpot values $<$1 °C do not result in cumulative heat stress leading to widespread coral bleaching. This high-pass clipping filter reduces run-away accumulations that can result when SSTs remain very close to the MMM for long periods of time, a condition often encountered in equatorial mid-Pacific regions. A DHW value of 2 °C-weeks is equivalent to 2 weeks of HotSpot values of 1 °C and also to 1 week of HotSpot values of 2 °C and so forth. CRW issues a coral bleaching alert via email to registered users when values at or near a reef reach DHW values $\geq$4 °C-weeks (Alert Level 1; see the next section for bleaching heat stress categories). Significant coral bleaching is expected to occur 1–3 weeks after the heat stress reaches this level. Mass bleaching and the onset of coral mortality are expected after reefs experience DHW values $\geq$8 °C-weeks (Alert Level 2). The DHW index has predicted bleaching of corals around the world (Eakin et al. 2017), including recently in the Commonwealth of the Northern Mariana Islands (2013, 2014; Heron et al. 2016b), the Hawaiian archipelago (2015; Couch et al. 2017), and the Great Barrier Reef (2016, 2017; Hughes et al. 2017b; Hughes and Kerry 2017).

4.2.2.4 Bleaching Alert Area

Both HotSpot and DHW products provide information needed to predict the likelihood of coral bleaching. However, the Bleaching Alert Area product was added in 2009 (Liu et al. 2013) to provide a single collective product where reef managers and others could quickly reference the information they needed most. The Bleaching Alert Area is a semiquantitative product that combines current and accumulated heat stress conditions into five categories: No Stress, Bleaching Watch, Bleaching Warning, Alert Level 1, and Alert Level 2. No Stress indicates the SST is below the MMM. A Bleaching Watch is issued once HotSpots are present (SST above MMM). A Bleaching Warning is issued once DHW accumulation begins (SST at or above bleaching threshold, MMM + 1 °C). At DHW values of 4 °C-weeks or above, significant bleaching is expected and an Alert Level 1 is issued. Finally, at 8 °C-weeks or above, widespread bleaching and significant mortality are expected and an Alert Level 2 is issued. Importantly, the three highest alert levels (Warning, Alert 1, and Alert 2) are only sustained when HotSpots reach or exceed the level at which heat stress is accumulated (i.e., HotSpot $\geq$ 1 °C).

4.2.2.5 Reprocessed and Heritage Satellite Observations of SST

In addition to the operational near-real-time satellite observations, retrospective analyses of satellite SST data have been produced and periodically updated. CRW has created multiple products based on the Pathfinder ver. 5.0–5.3 datasets (NOAA 2007; http://www.nodc.noaa.gov/sog/pathfinder4km/). Pathfinder provides daily global SST data at approximately 4-km resolution (Global Area Coverage) covering the period 1981–2014 (Kilpatrick et al. 2001). From the Pathfinder source data, SST archives were produced at 0.5° (50-km) and half-weekly resolution and 4-km and weekly resolution. These archives were applied to develop CRW historical product suites mimicking the methods for the near-real-time products (see Sects. 4.2.2.1–4.2.2.4) and were used to both support the development of earlier versions of CRW's near-real-time products (Heron et al. 2014; Liu et al. 2013, 2014) and examine coral impacts (e.g., Eakin et al. 2009; Heron et al. 2010, 2016b).

While these Pathfinder-based products have been valuable, the Pathfinder dataset has a very large percentage (87%) of data of too poor quality for use. This has limited the application of Pathfinder-based data to temporally or spatially composited applications and has resulted in significant and spatially variable biases. Such issues were especially large in tropical areas with monsoonal climates due to persistent high cloud cover. For this reason, CRW developed a new, long-term dataset at daily resolution from 1985 to present. The first version of this dataset, named CoralTemp (Skirving et al. 2018), was created in 2017 and merged together the OSTIA reanalysis from 1985 to 2002 (Roberts-Jones et al. 2012), a new NOAA/NESDIS reanalysis of geo-polar blended SSTs from 2002 to 2016 (A. Harris pers. comm.), and the geo-polar blended near-real-time data used in CRW's current operational products. CRW has already used a prototype of the CoralTemp dataset to compute the climatology for its latest near-real-time v3 products (see Sect. 4.2.2.1) and is currently recomputing all other historical data products formerly computed using Pathfinder-based SST.

4.2.2.6 Bleaching Prediction Using Climate Models

While CRW's satellite-based products have been very valuable to scientists and resource managers, most action to protect reef corals from bleaching required more lead time. In 2008 CRW, in collaboration with the Physical Sciences Division of the NOAA Earth System Research Laboratory (ESRL), released the world's first prediction tool for forecasting coral bleaching weeks to months in advance (Liu et al. 2009). That first system was based on a statistical global SST forecast model, the Linear Inverse Model (LIM) (Penland and Matrosova 1998), limiting the system to a single, deterministic subseasonal-to-seasonal-scale bleaching outlook. The Australian Bureau of Meteorology (BOM) released the first dynamical, ensemble-based coral bleaching prediction system built from their Predictive Ocean Atmosphere Model for Australia (POAMA) (Spillman et al. 2011). Simultaneously, CRW had partnered with the NOAA National Centers for Environmental Prediction (NCEP) to develop its next-generation global subseasonal-to-seasonal-scale bleaching outlook by applying SST predictions from NOAA's operational Climate Forecast System (CFS) in an ensemble prediction system. An example of CRW's Four-Month Bleaching Outlook and observed Bleaching Alert Area for the Pacific Ocean in 2015 is shown in Fig. 4.2 (CRW 2017; https://coralreefwatch.noaa.gov/satellite/bleachingoutlook_cfs/outlook_cfs.php).

These new systems from BOM and CRW provide dynamical, probabilistic subseasonal-to-seasonal-scale coral bleaching heat stress outlooks delivering advance warning to coral reef managers, scientists, stakeholders, and the public. Efforts at BOM and CRW have continued to advance, providing enhancements such

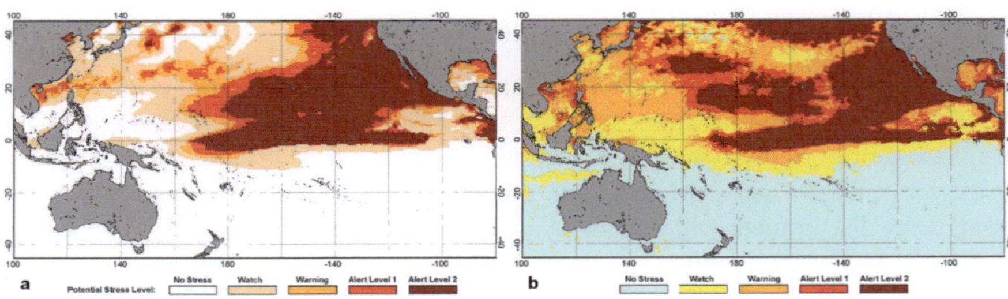

Fig. 4.2 NOAA Coral Reef Watch Pacific Ocean (**a**) 60% probability Four-Month Bleaching Outlook for July–October 2015 and (**b**) 5-km maximum Bleaching Alert Area for 2015 (full year)

as increased ensemble numbers, greater spatial resolution, skill assessments, and improved model physics (Griesser and Spillman 2016; Liu et al. 2017). High confidence in systems such as these provides assurance to managers to effectively undertake costly and/or controversial actions to protect corals when mass bleaching is anticipated (Maynard et al. 2009). Additional work has taken the same approach described here for seasonal outlooks to understand the impact a warming climate will have on coral reefs in future decades (Donner et al. 2005; IPCC 2014a; van Hooidonk et al. 2013, 2014, 2016).

4.2.2.7 Bleaching Weather: The Doldrums Product

While basin-scale coral bleaching occurs as a result of large-scale climatic phenomena, local weather patterns greatly influence heat stress and bleaching variability among sites within the basin. Three related factors that influence local bleaching patterns are temperature, light, and water-column mixing. One parameter that exerts a common influence on all of these is wind. When wind speeds drop, reductions in vertical mixing, evaporative cooling, and sensible heat transfer all increase the likelihood of higher temperatures and increased light penetration through the water column (Dunne and Brown 2001; Mumby et al. 2001a; Skirving and Guinotte 2001; Obura 2005). Additionally, low winds can increase stratification in the water column, resulting in enhanced photodegradation of colored dissolved organic material and, thereby, reducing shading (Manzello et al. 2006). CRW has developed an experimental Doldrums product using the NOAA/NESDIS Blended Sea Winds product derived from multiple satellites that provides 0.25° (approx. 25 km) resolution wind fields for the Earth's ocean surface (10-m height) every 6 hours (Zhang et al. 2006). The current experimental product identifies regions where mean daily wind speeds have remained below 3 m/s and records the persistence (doldrums-days) of such conditions (Fig. 4.3; http://coralreefwatch.noaa.gov/satellite/doldrums/). NOAA expects to use this product to augment SST- and light-based algorithms to help detect conditions conducive to coral bleaching.

4.2.2.8 Additional Products

To further refine its satellite product suite, CRW is developing products to monitor additional parameters that influence bleaching and coral reef health. Some parameters that further address the needs described above include ocean surface solar insolation, cloud cover, ocean color, and turbidity. These will directly address parameters that influence the quantity and quality of light that reaches reef corals at depth. Most of these products use both geostationary and polar-orbiting satellite data. Other products include SST-based indices to predict outbreaks of coral disease and connectivity among reefs estimated using oceanographic and hydrodynamic models.

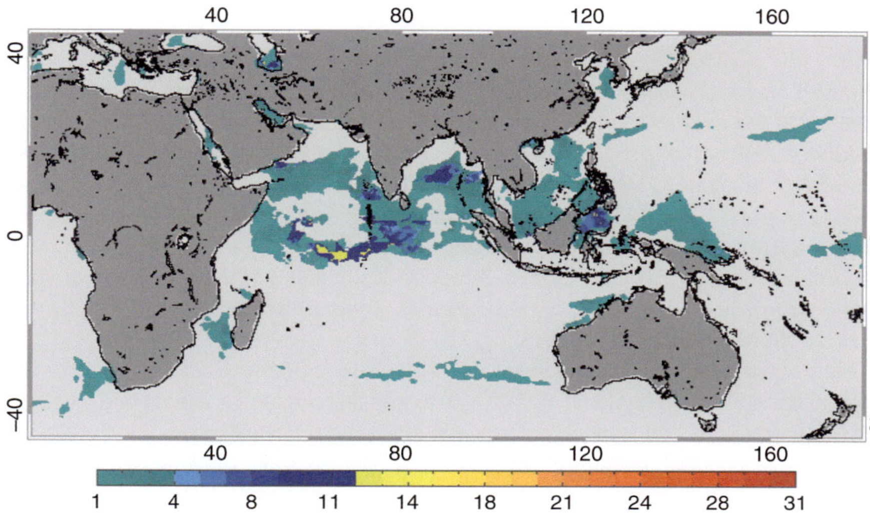

Fig. 4.3 NOAA Coral Reef Watch near-real-time satellite 25-km doldrums product for 24 April 2005 in the Eastern Hemisphere. The color scale indicates the number of consecutive days over which daily mean wind speed remained below 3 m/s

Also, CRW continues to work on improving the spatial and temporal resolution of its products. However, finer resolution in both space and time comes at a price, including an increased need to gap-fill cloudy regions and the greater influence of chaotic variability in coastal SST. Blended geostationary-polar-orbiting satellite data have provided the capacity for this in the global products at 5-km resolution, and higher-resolution regional products are currently being pursued, especially as new geostationary satellites with a higher spatial resolution of 2-km come online. These efforts complement regional collaborative developments, two of which are currently available for the Great Barrier Reef (GBR, Australia) and Florida regions. In 2007, a joint project between the Great Barrier Reef Marine Park Authority (GBRMPA) and the Australian Commonwealth Scientific and Industrial Research Organization (CSIRO) released an experimental 2-km polar-orbiting-satellite-based product suite for the GBR, ReefTemp, that includes SST, positive SST anomaly, Heating Rate, and Degree Heating Day products (Maynard et al. 2008; CSIRO at http://www.cmar.csiro.au/remotesensing/gbrmpa/ReefTemp.htm). More recently, the ReefTemp Next Generation system (Garde et al. 2014; http://www.bom.gov.au/environment/activities/reeftemp/reeftemp.shtml) expanded the suite of products and is now hosted by the Australian Bureau of Meteorology (BOM) with full operational support. The Institute for Marine Remote Sensing at the University of South Florida has developed 1-km polar-orbiting-satellite SST and Degree Heating Week products for Florida and the Caribbean (Vega-Rodriguez et al. 2015). These efforts will help define the requirements and limitations to developing global high-resolution SST-based monitoring products.

4.3 Tropical SST Trends Since the Nineteenth Century

The SST maxima reached during recent mass coral bleaching events were unprecedented in the available instrumental records at a range of coral reef sites (Hoegh-Guldberg and Salvat 1995; Brown et al. 1996; Winter et al. 1998; Lough 2000; Mumby et al. 2001b; Wellington et al. 2001; Eakin 2007; Wilkinson and Souter 2008; Lough et al. 2018). A number of correlated indices of local heat stress and other contributors to bleaching have been identified: absolute SST maximum, SST maximum anomaly, Mean Positive Summer Anomaly (MPSA), number of days above particular thresholds, etc. (Podesta and Glynn 1997; Liu et al. 2006; Strong et al. 2006; Maynard et al. 2008).

Here we used global compilations of 1871–2016 monthly HadISST1 SSTs with 1° spatial resolution (Rayner et al. 2003) to assess large-scale variations and trends in the tropical oceans and typical patterns associated with ENSO events. Such data are extremely useful for providing long-term perspectives on the changing tropical ocean climate in relation to coral bleaching events (Sheppard and Rayner 2002; Barton and Casey 2005; Sheppard and Rioja-Nieto 2005). It should, however, be recognized that data with such a coarse spatial resolution can disguise the considerable small-scale thermal variability on coral reefs (Potts and Swart 1984) and can significantly underestimate the real water temperature ranges experienced by corals. For example, monthly HadISST1 data in the vicinity of offshore Myrmidon Reef in the central GBR indicate an annual range of 5–6 °C, matching the 5 °C annual range of daily-average SSTs recorded by an automatic weather station (http://www.aims.gov.au/docs/data-centre/weatherstations.html); however, diurnal variation was observed to reach ~9.5 °C (Lough 2001). Furthermore, the modeling inherent within reconstructed SST datasets can also introduce errors in seasonal signals—a predecessor of the HadISST1 data (GISST 2.2) indicated an annual range of 2–3 °C in monthly SSTs for the same location.

4.3.1 Tropical SST and Global Temperature Trends

The instrumental record of global land and sea temperatures (Fig. 4.4a) illustrates the significant warming that has occurred since the end of the nineteenth century (Morice et al. 2012). Average temperatures for the most recent 30-year period, 1987–2016, are 0.7 °C warmer than the first 30-year period of the record, 1871–1900—a significant change in climate. This warming also has occurred in the coral reef regions of the tropical oceans (see Lough 2012) with maximum SSTs (Fig. 4.4b) averaging 0.5 °C warmer in the most recent 30-year period compared with the late nineteenth century. Paleoclimatic data from a pantropically distributed set of 57 coral records confirmed that the twentieth-century tropical oceans have significantly warmed and that 2016 was the warmest year in at least the past four centuries (Fig. 4.5; Tierney et al. 2015). There is also evidence from 32 coral isotopic

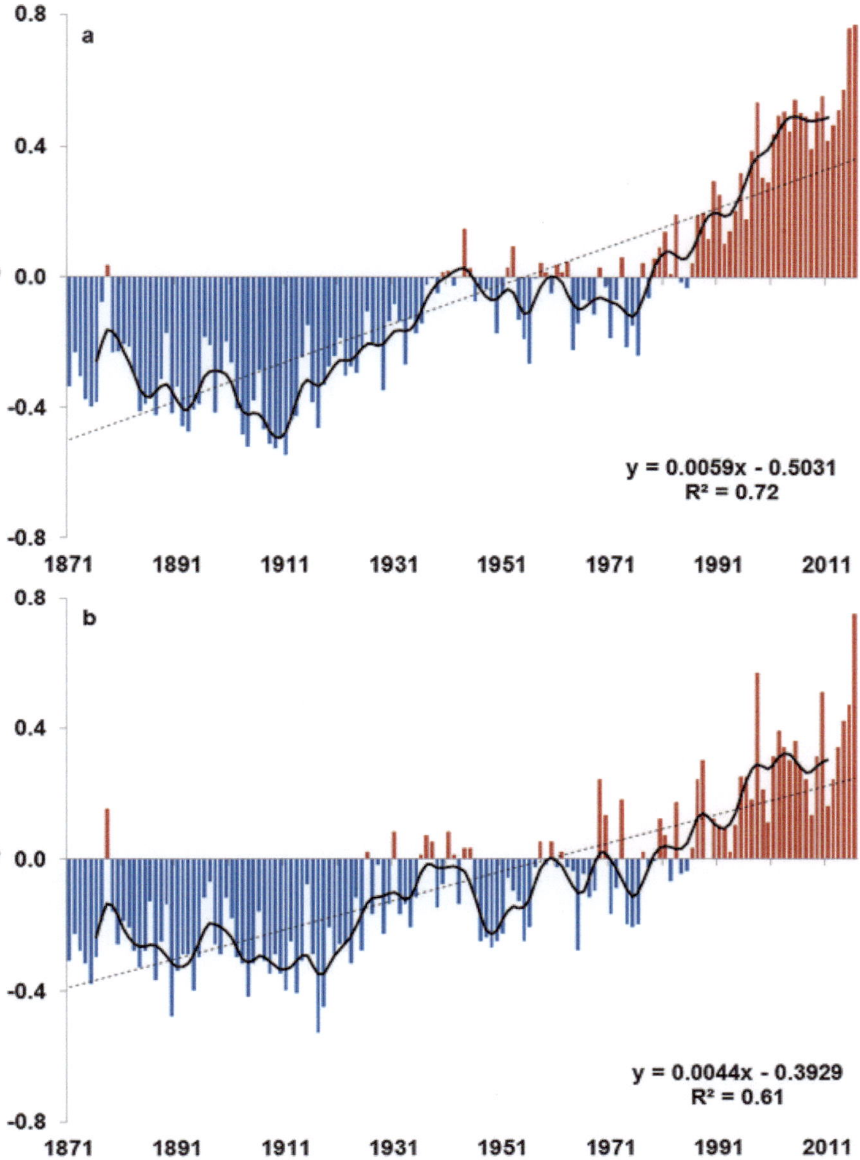

Fig. 4.4 (**a**) Global annual land and sea temperature (HadCRUT4, Morice et al. 2012). (**b**) Tropical coral reef (1670 1° boxes, see Lough 2012) annual maximum SST (HadISST1, Rayner et al. 2003). Both (**a**) and (**b**) show annual values, 1871–2016, as anomalies from 1961 to 1990 mean; *thick line* is 10-year Gaussian filter emphasizing decadal variability; linear trend is also shown

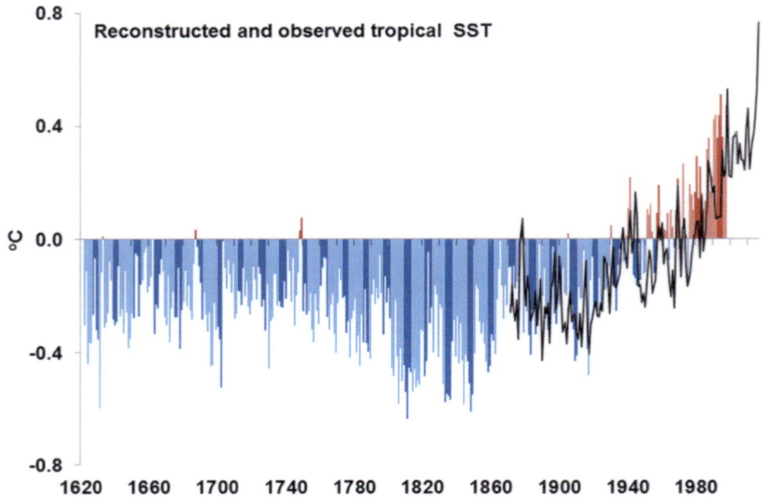

Fig. 4.5 Reconstructed annual tropical SST anomaly, 1622–1997 (bars), based on four regional SST reconstructions from coral $\delta^{18}O$, Sr/Ca, and growth records that are significantly correlated with SST (Tierney et al. 2015), and observed SST, 1871–2016 (black line) for tropical coral reef sites (HadISST1, Rayner et al. 2003). Annual values are based on April–March averages (dated by April year) and expressed as anomalies from 1961 to 1990 average. Over the period 1871–1997, the reconstructed and observational SST series are significantly correlated (at 1% level), $r = 0.75$

records that the tropical oceans have become fresher (less saline) than the previous two centuries at most sites (Grottoli and Eakin 2007).

Using an 18-year satellite time series, Strong et al. (2006) observed a pattern shift in decade-long SST trends during the latter half of the 1990s. Generally neutral trends across the Pacific Ocean transitioned to warming in the western Pacific and to cooling in the eastern Pacific; the tropical Indian Ocean moved from neutral to warming conditions; and warming trends in the northern and tropical Atlantic Ocean were enhanced. This pattern shift is likely a reflection of the superposition of short- and long-term climate variability.

Based on linear trend analysis (as used by IPCC 2013) over the period 1880–2016, the global surface land and sea temperature record (Fig. 4.4a) has significantly warmed by 0.9 °C and maximum annual SST at tropical reef sites (Fig. 4.4b) by 0.5 °C. For both series, 2016 was the warmest year on record. For maximum SST on coral reefs, the first, second, and third warmest years on record were 2016, 1998, and 2010—each associated with global mass coral bleaching events (Eakin et al. 2016, 2017; Hughes et al. 2018). Each of these bleaching events was associated with an El Niño which is typically linked with warmer than usual waters throughout much of the tropical oceans (Deser et al. 2010). Due to warming of baseline SST, however, the level of heat stress during these recent El Niño events was more than three times that of the "preindustrial," 1878 El Niño (Lough et al. 2018). Recent warming of the tropical oceans, home to the world's coral reefs, is inextricably linked to warming of the global climate system in

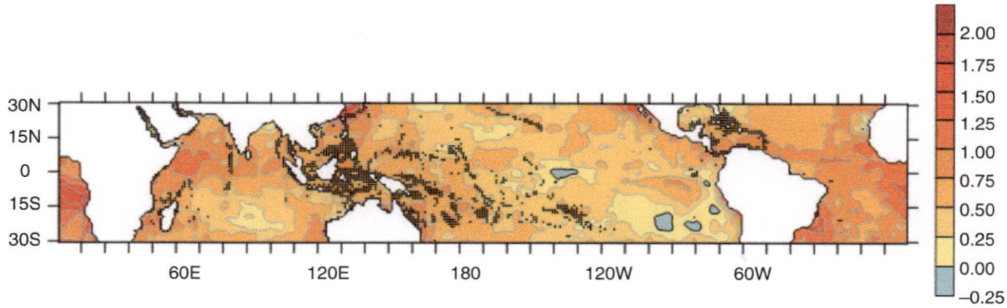

Fig. 4.6 Linear trend of annual maximum monthly SST, 1871–2016, using 0.25 °C contour interval. Black dots show 1° boxes containing coral reefs (HadISST1, Rayner et al. 2003)

response to increasing concentrations of greenhouse gases in the atmosphere. The human imprint on the global climate system was assessed as "clear" in the most recent IPCC report (Stocker et al. 2013). Analysis of the HadISST1 dataset indicated that warming of the tropical oceans alone represented ~70–80% of the total warming found in the global average land and sea data (Lough 2012). A rise in maximum SSTs has occurred throughout the tropical oceans, with greatest warming in the southern Atlantic, Indian Ocean, and parts of the northwestern and southwestern tropical Pacific and with less warming in the north-central and south-central Pacific (Fig. 4.5).

Evidence from reef cores covering the past three millennia (Aronson et al. 2002) and observations of the bleaching-induced death of centuries-old, slow-growing, and thermally more resistant *Porites* corals (Mumby et al. 2001b) strongly suggest that we have reached unprecedented thermal environmental conditions on coral reefs. Additionally, bleaching mortality of branching corals allowed corallivorous crown-of-thorns starfish, *Acanthaster planci*, to reach unprotected massive corals in Panama in 1983 for the first time in almost 200 years. The age of these massive corals was then used to estimate the return frequency of El Niño events to reefs in Pacific Panama (Glynn 1985). All of these data sources point to the likelihood that the recent severe El Niño events and the ocean temperature increases of the past 50 years are unique in at least the past few centuries (Tierney et al. 2015; Fig. 4.6).

4.3.2 Regional Trends in Heat Stress

The new 32-year CoralTemp satellite record (1985–2016; see Sect. 4.2.2.5) was used to examine the global and regional trends in SST anomalies at 100 coral reef sites with reliable coral bleaching histories (Hughes et al. 2018). SST anomaly values were averaged across reef-containing 5-km pixels for each 1° × 1° site, using reef locations as described in Heron et al. (2016a). Grouping these 100 sites into five geographic regions, annual average SST anomalies for each region are shown in Fig. 4.7. The 3 years of global coral beaching events (1998, 2010, 2016)

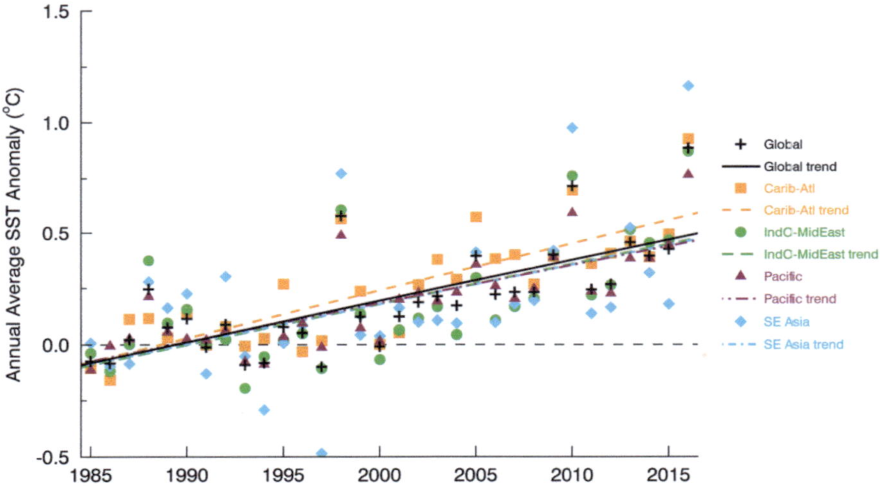

Fig. 4.7 Annual average SST anomalies and trends, 1985–2016, for the indicated sets of coral reef sites from the CoralTemp 32-year satellite record: 100 global coral reef sites with reliable coral bleaching histories (*black plus*, *solid line*), 28 sites in the Indian Ocean and Middle East (*green circle*, *dashed line*), 14 sites in Southeast Asia (*blue diamond*, *dash-dot line*), 36 sites in the Pacific Ocean (*purple triangle*, *dash-dot-dot line*), and 22 sites in the Caribbean and Atlantic Ocean (*orange square*, *dotted line*). The *black thin-dash line* shows the zero trend. Trend values for each line listed in Table 4.1

Table 4.1 Trends (and standard errors; *S.E.*) in SST anomalies from the CoralTemp 32-year satellite record, 1985–2016, at 100 global coral reef sites with reliable coral bleaching histories grouped into five geographic regions (as displayed in Fig. 4.7)

Region	Number of reef pixels	Trend in SST anomaly (°C/decade)	S.E. in trend (°C/decade)
Global	100	0.183	0.032
Indian Ocean and Middle East	28	0.179	0.037
Southeast Asia	14	0.177	0.055
Pacific Ocean	36	0.170	0.025
Caribbean and Atlantic Ocean	22	0.211	0.029

The SST anomaly values are averaged across specific reef pixels within each region and for each year

stand out as having the highest SST anomalies within the record period. The slopes of trend lines (Table 4.1) for each region over the 32-year record consistently range within 0.170–0.183 °C/decade, with the exception of reef sites in the Caribbean and Atlantic for which the trend is 0.211 °C/decade. The global rate of increase in SSTs since 1985 is significantly higher than the 1950–2009 average rate (IPCC 2014a) and has already reached the rate predicted for the twenty-first century by global climate models under Representative Concentration Pathway (RCP) 6.0 (average 2.2 °C/

century; IPCC 2013). Further considerations of the impacts of future climate change on the bleaching of corals are discussed in Chap. 13. Comparing the annual global and regional average SST anomalies (Fig. 4.7) shows that, while the values are fairly

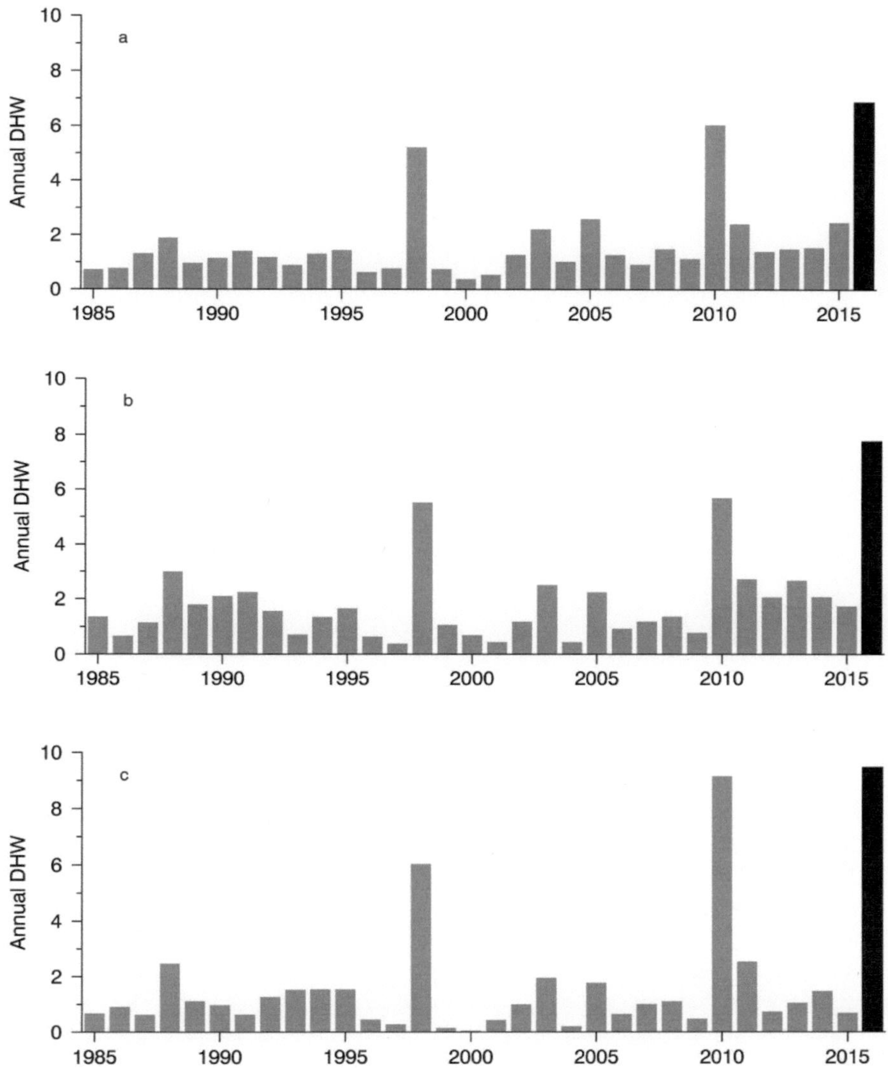

Fig. 4.8 Annual Degree Heating Week (*DHW*) indices, 1985–2016, averaged over the coral reef sites from the CoralTemp 32-year satellite record: (**a**) 100 global coral reef sites with reliable coral bleaching histories, (**b**) 28 sites in the Indian Ocean and Middle East, (**c**) 14 sites in Southeast Asia, (**d**) 36 sites in the Pacific Ocean, and (**e**) 22 sites in the Caribbean and Atlantic Ocean. Moderate bleaching has been shown to occur at DHW $\geq$ 4 °C-weeks; severe bleaching occurs at DHW $\geq$ 8 °C-weeks. *Black bar* shows most extreme year in 32-year record

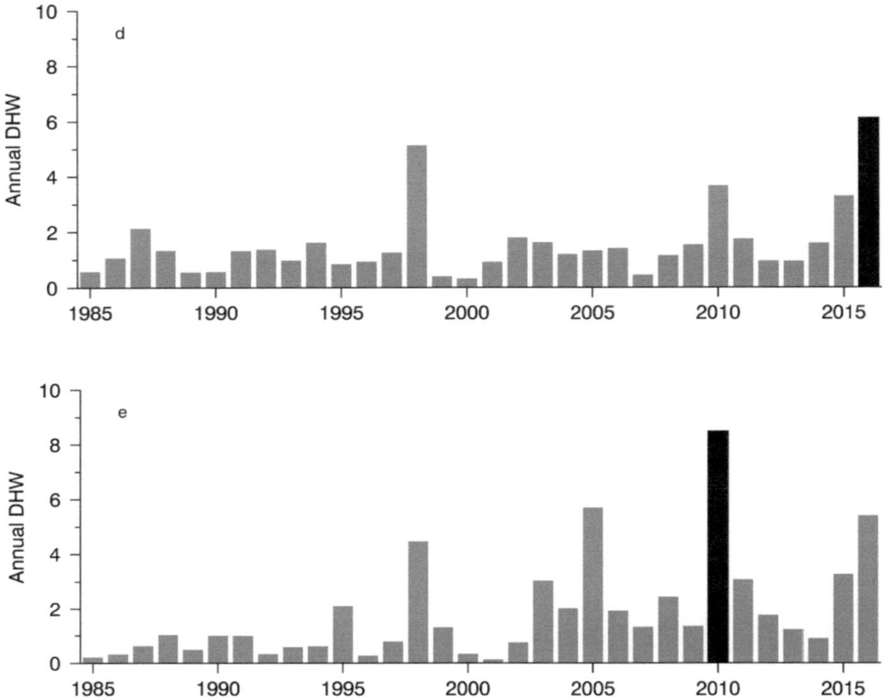

Fig. 4.8 (continued)

consistent across the globe within each year, there are important spatial patterns of SST anomaly that change through time. The role of large-scale patterns in regional bleaching events is discussed in Sect. 4.3.3.

Typical SST conditions that result in coral bleaching include high maximum SST sustained over a number of weeks, as measured by the DHW index. The record of CRW's DHWs over the 1985–2016 CoralTemp record (Sect. 4.2.2.5) for these 100 coral reef sites shows increasing heat stress throughout the 32-year period (Fig. 4.8). In this record, 2016 stands out as the year with the greatest cumulative reef heat stress both globally (Fig. 4.8a) and in the Indian Ocean, Southeast Asia, and the Pacific Ocean (Fig. 4.8b, c, d). The exception to this pattern is the Caribbean/Atlantic (Fig. 4.8e) where heat stress in 2010 greatly exceeded that of other years. The year 2016 was the peak of the continuous 2014–2017 global coral bleaching event (Eakin et al. 2017), the longest, most widespread, and perhaps most damaging bleaching event yet seen. Annual DHW values, and therefore the warming trends, were near identical for both the spatial average (as shown in Fig. 4.8) and for the 90th percentile value within each region (as used in CRW's Regional Virtual Stations), indicating the independence of the observed patterns with respect to the spatial statistic used.

The longer-term context of the increasing cumulative heat stress on coral reefs shown by the satellite data can be determined using a degree heating month index developed from multi-century records of reconstructed SST data. Similar to CRW's DHWs, degree heating months sum the monthly anomalies above the long-term average monthly maximum and have been used to analyze climate model outputs for future bleaching potential (Lough 2000; Donner et al. 2005). Using the HadISST1-based anomaly data (vs a 1961–1990 climatology), indices were created for the same 100 coral reef sites discussed above for the period January 1871–December 2016. With the exception of the 1877–1878 El Niño event, the regionally averaged degree-month values show very low heat stress until the latter half of the twentieth century (Fig. 4.9a). The heat stress in 2016 was the most extreme in terms of cumulative reef heat stress when averaged over all 100 reef locations, followed in magnitude by 1998 and 2010. Heat stress in 2016 equaled that in the previous record year, 1998, in the Indian Ocean and was the highest on record in Southeast Asia and the Pacific Ocean (Fig. 4.9a–d), consistent with CRW's satellite-based near-real-time record. Heat stress in the Caribbean region (Fig. 4.9e) showed significant heat stress beginning decades earlier than the other regions (c.f., Lough 2000; Lough et al. 2018; Barton and Casey 2005). It is unclear how much of the severe degradation of coral reefs in the Caribbean over recent decades resulted from this early increase in heat stress (Gardner et al. 2003).

4.3.3 Role of El Niño-Southern Oscillation and Other Large-Scale Patterns

ENSO events are the major source of short-term climatic variability within the tropical ocean-atmosphere system (McPhaden 2004). It was the major 1982–1983 El Niño event that first triggered warnings of a link between ENSO and mass coral bleaching events (Glynn 1983; Coffroth et al. 1990; Williams and Bunkley-Williams 1990). The 1997–1998 El Niño event (coinciding with what was then the warmest year on record; Fig. 4.4a) was one of the two most extreme El Niño events on record (Wolter and Timlin 1998; McPhaden 1999) and coincided with the greatest heat stress at many coral reef sites up to that time (Figs. 4.8a and 4.9a). It was estimated that over 15% of the world's reefs died and many reefs suffered over 90% bleaching in 1998 (Wilkinson 2000). Global bleaching was again seen in 2010 and most recently in 2014–2017 (Eakin et al. 2017).

ENSO events do not cause mass coral bleaching but instead increase the likelihood of anomalously warm SSTs in particular regions that result in coral bleaching. Major and minor ENSO years can be seen in the increased temperature anomalies in Fig. 4.7 (i.e., 1987–1988, 1994–1995, 1997–1998, 2002–2003, 2009–2010, 2015–2016; but note the absence of signal for the 1991–1992 event). Mass coral bleaching can occur in the absence of strong ENSO extremes when other climate anomalies cause regional warming, e.g., GBR in early 1982 (Coffroth et al. 1990), Moorea in 1994 (Hoegh-Guldberg and Salvat 1995), Hawaii in 1996 (Jokiel and Brown 2004), and the Caribbean in 2005 (Eakin 2007; Wilkinson and Souter 2008).

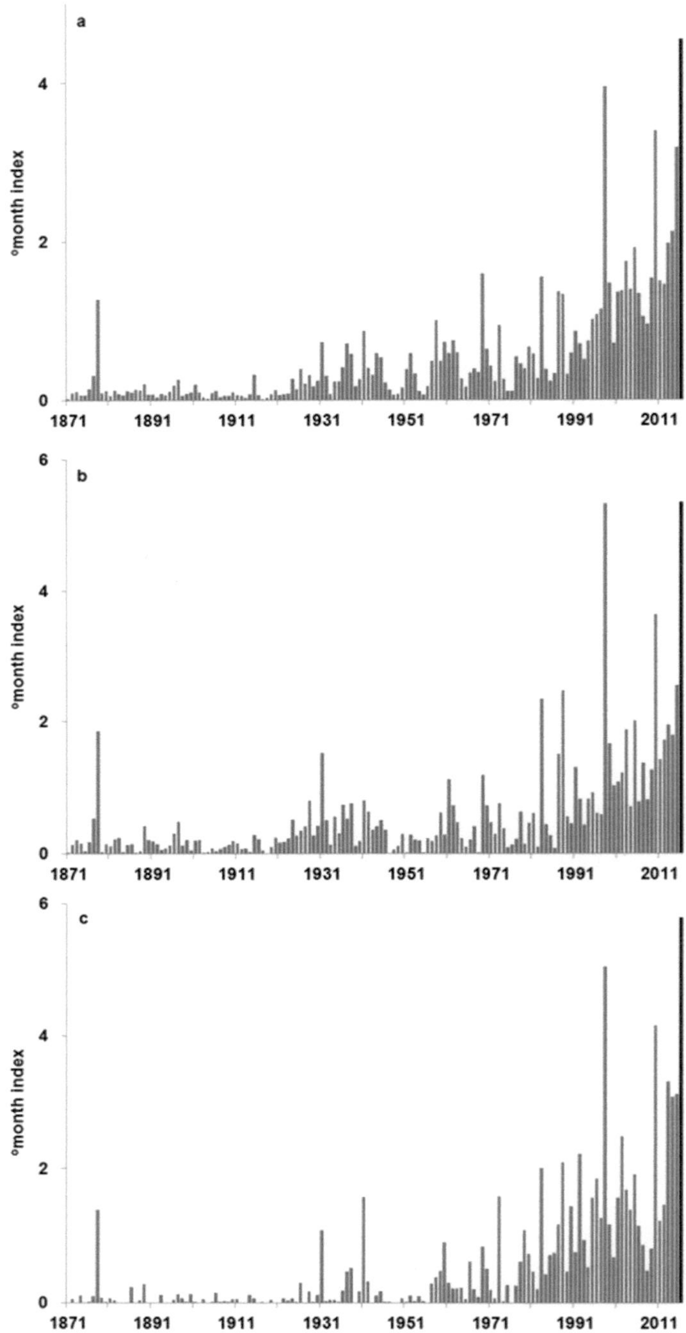

Fig. 4.9 Annual degree heating month indices (sum of anomalies above the long-term maximum; °C-*month*), 1871–2016 (HadISST1, Rayner et al. 2003) for: (**a**) 100 coral reef sites, (**b**) 28 sites in the Indian Ocean and Middle East, (**c**) 14 sites in Southeast Asia, (**d**) 36 sites in the Pacific Ocean, and (**e**) 22 sites in the Caribbean and Atlantic Ocean. *Black bar* shows most extreme year in 146-year record

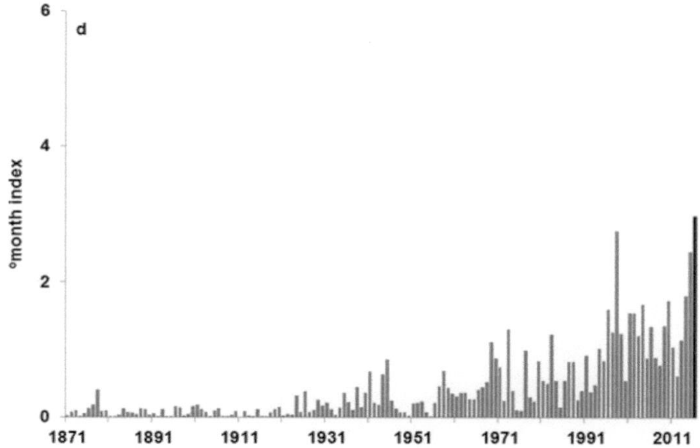

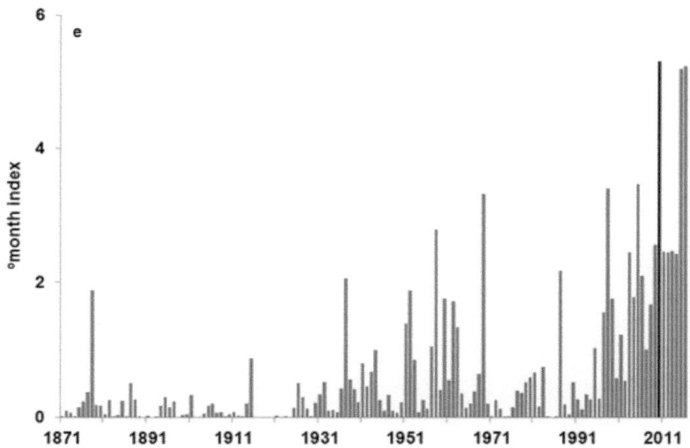

Fig. 4.9 (continued)

Here we define the phases of ENSO by the Oceanic Niño 3.4 SST index (ONI), calculated as the average SST anomaly for the near-equatorial central Pacific region (5 °N–5 °S, 170°–120 °W; obtained from http://www.cpc.ncep.noaa.gov/data/indices/ and extended back to 1871 using HadISST1). This index was used to identify 20 El Niño events, 20 La Niña events, and 20 years of ENSO-neutral conditions from the upper, lower, and middle percentiles, respectively, of the annual May–April values over the period 1871–2016. Monthly maxima were then averaged for each set of 20 years for both the target year and the following year (year t and year $t + 1$, e.g., 1982 and 1983) for the tropical oceans. For each 1° latitude-longitude box, the average values for the 20 El Niño years and 20 La Niña years were tested for significant difference from the average of the 20 ENSO-neutral years.

The two extreme phases of ENSO, El Niño and La Niña, typically evolve over 12–18 months and are associated with distinct and different ocean-atmosphere

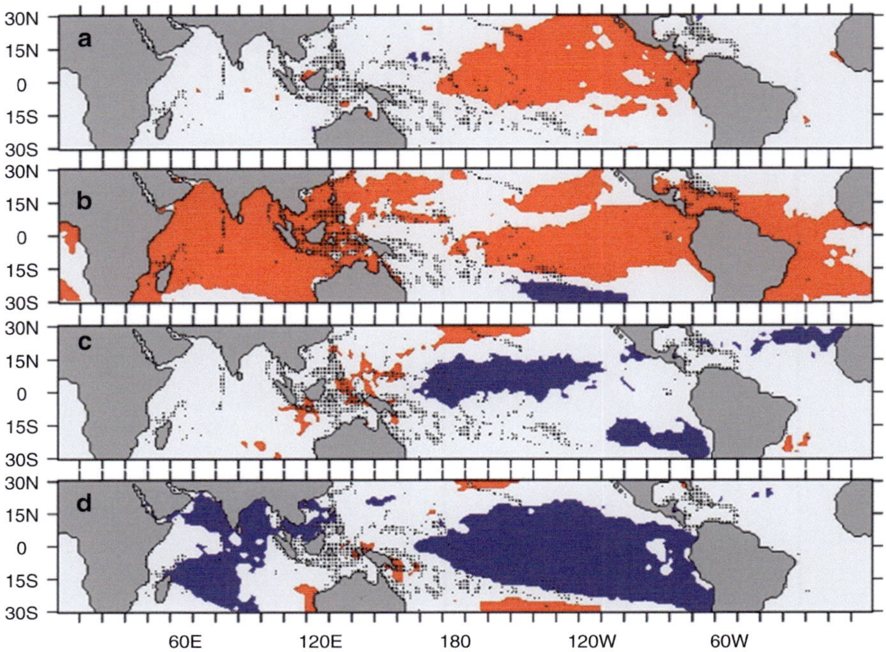

Fig. 4.10 Significantly (at 5% level) warmer (*red*) or cooler (*blue*) annual maximum SST difference: (**a**) El Niño year *t*, (**b**) El Niño year *t* + 1, (**c**) La Niña year *t*, and (**d**) La Niña year *t* + 1. Average values calculated for 20 El Niño events and 20 La Niña events and tested for significant differences from 20 ENSO-neutral years. The groups of years were identified from the Niño 3.4 SST ENSO index (http://www.cpc.ncep.noaa.gov/data/indices/). Black dots show 1° boxes containing coral reefs

circulation patterns in the core region of the central and eastern equatorial Pacific (McPhaden 2004). Typically for the 2 years spanning an El Niño event (e.g., 1982–1983, 1987–1988, 1997–1998, 2009–2010, 2015–2016), large areas of the tropical oceans have significantly warmer maximum SSTs than in ENSO-neutral years (Fig. 4.10a, b).

Figure 4.10b shows why, for example, many Caribbean bleaching events occur in the second year of an El Niño event. Conversely during the 2 years spanning a La Niña event, large areas of the tropical oceans have significantly cooler maximum SSTs than in ENSO-neutral years (Fig. 4.10c, d). An interesting exception is the region in the western equatorial Pacific lying under the South Pacific Convergence Zone where waters tend to be warmer during La Niña years and coral reefs experienced mass bleaching not in 1997–1998 but during the ensuing 1998–1999 La Niña (Wilkinson 2002). Reefs in the southeastern Indian Ocean off Western Australia are also more likely to be warmer than normal during the La Niña phase (Zhang et al. 2017), and this region experienced mass bleaching during the major 2010–2011 La Niña event (Moore et al. 2012; Wernberg et al. 2012). SST reconstructions from Western Australian corals highlight how these "marine heat waves" have become more frequent in recent decades (Zinke et al. 2015). From a global perspective, the risk of warmer than normal maximum SSTs and thus bleaching is greater during El Niño events for many of the world's coral reefs, and this risk is much lower during La Niña and ENSO-neutral years. Some studies have indicated that El Niño events are projected to become more frequent and more severe as the world continues to warm (Power et al. 2013; Cai et al. 2014). The important point for future mass bleaching on coral reef ecosystems is that even if temperature anomalies resulting from ENSO events remain within the range observed during the past 150 years, the warming of baseline maximum SSTs (Figs. 4.4b, 4.5, and 4.7) increases the probability that waters overlying reefs will reach or exceed critical temperature thresholds for bleaching.

The ENSO system is not the only climatic pattern that influences ocean temperatures and, accordingly, the risk of coral bleaching. Oceanic SST variations also result from other longer-term modes including the Interdecadal Pacific Oscillation (Folland et al. 2002) and the Atlantic Multidecadal Oscillation (AMO; Schlesinger and Ramankutty 1994), which vary on timescales of 25 years and 65 years, respectively. These oscillations have typically been observed in periodic variations of temperature anomalies in the northern parts of their respective oceans. The Indian Ocean Dipole (Saji et al. 1999) is considered an aperiodic variation of SST anomalies between the eastern and western tropical Indian Ocean that has been found to be often intertwined with ENSO (Luo et al. 2010). Especially during times that ENSO and the Indian Ocean Dipole are in phase, bleaching across much of the Indian Ocean can be severe (McClanahan 2004). However, patterns of bleaching across the Indian Ocean can be spatially variable with different portions of the region influenced by different climatic forcing (Lix et al. 2016; Zhang et al. 2017).

Many authors have suggested that SST increases due to global climate change, especially in combination with natural variability such as ENSO, have and will continue to increase the frequency and severity of coral bleaching events (Williams and Bunkley-Williams 1990; Hoegh-Guldberg 1999; Eakin 2007). However, as global temperatures have increased, major bleaching during non-ENSO years also has also increased. Global temperatures in 2005 were the first to surpass 1998 as the warmest year on record (Shein 2006). This resulted in the most severe and widespread coral bleaching and mortality ever seen in the Caribbean despite only mild El Niño conditions the preceding winter (Eakin 2007; Wilkinson and Souter 2008). Trenberth and Shea (2006) found that 0.45 °C of the 0.9 °C warming during the boreal summer of 2005 was due to the monotonic rise in global SSTs, 0.2 °C due to after effects of the 2004–2005 mild El Niño and 0.1 °C due to AMO. Donner et al. (2007) suggest that twentieth-century anthropogenic warming, on top of the natural modes of variability, increased the probability that the Caribbean would experience the level of heat stress observed in 2005 by an order of magnitude, compared with the natural modes alone. Their projections of future SST (including anthropogenic warming) suggest that, in the absence of acclimatization or adaptation by coral organisms, stress levels like those seen in 2005 will be experienced almost biennially by the 2030s.

These warming trends have continued, with the 12 warmest years on record occurring since 1998, and the five warmest since 2010. Most importantly for corals was the series of three warmest years in a row from 2014 to 2016 (and global temperature during the first three-quarters of 2017 was comparable with these 3 years). These contributed to the longest, most widespread, and probably most damaging coral bleaching event on record, which lasted from June 2014 to May 2017 (Eakin et al. 2017). This global bleaching began before the onset of the 2015–2016 El Niño and lasted for a full year after it collapsed. The heat of these warmest years now resides in the upper ocean, pushing baseline temperatures for corals that are now much closer to their upper thermal limits.

4.4 Local Environmental Variables

Large-scale oceanic temperature anomalies in the tropical oceans (Sect. 4.3.1) are the principal drivers of conditions that result in mass coral bleaching. At the local scale, the occurrence and intensity of bleaching can be highly variable both within a coral colony, between coral colonies, within a reef, and between reefs in a region (Hoegh-Guldberg 1999). These variations are in addition to the differential susceptibility of different coral species to heat stress (Marshall and Baird 2000). Other physical factors that operate locally can either enhance or suppress the impacts of higher-than-normal regional SSTs and thus the intensity of coral bleaching. Observations that corals often bleach more on their upper surface than at the sides clearly implicate light as an additional factor, and frequently the local weather conditions that cause intense warming of the water column (low winds, low cloud amount, still waters; Skirving et al. 2006a) allow increased light penetration to the coral's surface (Coles and Jokiel 1978; Salm and Coles 2001). Increased cloudiness can mitigate bleaching even when SSTs are unusually warm (Mumby et al. 2001a). Lowered salinity due to a major flood event appeared to increase the intensity of coral bleaching on nearshore reefs of the central GBR in 1998 (Berkelmans and Oliver 1999). There can also be considerable local-scale variations in SSTs within and between reefs that can affect bleaching occurrence and intensity (Nadaoka et al. 2001; Berkelmans 2002; Berkelmans et al. 2004). Such local-scale SST variations can be related to water movements such as upwelling, mixing, tidal range and wave energy, shading, and exposure that reduce the local heat stress (Salm and Coles 2001; Skirving and Guinotte 2001; Skirving et al. 2006b). Although often small in scale, identification and enhanced protection of such bleaching-resistant sites may be critical for recovery of adjacent bleaching-damaged coral populations (Marshall and Schuttenberg 2006; Skirving et al. 2006b).

Many characteristic features of coral locations have been identified as increasing the capacity of corals to resist and/or recover from disturbance impacts. While the importance of these "resilience factors" may vary from site to site, 11 were identified as key factors to support and inform management decisions, including the presence of heat-tolerant coral taxa, coral recruitment, and the level of nutrients (McClanahan et al. 2012). For example, those Seychelles coral reefs that recovered from the severe bleaching in 1998 were associated with complex reef structures that had high density of juvenile corals and were located in relatively deep water with plentiful herbivore populations and low nutrient loads (Graham et al. 2015). In addition to the variability across sites as to which factors are most important to support resilience is the variation in tools available to enhance resilience. A framework to support resilience-based management by Anthony et al. (2015) describes opportunities to effectively influence the level of resilience. Guiding principles with respect to resilience continue to be developed to assist reef managers in translating the ideals of resilience into conservation planning, monitoring, and management strategies.

4.5 Summary

We now have strong evidence of how global climate change due to the increased concentration of atmospheric greenhouse gases already has caused significant coral bleaching. This threat of future warming is so strong that coral bleaching has been highlighted among the greatest threats to ecosystems brought about by anthropogenic climate change (IPCC 2014a, b). Warming of the tropical oceans has raised the baseline SSTs where coral reefs live much closer to their upper thermal limits, so that weather conditions and interannual variability are more likely to raise SSTs above these limits than similar weather conditions did 100, or even 30, years ago. The risk of unusually warm maximum SSTs also varies with the ENSO cycle, so that the risk of conditions conducive to bleaching further increases for much of the tropical oceans during ENSO events. This has allowed us to use El Niño conditions as an analogue for potential future climatic conditions. However, in recent years mass bleaching events have occurred in all phases of ENSO—fulfilling earlier projections (e.g., Hoegh-Guldberg 1999; Hughes et al. 2018). What remains is to determine if corals are capable of evolving physiological adaptations to heat stress rapidly enough to cope with the combination of natural variability and climate change—a now ongoing, uncontrolled experiment that could result in the extinction of many coral species and destruction of many of the world's reefs.

Our ability to identify, monitor, and predict SST conditions that can lead to coral bleaching has improved dramatically since the first reports of mass bleaching events in the early 1980s. This allows near-real-time monitoring and identification of potential bleaching conditions throughout the world's coral reefs. Of course, monitoring alone cannot prevent coral bleaching or mortality. However, monitoring, combined with recent advances in bleaching forecasting, enables scientists and managers to be alert to bleaching and to document the extent, intensity, impact, and follow-on effects more comprehensively than was possible 10–20 years ago (Chap. 5). It also allows managers to take actions to help protect reefs at times when bleaching makes them more vulnerable to other stressors (Marshall and Schuttenberg 2006; Obura et al. 2006). Understanding linkages between the physical environment and biological processes on coral reefs improves our knowledge of the bleaching phenomenon, its ramifications, and potential management responses. This helps managers develop and test management approaches, even controversial or "high-risk" actions, to protect corals from mortality at the time of bleaching events. It also allows identification of "bleaching-resistant" corals, reefs, and regions that should be targeted for enhanced protection. Resilient reefs may provide important refugia for coral reef organisms as climate continues to change and increasingly stress the world's reefs.

Many of the world's coral reefs already have been severely degraded by local human-induced impacts, and most conservation practitioners used to believe reducing local impacts was all that was needed to preserve reefs. Unfortunately, a lack of local stressors on reefs, often through sustained conservation efforts, has repeatedly proven insufficient to save reefs from the extreme impacts of heat stress in many locations around the globe (Graham et al. 2015; Obura and Mangubhai 2011; Couch et al. 2017; Hughes et al. 2017b). Climate change impacts now have surpassed these local impacts and have been seen to overwhelm even the most comprehensive conservation efforts. In addition to the other consequences of a rapidly warming, increased greenhouse world (ocean acidification, more intense tropical storms, etc.), the observed increase in mass coral bleaching events does not bode well for the near- and long-term future of these vital ecosystems.

Acknowledgments C.M.E. thanks NOAA, including Coral Reef Watch and the Coral Reef Conservation Program, for work that contributed to this manuscript. C.M.E. thanks D. Gledhill for providing figures and text for this chapter. The scientific results and conclusions, as well as any views or opinions expressed herein, are those of the author(s) and do not necessarily reflect the views of NOAA or the Department of Commerce.

References

Anthony K, Marshall PA, Abdulla A, Beeden R, Bergh C, Black R et al (2015) Operationalizing resilience for adaptive coral reef management under global environmental change. Glob Chang Biol 21:48–61. https://doi.org/10.1111/gcb.12700

Aronson RB, Precht WF, Toscano MA, Koltes KH (2002) The 1998 bleaching event and its aftermath on a coral reef in Belize. Mar Biol 141:435–447. https://doi.org/10.1007/s00227-002-0842-5

Atwood DK, Hendee JC, Mendez A (1992) An assessment of global warming stress on Caribbean coral reef ecosystems. Bull Mar Sci 51:118–130

Barkley HC, Cohen AL (2016) Skeletal records of community-level bleaching in *Porites* corals from Palau. Coral Reefs 35:1407–1417. https://doi.org/10.1007/s00338-016-1483-3

Barton AD, Casey KS (2005) Climatological context for large-scale coral bleaching. Coral Reefs 24:536–554. https://doi.org/10.1007/s00338-005-0017-1

Berkelmans R (2002) Time-integrated thermal bleaching thresholds of reefs and their variation on the Great Barrier Reef. Mar Ecol Prog Ser 229:73–82. https://doi.org/10.3354/meps229073

Berkelmans R, Oliver JK (1999) Large-scale bleaching of corals on the Great Barrier Reef. Coral Reefs 18:55–60. https://doi.org/10.1007/s003380050154

Berkelmans R, De'ath G, Kininmonth S, Skirving W (2004) A comparison of the 1998 and 2002 coral bleaching events on the Great Barrier Reef: spatial correlation, patterns and predictions. Coral Reefs 23:74–83. https://doi.org/10.1007/s00338-003-0353-y

Brown BE (1987) Worldwide death of corals – natural cyclical events or man-made pollution? Mar Pollut Bull 18:9–13. https://doi.org/10.1016/0025-326X(87)90649-7

Brown BE (1997) Coral bleaching: causes and consequences. Coral Reefs 16:S129–S138. https://doi.org/10.1007/s003380050249

Brown BE, Dunne RP, Chansang H (1996) Coral bleaching relative to elevated seawater temperature in the Andaman Sea (Indian Ocean) over the last 50 years. Coral Reefs 15:151–152. https://doi.org/10.1007/BF01145885

Buddemeier RW, Kinzie RA (1976) Coral growth. Oceanogr Mar Biol Annu Rev 14:183–225

Buddemeier RW, Kleypas JA, Aronson RB (2004) Coral reefs and global climate change. Potential contributions of climate change to stresses on coral reef ecosystem. Pew Center on Global Climate Change, Arlington

Cai W, Borlace S, Lengaigne M, van Rensch P, Collins M, Vecchi G et al (2014) Increasing frequency of extreme El Niño events due to greenhouse warming. Nat Clim Chang 4:111–116. https://doi.org/10.1038/nlimate2100

Cantin NE, Lough JM (2014) Surviving coral bleaching events: *Porites* growth anomalies on the Great Barrier Reef. PLoS One 9:e88720. https://doi.org/10.1371/journal.pone.0088720

Carilli JE, Norris RD, Black BA, Walsh SM, McField M (2009) Local stressors reduce coral resilience to bleaching. PLoS One 4:e6324. https://doi.org/10.1371/journal.pone.0006324

Carilli JE, Norris RD, Black B, Walsh SM, McField M (2010) Century-scale records of coral growth rates indicate that local stressors reduce coral thermal tolerance threshold. Glob Chang Biol 16:1247–1257. https://doi.org/10.1111/j.1365-2486.2009.02043.x

Coffroth MA, Lasker HR, Oliver JK (1990) Coral mortality outside of the eastern Pacific during 1982–1983: relationship to El Niño. In: Glynn PW (ed) Global ecological consequences of the 1982–83 El Niño–Southern Oscillation. Elsevier, New York, pp 141–182

Coles SL, Jokiel PL (1978) Synergistic effects of temperature, salinity and light on the hermatypic coral *Montipora verrucosa*. Mar Biol 49:187–195. https://doi.org/10.1007/BF00391130

Coles SL, Jokiel PL, Lewis CR (1976) Thermal tolerance in tropical versus subtropical Pacific reef corals. Pac Sci 30:159–166

Cooper TF, O'Leary RA, Lough JM (2012) Growth of Western Australian corals in the Anthropocene. Science 335:593–596. https://doi.org/10.1126/science.1214570

Couch CS, Burns JHR, Liu G, Steward K, Gutlay TN, Kenyon J et al (2017) Mass coral bleaching due to unprecedented marine heatwave in the Northwestern Hawaiian Islands. PLoS One 12: e0185121. https://doi.org/10.1371/journal.pone.0185121

De'ath G, Lough JM, Fabricius KE (2009) Declining coral calcification on the Great Barrier Reef. Science 323:116–119. https://doi.org/10.1126/science.1165283

De'ath G, Lough JM, Fabricius KE (2013) Corrigendum: declining coral calcification on the Great Barrier Reef. Science 342:559

Deser C, Alexander MA, Xie SP, Phillips AS (2010) Sea surface temperature variability: patterns and mechanisms. Annu Rev Mar Sci 2:115–143. https://doi.org/10.1146/annurev-marine-120408-151453

D'Olivo JP, McCulloch MT, Judd K (2013) Long-term records of coral calcification across the central Great Barrier Reef: assessing the impacts of river runoff and climate change. Coral Reefs 32:999–1012. https://doi.org/10.1007/s00338-013-1071-8

Donner SD, Skirving WJ, Little CM, Oppenheimer M, Hoegh-Guldberg O (2005) Global assessment of coral bleaching and required rates of adaptation under climate change. Glob Chang Biol 11:2251–2265. https://doi.org/10.1111/j.1365-2486.2005.01073.x

Donner SD, Knutson TR, Oppenheimer M (2007) Model-based assessment of the role of human-induced climate change in the 2005 Caribbean coral bleaching event. Proc Natl Acad Sci U S A 104:5483–5488. https://doi.org/10.1073/pnas.0610122104

Dunne R, Brown B (2001) The influence of solar radiation on bleaching of shallow water reef corals in the Andaman Sea, 1993–1998. Coral Reefs 20:201–210. https://doi.org/10.1007/s003380100160

Eakin CM (2007) Written testimony of C. Mark Eakin, Ph.D., Coordinator, Coral Reef Watch, National Environmental Satellite, Data and Information Service, National Oceanic and Atmospheric Administration, US Department of Commerce, 17 April 2007. (Oversight hearing on wildlife and oceans in a changing climate before the Committee on Natural Resources Subcommittee on Fisheries, Wildlife and Oceans) US House of Representatives, Washington, DC

Eakin CM, Lough JM, Heron SF (2009) Climate variability and change: monitoring data and evidence for increased coral bleaching stress. In: van Oppen MJH, Lough JM (eds) Coral bleaching. Patterns, processes, causes and consequences, Ecological Studies, vol 205. Springer, Berlin, pp 41–67

Eakin CM, Liu G, Gomez AM, De La Cour JL, Heron SF, Skirving WJ et al (2016) Global coral bleaching 2014-2017. Reef Encount 31:20–26

Eakin CM, Liu G, Gomez AM, De La Cour JL, Heron SF, Skirving WJ et al (2017) Ding, dong, the witch is dead (?) — three years of global coral bleaching 2014-2017. Reef Encount 32:31–36

Evans MN, Kaplan A, Cane MA (2002) Pacific sea surface temperature field reconstruction from coral $\delta^{18}O$ data using reduced space objective analysis. Paleoceanography 17:7-1–7-13. https://doi.org/10.1029/2000PA000590

Felis T, Patzold J (2003) Climate records from corals. In: Wefer G, Lamy F, Mantoura F (eds) Marine science frontiers for Europe. Springer, New York, pp 11–27

Folland CK, Renwick JA, Salinger MJ, Mullan AB (2002) Relative influences of the Interdecadal Pacific Oscillation and ENSO on the South Pacific convergence zone. Geophys Res Lett 29:1643. https://doi.org/10.1029/2001GL014201

Gagan MK, Abram NJ (2011) Stable isotopes and trace elements. In: Hopley D (ed) Encyclopedia of modern coral reefs – structure, form and process. Springer, Dordrecht, pp 1034–1043

Garde LA, Spillman CM, Heron SF, Beeden RJ (2014) ReefTemp next generation: a new operational system for monitoring reef thermal stress. J Oper Oceanog 7:21–33. https://doi.org/10.1080/1755876X.2014.11020150

Gardner TA, Cote IM, Gill JA, Grant A, Watkinson AR (2003) Long-term region-wide declines in Caribbean corals. Science 301:958–960. https://doi.org/10.1126/science.1086050

Gleeson MW, Strong AE (1995) Applying MCSST to coral reef bleaching. Adv Space Res 16:151–154. https://doi.org/10.1016/0273-1177(95)00396-V

Glynn PW (1983) Extensive “bleaching” and death of reef corals on the Pacific coast of Panama. Environ Conserv 10:149–154. https://doi.org/10.1017/S0376892900012248

Glynn PW (1984) Widespread coral mortality and the 1982–83 El Niño warming event. Environ Conserv 11:133–146. https://doi.org/10.1017/S0376892900013825

Glynn PW (1985) El Niño-associated disturbance to coral reefs and post disturbance mortality by *Acanthaster planci*. Mar Ecol Prog Ser 26:295–300. https://doi.org/10.3354/meps026295

Glynn PW (1990) Coral mortality and disturbances to coral reefs in the tropical eastern Pacific. In: Glynn PW (ed) Global ecological consequences of the 1982–83 El Nino–Southern Oscillation. Elsevier, New York, pp 55–126

Glynn PW (1991) Coral reef bleaching in the 1980s and possible connections with global warming. Trends Ecol Evol 6:175–179. https://doi.org/10.1016/0169-5347(91)90208-F

Glynn PW (1993) Coral reef bleaching: ecological perspectives. Coral Reefs 12:1–17. https://doi.org/10.1007/BF00303779

Glynn PW (1996) Coral reef bleaching: facts, hypotheses and implications. Glob Chang Biol 2:495–509. https://doi.org/10.1111/j.1365-2486.1996.tb00063.x

Glynn PW, D’Croz L (1990) Experimental evidence for high temperature stress as the cause of El Niño-coincident coral mortality. Coral Reefs 8:181–191. https://doi.org/10.1007/BF00265009

Goreau TJ, Hayes RL (1994) Coral bleaching and ocean “hot spots”. Ambio 23:176–180

Goreau T, McClanahan T, Hayes R, Strong A (2000) Conservation of coral reefs after the 1998 global bleaching event. Conserv Biol 14:5–15. https://doi.org/10.1046/j.1523-1739.2000.00011.x

Graham NAJ, Jennings S, MacNeil MA, Mouillot D, Wilson SK (2015) Predicting climate-driven regime shifts versus rebound potential in coral reefs. Nature 518:94–97. https://doi.org/10.1038/nature14140

Griesser AG, Spillman CM (2016) Assessing the skill and value of seasonal thermal stress forecasts for Coral bleaching risk in the Western Pacific. J Appl Meteorol Climatol 55:1565–1578. https://doi.org/10.1175/JAMC-D-15-0109.1

Grottoli AG, Eakin CM (2007) A review of modern coral $\delta^{18}O$ and $\Delta^{14}C$ proxy records. Earth Sci Rev 18:67–91. https://doi.org/10.1016/j.earscirev.2006.10.001

Heron SF, Steinberg CR, Heron ML, Mantovanelli A, Jaffrés JBD, Skirving WJ et al (2010) Advancing ocean monitoring near coral reefs. Eos 91:369–380. https://doi.org/10.1029/2010E0410001

Heron SF, Liu G, Rauenzahn JL, Christensen TRL, Skirving WJ, Burgess TFR et al (2014) Improvements to and continuity of operational global thermal stress monitoring for coral bleaching. J Oper Oceanogr 7:3–11. https://doi.org/10.1080/1755876X.2014.11020154

Heron SF, Liu G, Eakin M, Skirving WJ, Muller-Karger FE, Vega-Rodriguez M et al. (2015) Climatology development for NOAA Coral Reef Watch's 5-km product suite. NOAA Technical Report NESDIS 145. NOAA/NESDIS, College Park, MD, p 21

Heron SF, Maynard J, van Hooidonk R, Eakin CM (2016a) Warming trends and bleaching stress of the world's coral reefs 1985–2012. Sci Rep 6:38402. https://doi.org/10.1038/srep38402

Heron SF, Johnston L, Liu G, Geiger EF, Maynard JA, De La Cour JL et al (2016b) Validation of reef-scale thermal stress satellite products for coral bleaching monitoring. Remote Sens 8:59. https://doi.org/10.3390/rs8010059

Hoegh-Guldberg O (1999) Climate change, coral bleaching and the future of the world's coral reefs. Mar Freshw Res 50:839–866. https://doi.org/10.1071/MF99078

Hoegh-Guldberg O (2005) Low coral cover in a high-CO_2 world. J Geophys Res 110. https://doi.org/10.1029/2004JC002528

Hoegh-Guldberg O, Salvat B (1995) Periodic mass-bleaching and elevated sea temperatures: bleaching of outer reef slope communities in Moorea, French Polynesia. Mar Ecol Prog Ser 121:181–190. https://doi.org/10.3354/meps121181

Hoegh-Guldberg O, Mumby PJ, Hooten AJ, Steneck RS, Greenfield P, Gomez E et al (2007) Coral reefs under rapid climate change and ocean acidification. Science 318:1737–1742. https://doi.org/10.1126/science.1152509

Huang B, Thorne PW, Banzon VF, Boyer T, Chepurin G, Lawrimore JH et al (2017) Extended reconstructed sea surface temperature, Version 5 (ERSSTv5): upgrades, validations and intercomparisons. J Clim 30:8179–8205. https://doi.org/10.1175/JCLI-D-16-0836.1

Hughes TP, Kerry JT (2017) Back-to-back bleaching has now hit two-thirds of the Great Barrier Reef. The Conversation, published online April 12 2017. http://theconversation.com/back-to-back-bleaching-has-now-hit-two-thirds-of-the-great-barrier-reef-76092

Hughes TP, Baird AH, Bellwood DR, Card M, Connolly SR, Folke C et al (2003) Climate change, human impacts, and the resilience of coral reefs. Science 301:929–933. https://doi.org/10.1126/science.1085046

Hughes TP, Barnes ML, Bellwood DR, Cinner JE, Cumming GS, Jackson JBC et al (2017a) Coral reefs in the Anthropocene. Nature 546:82–90. https://doi.org/10.1038/nature22901

Hughes TP, Kerry J, Álvarez-Noriega M, Álvarez-Romero JG, Anderson KD, Baird AH et al (2017b) Global warming and recurrent mass bleaching of corals. Nature 543:373–378. https://doi.org/10.1038/nature21707

Hughes TP, Anderson KD, Connolly SR, Heron SF, Kerry JT, Lough JM et al (2018) Spatial and temporal patterns of mass bleaching of corals in the Anthropocene. Science 359:80–83. https://doi.org/10.1126/science.aan8048

IPCC (2013) In: Stocker TF, Qin D, Plattner G-K, Tignor M, Allen SK, Boschung J et al (eds) Climate change 2013: the physical science basis. Contribution of working group I to the fifth assessment report of the intergovernmental panel on climate change. Cambridge University Press, Cambridge

IPCC (2014a) In: Barros VR, Field CB, Dokken DJ, Mastrandrea MD, Mach KJ, Bilir TE et al (eds) Climate change 2014: impacts, adaptation, and vulnerability. Part B: regional aspects. Contribution of working group II to the fifth assessment report of the intergovernmental panel on climate change. Cambridge University Press, Cambridge, p 688

IPCC (2014b) In: Field CB, Barros VR, Dokken DJ, Mach KJ, Mastrandrea MD, Bilir TE et al (eds) Climate change 2014: impacts, adaptation, and vulnerability. Part A: global and sectoral aspects. Contribution of working group II to the fifth assessment report of the intergovernmental panel on climate change. Cambridge University Press, Cambridge, p 1132

Jokiel PL, Coles SL (1977) Effects of temperature on the mortality and growth of Hawaiian reef corals. Mar Biol 43:201–208. https://doi.org/10.1007/BF00402312

Jokiel PL, Brown EK (2004) Global warming, regional trends and inshore environmental conditions influence coral bleaching in Hawaii. Glob Chang Biol 10:1627–1641. https://doi.org/10.1111/j.1365-2486.2004.00836.x

Jones PD, Mann ME (2004) Climate over past millennia. Rev Geophys 42:RG2002. https://doi.org/10.1029/2003RG000143

Jones PD, Briffa KR, Osborn TJ, Lough JM, van Ommen TD, Vinther BM et al (2009) High-resolution palaeoclimatology of the last millennium: a review of current status and future prospects. The Holocene 19:3–49. https://doi.org/10.1177/0959683608098952

Kilpatrick KA, Podesta GP, Evans R (2001) Overview of the NOAA/NASA advanced very high resolution radiometer Pathfinder algorithm for sea surface temperature and associated matchup database. J Geophys Res 106:9179–9197. https://doi.org/10.1029/1999JC000065

Kleypas JA, McManus JW, Menez LAB (1999) Environmental limits to coral reef development: where do we draw the line? Am Zool 39:146–159. https://doi.org/10.1093/icb/39.1.146

Liu G, Skirving W, Strong AE (2003) Remote sensing of sea surface temperatures during 2002 Barrier Reef coral bleaching. Eos 84:137–144. https://doi.org/10.1029/2003EO150001

Liu G, Strong AE, Skirving W, Arzayus LF (2006) Overview of NOAA Coral Reef Watch program's near-real-time satellite global coral bleaching monitoring activities. In: Proceedings of the 10th internation coral reef symposium, Japan, vol 10, pp 1783–1793

Liu G, Matrosova LE, Penland C, Gledhill DK, Eakin CM, Webb RS et al. (2009). NOAA Coral Reef Watch coral bleaching outlook system. In: Proceedings of the 11th international coral reef symposium, Florida, vol 2, pp 956–960

Liu G, Rauenzahn JL, Heron SF, Eakin CM, Skirving WJ, Christensen TRL et al. (2013) NOAA Coral Reef Watch 50 km satellite sea surface temperature-based decision support system for Coral Bleaching Management. NOAA technical report NESDIS 143. NOAA/NESDIS. College Park, MD

Liu G, Heron SF, Eakin CM, Muller-Karger FE, Vega-Rodriguez M, Guild LS et al (2014) Reef-scale thermal stress monitoring of coral ecosystems: new 5-km global products from NOAA Coral Reef Watch. Remote Sens 6:11579–11606. https://doi.org/10.3390/rs61111579

Liu G, Skirving WJ, Geiger EF, De La Cour JL, Marsh BL, Heron SF et al (2017) NOAA Coral Reef Watch's 5km satellite coral bleaching heat stress monitoring product suite version 3 and four-month outlook version 4. Reef Encount 32:37–43

Lix JK, Venkatesan R, Grinson G, Rao RR, Jineesh VK, Muthiah MA et al (2016) Differential bleaching of corals based on El Niño type and intensity in the Andaman Sea, southeast Bay of Bengal. Environ Monit Assess 188:175. https://doi.org/10.1007/s10661-016-5176-8

Lough JM (2000) 1997–98: unprecedented thermal stress to coral reefs? Geophys Res Lett 27:3901–3904. https://doi.org/10.1029/2000GL011715

Lough JM (2001) Climate variability and change on the Great Barrier Reef. In: Wolanski E (ed) Oceanographic processes of coral reefs. Physical and biological links in the Great Barrier Reef. CRC, Boca Raton, pp 269–300

Lough JM (2004) A strategy to improve the contribution of coral data to high-resolution paleoclimatology. Palaeogeogr Palaeoclimatol Palaeoecol 204:115–143. https://doi.org/10.1016/S0031-0182(03)00727-2

Lough JM (2010) Climate records from corals. Wiley interdisciplinary reviews – climate change. https://doi.org/10.1002/wcc.39

Lough JM (2012) Small change, big difference: sea surface temperature distributions for tropical coral reef ecosystems, 1950–2011. J Geophys Res Oceans 117. https://doi.org/10.1029/2012JC008199

Lough JM, Cantin NE (2014) Perspectives on massive coral growth rates in a changing ocean. Biol Bull 226:187–202. https://doi.org/10.1086/BBLv226n3p187

Lough JM, Anderson KD, Hughes TP (2018) Increasing thermal stress for tropical coral reefs. Sci Rep 8:1871–2017. https://doi.org/10.1038/s41598-018-24530-9

Luo J-J, Zhang R, Behera SK, Masumoto Y (2010) Interaction between El Niño and extreme Indian Ocean dipole. J Clim 23:726–742. https://doi.org/10.1175/2009JCLI3104.1

Manzello DP (2010) Coral growth with thermal stress and ocean acidification: lessons from the eastern tropical Pacific. Coral Reefs 29:749–758. https://doi.org/10.1007/s00338-010-0623-4

Manzello D, Hendee JC, Ward D, Hillis-Starr Z (2006) An evaluation of environmental parameters coincident with the partial bleaching event in St Croix, US Virgin Islands 2003. In: Proceedings of the 10th international coral reef symposium, Japan, vol 10, pp 709–717

Manzello DP, Enochs IC, Kolodziej G, Carlton R (2015) Recent decade of growth and calcification of *Orbicella faveolata* in the Florida Keys: an inshore-offshore comparison. Mar Ecol Prog Ser 521:81–89. https://doi.org/10.3354/meps11085

Marshall PA, Baird AH (2000) Bleaching of corals on the Great Barrier Reef: differential susceptibilities among taxa. Coral Reefs 19:155–163. https://doi.org/10.1007/s003380000086

Marshall P, Schuttenberg H (2006) A reef manager's guide to coral bleaching. Great Barrier Reef Marine Park Authority, Townsville

Maturi E, Harris A, Mittaz J, Sapper J, Wick G, Zhu X et al (2017) A new high-resolution sea surface temperature blended analysis. Bull Am Meteorol Soc 98:1015–1026. https://doi.org/10.1175/BAMS-D-15-00002.1

Maynard JA, Turner PJ, Anthony KRN, Baird AH, Berkelmans R, Eakin CM et al (2008) ReefTemp: an interactive monitoring system for coral bleaching using high-resolution SST and improved stress predictors. Geophys Res Lett 35:L05603. https://doi.org/10.1029/2007GL032175

Maynard JA, Johnson JE, Marshall PA, Eakin CM, Goby G et al (2009) A strategic framework for responding to coral bleaching events in a changing climate. Environ Manag 44:1–11. https://doi.org/10.1007/s00267-009-9295-7

McClanahan T (2004) Coral bleaching, diseases and mortality in the Western Indian Ocean. In: Rosenberg E, Loya Y (eds) Coral health and disease. Springer, Berlin, pp 157–176

McClanahan TR, Donner SD, Maynard JA, MacNeil MA, Graham NAJ, Maina J et al (2012) Prioritizing key resilience indicators to support coral reef management in a changing climate. PLoS One 7:e42884. https://doi.org/10.1371/journal.pone.0042884

McGregor HV (2011) Paleoclimate from corals. In: Hopley D (ed) Encyclopedia of modern coral reefs – structure, form and process. Springer, Dordrecht, pp 777–785

McPhaden MJ (1999) Genesis and evolution of the 1997–98 El Niño. Science 83:950–954. https://doi.org/10.1126/science.283.5404.950

McPhaden MJ (2004) Evolution of the 2002/03 El Niño. Bull Am Meteorol Soc 85:677–695. https://doi.org/10.1175/BAMS-85-5-677

Montgomery RS, Strong AE (1994) Coral bleaching threatens oceans, life. Eos 75:145–147. https://doi.org/10.1029/94EO00837

Moore JAY, Bellchambers LM, Depczynski MR, Evans RD, Evans SN, Field SN et al (2012) Unprecedented mass bleaching and loss of coral across 12° of latitude in Western Australia in 2010–11. PLoS One 7:e51807. https://doi.org/10.1371/journal.pone.0051807

Morice CP, Kennedy JJ, Rayner NA, Jones PD (2012) Quantifying uncertainties in global and regional temperature change using an ensemble of observational estimates: the HadCRUT4 data set. J Geophys Res Atmos 117. https://doi.org/10.1029/2011JD017187

Mumby PJ, Chisholm JRM, Edwards AJ, Andrefouet S, Jaubert J (2001a) Cloudy weather may have saved Society Island reef corals during the 1998 ENSO event. Mar Ecol Prog Ser 222:209–216. https://doi.org/10.3354/meps222209

Mumby PJ, Chisholm JRM, Edwards AJ, Clark CD, Roark EB, Andrefouet S et al (2001b) Unprecedented bleaching-induced mortality in *Porites* spp at Rangiroa Atoll, French Polynesia. Mar Biol 139:183–189. https://doi.org/10.1007/s002270100575

Nadaoka K, Nihei Y, Wakaki K, Kumano R, Kakuma S, Moromizato S et al (2001) Regional variation of water temperature around Okinawa coasts and its relationship to offshore thermal environments and coral bleaching. Coral Reefs 20:373–384. https://doi.org/10.1007/s00338-001-0187-4

Obura DO (2005) Resilience and climate change: lessons from coral reefs and bleaching in the Western Indian Ocean. Estuar Coast Shelf Sci 63:353–372. https://doi.org/10.1016/j.ecss.2004.11.010

Obura D, Mangubhai S (2011) Coral mortality associated with thermal fluctuations in the Phoenix Islands, 2002–2005. Coral Reefs 30:607–619. https://doi.org/10.1007/s00338-011-0741-7

Obura D, Causey B, Church J (2006) Management response to a bleaching event. In: Phinney JT, Hoegh-Guldberg O, Kleypas J, Skirving W, Strong A (eds) Coral reefs and climate change science and management, Coastal and estuarine studies, vol 61. American Geophysical Union, Washington, DC, pp 181–206

Penland C, Matrosova L (1998) Prediction of tropical Atlantic sea surface temperatures using linear inverse modeling. J Clim 11:483–496. https://doi.org/10.1175/1520-0442(1998)011<0483:POTASS>2.0.CO;2

Podesta GP, Glynn PW (1997) Sea surface temperature variability in Panama and Galapagos: extreme temperatures causing coral bleaching. J Geophys Res 102:15749–15759. https://doi.org/10.1029/96JC03557

Potts DC, Swart PK (1984) Water temperature as an indicator of environmental variability on a coral reef. Limnol Oceanogr 29:504–516

Power S, Delage F, Chung C, Kociuba G, Keay K (2013) Robust twenty-first-century projections of El Niño and related precipitation variability. Nature 502:541–545. https://doi.org/10.1038/nature12580

Rayner NA, Parker DE, Horton EB, Folland CK, Alexander LP, Rowell DP et al (2003) Global analyses of SST, sea ice and night marine air temperature since the late nineteenth century. J Geophys Res 108. https://doi.org/10.1029/2002JD002670

Reynolds RW, Smith TM (1994) Improved global sea surface temperature analyses using optimum interpolation. J Clim 7:929–948. https://doi.org/10.1175/1520-0442(1994)007<0929:IGSSTA>2.0.CO;2

Reynolds RW, Rayner NA, Smith TM, Stokes DC, Wang W (2002) An improved in situ and satellite SST analysis for climate. J Clim 15:1609–1625. https://doi.org/10.1175/1520-0442(2002)015<1609:AIISAS>2.0.CO;2

Roberts-Jones J, Fiedler EK, Martin MJ (2012) Daily, global, high-resolution SST and sea ice reanalysis for 1985–2007 using the OSTIA system. J Clim 25:6215–6232. https://doi.org/10.1175/JCLI-D-11-00648.1

Saji NH, Goswami BN, Vinayachandran PN, Yamagata T (1999) A dipole mode in the tropical Indian Ocean. Nature 401:360–363. https://doi.org/10.1038/43854

Salm RV, Coles SL (eds) (2001) Coral bleaching and marine protected areas. Proceedings of the workshop on mitigating coral bleaching impact through MPA design, Bishop Museum, Honolulu, Hawaii, 29–31 May 2001. Asia Pacific coastal marine program report 0102. The Nature Conservancy, Honolulu

Schlesinger ME, Ramankutty N (1994) An oscillation in the global climate system of period 65–70 years. Nature 367:723–726. https://doi.org/10.1038/367723a0

Shein KA (ed) (2006) State of the climate in 2005. Bull Am Meteorol Soc 87:S1–S102. https://doi.org/10.1175/BAMS-87-6-shein

Sheppard C, Rayner NA (2002) Utility of the Hadley sea-ice and sea surface temperature data set (HadISST1) in two widely contrasting coral reef areas. Mar Pollut Bull 44:303–308. https://doi.org/10.1016/S0025-326X(02)00059-0

Sheppard C, Rioja-Nieto R (2005) Sea surface temperature 1871–2099 in 38 cells in the Caribbean region. Mar Environ Res 60:389–396. https://doi.org/10.1016/j.marenvres.2004.12.006

Skirving W, Guinotte J (2001) The sea surface temperature story on the Great Barrier Reef during the coral bleaching event of 1998. In: Wolanski E (ed) Oceanographic processes of coral reefs. Physical and biological Links in the Great Barrier Reef. CRC, Boca Raton, pp 301–313

Skirving WJ, Strong AE, Liu G, Liu C, Arzayus F, Sapper J (2006a) Extreme events and perturbations of coastal ecosystems: sea surface temperature change and coral bleaching. In: Richardson LL, LeDrew EF (eds) Remote sensing of aquatic coastal ecosystem processes. Springer, Dordrecht, pp 11–25

Skirving W, Heron M, Heron S (2006b) The hydrodynamics of a bleaching event: implications for management and monitoring. In: Phinney JT, Hoegh-Guldberg O, Kleypas J, Skirving W, Strong A (eds) Coral reefs and climate change science and management, Coastal and estuarine studies, vol 61. American Geophysical Union, Washington, DC, pp 145–161

Skirving WJ, Marsh B, Heron SF, Liu G, De La Cour JL, Maturi E, Harris A, Geiger EF, Tirak K, Strong AE, Eakin CM (2018) CoralTemp: A daily global 5km satellite sea surface temperature product (1985 to present). Abstract OD43A-05 presented at 2018 Ocean Sciences Meeting, Portland, OR, 12–16 Feb

Smith SV, Buddemeier RW (1992) Global change and coral reef ecosystems. Annu Rev Ecol Syst 23:89–118. https://doi.org/10.1146/annurev.es.23.110192.000513

Smith TM, Reynolds RW (2003) Extended reconstruction of global sea surface temperatures based on COADS data (1854–1997). J Clim 16:1495–1510. https://doi.org/10.1175/1520-0442-16.10.1495

Smith TM, Reynolds RW (2004) Improved extended reconstruction of SST (1854–1997). J Clim 17:2466–2477. https://doi.org/10.1175/1520-0442(2004)017<2466:IEROS>2.0.CO;2

Spillman C, Alves O, Hudson DA (2011) Seasonal prediction of thermal stress accumulation for coral bleaching in the tropical oceans. Mon Weather Rev 139:317–331. https://doi.org/10.1175/2010MWR3526.1

Stocker TF, Qin D, Plattner G-K, Alexander LV, Allen SK, Bindoff NL et al (2013) Technical summary. In: Stocker TF, Qin D, Plattner G-K, Tignor M, Allen SK, Boschung J et al (eds) Climate change 2013: the physical science basis. Contribution of working group I to the fifth assessment report of the intergovernmental panel on climate change. Cambridge University Press, Cambridge

Strong AE, Barrientos CS, Duda C, Sapper J (1997) Improved satellite techniques for monitoring coral reef bleaching. In: Lessios HA, Macintyre IG (eds) Proceedings of the 8th international coral reef symposium, Panama, 8-2, pp 1495–1498

Strong AE, Arzayus F, Skirving W, Heron S (2006) Identifying coral bleaching remotely via coral reef watch – improved integration and implications for changing climate. In: Phinney JT, Hoegh-Guldberg O, Kleypas J, Skirving W, Strong A (eds) Coral reefs and climate change science and management, Coastal and estuarine studies, vol 61. American Geophysical Union, Washington, DC, pp 163–180

Tanzil JTI, Brown BE, Tudhope AW, Dunne RP (2009) Decline in skeletal growth of the coral *Porites lutea* from the Andaman Sea, South Thailand between 1984 and 2005. Coral Reefs 28:519–528. https://doi.org/10.1007/s00338-008-0457-5

Tanzil JTI, Brown BE, Dunne RP, Lee JN, Kaandorp JA, Todd PA (2013) Regional decline in growth rates of massive *Porites* corals in southeast Asia. Glob Chang Biol 19:3011–3023. https://doi.org/10.1111/gcb.12279

Tierney JE, Abram NJ, Anchukaitis KJ, Evans MN, Giry C, Kilbourne KH et al (2015) Tropical sea surface temperatures for the past four centuries reconstructed from coral archives. Paleoceanography:30, 226–252. https://doi.org/10.1002/2014PA002717

Trenberth KE, Shea DJ (2006) Atlantic hurricanes and natural variability in 2005. Geophys Res Lett 33:L12704. https://doi.org/10.1029/2006GL026894

van Hooidonk R, Maynard JA, Planes S (2013) Temporary refugia for coral reefs in a warming world. Nat Clim Chang 3:508–511. https://doi.org/10.1038/nclimate1829

van Hooidonk R, Maynard JA, Manzello D, Planes S (2014) Opposite latitudinal gradients in projected ocean acidification and bleaching impacts on coral reefs. Glob Chang Biol 20:103–112. https://doi.org/10.1111/gcb.12394

van Hooidonk R, Maynard J, Tamelander J, Gove J, Ahmadia G, Raymundo L et al (2016) Local-scale projections of coral reef futures and implications of the Paris agreement. Sci Rep 6:39666. https://doi.org/10.1038/srep39666

Vega-Rodriguez M, Müller-Karger FE, Hallock P, Quiles-Perez GA, Eakin CM, Colella M et al (2015) Influence of water-temperature variability on stony coral diversity in Florida Keys patch reefs. Mar Ecol Prog Ser 528:173–186. https://doi.org/10.3354/meps11268

Wellington GM, Strong AE, Merlen G (2001) Sea surface temperature variation in the Galapagos archipelago: a comparison between AVHRR nighttime satellite data and in situ instrumentation (1982–1998). Bull Mar Sci 69:27–42

Wernberg T, Smale DA, Tuya F, Thomsen MS, Langlois TJ, de Bettignies T et al (2012) An extreme climatic event alters marine ecosystem structure in a global biodiversity hotspot. Nat Clim Chang 3:78–82. https://doi.org/10.1038/nclimate1627

Wilkinson C (2000) Status of coral reefs of the world: 2000. Global Coral Reef Monitoring Network, Australian Institute of Marine Science, Townsville

Wilkinson C (2002) Status of coral reefs of the world: 2002. Global Coral Reef Monitoring Network, Australian Institute of Marine Science, Townsville

Wilkinson C (2004) Status of coral reefs of the world: 2004. Global Coral Reef Monitoring Network, Australian Institute of Marine Science, Townsville

Wilkinson C, Souter D (2008) Status of Caribbean coral reefs after bleaching and hurricanes in 2005. Global Coral Reef Monitoring Network and Reef and Rainforest Research Centre, Townsville

Williams EH, Bunkley-Williams L (1990) The world-wide coral reef bleaching cycle and related sources of coral mortality. Atoll Res Bull 335:1–71. https://doi.org/10.5479/si00775630.335.1

Winter A, Appeldoorn RS, Bruckner A, Williams EH, Goenaga C (1998) Sea surface temperatures and coral reef bleaching off La Parguera, Puerto Rico (northeastern Caribbean Sea). Coral Reefs 17:377–382. https://doi.org/10.1007/s003380050143

Wolter K, Timlin MS (1998) Measuring the strength of ENSO events – how does 1997/98 rank? Weather 53:315–324. https://doi.org/10.1002/j.1477-8696.1998.tb06408.x

Zhang H-M, Reynolds RW, Bates JG (2006) Blended and gridded high resolution global sea surface wind speed and climatology from multiple satellites: 1987–present. 14th conference on satellite meteorology and oceanography, Atlanta, GA, American Meteorological Society, Paper 100004. Available at https://ams.confex.com/ams/Annual2006/webprogram/Paper100004

Zhang N, Feng M, Hendon HH, Hobday AJ, Zinke J (2017) Opposite polarities of ENSO drive distinct patterns of coral bleaching potentials in the southeast Indian Ocean. Sci Rep 7:2443. https://doi.org/10.1038/s41598-017-02688-y

Zinke J, Hoell A, Lough JM, Feng M, Kuret AJ, Clarke H et al (2015) Coral record of southeast Indian Ocean marine heatwaves with intensified western Pacific temperature gradient. Nat Commun 6:8562. https://doi.org/10.1038/ncomms9562

Chapter 5
Detecting and Monitoring Coral Bleaching Events

N. E. Cantin and M. Spalding

5.1 Introduction

Over the past 30 years, mass coral bleaching has become a widespread phenomenon and is now considered as one of the most distinct visual manifestations of climate change impacts on tropical coral reef ecosystems. Oliver et al. (Chap. 3) provide evidence that the global occurrence of widespread coral bleaching and mortality events has increased since the 1980s. This was most dramatically illustrated for the first time in 1998, linked to a major El Niño event, which led to bleaching in almost every coral reef region and to widespread coral mortality in some areas (Chap. 3). Since 2014, the three warmest years on record have been observed throughout most tropical coral reef regions, and, as a result, the most severe coral bleaching event impacted most of the world's reefs from 2014 to 2017 (Heron et al. 2017).

Quantifying the scale of such events presents particular challenges. In situ underwater observation is clearly limited by the magnitude of observations required—few reefs are within striking distance of research institutions and remote research stations. Significant planning and preparation are required to document coral bleaching events across large regional coral reef ecosystems and remote, distant coral reefs.

As an extreme example of the problem, the 1998 coral bleaching event caused mass mortality throughout the Seychelles (Spencer et al. 2000) and the Maldives (McClanahan 2000; Edwards et al. 2001). The Chagos Archipelago (British Indian Ocean Territory) consists of a vast area of coral reefs in the same broad vicinity as these island groups, but no underwater observations were made in 1998. Almost a

N. E. Cantin (✉)
Australian Institute of Marine Science, Townsville MC, QLD, Australia
e-mail: n.cantin@aims.gov.au

M. Spalding
Global Conservation Approach Team, The Nature Conservancy, Newmarket, UK

M. J. H. van Oppen, J. M. Lough (eds.), *Coral Bleaching*, Ecological Studies 233,
https://doi.org/10.1007/978-3-319-75393-5_5

year later the UK government and their senior fisheries advisors (who have a boat on permanent patrol in the islands) suggested that the Chagos reefs had not bleached. It was only the observation of a bleached coral in the background of a photograph taken by an engineer undertaking work on a military base that enabled the case to be made for a post hoc assessment (pers. obs. M. Spalding). The subsequent survey noted "near-total" mortality of hard corals down to 15 m in the northern atolls and to >35 m in the central and southern atolls throughout the Chagos Archipelago (Sheppard et al. 2002).

Due to the increased frequency and severity of widespread coral bleaching events over the past three decades, researchers and reef managers are better prepared to respond to ocean heat waves to document the response and impacts of thermal stress on local, regional and remote coral reef locations. While detailed and frequent observations are necessary to fully document the impacts of a bleaching event, well-planned and co-ordinated response efforts are necessary to assess the full spatial severity of widespread coral bleaching events as heat waves unfold throughout coral reef ecosystems. Integration of coral reef researchers with local communities is important to accurately assess the extent of bleaching in populated and remote locations. For example, the severe bleaching in the remote central Pacific Phoenix Islands group of the Republic of Kiribati was fully documented in 2002–2005 through repeated surveys over time to detect a 60% decline in live coral cover and then again with observations at the peak of severe bleaching in 2015 (Obura and Mangubhai 2011).

There remains a need, however, for remote and, ideally, near-continuous observation of tropical coral reefs to increase the spatial assessment capability and to ensure that an event as dramatic as the near-total loss of reef corals in 1998 throughout the Chagos Archipelago, a region the size of England, cannot pass unnoticed in the future. At the same time, finer-scale observations of bleaching events across space and time will always prove to be of critical value in understanding mechanisms of thermal tolerance and susceptibility and in determining management responses (West and Salm 2003). Fine-scale observations can clarify patterns of differential survival (deep versus shallow, lagoon or sheltered bays versus current-swept reef slopes) and support a better understanding of bleaching processes within coral communities. At even finer resolutions, the patterns of survival and recovery also require observation at levels right down to the individual colony, the polyp and indeed within the cell.

In this chapter we discuss current approaches for monitoring bleaching events which have been used by coral reef researchers. We highlight the importance of consistent, comparable observations to synthesise the global impacts of coral bleaching for coral reef communities. As technology develops, remote-sensing applications from satellite platforms will enhance our capacity to monitor remote coral reef locations. We review advances in these remote-sensing applications as well as the aerial and in-water survey methods that have rapidly advanced to document the spatial impacts and severity of coral bleaching in recent years.

5.2 Broad-Scale Approaches

5.2.1 *Remote Sensing of Bleaching Events*

At the broadest scales, there is a need to be able to observe coral bleaching over very wide areas, across all tropical coral reef ecosystems. The only effective means of "seeing" reefs at these broad scales is using remote sensing with satellite platforms or aerial surveys from fixed-wing aircraft and helicopters and possibly long-range, fixed-wing drones in the future. Bleached corals can hardly be mistaken in the field, but this also translates to a very distinctive spectral signature that, theoretically, should be visible from remote platforms (Holden and LeDrew 1998; Call et al. 2003).

The practical challenges of remote detection of coral bleaching remain considerable. Coral reefs present highly heterogeneous substrates—even up close, most are a complex patchwork of coral, algae, sponges, sand, reef rock and other surface cover. Thus, all but the highest resolution remote-sensing platforms are sampling areas (pixels) that include a very broad mix of reflecting surfaces. This challenge is further compounded by the influence of differing depths of the water column—in clear waters it is possible to differentiate marine features and detect coral bleaching to depths as deep as 15 m (Elvidge et al. 2004), but it is a challenge to differentiate many features beyond that depth. In addition, as most reef corals grow on sloping substrates, considerable variation in depth and reflectance can occur even within the space of individual pixels. Current satellite imagery capability is largely restricted to the upper reef slopes and reef flat regions. The descriptive resolution derived from a remote-sensing platform can range from broad benthic community cover categories (e.g. coral, sand, rubble, reef matrix) to benthic types (e.g. branching coral, tabulate coral, massive coral).

Historically the most widely used remote platforms for general reef mapping are those that allow coverage of relatively large areas relatively cheaply. The most commonly used was Landsat 8 (USGS and NASA), while the recently launched Sentinel-2A and Sentinel-2B (ESA) continue this family of sensors. In the last 10 years, commercial higher-resolution satellite data (pixel sizes less than 10 m) from IKONOS, GeoEye, Quickbird, WorldView 2 and 3 and Pleiades clearly enable more accurate feature assessment and classification from pixel scales more closely related to reef substrate patch sizes (Andréfouët et al. 2003; Mumby et al. 2004a). This is, however, at the increased cost of image acquisition and pre- and post-processing corrections and remains challenging to apply consistently across large spatial scales (Hedley et al. 2016).

Coral bleaching can be a short-lived phenomenon, and its spatial distribution can vary considerably across broad latitudinal spatial scales and as the thermal stress progresses throughout the warm season. Conversely, many bleaching events are becoming more extensive and tightly synchronised (i.e. more species are fully bleached at the same time) and these are likely to be easier to detect with remote-sensing techniques, particularly in areas of high coral cover. For reef locations where a low proportion of colonies are bleached, or where the loss of colour in bleached colonies is only partial, or reefs that have low live coral cover, detection of bleaching occurrence from remote-sensing platforms remains challenging.

High spatial resolution is undoubtedly the most critical factor in helping to disaggregate the complex patchwork of substrate which typifies most coral reefs (Hedley et al. 2012a). The successful detection of coral bleaching from remote-sensing satellites can be limited by several environmental factors such as cloud cover, wave action, solar glint and satellite path over particular reefs of interest (i.e. reefs with in-water survey data). Temporal return is also critical: corals can shift from bleached to recovered in just a few weeks, and dead corals can become overgrown with algae in even shorter timeframes. Differential susceptibilities of different coral species or at different depths mean that the "peak" of a bleaching event may only last a few days, although more typically it will be from 2 to 4 weeks. With regular cloud cover in the tropics, a return of, say, 2–3 weeks for a sensor may be insufficient to capture a bleaching event at particular reefs. The Sentinel-2 satellite pair has short re-visit times and will eventually resample each location around the globe every 5 days, which will enable the use of image time series analysis, making this platform better suited to detect spectral changes in the benthic community at scales of 10 m or more due to mass coral bleaching, but only where coral cover is sufficiently high.

The challenges of using satellite platforms to detect and map bleaching events, even high-resolution systems, are thus considerable. Andréfouët et al. (2002) tried to assess the optimum resolution for remote sensing of bleached corals using aerial photographs taken during the 1998 bleaching event and subjecting these to interpretation at varying resolutions. They noted a rapid, linear decrease in the detection capability at coarser pixel resolution, with as much as 50% of the 10 cm resolution signal lost at 1 m resolution. As a general guide, resolution closest to that of the mean colony size will be most accurate (40–80 cm), but resolution up to 1 m may still give some ability both to detect bleaching and estimate variance between locations. At their seven reef locations in the central and southern Great Barrier Reef (GBR), Andréfouët et al. (2002) also compared satellite-derived images taken before, during and after the 1998 bleaching using 20 m and 10 m resolution imagery, but showed a complete inability to detect the fairly major bleaching event of 1998.

Philipson and Lindell (2003) showed at least a basic detection of a very large-scale bleaching event in Belize using the 24 m resolution IRS LISS-III platform and suggested that much better detection should be possible with higher-resolution sensors (e.g. IKONOS 4 m resolution). Elvidge et al. (2004) showed very good detection with IKONOS imagery on the GBR, but point to the need for pre-bleaching reference images. Yamano and Tamura (2004) were also able to show detection of severe bleaching at Ishigaki Island in Japan, using 30 m resolution imagery from Landsat TM, but only in shallow, coral rich areas. Mumby et al. (2004b) were able to distinguish living massive *Porites* and branching *Pocillopora* colonies from various stages of recently dead and old dead colonies (>6 months). They used multispectral (10 bands) imagery from an airborne sensor (CASI) at 1 m resolution and concluded that the precision of sampling benthic cover was similar

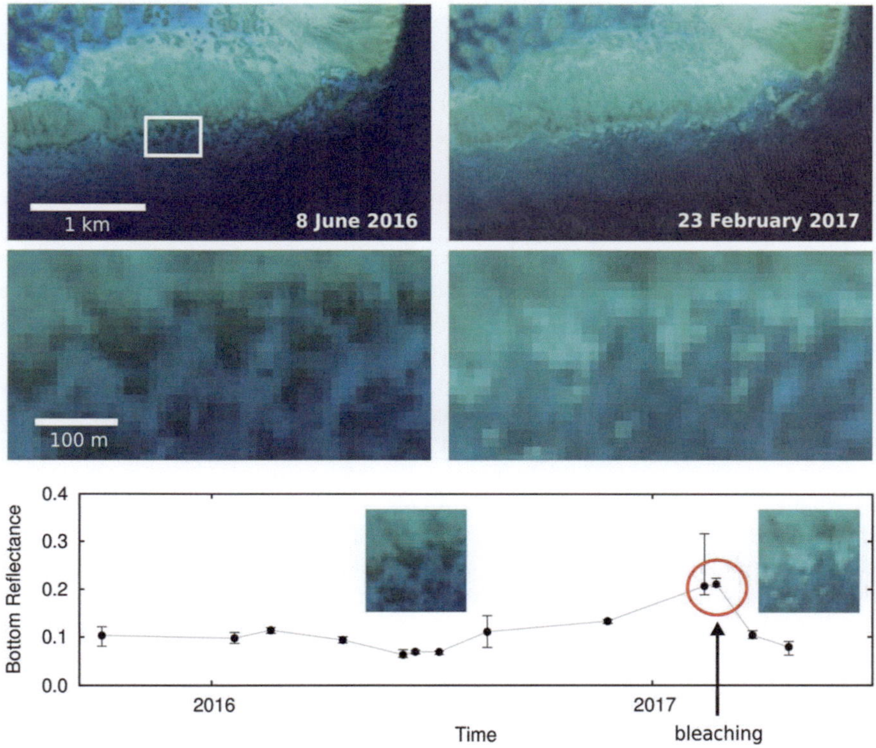

Fig. 5.1 Images from the Copernicus Sentinel-2A satellite captured visual signs of widespread community-level coral bleaching and increased bottom reflectance signatures at Adelaide Reef from the Great Barrier Reef on 23 February 2017, through time series comparisons of frequent geo-located images (Hedley and Roelfsema 2017)

between the remote-sensing and field-based quadrat surveys, provided that bathymetry is known independently, water is shallow and clear and brown fleshy macroalgae are scarce (Mumby et al. 2004b). Prelaunch tests of Sentinel-2 capabilities suggested that detection of bleaching signals was a possibility in shallow water (2–3 m; Hedley et al. 2012b) and, through further development, bleaching signatures were detected by Sentinel-2a on the GBR in 2017 (Fig. 5.1).

The overall conclusion from these efforts is that bleaching detection at regional to global scales from satellite platforms and hyperspectral airborne sensors remains limited and challenging due to the poor temporal availability of quality data that is often compromised by cloud and sea state optical distortions at the air-water interface during the peak bleaching window. New technologies called fluid lensing on unmanned aerial vehicle (UAV) drone platforms are under development. These appear capable of reducing surface wave distortion and producing high-resolution 2D and 3D reconstructions of the reef at centimetre scales to a depth of ~10 m (Chirayath and Earle 2016). This technique offers a new option for large-scale surveys of shallow reef environments, if the imaging and high-performance computing technologies can be made widely available. However, UAV platforms will still restrict the spatial scale over which observations are consistently acquired. Improvements in availability of high-resolution imagery, notably reduction in the costs, have helped. However, unless mass bleaching is widespread across hundreds of metres of reef, detection of change in the benthic community due to coral bleaching through time series image comparisons remains limited and an avenue that requires further development.

Under current development is a programme called Sen2Coral, by the European Space Agency (www.sen2coral.org). Sen2Coral is aimed at using the open data policy of the Sentinel-2 Copernicus programme and the improved capabilities of Sentinel-2A and 2B (multispectral, high-resolution, near-global coverage and frequent return times) to develop and implement a suite of practical tools that will likely lead to better detection capabilities of coral bleaching from satellite remote-sensing platforms. Tools under development that are crucial to the detection of bleaching signals and benthic classification from multispectral imagery relate to the visual interpretation and data processing of time series and detecting pixel-based differences. These include a combination of accessible algorithms that can process sequences of images that are spatially aligned, exclude land, clouds, cloud shadows and breaking waves (white caps) and atmospheric corrections and proprietary algorithms related to solar surface glint removal, water column corrections and bathymetry. Studying Sentinel-2 images captured over the GBR between January and April 2017, the Sen2Coral science team captured clear visual evidence of widespread bleaching and increasing bottom reflectance from a series of images (Fig. 5.1; Hedley and Roelfsema 2017). While the resolution of these observations does not replace in-water survey assessments, this technological advance certainly has significant implications for documenting the presence or absence of community-scale bleaching and will enable wide spatial assessments of the early onset of the heat stress response. Collaboration between reef ecologists documenting bleaching and remote-sensing teams following the recent coral bleaching events in 2014–2017 will certainly drive effective developments in remote sensing of coral bleaching in the coming years. Clearly the capability for detecting the early onset and presence of bleached and non-bleached reef communities is becoming available with targeted approaches.

The finer spatial resolution of aerial photography offers increased pixel resolution down to the scale of individual coral colonies, at similar costs to remote satellite data acquisition and processing: however, considerable challenges in surveying more remote reef systems remain. Some of the most extensive aerial survey work has been conducted along the length of the GBR in 1998 and 2002 (Berkelmans and Oliver 1999; Berkelmans et al. 2004) and in 2016 (Hughes et al. 2017) and 2017 (https://www.coralcoe.org.au/for-managers/coral-bleaching-and-the-great-barrier-reef). These surveys were able to document bleaching at a total of 25–35% of the >3000 individual reefs that make up the GBR during four mass bleaching events. This work was undertaken from a combination of fixed-wing aircraft and helicopters flying at an altitude of 150 m. This aerial survey approach covered over 2000 km of coastline and provided critical information to compare spatial patterns of recent mass bleaching events (Hughes et al. 2017). Broad categories are used to score the bleaching severity by a pair of observers. The categories match in-water transect-based bleaching categories [0, no bleaching (<1%); 1, minor (1–10%); 2, moderate (10–30%); 3, major (30–60%) and 4, severe (>60%)], based on the percentage of live coral that appears bleached and is visible to the observer. Bleached coral colonies and early mortality are easily differentiated from reef substrate within high-resolution images captured with telephoto digital camera systems (e.g. single lens reflex (SLR) or mirrorless options) during flight (Fig. 5.2). Paired observers simultaneously score the bleaching severity.

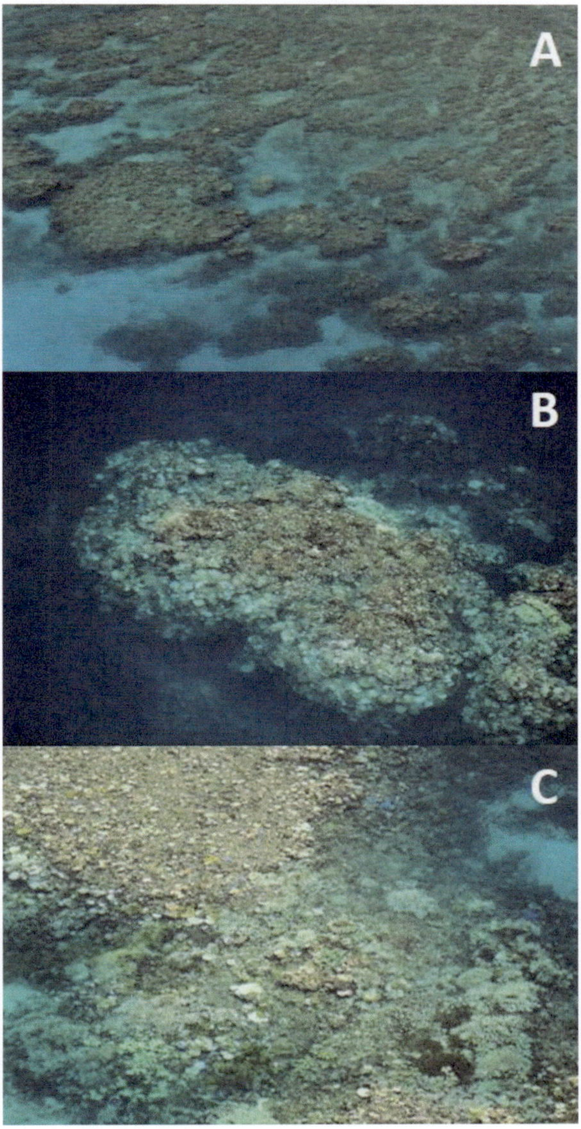

Fig. 5.2 Aerial images of the 2017 Great Barrier Reef bleaching event, clearly showing the difference between a reef that is not bleached [Category 0, (**a**)] and reefs suffering severe bleaching [Category 4, >60% corals bleached, pale or fluorescent (**b**)] and recent stages of mortality of staghorn and tabulate *Acropora* colonies [patches of green and brown algae forming on white coral skeletons (**c**)]. Image credit: R. Berkelmans (**a**) and N. Cantin (**b** and **c**) AIMS

Aerial surveys provide a rapid assessment of coral bleaching severity across wide spatial scales that, to date, has not been gathered either from satellite or field-based in-water observations.

Timing of aerial bleaching surveys is critical (Berkelmans et al. 2004). Aerial surveys should be planned for when bleaching is at its peak (most intense) but before major mortality sets in. This will be informed by early warning systems (e.g. satellite-derived temperature or bleaching products (Chap. 4), real-time temperature monitoring, bleaching thresholds, etc.), in-water surveys by researchers, tourism and industry partners, and anecdotal reports from the general public. Wind, weather (rain, storms and cloud cover) and tides are also key attributes that should be considered in planning the schedule of aerial bleaching surveys. Clear skies, calm winds (<10–15 knots) and low tides that coincide with midday direct overhead sunlight (10a.m.–3p.m.) provide the best water clarity conditions for aerial imagery and observations.

Aerial surveys are a cost-effective approach to assess the severity of widespread mass coral bleaching across wide spatial areas. Data gathered provides a categorical assessment of community-level bleaching severity in the upper 4–6 m of the reef flat, back reef and slope habitats. In-water surveys are still required to assess the taxonomic spectrum of bleaching responses, to quantify bleaching-related mortality and to assess bleaching depth distributions.

5.2.2 Remote Sensing of Indicators of Bleaching Likelihood

Proxy measures of likelihood of bleaching have been available since the 1990s using both low- and now high-resolution sea surface temperatures (SST) data. Originally, these products were produced from night-time only SST records at 50 km resolution, gathered twice weekly by the National Oceanic and Atmospheric Administration's (NOAA) Coral Reef Watch (CRW) programme (Strong et al. 1997, 2004; NOAA-NESDIS 2006). Using data generated from an AVHRR (Advanced Very High Resolution Radiometer) satellite, SST measurments were gathered twice weekly in near real time. NOAA CRW has since developed a daily global 5 km product suite derived from a combination of satellite SST data from US, Japanese and European geostationary infrared imagers and low-earth orbiting infrared SST data (Liu et al. 2014; Chap. 4). The new version 3 climatology combines the NOAA/NESDIS 2002–2015 daily 5 km geo-polar blended night-only SST analysis and the UK Met Office Operational SST and Sea Ice Analysis (OSTIA) 1985–2007 daily global 5 km SST reanalysis and employs a rigorous multi-scale optimal interpolation (OI) methodology to improve the resolution of SST features in areas of persistent cloud cover. The increased resolution of the new product suite provides service at near reef scales and preserves oceanographic features (e.g. eddies) that the previous 50 km product could not. These product suites have been used to generate a range of coral bleaching predictive risk tools which include Doldrums (persistent low wind regions), HotSpots (areas where SST are at least 1 °C above the mean maximum summer temperature) and Degree Heating Weeks (DHW)—an index of accumulated heat stress that summarises both longevity of an SST anomaly and its magnitude (size of temperature deviation) over a 12-week period (Chap. 4).

It must be recognised that real-time monitoring of SST, SST anomalies, HotSpots and DHW will only ever provide an approximate estimate highlighting locations that are exceeding the normal summer maximum to guide where coral bleaching is likely. Based on the severity and duration of the anomaly, typically when alert level 1 is reached (DHW > 4), bleaching and mortality of some species are likely. The NOAA Coral Reef Watch set of bleaching alert tools have shown themselves to be valuable predictors. Numerous studies have shown their general validity (Sheppard 1999; Spencer et al. 2000) and a number of more extensive observations during the recent 2014–2017 bleaching events to assess the DHW thresholds and improve and refine the predictive capacity of the Coral Reef Watch products (pers. comm. N. Cantin), and experimental products incorporating the influence of light on the bleaching response are under development (https://coralreefwatch.noaa.gov/satellite/lsd/index.php; http://coralreefwatch.noaa.gov/satellite/publications.html). As our knowledge of the thresholds for bleaching in different areas improves, the predictive capacity of such measures is also likely to improve.

During thermal stress events, the occurrence and/or intensity of bleaching will also be influenced by a range of other factors such as solar insolation, wind strength, tidal mixing and sea state. The important role of solar insolation was further corroborated by observations of near-continuous cloud cover during critical periods of high temperature, which may have prevented bleaching in the Society Islands (Mumby et al. 2001) and in Mauritius (Turner 1999). Different coral taxa also typically show different susceptibilities to bleaching (Marshall and Baird 2000; McWilliams et al. 2005), although recent widespread bleaching observations indicate that the disparity in taxonomic bleaching resistance diminishes as the severity of heat stress increases (Hughes et al. 2017). This trend in bleaching severity could mean that detection of coral bleaching with remote-sensing platforms could become easier in a warming ocean, as the spatial footprint of white reef tops from coral bleaching becomes more widespread and uniform with increasingly extreme thermal anomalies. However, this will only remain true if coral cover is high prior to the thermal anomaly and will vary regionally with habitat complexity.

SST data at 50 km resolution were clearly insufficient to show the fine-scale patterns of variation in bleaching which certainly occur. The aerial surveys of Berkelmans et al. (2004) observed changes at scales of ~10s of km, indicating local-scale variance, perhaps related to oceanographic mixing or weather patterns (e.g. upwelling and tidal mixing or persistent cloud cover adjacent to islands). Proximity to land can also further influence water temperatures and bleaching likelihood—runoff and water circulation—but these will not be picked up in very low-resolution SST data, which highlights the need to incorporate in situ temperature measurements with satellite-based SST products. Other studies have noted even finer-scale variance in bleaching tolerance or survivorship linked to shading, aspect or water flows (Spencer et al. 2000; West and Salm 2003). Analyses of recent bleaching observations will advance the predictive capacity of SST anomalies and bleaching stress indicators.

At very broad scales, satellite-derived SST data provide a useful indicator of bleaching likelihood, and the various refinements that have been made in the past decade have further improved this suite of predictive tools (Chap. 4). At the same time, however, the finer-scale variance in bleaching severity during thermal stress events needs to be documented and understood to determine how coral reefs will respond in a warming ocean. Current thresholds for bleaching response may change as the frequency of thermal stress events increases in the future.

5.2.3 Summary of Remote-Sensing Tools

- There are considerable challenges in using current satellite sensors to document bleaching. Generally, very fine spatial resolution is critical, and detection is better in areas where coral cover and water clarity are both high. There are no broadly accepted tools as yet, but advances are under development.
- Detection of bleaching from satellites is more likely in the near future due to technological developments and increased return times of current satellite sensors which have increased the capability to detect the early onset of community-scale coral bleaching.
- Low-level (~150 m altitude) aerial observations and photography have been highly successful in detection of bleaching and also in rapidly quantifying and mapping bleaching impacts across large spatial scales (>2000 km).
- Co-ordinated planning and the availability of archived metadata records have enabled comparisons of recent bleaching events on the GBR (2016 and 2017) to prior ones (1998 and 2002)—a detailed approach that has only been conducted along the GBR.
- SST anomalies from satellite sensors continue to be the most reliable and widely applied remote-sensing tools to predict coral bleaching risk across global spatial scales. NOAA's Coral Reef Watch bleaching HotSpots and Degree Heating Week products have proved powerful predictive tools that are consistent and comparable between regions and over time.
- Satellite-derived SST data sets have also been used in post hoc studies as a proxy measure of thermal stress and possible impacts.
- Ground-truthing remote-sensing platforms against recent in-water and aerial observations from the 2015 to 2017 bleaching event will likely lead to progress in the analytical techniques required to detect and quantify coral bleaching from remote-sensing platforms.

5.3 Field-Based Observations

Almost all bleaching monitoring is based on in situ observations by scientists and trained observers through volunteer citizen science programmes (e.g. Reef Check; www.reefcheck.org, Done et al. 2017). Standardised reef monitoring techniques provide the basis for most assessments to document how much coral within the community has bleached, which species, the severity level of each individual colony's response and the reef location where bleaching observations are recorded. Such in situ surveys are also important for validating remote-sensing bleaching assessments.

More challenging, but at least as important as documenting the severity of a bleaching event, are differential survivorship among coral taxa in the community; the identification of resistant locations, species or colonies; and the gathering of sufficient environmental information to understand the processes and mechanisms of bleaching, subsequent survival and bleaching tolerance in a warming ocean. These questions cannot be addressed through today's remote-sensing tools, and thus the need for field-based, underwater observation remains.

Good resources describing the broad array of coral reef monitoring techniques are widely available (e.g. English et al. 1997; Hill and Wilkinson 2004), while subtle themes and variations are devised for almost every study that is published. Broadly, it can be seen that approaches span a spectrum of broad to narrow spatial coverage which is countered by a reverse spectrum of low to high accuracy and level of taxonomic detail:

1. *Rapid semi-quantitative assessment.* This might be gathered on directed or undirected swims or using manta tows. These enable assessments of large areas and the wider patterns of bleaching versus non-bleaching. This approach enables detection of low-level bleaching and for obtaining a general picture with respect to the onset of a bleaching event in early summer, but it is inadequate for numerical studies and spatial or temporal comparisons.
2. *Fixed transects.* Fixed permanent transects (recommended length of 100–500 m) offer some key advantages of relatively large spatial coverage while allowing more accurate quantification of bleaching and mortality following the peak of the event. Accuracy of quantification is greatly increased using video or photo-based transects although processing time is increased. A disadvantage of this approach is that the initial deployment and maintenance of fixed transects increase the time investment per reef and will reduce the spatial coverage during a field campaign. However, the gains acquired by following a standardised, long-term approach that follows permanently marked transect locations or at the very least strictly controlled fixed depths and reef zones with GPS co-ordinates will enable comparison over time to document bleaching mortality and recovery processes.
3. *Quadrats, photo quadrats and line-point intercept approaches.* These give highest levels of accuracy, but there are considerable costs in terms of reductions in sampling area.

In developing a monitoring protocol, researchers must consider available resources and funding (over time) while keeping in mind repeatability and ease of integrating observations with other programmes and researchers. Ideally a nested sampling approach may be devised with some elements of monitoring across a range of spatial and temporal scales. This would allow more qualitative statements to be made about the timing and onset of bleaching and the overall impacts on the entire reef system while allowing more quantitative rigour in fixed localities.

As with all monitoring, there is a requirement for repeat surveys, and researchers must build continuing observation into their plans. Repeat surveys after peak bleaching and before the next summer season can be critical for ascertaining levels of recovery and or mortality, but ongoing mortality may continue even longer from associated coral disease impacts. Monitoring should be designed to assess not only immediate bleaching impacts but also ongoing recovery processes following major bleaching events, including coral recruitment and growth.

The development of fixed sampling sites is essential to assess trends over time following the initial bleaching event. As the frequency of coral bleaching events is likely to increase in a warming ocean (Donner et al. 2005; Frieler et al. 2013; Chap. 13), initial design of the survey locations should consider the need to return to permanent transects in the near future. Representative, broad spatial coverage of sites across an entire reef system is always important to reliably assess the full extent of the impact of coral bleaching for each coral reef region; thus co-ordinated collaboration should be of particular concern to ensure reliable observations capture the full picture of the event.

5.3.1 Describing Bleaching

During a bleaching event, a coral's colours will fade gradually as the summer progresses, but severe bleaching responses within the community can escalate rapidly with just 5–10 days of thermal stress. A diver's visual perception of coral bleaching and the loss in colour due to stress, generally, is not detectable until 60–80% of the *Symbiodinium* cells and chlorophyll have been lost by the coral host (Enriquez et al. 2005; Siebeck et al. 2006). As a result the timing of bleaching observations is critical. Early in the development of a mass bleaching event or at reefs exposed to less severe accumulation of heat stress, individual colonies will show variation in bleaching intensity across their surface; for example, those surfaces receiving higher exposure to solar irradiance often show more bleaching on the upper surfaces than on the shaded side of the colony which may be unbleached (Hoegh-Guldberg 1999). Depending on the magnitude and duration of the heat stress, a bleaching event may end with the gradual recovery of the coral, with subtle increases in colour over time. In other cases, when the anomaly is severe in magnitude and/or prolonged over time, part or all of the colony may die due to either oxidative stress and cell death or insufficient energy reserves to sustain the coral host for prolonged periods without the algal symbiosis (Chap. 8). In the field it

Fig. 5.3 (**a**) Tabulate and digitate *Acropora* colonies and (**b**) staghorn *Acropora* colonies; both displaying severe levels of bleached and fluorescent host tissue and early signs of bleaching-related mortality as the skeleton becomes overgrown with green and brown filamentous algae. Image credit: N. Cantin AIMS

can be challenging to identify the subtle differences in colour and texture between a very recently dead coral due to bleaching stress, with a fine haze of filamentous algae, and a recovering coral or a dead coral due to predators such as the crown-of-thorns starfish. Consistent categories need to be maintained to properly identify bleaching-related mortality. Ubiquitous bleaching of similar taxa within the community and signs of bleached sections of the colony that are becoming overgrown with filamentous algae (Fig. 5.3) should be observed when attributing mortality to bleaching.

During the 2016–2017 bleaching event on the GBR, the Australian National Bleaching Taskforce used transect-based 1 m belt survey approaches and categorised the bleaching severity of every colony within 1 m × 1 m photo-based quadrats (Fig. 5.3) or with in-water observations only. Bleaching severity categories were based on the categories used historically on the GBR (Baird and Marshall 2002) and severity of the bleaching response scored in relation to the total area of bleached tissue for individual colonies. Categories were 1, no bleaching; 2, pale; 3, 1–50% (mostly upper surfaces); 4, 51–99%, 5, 100% completely white and fluorescent; and 6, early partial or full mortality (Fig. 5.4).

In describing the bleaching response at the community level, the most common metric is simply the total percentage of the benthic community that has bleached. In contrast to remotely sensed studies, this term almost always refers to the proportion of the hard and soft coral cover which is bleached rather than the proportion of the total substrate. Comparison between studies is sometimes challenged by poor reporting and inconsistency in categorising the variation in levels of individual bleaching responses. Such problems may be avoided by differentiating and quantifying the proportion of bleaching across the entire colony from patchy bleaching

Fig. 5.4 Individual colony bleaching severity based on percent area of the colony bleached. (**a**) No bleaching, Category 1; (**b**) 1–50% (mostly upper surfaces), Category 3; (**c**) 50–99%, Category 4; (**d**) >95% completely white and/or fluorescent, Category 5; and (**e**) early mortality, Category 6. Image credit: N. Cantin AIMS

(including evidence of photosensitivity, which is captured by the upper surface bleaching Category 3, Fig. 5.4) and partial bleaching where some symbiont colour is still maintained. Employing the broad five-category scale of colony bleaching severity described here captures the severity of bleaching responses and can help to standardise observations and allow for comparisons between geographic locations and to historical datasets.

Attention is typically focused on scleractinian corals; however, many zooxanthellate invertebrates have shown bleaching responses to thermal stress, and it is worth including observations of bleaching in other benthic groups (*Millepora*, *Heliopora*, soft corals, giant clams, sponges and sea anemones). Since these taxa will contribute to the overall aerial white appearance of the reef, they should be included in more general estimates of percentage bleaching when linking in-water estimates of community bleaching responses with aerial and remote-sensing indicators of coral reef community bleaching.

Finally, it is worth noting that as heat stress intensifies with record-breaking summer thermal anomalies, partial and severe bleaching in some species (e.g. *Platygyra*) can be prolonged (Lang et al. 1992, Fig. 5.5). Continued monitoring and repeat surveys prior to the next summer should highlight this phenomenon and document the full extent of mortality resulting from annual bleaching events.

Fig. 5.5 (**a**) Long-lived century-old, massive *Platygyra* sp. from the Great Barrier Reef exhibiting prolonged periods of severe bleaching in June 2016, 3 months after the peak of the heat stress. (**b**) The same massive *Platygyra* sp. colony suffering complete mortality in September 2016, highlighting the delayed progression of final mortality or survivorship outcomes throughout the winter. Image credit: N. Cantin AIMS

5.3.2 Assessing Mortality

Patterns of community-level responses have shown that coral mortality during the peak of the heat stress has become an important component of recent bleaching events to assess the potential for adaptive responses and shifts in susceptibility among taxa (Guest et al. 2012). Recent studies have differentiated bleaching severity among reefs based not only on the proportion of bleached corals within the community but also on the extent of coral mortality and if these patterns are shifting during subsequent repetitive bleaching stress (Guest et al. 2012). Bleaching-related mortality should be incorporated into bleaching response monitoring programmes to clearly differentiate the minor impacts of coral bleaching (i.e. recovery) from the severe impacts of coral bleaching (i.e. mortality) that may result in a shift in the community composition of the reef.

Mortality may also be important in determining the extent of the bleaching impact, especially in locations where it may be difficult to time a survey to coincide with peak bleaching. There is a risk that mortality assessed after peak bleaching may be non-bleaching related, and here the contextual framework of the surrounding reef environment, presence of widespread bleaching responses and evidence of recently bleached sections within the colony (Fig. 5.6), coupled with the experience of the observer, are critical. Dead corals can be rapidly overgrown within a matter of weeks (Lang et al. 1992). In the context of an ongoing bleaching event, it is, therefore, relatively simple to differentiate recently dead corals due to bleaching from corals

Fig. 5.6 Comparison of signs of long dead coral colonies, likely from crown-of-thorns starfish, that would not be attributed to bleaching (**a**) and to recently dead coral colonies due to bleaching (**b**). Image credit: N. Cantin AIMS

that have been dead for some months or years, and there are many instances when it may be reasonable to attribute recent death to bleaching-related heat stress.

Bleaching-related mortality can be estimated even some time after the event in the more extreme cases, as was the case in the Chagos Archipelago, which was only surveyed 1 year after a suspected major bleaching event—here the context was provided both by SST observations and the impacts in adjacent reef systems including the Maldives and Seychelles. It was clear that the "near-total" loss of living hard corals on these reefs was directly related to an unobserved massive bleaching event (Sheppard et al. 2002).

Semi-quantitative approaches may also be important, and the semi-quantitative categorisation of bleaching severity and mortality results gathered from different surveys is, in many areas, the only means available to place bleaching and other impacts into a historical context (Winter et al. 1998). As bleaching events become more frequent in a warming ocean, records of mortality after the peak of the thermal heat stress will become a critical data set for assessments of bleaching impacts.

5.3.3 *Broader Patterns Across the Reef*

Many of the traditional coral monitoring approaches give little space for a standardised approach to documenting patterns of survival over broader scales than can be provided by the transect or quadrat methods. This problem is exacerbated by the otherwise quite sensible desire among reef ecologists to standardise sampling techniques. This has led to most ongoing reef monitoring programmes, many of which have been usefully adapted to look at bleaching, remaining fixed on "standard" reef zones, at fixed (usually quite shallow) depths down the reef front. Sampling in channels, on isolated or lagoon bommies, on current-swept "headlands", on deeper reef slope areas (below 30 m) or on coral patches across the reef flat remains the exception. A growing number of anecdotal reports suggest that there are indeed differential patterns of impact in these locations and that these may offer critical refugia or places for adaptation and recovery. Effort should thus be made to incorporate such locations into monitoring systems, and even post hoc descriptions of differential bleaching patterns can be valuable. Researchers should be encouraged to note any variance in bleaching intensity ("although not quantified, bleaching appeared to be more or less severe in lagoons vs. reef channels vs. deeper waters vs. close to land, etc.") or even the lack of such variance ("patterns of bleaching appeared to be very similar in. . ."). Such information may provide a valuable guide to future research and to managers. Such issues can be solved with the advancement of remote-sensing, aerial drone and autonomous underwater video-/photo-based survey technologies combined with the necessary automated image analysis capability to expand the spatial scale over which reefs can be rapidly surveyed during a brief bleaching response research monitoring campaigns.

5.3.4 *Summary of In Situ Field Protocols to Document Coral Bleaching*

- Qualitative broad-scale methods are effective to document the early onset of a widespread bleaching event and to inform large-scale coral bleaching research response efforts.
- Quantitative approaches which document community composition, bleaching severity and mortality are recommended for major regional bleaching events.
- Line intercept transects (LIT) provide assessments of coral cover, taxonomic composition and colony size demographics.
- Photo-based quadrats conducted along with the LIT transect as a belt survey assessment provide broad community detail of bleaching severity.
- Bleaching severity (ranging from no bleaching to 100% white and fluorescent) should be assessed based on the percent area of the colony bleached.
- Bleaching-related mortality should be carefully included using key indicators that include the presence of widespread bleaching responses, intact corallites and evidence of recently bleached sections within the colony.

5.3.5 *Colony Scale and Finer*

At the level of the coral colony, there can be variation in the intensity of bleaching. At these scales it may be important to establish repeat sampling of colonies, and tagging of individual colonies is thus required. Baird and Marshall (2002) used a simple categorisation of degree of colony bleaching as follows: no bleaching, 1–10%, 11–50%, 51–99%, 100% bleached, and dead and placed all pale colonies into the 1–10% category. Such a system should be more widely applied to enable comparison between coral reef regions, although it may need some revision for Caribbean reefs where partial loss of colour appears to be much more widespread. Siebeck et al. (2006) provide a useful tool for describing the degree of colour change which may improve consistency of monitoring through time; it should be noted that this has so far only been used on five species and there was some degree of interobserver error.

Describing patterns across a colony can also be important and can lead to advances in understanding processes and patterns of resilience. Patchy bleaching responses throughout individual colonies are quite common and can often be linked to exposure to solar radiation. Typically, bleaching may commence on those coral surfaces receiving the highest levels of solar radiation. More complex patterns have been observed, notably by Brown et al. (2002), where faces of corals that had undergone prior exposure to high solar insolation were subsequently resilient to a high temperature-related bleaching event.

At the level of colonies, it is also possible to measure and assess recovery, including sublethal impacts on growth and reproduction. Baird and Marshall (2002) examined growth rates in four coral species, by regular return sampling to over 100 separate colonies during and after a bleaching event. They also assessed the reproductive potential, taking coral samples for laboratory analysis both before and after the known spawning period and assessing development of gametes and presence of fertilised eggs. Long-term impacts of bleaching stress on the fecundity of the survivors remain an unanswered question, but this study showed that 12 and 55% of *Acropora millepora* and *A. hyacinthus* colonies, respectively, failed to develop eggs in the following year after bleaching stress in 1998, highlighting the long-term energetic impacts of coral bleaching for the coral host.

Measurements of photosynthetic activity [of photosystem II (PSII)] can be undertaken with relatively non-invasive techniques. Although many studies still involve the removal of specimens to adjacent laboratories to quantify symbiont densities or to run photosynthesis/respiration (P/R) incubations (Brown et al. 2000), submersible diving pulse-amplitude-modulated (PAM) fluorometers have been used (Ralph et al. 1999; Warner et al. 1999; Fitt et al. 2001) for in situ measurement of PSII before, during and after bleaching events. Lombardi et al. (2000) used a fast repetition rate fluorometer to show variability in fluorescence yields from corals at different stages of bleaching or recovery. The PAM fluorometer is capable of detecting the photoinhibition processes of the algal symbiont within the coral host that lead to widespread coral bleaching. Coral bleaching on regional scales is caused primarily by thermal heat stress and amplified by high light exposure (Jones et al. 1998). As the enzymes degrade that drive the dark reactions of carbon fixation in PSI, the light-harvested energy in PSII builds up at the Q_a binding site prior to entering the plastoquinone pool and generates an increase in fluorescence and toxic reactive oxygen (Jones et al. 1998; Hoegh-Guldberg 1999). The PAM fluorometer detects this process of photoinhibition during thermal stress by measuring the increased fluorescence emitted and quantifies the decrease in the photosynthetic efficiency (F_v/F_m) and the photoinhibitory damage to the D1 protein of PSII as prolonged decreases in dark-adapted yields at night for hours to days depending on the level of damage and rate of repair of the D1 protein (Hill and Ralph 2005).

Working at the scale of polyps, cells and algal symbionts always require a shift of effort from the field to the laboratory, and here we only provide a brief review of typical methods used on wild-sampled corals. Considerably more work of this nature has been developed in laboratory-cultured corals, and this work is clearly important in refining techniques, but it does not constitute monitoring of natural bleaching events.

Assessing *Symbiodinium* densities can be undertaken on tissue prepared for microscope counts. Some studies have used a water pik to remove soft tissue, with the subsequent isolation of *Symbiodinium* cells by centrifugation (Warner et al. 1999). Others have assessed chlorophyll content from frozen samples and symbiont densities from preserved tissue in 10% formalin and seawater (Le Tissier and Brown 1996) based on small cores collected from massive coral colonies.

Histological examination enables further investigation of the distribution of *Symbiodinium* cells within a polyp and provides some picture of the internal mechanisms of bleaching, with differential patterns of *Symbiodinium* loss and degradation typically observed in different tissues and at different stages of bleaching. Such patterns (e.g. the existence of a reservoir of algae in the base of the polyp) may give some indication of the ability of polyps to recover when environmental conditions improve (Brown et al. 1995; Le Tissier and Brown 1996).

Assessing chlorophyll concentrations is another useful measure, and field studies have used both spectrophotometry (Le Tissier and Brown 1996) and high-performance liquid chromatography (HPLC) (Ambarsari et al. 1997; McDougall et al. 2006; Apprill et al. 2007). Unlike HPLC, spectrophotometry cannot discern chlorophyll-a-like compounds and can thus lead to higher chlorophyll readings than would be expected (McDougall et al. 2006).

It is now clear that different types of *Symbiodinium* offer differing levels of thermal tolerance to their coral host and hence influence bleaching resistance (Rowan et al. 1997; Baker et al. 2004; Chap. 6). Identification and quantification of different symbiont types requires extraction of *Symbiodinium* DNA and sequence analysis (Chap. 6). The genotype of the coral host itself also plays a role in determining the relative thermal tolerance of a colony (Bay and Palumbi 2014; Dixon et al. 2015; Jin et al. 2016).

From an environmental perspective, the monitoring of these different algal types may provide critical understanding of existing resistance to future bleaching events and ongoing patterns of adaptation/acclimatisation in recovering corals (LaJeunesse 2002; Baker et al. 2004; Thornhill et al. 2006; Boulotte et al. 2016). Such work requires not only very fine-scale molecular techniques in assessing *Symbiodinium* diversity, but also broad-scale sampling to understand the distribution of types between species, across the reef profile, between adjacent reefs and across much wider spatial scales (van Oppen et al. 2005; Garren et al. 2006; Chap. 6).

5.3.6 *Temporal Context*

Although the primary purpose of this chapter is an overview of monitoring techniques during individual bleaching events, it should be noted that long-term monitoring before and after a bleaching event is invaluable especially as repeated bleaching events are more likely in the future (Chap. 13). Designing bleaching survey approaches so that the transects or quadrats can be repeated will increase the ability to detect future impacts of heat stress, recovery of the community and shifts in bleaching tolerance in a warming ocean. Knowledge of pre-bleaching state—including reef conditions of coral cover; variation across the reef system; prior bleaching impacts and severity per reef; "natural", seasonal or recurring bleaching observations; and, at finer scales, "natural" backgrounds of *Symbiodinium* types—can provide an invaluable reference in understanding thermal tolerance and susceptibility to coral bleaching. Similar monitoring protocols can and should be used post hoc in an attempt to document rates of recovery and patterns of recovery and further to document shifts in community structures or *Symbiodinium* communities.

Monitoring changes in calcification, growth and reproductive success (Baird and Marshall 2002; Cantin and Lough 2014) may identify sublethal impacts of prior bleaching events or conversely may uncover other processes that could hinder or expedite recovery. Changes in disease incidence may also provide some measure of underlying sublethal effects of heat stress events that can be documented with follow-up surveys after the peak bleaching response effort (Miller et al. 2006; Bruno et al. 2007). Others have looked at recruitment itself—counting densities of new coral recruits (<5 cm in size) and identifying (where possible) species (Sheppard et al. 2002)—as this may provide an indicator of rates and patterns of recovery in different management areas; the influence of substrate (bare rock, dense algal cover, loose rubble, etc.) and connectivity to neighbouring non-bleached reefs on recovery dynamics remain poorly understood and should continue to be a focus of research efforts following the global bleaching event of 2014–2017, as some remote isolated reef locations have suffered extensive mortality.

5.4 Summary

We have described the various means of assessing and monitoring bleaching from global scales to patterns within individual colonies, and each of these approaches is critical. It is also important, however, to be able to move between scales. Global and regional patterns are helping to inform the science of global change and are also strengthening the hands of policy-makers in seeking to address climate change.

Broad scales also help provide context for more localised in-field observations and support informed extrapolation to areas where there have been no field studies. Regional field studies provide context to the global patterns, enabling a better understanding of patterns of resistance and resilience. These in turn can help with the wider interpretation of coral reef futures and are also helping policy-makers in planning for uncertain futures. Clear examples of such applications for planning and conservation are being provided with the promotion of resilient networks of marine protected areas (Grimsditch and Salm 2006; Marshall and Schuttenberg 2006; see also http://www.reefresilience.org/). But it is only with the finest scale studies that we can begin to understand the mechanisms of bleaching and the variability in responses which will be critical in understanding future impacts. Once again, connecting across scales provides further evidence to the growing picture of the many facets of the phenomenon of coral bleaching. Perhaps most striking in this regard have been the observations of variance in *Symbiodinium* community composition over both space and time, providing what may be critical clues into the future for coral reefs in a warming world (see Chap. 6).

It is also critical to consider patterns across timescales. "Monitoring" implies ongoing observation, and understanding the temporal component is important in determining long-term futures both for coral reefs and those who rely on them. It is important that bleaching events are not seen in isolation. Bleaching impacts appear to be significantly affected by the location and severity of prior events. There also appear to be quite significant non-lethal impacts affecting long-term survivorship. Mortality is a frequent component of bleaching events and must be accurately assessed, alongside growth and recruitment of remaining corals.

Finally, bleaching as a global phenomenon should not only be reported in one-off reports in local or national contexts. Although valuable, such reports need to be given a regional and global context to raise awareness of the scale of the problem. Researchers should consider existing methodologies before devising new ones and strive to make survey techniques as comparable as possible to make global assessments more feasible. Should researchers choose to develop or refine their own approaches, they should at least consider how their bleaching severity and mortality categories can be directly compared to other previously published research protocols. In the reporting process they should indeed make such comparisons, providing a regional and temporal context for their own work. In this chapter we have provided the bleaching and mortality categories that have been employed on the GBR in 1998, 2002, 2016 and 2017 bleaching response campaigns. We encourage monitoring programmes to adopt similar bleaching and mortality categories in the future to facilitate global comparisons.

The challenges of summarising a vast literature and burgeoning array of methods into wider regional and global assessments are considerable (Chap. 3), but such reporting needs to be continued and indeed strengthened, and again researchers should be encouraged to contribute to global monitoring schemes, such as the NOAA Coral Reef Watch Bleaching Observations reporting tool (https://coralreefwatch.noaa.gov/satellite/research/coral_bleaching_report.php), or the newly developed public historical bleaching observation database and modelling tools (http://simondonner.com/bleachingdatabase/) (Donner et al. 2017). Such broader review greatly helps in developing our understanding of the global threats posed by coral bleaching; it will support more informed considerations of management responses; and it will provide yet stronger arguments to governments and to civil society on the urgent need to aggressively tackle the ultimate causes of climate change and ocean warming that are driving the patterns of widespread coral bleaching.

Acknowledgements We would like to thank the editors Janice Lough and Madeleine van Oppen for the invitation to contribute this book chapter and for comments on earlier drafts. Sincere thanks to John Hedley and Bernardo Vargas-Angel for valuable comments and review of this chapter. Experiences gained as a member of the Australian National Bleaching Taskforce and with the Great Barrier Reef Marine Park Authority during the 2016–2017 bleaching events were instrumental in developing some of the insights in this chapter.

References

Ambarsari I, Brown BE, Barlow RG, Britton G, Cummings D (1997) Fluctuations in algal chlorophyll and carotenoid pigments during solar bleaching in the coral Goniastrea aspera at Phuket, Thailand. Mar Ecol Prog Ser 159:303–307. https://doi.org/10.3354/meps159303

Andréfouët S, Berkelmans R, Odriozola L, Done T, Oliver J, Müller-Karger F (2002) Choosing the appropriate spatial resolution for monitoring coral bleaching events using remote sensing. Coral Reefs 21:147–154. https://doi.org/10.1007/s00338-002-0233-x

Andréfouët S, Kramer P, Torres-Pulliza D, Joyce KE, Hochberg EJ, Garza-Pérez R et al (2003) Multi-site evaluation of IKONOS data for classification of tropical coral reef environments. Remote Sens Environ 88:128–143. https://doi.org/10.1016/j.rse.2003.04.005

Apprill AM, Bidigare RR, Gates RD (2007) Visibly healthy corals exhibit variable pigment concentrations and symbiont phenotypes. Coral Reefs 26:387–397. https://doi.org/10.1007/s00338-007-0209-y

Baird AH, Marshall PA (2002) Mortality, growth and reproduction in scleractinian corals following bleaching on the Great Barrier Reef. Mar Ecol Prog Ser 237:133–141. https://doi.org/10.3354/meps237133

Baker AC, Starger CJ, McClanahan TR, Glynn PW (2004) Corals' adaptive response to climate change. Nature 430:741. https://doi.org/10.1038/430741a

Bay RA, Palumbi SR (2014) Multilocus adaptation associated with heat resistance in reef-building corals. Curr Biol 24:2952–2956. https://doi.org/10.1016/j.cub.2014.10.044

Berkelmans R, Oliver JK (1999) Large-scale bleaching of corals on the Great Barrier Reef. Coral Reefs 18:55–60. https://doi.org/10.1007/s003380050154

Berkelmans R, De'ath G, Kininmonth S, Skirving WJ (2004) A comparison of the 1998 and 2002 coral bleaching events on the Great Barrier Reef: spatial correlation, patterns, and predictions. Coral Reefs 23:74–83. https://doi.org/10.1007/s00338-003-0353-y

Boulotte NM, Dalton SJ, Carroll AG, Harrison PL, Putnam HM, Peplow LM et al (2016) Exploring the *Symbiodinium* biosphere provides evidence for symbiont switching in reef-building corals. ISME J 10:2693–2701. https://doi.org/10.1038/ismej.2016.54

Brown BE, Tissier MDAL, Bythell JC (1995) Mechanisms of bleaching deduced from histological studies of reef corals sampled during a natural bleaching event. J Mar Biol 122:655–663. https://doi.org/10.1007/BF00350687

Brown BE, Dunne RP, Warner ME, Ambarsari I, Fitt WK, Gibb SW et al (2000) Damage and recovery of Photosystem II during a manipulative field experiment on solar bleaching in the coral *Goniastrea aspera*. Mar Ecol Prog Ser 195:117–124. https://doi.org/10.3354/meps195117

Brown BE, Dunne RP, Goodson MS, Douglas AE (2002) Experience shapes the susceptibility of a reef coral to bleaching. Coral Reefs 21:119–126. https://doi.org/10.1007/s00338-002-0215-z

Bruno JF, Selig ER, Casey KS, Page CA, Willis BL, Harvell CD et al (2007) Thermal stress and coral cover as drivers of coral disease outbreaks. PLoS Biol 5. https://doi.org/10.1371/journal.pbio.0050124

Call KA, Hardy JT, Wallin DO (2003) Coral reef habitat discrimination using multivariate spectral analysis and satellite remote sensing. Int J Remote Sens 24:2627–2639. https://doi.org/10.1080/0143116031000066990

Cantin NE, Lough JM (2014) Surviving coral bleaching events: *Porites* growth anomalies on the Great Barrier Reef. PLoS One 9:e88720. https://doi.org/10.1371/journal.pone.0088720

Chirayath V, Earle SA (2016) Drones that see through waves – preliminary results from airborne fluid lensing for centimetre-scale aquatic conservation. Aquat Conserv Mar Freshw Ecosyst 26:237–250. https://doi.org/10.1002/aqc.2654

Dixon GB, Davies SW, Aglyamova GA, Meyer E, Bay LK, Matz MV (2015) Genomic determinants of coral heat tolerance across latitudes. Science 348:1460–1462. https://doi.org/10.1126/science.1261224

Done T, Roelfsema C, Harvey A, Schuller L, Hill J, Schläppy M-L et al (2017) Reliability and utility of citizen science reef monitoring data collected by Reef Check Australia, 2002–2015. Mar Pollut Bull 117:148–155. https://doi.org/10.1016/j.marpolbul.2017.01.054

Donner SD, Skirving WJ, Little CM, Oppenheimer M, Hoegh-Guldberg O (2005) Global assessment of coral bleaching and required rates of adaptation under climate change. Glob Chang Biol 11:2251–2265. https://doi.org/10.1111/j.1365-2486.2005.01073.x

Donner SD, Rickbeil GJM, Heron SF (2017) A new, high-resolution global mass coral bleaching database. PLoS One 12:e0175490. https://doi.org/10.1371/journal.pone.0175490

Edwards AJ, Clark S, Zahir H, Rajasuriya A, Naseer A, Ruben J (2001) Coral bleaching and mortality on artificial and natural reefs in Maldives in 1998, sea surface temperature anomalies and initial recovery. Mar Pollut Bull 42:7–15. https://doi.org/10.1016/S0025-326X(00)00200-9

Elvidge CD, Dietz JB, Berkelmans R, Andréfouët S, Skirving W, Strong AE et al (2004) Satellite observation of Keppel Islands (Great Barrier Reef) 2002 coral bleaching using IKONOS data. Coral Reefs 23:123–132. https://doi.org/10.1007/s00338-003-0364-8

English S, Wilkinson C, Baker V (1997) Survey manual for tropical marine resources. Australian Institute of Marine Science, Townsville

Enríquez S, Mendez ER, Iglesias-Prieto R (2005) Multiple scattering on coral skeletons enhances light absorption by symbiotic algae. Limnol Oceanogr 50:1025–1032. https://doi.org/10.4319/lo.2005.50.4.1025

Fitt WK, Brown BE, Warner ME, Dunne RP (2001) Coral bleaching: interpretation of thermal tolerance limits and thermal thresholds in tropical corals. Coral Reefs 20:51–65. https://doi.org/10.1007/s003380100146

Frieler K, Meinshausen M, Golly A, Mengel M, Lebek K, Donner SD et al (2013) Limiting global warming to 2 °C is unlikely to save most coral reefs. Nat Clim Chang 3:165–170. https://doi.org/10.1038/nlcimate1674

Garren M, Walsh SM, Caccone A, Knowlton N (2006) Patterns of association between *Symbiodinium* and members of the *Montastraea annularis* species complex on spatial scales ranging from within colonies to between geographic regions. Coral Reefs 25:503–512. https://doi.org/10.1007/s00338-006-0146-1

Grimsditch GD, Salm RV (2006) Coral reef resilience and resistance to bleaching. IUCN, Gland

Guest JR, Baird AH, Maynard JA, Muttaqin E, Edwards AJ, Campbell SJ et al (2012) Contrasting patterns of coral bleaching susceptibility in 2010 suggest an adaptive response to thermal stress. PLoS One 7:e33353. https://doi.org/10.1371/journal.pone.0033353

Hedley J, Roelfsema C (2017) Sentinel-2 captures coral bleaching of Great Barrier Reef. European Space Agency. http://www.esa.int/Our_Activities/Observing_the_Earth/Copernicus/Sentinel-2/Sentinel-2_captures_coral_bleaching_of_Great_Barrier_Reef

Hedley JD, Roelfsema C, Phinn SR, Mumby PJ (2012a) Environmental and sensor limitations in optical remote sensing of coral reefs: implications for monitoring and sensor design. Remote Sens 4:271–302. https://doi.org/10.3390/rs4010271

Hedley J, Roelfsema C, Koetz B, Phinn S (2012b) Capability of the Sentinel 2 mission for tropical coral reef mapping and coral bleaching detection. Remote Sens Environ 120:145–155. https://doi.org/10.1016/j.rse.2011.06.028

Hedley J, Roelfsema C, Chollett I, Harborne A, Heron S, Weeks S et al (2016) Remote sensing of coral reefs for monitoring and management: a review. Remote Sens 8:118. https://doi.org/10.3390/rs8020118

Heron SF, Eakin CM, Douvere F, Anderson KD, Day JC, Geiger E et al. (2017) Impacts of climate change on world heritage coral reefs: a first global scientific assessment. UNESCO World Heritage Centre

Hill R, Ralph P (2005) Diel and seasonal changes in fluorescence rise kinetics of three scleractinian corals. Funct Plant Biol 32:549–555. https://doi.org/10.1071/FP05017

Hill J, Wilkinson C (2004) Methods for ecological monitoring of coral reefs. A resource for managers. Version 1. Australian Institute of Marine Science, Townsville

Hoegh-Guldberg O (1999) Climate change, coral bleaching and the future of the world's coral reefs. Mar Freshw Res 50:839–866. https://doi.org/10.1071/MF99078

Holden H, LeDrew E (1998) Spectral discrimination of healthy and non-healthy corals based on cluster analysis, principal components analysis, and derivative spectroscopy. Remote Sens Environ 65:217–224. https://doi.org/10.1016/S0034-4257(98)00029-7

Hughes TP, Kerry JT, Álvarez-Noriega M, Álvarez-Romero JG, Anderson KD, Baird AH et al (2017) Global warming and recurrent mass bleaching of corals. Nature 543:373–377. https://doi.org/10.1038/nature21707

Jin YK, Lundgren P, Lutz A, Raina JB, Howells EJ, Paley AS et al (2016) Genetic markers for antioxidant capacity in a reef-building coral. Sci Adv 2. https://doi.org/10.1126/sciadv.1500842

Jones RJ, Hoegh-Guldberg O, Larcum AWD, Schreiber U (1998) Temperature-induced beaching of corals begins with impairment of the CO_2 fixation mechanism in zooxanthellae. Plant Cell Environ 21:1219–1230. https://doi.org/10.1046/j.1365-3040.1998.00345.x

LaJeunesse T (2002) Diversity and community structure of symbiotic dinoflagellates from Caribbean coral reefs. Mar Biol 141:387–400. https://doi.org/10.1007/s00227-002-0829-2

Lang JC, Lasker HR, Gladfelter EH, Hallock P, Jaap WC, Losada FJ et al (1992) Spatial and temporal variability during periods of "recovery" after mass bleaching on Western Atlantic coral reefs. Integr Comp Biol 32:696–706. https://doi.org/10.1093/icb/32.6.696

Le Tissier MDAA, Brown BE (1996) Dynamics of solar bleaching in the intertidal reef coral *Goniastrea aspera* at Ko Phuket, Thailand. Mar Ecol Prog Ser 136:235–244. https://doi.org/10.3354/meps136235

Liu G, Heron S, Eakin C, Muller-Karger F, Vega-Rodriguez M, Guild L et al (2014) Reef-scale thermal stress monitoring of coral ecosystems: new 5-km global products from NOAA Coral Reef Watch. Remote Sens 6:11579–11606. https://doi.org/10.3390/rs61111579

Lombardi MR, Lesser MP, Gorbunov MY (2000) Fast repetition rate (FRR) fluorometry: variability of chlorophyll a fluorescence yields in colonies of the corals, Montastraea faveolata (w.) and Diploria labyrinthiformes (h.) recovering from bleaching. J Exp Mar Biol Ecol 252:75–84. https://doi.org/10.1016/S0022-0981(00)00238-0

Marshall PA, Baird AH (2000) Bleaching of corals on the Great Barrier Reef: differential susceptibilities among taxa. Coral Reefs 19:155–163. https://doi.org/10.1007/s003380000086

Marshall P, Schuttenberg H (2006) A reef manager's guide to coral bleaching. Great Barrier Reef Marine Park Authority, Townsville

McClanahan TR (2000) Bleaching damage and recovery potential of Maldivian Coral Reefs. Mar Pollut Bull 40:587–597. https://doi.org/10.1016/S0025-326X(00)00064-3

McDougall KE, Gibb SW, Boyd KG, Brown BE (2006) 'Chlorophyll-like' compounds as novel biomarkers of stress in corals. Mar Ecol Prog Ser 325:137–144. https://doi.org/10.3354/meps325137

McWilliams JP, Côté IM, Gill JA, Sutherland WJ, Watkinson AR (2005) Accelerating impacts of temperature-induced coral bleaching in the Caribbean. Ecology 86:2055–2060. https://doi.org/10.1890/04-1657

Miller J, Waara R, Muller E, Rogers C (2006) Coral bleaching and disease combine to cause extensive mortality on reefs in US Virgin Islands. Coral Reefs 25:418. https://doi.org/10.1007/s00338-006-0125-6

Mumby PJ, Chisholm JRM, Edwards AJ, Andréfouët S, Jaubert J (2001) Cloudy weather may have saved Society Island reef corals during the 1998 ENSO event. Mar Ecol Prog Ser 222:209–216. https://doi.org/10.3354/meps222209

Mumby PJ, Skirving W, Strong AE, Hardy JT, LeDrew EF, Hochberg EJ et al (2004a) Remote sensing of coral reefs and their physical environment. Mar Pollut Bull 48:219–228. https://doi.org/10.1016/j.marpolbul.2003.10.031

Mumby P, Hedley J, Chisholm J, Clark C, Ripley H, Jaubert J (2004b) The cover of living and dead corals from airborne remote sensing. Coral Reefs 23:171–183. https://doi.org/10.1007/s00338-004-0382-1

NOAA-NESDIS (2006) Operational satellite coral bleaching monitoring products methodology. National Oceanic and Atmospheric Administration, National Environmental Satellite Data and Information Service

Obura D, Mangubhai S (2011) Coral mortality associated with thermal fluctuations in the Phoenix Islands, 2002–2005. Coral Reefs 30:607–619. https://doi.org/10.1007/s00338-011-0741-7

Philipson P, Lindell T (2003) Can coral reefs be monitored from space? Ambio 32:586–593. https://doi.org/10.1579/0044-7447-32.8.586

Ralph P, Gademan R, Larkum AWD, Schreiber U (1999) In situ underwater measurements of photosynthetic activity of coral zooxanthellae and other reef-dwelling dinoflagellate endosymbionts. Mar Ecol Prog Ser 180:139–147. https://doi.org/10.3354/meps180139

Rowan R, Knowlton N, Baker A, Jara J (1997) Landscape ecology of algal symbionts creates variation in episodes of coral bleaching. Nature 388:265–267. https://doi.org/10.1038/40843

Sheppard CRC (1999) Coral decline and weather patterns over 20 years in the Chagos Archipelago, central Indian Ocean. Ambio 28:472–478

Sheppard CRC, Spalding M, Bradshaw C, Wilson S (2002) Erosion *vs.* recovery of coral reefs after 1998 El Niño: Chagos reefs, Indian Ocean. Ambio 31:40–48. https://doi.org/10.1579/0044-7447-31.1.40

Siebeck UE, Marshall NJ, Klüter A, Hoegh-Guldberg O (2006) Monitoring coral bleaching using a colour reference card. Coral Reefs 25:453–460. https://doi.org/10.1007/s00338-006-0123-8

Spencer T, Teleki KA, Bradshaw C, Spalding MD (2000) Coral bleaching in the southern Seychelles during the 1997–1998 Indian Ocean warm event. Mar Pollut Bull 40:569–586. https://doi.org/10.1016/S0025-326X(00)00026-6

Strong AE, Barrientos CS, Duda C, Sapper J (1997) Improved satellite techniques for monitoring coral reef bleaching. In: Proceedings of the 8th international coral reef symposium, Panama, vol 2, pp 1495–1498

Strong AE, Liu G, Meyer J, Hendee JC, Sasko D (2004) Coral Reef Watch 2002. Bull Mar Sci 75:259–268

Thornhill DJ, LaJeunesse TC, Kemp DW, Fitt WK, Schmidt GW (2006) Multi-year, seasonal genotypic surveys of coral-algal symbioses reveal prevalent stability or post-bleaching reversion. Mar Biol 148:711–722. https://doi.org/10.1007/s00227-005-0114-2

Turner JR (1999) Status report Mauritius. In: Lindén O, Sporrong N (eds) Coral reef degradation in the Indian Ocean Status Reports and Project Presentations. CORDIO, Stockholm, pp 60–62

van Oppen MJH, Mahiny AJ, Done TJ (2005) Geographic distribution of zooxanthella types in three coral species on the Great Barrier Reef sampled after the 2002 bleaching event. Coral Reefs 24:482–487. https://doi.org/10.1007/s00338-005-0487-1

Warner ME, Fitt WK, Schmidt GW (1999) Damage to photosystem II in symbiotic dinoflagellates: a determinant of coral bleaching. Proc Natl Acad Sci U S A 96:8007–8012. https://doi.org/10.1073/pnas.96.14.8007

West JM, Salm RV (2003) Resistance and resilience to coral bleaching: implications for coral reef conservation and management. Conserv Biol 17:956–967. https://doi.org/10.1046/j.1523-1739.2003.02055.x

Winter A, Appeldoorn RS, Bruckner A, Williams EH, Goenaga C (1998) Sea surface temperatures and coral reef bleaching off La Parguera, Puerto Rico (northeastern Caribbean Sea). Coral Reefs 17:377–382. https://doi.org/10.1007/s003380050143

Yamano H, Tamura M (2004) Detection limits of coral reef bleaching by satellite remote sensing: Simulation and data analysis. Remote Sens Environ 90:86–103. https://doi.org/10.1016/j.rse.2003.12.005

Chapter 6
Bleaching Resistance and the Role of Algal Endosymbionts

K. M. Quigley, A. C. Baker, M. A. Coffroth, B. L. Willis, and M. J. H. van Oppen

6.1 Introduction

Scleractinian corals and octocorals form mutualistic endosymbioses with single-celled dinoflagellate algae in the genus *Symbiodinium*. This association is obligate in most coral species, with the coral host capable of deriving over 90% of its energy budget from its algal endosymbionts (Muscatine and Porter 1977), although nutritional dependence on *Symbiodinium* is generally lower in octocorals (Fabricius and Klumpp 1995). These endosymbionts also play a vital role in the light-enhanced calcification of scleractinian corals (Barnes and Chalker 1990; Moya et al. 2006). In healthy corals, *Symbiodinium* typically occur at extremely high densities ($>10^6$ cells per cm^2 coral tissue). During bleaching events, symbiont photosynthetic pathways become impaired (Chap. 8), leading to a breakdown in the symbiosis resulting in the

K. M. Quigley (✉)
Australian Institute of Marine Science, Townsville, QLD, Australia

ARC Centre of Excellence for Coral Reef Studies, James Cook University, Townsville, QLD, Australia
e-mail: katemarie.quigley@my.jcu.edu.au

A. C. Baker
Department of Marine Biology and Ecology, Rosenstiel School of Marine and Atmospheric Science, University of Miami, Miami, FL, USA

M. A. Coffroth
Department of Biological Sciences, State University of New York at Buffalo, Buffalo, NY, USA

B. L. Willis
ARC Centre of Excellence for Coral Reef Studies, James Cook University, Townsville, QLD, Australia

M. J. H. van Oppen
Australian Institute of Marine Science, Townsville, QLD, Australia

School of BioSciences, The University of Melbourne, Parkville, Victoria, Australia

M. J. H. van Oppen, J. M. Lough (eds.), *Coral Bleaching*, Ecological Studies 233,
https://doi.org/10.1007/978-3-319-75393-5_6

expulsion of *Symbiodinium* and/or a loss of photosynthetic pigment. Until recently, it was unclear whether the loss of *Symbiodinium* was due to death or detachment of host cells surrounding the symbionts or to expulsion or degradation of the symbionts themselves. Recent evidence suggests that bleaching is predominantly caused by the expulsion of intact *Symbiodinium* (Bieri et al. 2016), likely due to the production of reactive oxygen species (ROS) in the symbiont (Weis 2008). The production of ROS and the resulting cellular cascade leads to a breakdown in host–symbiont signalling and is a key trigger of bleaching responses (Baird et al. 2009a). Episodes of mass coral bleaching are predominantly driven by elevated temperature and irradiance (Hoegh-Guldberg 1999) but may also result from a variety of stressors, including disease (i.e. bacteria; reviewed in Harvell et al. 2007), viral infections (Lohr et al. 2007; Marhaver et al. 2008; van Oppen et al. 2009; Levin et al. 2017a; Weynberg et al. 2017), ocean acidification (Anthony et al. 2008; Chap. 12), salinity, heavy metals, cyanide, herbicides and other factors (reviewed in Baker and Cunning 2016). Severely bleached corals typically starve and die unless symbiont densities recover sufficiently rapidly to meet minimal phototrophic requirements and/or the coral has the ability to supplement its energy demands through increased heterotrophy (Grottoli et al. 2006; Anthony et al. 2009; Hoogenboom et al. 2012).

Mutualisms with *Symbiodinium* are established via horizontal (i.e. from the environment), vertical (i.e. maternal), or mixed transmission (a combination of both mechanisms). Approximately 85% of scleractinian corals are horizontal transmitters that produce gametes free of *Symbiodinium* (Fadlallah 1983; Babcock and Heyward 1986; Harrison and Wallace 1990; Baird et al. 2009b). In these cases, *Symbiodinium* must be acquired de novo from the environment in each generation, at the larval or early juvenile stage. Most broadcast-spawning corals show horizontal symbiont transmission, although there are some that transmit *Symbiodinium* directly to the oocytes (e.g. *Montipora* spp., *Porites* spp.). In contrast, the majority of brooding corals show vertical transmission, releasing internally brooded larvae that already harbour *Symbiodinium*. A small proportion of brooding corals have horizontal symbiont transmission, an example being *Isopora palifera* and several surface-brooding octocorals [e.g. *Briareum asbestinum*, *Antillogorgia elisabethae* (Kinzie 1974; Brazeau and Lasker 1990; Gutiérrez-Rodríguez and Lasker 2004; Poland et al. 2013)]. It is currently unclear how many species have mixed-mode transmission or if both vertical and horizontal infection occur simultaneously or in succession. For example, eggs or larvae may first undergo vertical transmission and then subsequently acquire *Symbiodinium* horizontally. Mixed-mode transmission has been documented for the eggs of *Montipora digitata* (Quigley et al. 2017a) and *Montipora capitata* (Padilla-Gamiño et al. 2012) and the planulae of *Seriatopora hystrix* (Quigley et al. 2018), and it has been hypothesized to occur in *Stylophora pistillata*, *Pocillopora damicornis* and *Porites astreoides* (Byler et al. 2013; Boulotte et al. 2016; Reich et al. 2017). Winnowing (i.e. the stepwise elimination of certain *Symbiodinium* during ontogeny) occurs in hard (Abrego et al. 2009a; Dunn and Weis 2009) and soft corals (Poland and Coffroth 2017) with horizontal transmission and likely also occurs in corals that have strictly vertical or mixed-mode transmission (Byler et al. 2013). This has led to the notion that some

Symbiodinium acquired during early ontogeny may be opportunistic or ephemeral, but these properties may themselves be dependent on environmental or ontogenetic factors.

Symbiodinium was long considered to be a monospecific genus with a worldwide distribution (Taylor 1974). However, in the early 1980s, it became evident that the species *Symbiodinium microadriaticum* Freudenthal comprises a highly diverse group of organisms (Schoenberg and Trench 1980; Blank and Trench 1985, 1986). This was confirmed more recently by molecular analyses (for reviews see Baker 2003; Coffroth and Santos 2005; Stat et al. 2006; Pochon et al. 2012) and detailed morphological analyses (Trench and Blank 1987; LaJeunesse 2001; Hansen and Daugbjerg 2009; Jeong et al. 2014; Lee et al. 2014, 2015). Despite the existence of subtle morphological differences between some *Symbiodinium* species, most *Symbiodinium* are morphologically extremely similar when examined using light microscopy. Therefore, the identification of *Symbiodinium* has relied primarily on genetic methods. These methods have resulted in the relatively recent introduction of a number of formal taxonomic descriptions of *Symbiodinium* species that have resolved substantial diversity within taxa from clades B (Lajeunesse et al. 2012; Parkinson et al. 2015a), C (Thornhill et al. 2014; Hume et al. 2015), and D (Wham et al. 2011; LaJeunesse et al. 2014). It is now clear that the original clades of *Symbiodinium* first described by Rowan and Powers (1991) represent major phylogenetic lineages, each probably deserving its own generic status (with the genus *Symbiodinium* being reserved for clade A, which contains the original species *S. microadriaticum* first described by Freudenthal in 1962). However this substantial taxonomic revision has not yet been undertaken.

Early molecular techniques using DNA/DNA hybridization estimated only 36% homology among symbionts representing *S. microadriaticum*, *S. kawagutii* and *S. pilosum* extracted from different hosts, highlighting the high diversity present within *Symbiodinium* (Blank and Huss 1989). Subsequent restriction fragment length polymorphism (RFLP) analyses and direct sequencing detected one to two predominant clades in corals (Rowan and Powers 1991; Baker and Rowan 1997; Burnett 2002), demonstrating the low resolving and/or sensitivity power of these early methods (Rowan et al. 1997). Other gel-based fingerprinting methods such as DGGE (denaturing gradient gel electrophoresis) of the Internal Transcribed Spacer 2 (ITS2) rDNA (LaJeunesse 2002) are still prevalent but are problematic because of extensive intra-genomic variation within *Symbiodinium* ITS2 (Thornhill et al. 2007) and the propensity for co-migration (Fabricius et al. 2004; Stat et al. 2009) and technical biases (Thornhill et al. 2010) inherent in methods that rely on differences in electrophoretic movement of distinct sequences (Coffroth and Santos 2005). However, these methods may still have value in systems where the diversity of *Symbiodinium* types is relatively well understood (i.e. controlled experiments involving well-studied coral species).

Quantitative polymerase chain reaction (qPCR) and next-generation sequencing (NGS) techniques offer increased sensitivity, especially for low-abundance *Symbiodinium* (Ulstrup and van Oppen 2003; Mieog et al. 2007; Correa et al. 2009; Silverstein et al. 2012; Cunning and Baker 2013; Arif et al. 2014; Quigley et al. 2014), and have the power to resolve diversity undetected by earlier methodologies (Kennedy et al. 2015; Boulotte et al. 2016; Quigley et al. 2017a, b). It was suggested that the increased detection ability of qPCR and NGS might also increase the risk of false positives due to the detection of non-symbiotic surface contaminants (Loram et al. 2007; Boldt et al. 2008; Correa et al. 2009); however, more recently, rates of false positives have been found to be low (0.07–9%) for both methods (Silverstein et al. 2012; Quigley et al. 2014). Both techniques may potentially overestimate genetic diversity if the genetic region quantified is multicopy and/or displays intra-genomic variation (Thornhill et al. 2007; Stat et al. 2009). Recent developments in whole genome sequencing (Shoguchi et al. 2013, 2015; Barbrook et al. 2014; Lin et al. 2015; Aranda et al. 2016) should accelerate the discovery of single-copy markers, with 261–1792 single-copy orthologs having been identified as candidates (Levin et al. 2017b). Computational methods can be applied to identify and account for intra-genomic variation, including identifying clusters of operational taxonomic units (OTUs) based on co-occurrence and proportionality (Kenkel et al. 2013; Quigley et al. 2014, 2016; Stat et al. 2015), applying 97% clustering across samples (Cunning et al. 2017a, b) and metahaplotype clustering (Smith et al. 2017). Another significant advantage of qPCR and NGS approaches, in addition to their high sensitivity, is the quantitative nature of the data generated, which lend themselves to relatively sophisticated analytical methods and statistical modelling approaches that cannot be applied to earlier, gel-based methods. These approaches have particularly helped understand the dynamics of mixed communities and their response to environmental changes (Cunning et al. 2015a, b, 2017a; Silverstein et al. 2017). Microsatellite markers are increasingly used to differentiate diversity within species and among populations of *Symbiodinium* (Howells et al. 2013; Prada et al. 2014; Davies et al. 2016; Wham and LaJeunesse 2016), although the interpretation of microsatellite data to delineate species boundaries is still in dispute (Howells et al. 2016a; Wirshing and Baker 2016).

In the following sections, we review what is currently known about *Symbiodinium* taxonomic and physiological diversity and patterns in the genetic diversity of *Symbiodinium*, both within individual coral colonies and among populations. We further discuss patterns in spatial (i.e. biogeographic and bathymetric gradients) and temporal (i.e. ontogenetic changes and shifts in response to environmental change) variability in *Symbiodinium* communities. We finish with a summary of recent advances in targeted research areas and identify new research fronts still to be explored.

6.2 Genetic Diversity of *Symbiodinium*

Nine phylogenetic lineages or clades (A–I) have been distinguished based on nuclear small and large subunit ribosomal DNA (nrDNA), as well as chloroplast large subunit ribosomal DNA (Pochon and Gates 2010). Relationships among seven of these clades have been verified by analyses of mitochondrial *cox1* and chloroplast *psbA* DNA (Baker 2003; Takabayashi et al. 2004; Coffroth and Santos 2005; Barbrook et al. 2006; Pochon et al. 2006, 2012; Sampayo et al. 2009; Pochon and Gates 2010). Estimates from molecular clocks calibrated with a "universal" dinoflagellate rate of sequence evolution for large subunit nrDNA (Tchernov et al. 2004) or based on vicariant events and host fossil dating (Pochon et al. 2006) suggest that the genus originated between 65 and 50 Mya, with subsequent diversification events, at least within clade C, occurring as recently as 12–13 Mya as the global climate cooled (Thornhill et al. 2014). Corals mainly associate with a suite of evolutionarily highly divergent symbionts in six of the nine known clades (A–D, F, G) of *Symbiodinium*, including the most basal lineage A and several derived lineages. More limited records of hard corals associating with members of clade E exist (Lee et al. 2016), whereas clades H and I associate predominantly with soritid Foraminifera (Pochon and Gates 2010; Pochon et al. 2012). However, recent genotyping of coral juveniles with horizontal transmission has revealed symbionts from all known clades, including members of E, H and I (Quigley et al. 2017a).

Each of the nine *Symbiodinium* clades contains some degree of diversity. Some clades with very distinct phylogenetic structure, such as F and G, have been subdivided into numbered "subclades", which themselves also contain further diversity (Pochon et al. 2006; Pochon and Gates 2010; Bo et al. 2011). Here we will use the term "type" to indicate taxonomic entities below the level of the nine known clades. Types have been generally characterized as operational taxonomic units (OTUs) and in some cases have been formally described as distinct species. We use the term "genotype" to denote genetic diversity within types. Genotypes may replicate asexually to form clones or genets (strains) within a type.

Diversity below the level of the clade is usually assessed using the internal transcribed spacers (ITS1, ITS2) of the nrDNA (Hunter et al. 1997; Baillie et al. 2000a; LaJeunesse 2001; van Oppen et al. 2001; Pochon et al. 2012), the chloroplast rDNA 23S gene (cpDNA) (Santos et al. 2002) or the *psbA* non-coding region (Moore et al. 2003; Thornhill et al. 2014) of chloroplast DNA, or microsatellite flanking regions (Santos et al. 2004). The delineation of species boundaries in *Symbiodinium* is much debated and not easily resolved. Initially several authors proposed that distinct ITS or cpDNA sequences represented different species (LaJeunesse 2001; Coffroth and Santos 2005); however, more recent studies have demonstrated that these markers do not resolve all genetic diversity, with types that share ITS2 or cpDNA sequences sometimes representing distinct taxa based on other molecular markers (Santos et al. 2004; Parkinson et al. 2015b). Investigations of genotypic diversity within these types or species have used allozymes (Schoenberg and Trench 1980), DNA fingerprinting (Goulet and Coffroth 2003a), random amplified polymorphic DNA (Baillie et al. 2000b) and microsatellites (Santos et al. 2001, 2003; Magalon et al. 2006; Pettay and LaJeunesse 2009; Pinzón et al. 2011; Wham et al. 2011; Parkinson et al. 2015b), with the most recent papers focusing on identifying species boundaries but also beginning to point to the functional importance of different *Symbiodinium* genotypes in understanding coral performance (Parkinson et al. 2015a).

Studies using high-resolution markers to examine the diversity and spatial structure of *Symbiodinium* populations have produced very dissimilar results. In the Caribbean gorgonian *Antillogorgia* (previously *Pseudopterogorgia*) *elisabethae*, less than 5% of the 575 colonies surveyed in 12 populations harboured more than one *Symbiodinium* clone (Santos et al. 2003). Furthermore, distinct population structure was observed over spatial scales ranging from a few to ~450 km, with 66–100% of colonies within 10 of the 12 surveyed populations harbouring a single genotype; only in a few cases were genotypes shared among populations. Similarly, most genotypes found for *Gorgonia ventalina* were restricted to a single or a few populations across the entire host range (Kirk et al. 2009; Andras et al. 2011). A global analysis of clade D *Symbiodinium* isolated from symbiotic cnidarians also revealed that a majority of hosts harboured a single genotype (Pettay and LaJeunesse 2009; Wham et al. 2011), as did *Acropora palmata* from the Caribbean, which harboured *Symbiodinium* A3 (*S. fitti*) (Pinzón et al. 2011), and *Orbicella* spp. collected across the Caribbean, which had single clones of C1 and C3 per host (Thornhill et al. 2007). In contrast to only harbouring a single genotype, up to four *Symbiodinium* clones from clade C were found within individual colonies of the scleractinian coral *Pocillopora meandrina* from the Tonga and Society archipelagos in the South Pacific (Magalon et al. 2006), up to 15 clones within type C2 were associated with *Acropora millepora* on the Great Barrier Reef (GBR; Howells et al. 2013), and up to six clones were found within *Pocillopora* type 1 (*sensu* Pinzón and LaJeunesse 2011) from the Gulf of California (Pettay et al. 2011). Furthermore, *P. meandrina* commonly shared alleles among South Pacific populations, with significant genetic differentiation detected only at scales greater than ~200 km (Magalon et al. 2006). *Symbiodinium* populations harboured by colonies of the soft coral *Sinularia flexibilis* from the GBR were comprised of between 2 and 6 clones; and up to 14 clones were found within colonies from Torres Strait populations, with significant genetic differentiation existing among populations at scales ranging from 10s to 100s of kilometres (Howells et al. 2009). Single clones of B1 in two *Orbicella* species in the Caribbean (~40%) (Thornhill et al. 2009) and of C3 and C40 in *Acropora digitifera* and *Acropora hyacinthus* across Micronesia (~67% of colonies tested) were less common (Davies et al. 2016).

There appears to be no relationship between the number of *Symbiodinium* genotypes present within individual colonies and either the mode of symbiont transmission, the specificity of the symbiosis or the extent of spatial structure observed. However, specificity for a symbiont type may be related to either the transmission mode (Fabina et al. 2012) or a genetically determined host trait (Quigley et al. 2016, 2017a, b; Poland and Coffroth 2017). For example, strong symbiont specificity is shown by both *Sinularia flexibilis*, which has horizontal transmission, and *P. meandrina*, which transmits algal symbionts directly to its offspring and exhibits some symbiont specificity for at least two distinct symbiont C1 types in the South Pacific (Magalon et al. 2006) and on the GBR (Hirose et al. 2001; LaJeunesse et al. 2004). The pattern emerging is that a single clone per *Symbiodinium* type is the most common association in Caribbean hard and soft corals, but not in Indo-Pacific corals (reviewed in Thornhill et al. 2017), although in some instances individual Caribbean colonies may harbour multiple types simultaneously. These patterns likely reflect the fact that Caribbean corals commonly host symbionts across four clades of *Symbiodinium* (A, B, C and D), whereas *Symbiodinium* communities associated with Indo-Pacific corals tend to be dominated by diverse members of clade C, perhaps driving diversification within this group in the Indo-Pacific. Genetic diversity within *Symbiodinium* types in the Caribbean also appears lower than within types in the Pacific. However, studies of a much wider range of taxa are required to confirm these patterns.

6.3 Biogeographic Patterns in *Symbiodinium* Diversity: In Hospite and Ex Situ

The distributions of *Symbiodinium* types associated with scleractinian corals vary considerably between the Caribbean and the Indo-Pacific (Baker and Rowan 1997; Baker 2003; LaJeunesse et al. 2003; LaJeunesse 2005), and between the Indian Ocean and western Pacific (LaJeunesse et al. 2010a), as well as within both the Caribbean (Finney et al. 2010; Kennedy et al. 2015, 2016) and Indo-Pacific (Loh et al. 2001; LaJeunesse et al. 2004). Scleractinian corals often host multiple symbionts simultaneously (Baker and Romanski 2007), particularly when the low-abundance background symbionts are included (Silverstein et al. 2012). Nevertheless, clear biogeographic patterns exist. At shallow depths (<7 m) in the tropical western Atlantic (Caribbean), *Symbiodinium* clades A–D are common, with individual coral species typically hosting members of one of these clades, but sometimes hosting numerous ITS-types from up to three additional clades, although more often each species is dominated by one specific ITS-type (LaJeunesse et al. 2003; Fabina et al. 2012; Kennedy et al. 2015). In contrast, at similar depths in the tropical Indo-Pacific, scleractinian corals are dominated by a lower phylogenetic diversity of ITS-types, mainly in clade C and a few in clade D. Similar patterns are found in the Red Sea, Sea of Oman and Arabian Gulf, where scleractinian corals are also dominated by clades C and D (Hume et al. 2015; Ziegler et al. 2017) [Fig. 6.1; see also online databases of *Symbiodinium* biogeography: Geosymbio (Franklin et al. 2012) and SymbioGBR (LaJeunesse et al. 2003; Fabina et al. 2012)]. The generalist symbiont types within clade C [i.e. the more ancestral ITS2-types C1 and C3 (*sensu* LaJeunesse 2002) or ITS1-type C2 (*sensu* van Oppen et al. 2001, which is equivalent to ITS2-type C3)] are present in both Caribbean and Pacific hosts, suggesting that they arose prior to the closure of the Isthmus of Panama (LaJeunesse 2005) (Fig. 6.1). The high diversity within clade C found in both oceans is thought to have evolved independently through a series of adaptive radiation events driven by post mid-Miocene cooling (LaJeunesse 2005; Thornhill et al. 2014).

Apparent similarity between tropical Caribbean scleractinian symbioses and their counterparts from the temperate Indo-Pacific and Mediterranean led to speculation that the closure of the Isthmus of Panama and subsequent cycles of glaciations in the

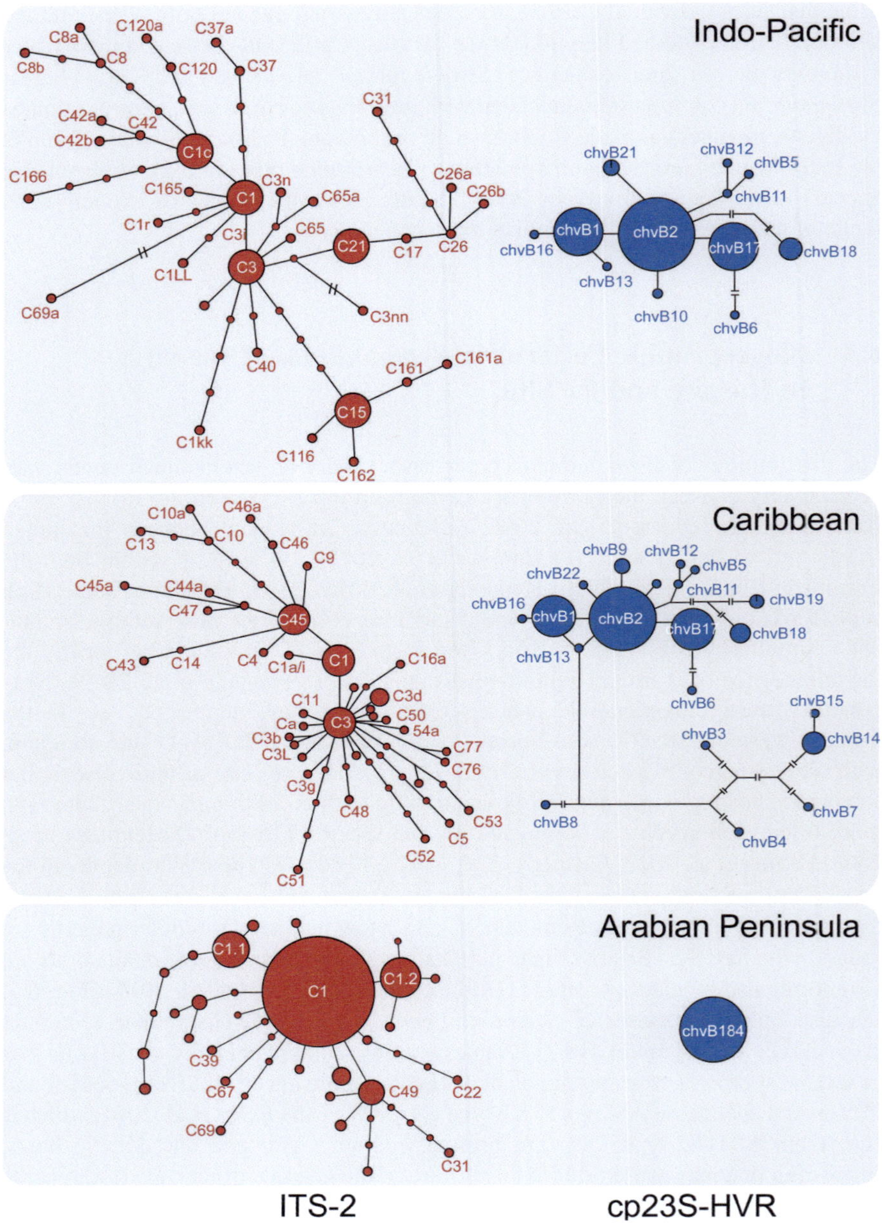

Fig. 6.1 Haplotype networks depicting the variability in in hospite and ex situ diversity detected in *Symbiodinium* clades C using ITS2 (red networks) (LaJeunesse 2005; Tonk et al. 2014; Reimer et al. 2016) and B using cp23S (blue networks) (Takabayashi et al. 2012; Ziegler et al. 2017) within the Indo-Pacific, Caribbean and Arabian Peninsula (Sea of Oman and Persian/Arabian Gulf). Each node represents a distinct ITS2 or cp23S sequence and potential variant or "type". For clarity, not every node has been labelled. The size of each haplotype reflects the relative abundance of the type

Northern Hemisphere led to cooler and more variable temperature regimes in the tropical western Atlantic which, in turn, selected for *Symbiodinium* that are more typical of temperate hosts (Baker and Rowan 1997; Baker 2003). A Caribbean radiation of clade B also dates to this period (LaJeunesse 2005), supporting the idea that environmental change during the Plio–Pleistocene transition drove these patterns of association (Baker and Rowan 1997; Baker 2003; LaJeunesse et al. 2003) (Fig. 6.1). These general hypotheses are somewhat further supported by the unusual cold tolerance of *Symbiodinium* B2 in coral hosts from northerly habitats of the western Atlantic (Thornhill et al. 2008).

As documented in scleractinians, the distributions of *Symbiodinium* clades and types associated with octocorals vary between the major ocean basins (van Oppen et al. 2005a; Goulet et al. 2008). Overall diversity is greatest in the western Pacific and the GBR, where, at the clade level, the symbiosis is dominated primarily by *Symbiodinium* clade C (van Oppen et al. 2005a; Howells et al. 2009; FitzPatrick et al. 2012), with a substantial number of hosts also having clade D. Clades A, B and G are also reported from soft corals on the GBR, although occurrences of clade B and of clade A and G symbionts are rare on the GBR and Caribbean, respectively (Franklin et al. 2012; Tonk et al. 2013). Within the Red Sea, soft coral symbioses are dominated by *Symbiodinium* clade C and, to a lesser extent, clade A (Barneah et al. 2004). In the Caribbean and Bermuda, soft corals harbour predominantly *Symbiodinium* clade B, with some species also engaging in symbiosis with *Symbiodinium* clade C. Within octocorals, symbiont distribution does not appear to vary with depth, in contrast to patterns observed in some scleractinians. Both general reef surveys (LaJeunesse 2002; LaJeunesse et al. 2003; van Oppen et al. 2005a) and studies that specifically examined symbiont diversity over depth (Goulet and Coffroth 2003a, 2004) failed to detect any depth-related pattern in symbiont distribution. These patterns suggest that, in contrast to scleractinian corals, many octocoral species may be less flexible in their associations with *Symbiodinium*, perhaps due to their greater dependence on heterotrophy (Baker and Romanski 2007).

Patterns of host and symbiont specificity are influenced by differences in symbiont diversity between the Atlantic and Pacific. LaJeunesse et al. (2003) suggested that, although the total number of *Symbiodinium* types may be greater in the Indo-Pacific than in the Caribbean (due in large part to the enormous diversity of *Symbiodinium* in clade C) (Fig. 6.1), the ratio of symbiont diversity to host diversity may be higher in the Atlantic than in the Pacific. This effect is mostly a result of the

Fig. 6.1 (continued) in that region. Clades B and C are the most common clades in the Caribbean and Indo-Pacific, respectively; and clade C is more diverse compared to clade B (LaJeunesse et al. 2003). The diversity of clade B is also higher in the Caribbean relative to the Indo-Pacific, whereas the converse is true for clade C (LaJeunesse et al. 2003). Indo-Pacific and Caribbean B haplotypes were constructed using cp23S sequences with the exception of chvB184, which was translated from the B1 ITS2 type reported in Ziegler et al. (Santos et al. 2003; Lajeunesse et al. 2012)

additional diversity found within clades A and B, which both contribute more commonly to symbiont pools in the Atlantic.

Less is known about the worldwide biogeography of free-living *Symbiodinium* (although some locations are well studied: Takabayashi et al. 2012). However, similar to patterns found for *Symbiodinium* associated with scleractinians and octocorals, the distributions of free-living *Symbiodinium* vary between major ocean basins. In Florida, water samples and sediments were dominated by clade B, with clade A types also present in high abundance (Coffroth et al. 2006; Takabayashi et al. 2012). In Curaçao, sediments, vegetative surfaces, rubble and water column samples were dominated by clade A, with the water column being particularly speciose, containing types from clades A, B, C, F, G and H (Granados-Cifuentes et al. 2015). Clades B, C and D dominated water samples collected from Mexico (Manning and Gates 2008). An exception to the general dominance of clades A and B in the Caribbean was observed in Colombian reefs, where clade C was dominant, although diverse types in clades A and B were also detected in macroalgal beds and sediments (Porto et al. 2008).

Diversity within the Indo-Pacific region appears to be greater than in the Caribbean, with initial surveys reporting clades A, B, C, D, G, H (Huang et al. 2013) and F (Gou et al. 2003) in the water column around Chinese reefs. Japanese reefs have been found to have members of clades A, C, D and E in the water column (Yamashita and Koike 2013) and an array of A types, C types and novel D types in the sediments (Hirose et al. 2008; Reimer et al. 2010; Yamashita and Koike 2013; Yamashita et al. 2013). Studies in the southern GBR report that macroalgal thalli, cyanobacterial mats, crustose coralline algae and water samples were dominated by clade C, but no *Symbiodinium* sequences were retrieved from the sediment (Venera-Ponton et al. 2010). However, later genotyping at the same location discovered clade A in the sediments, as well as members of clades A, C and E in the water column (Sweet 2013). The dominance of clade C in the western Pacific also extends to central Pacific (Hawaiian) waters, which were dominated almost exclusively by clade C (Manning and Gates 2008). Further sampling also detected types from clades A, B and D in the water column. Sediment samples were dominated by a range of types, particularly from clades A and B and, to a lesser extent, C and G (Carlos et al. 1999; Pochon et al. 2010; Takabayashi et al. 2012).

As with in hospite *Symbiodinium*, the biogeographical distributions of free-living *Symbiodinium* appear to be influenced by temperature and light gradients. In the water column, clade A was more abundant closer to the shore in American Samoa, but the opposite trend was documented in sediments (Cunning et al. 2015c), suggesting A types partition their distributions based on temperature or turbidity differences. *Symbiodinium* found in sediments collected from the northern and central GBR also showed strong partitioning along thermal and water quality gradients, with types within clades A and C dominating samples from warmer areas, whereas types within clade D dominated samples from more turbid inshore sites (Quigley et al. 2017b). Continued sampling is needed to determine the full extent of niche partitioning among *Symbiodinium* types, as well as key environmental drivers of their distributions.

6.4 Physiological Differences Among Genetically Distinct *Symbiodinium* Types

The influence of symbiont identity and diversity on fitness of the coral host has been increasingly recognized. To a large extent, physiological characteristics of distinct symbiont types have been inferred from correlative studies. For example, zonation of *Symbiodinium* types over light gradients within colonies and between shallow and deep colonies of *Orbicella annularis*, *O. faveolata* and *O. franksi* suggests that these symbionts have distinct light sensitivities (Rowan and Knowlton 1995; Rowan et al. 1997; Toller et al. 2001a; Kemp et al. 2008, 2015). Observations of patchy bleaching within *Orbicella* colonies during a natural bleaching event further suggest that bleaching tolerances of different *Symbiodinium* types within clades A–C differ substantially (Rowan et al. 1997). For example, bleached patches with lower photosynthetic values had higher abundances of B17 and C7 compared to non-bleached patches that harboured A3 and D1a (Kemp et al. 2014). Toppled *O. annularis* colonies showing a cline of *Symbiodinium* types within B and C along an irradiance gradient re-established their original zonation patterns with respect to light in the newly oriented colonies over a 6-month period (Rowan et al. 1997). Other environmental factors such as temperature, substrate and water quality also shape distributions of in hospite *Symbiodinium* (Toller et al. 2001b; Cooper et al. 2011; Silverstein et al. 2011; Kemp et al. 2015; Davies et al. 2016; Howells et al. 2016b; Ziegler et al. 2017) but may also depend on the scale of environmental variation across spatial gradients (Cunning et al. 2017b). In combination, these studies suggest that coral–algal associations can vary spatially and be controlled by environmental factors (see also Sect. 6.5).

Similar patterns of depth zonation in *Symbiodinium* types have been observed for Caribbean (i.e. *Acropora cervicornis*, *Stephanocoenia intersepta*, *Porites astreoides*, *Diploria strigosa*, *Acropora lamarcki*, *Madracis pharensis* and *Orbicella faveolata*) and GBR (*Seriatopora hystrix*) scleractinian corals, and for the Japanese zoanthid *Zoanthus sansibaricus* (Baker et al. 1997; Baker 2001; Frade et al. 2008; Bongaerts et al. 2010, 2015a; Kamezaki et al. 2013; Kemp et al. 2015; Lucas et al. 2016), among others (Bongaerts et al. 2015b), although important exceptions exist (*Agaricia* and *Madracis* spp., Frade et al. 2008; Bongaerts et al. 2015a; *Montastraea cavernosa*, Serrano et al. 2014). When corals were transplanted from deep to shallow habitats (Baker 2001) or inversely (Bongaerts et al. 2015a), changes in light levels (low to high or high to low, respectively) following transplantation caused a portion of the transplanted colonies to bleach. When transplanted deep to shallow, corals that showed depth zonation in symbiont types recovered with the symbiont types typical of shallow conspecifics, rather than their original deep-water types, whereas coral species that showed no depth zonation also showed no change in symbionts following recovery (changes in transplanted colonies from shallow to deep were not assessed). These results suggest that bleaching may provide an opportunity to change *Symbiodinium* communities inside host tissues in favour of a community that is better adapted to the changed environmental conditions (Buddemeier and Fautin 1993; Baker 2001; Baker et al. 2004a; Buddemeier et al. 2004). However, communities in some colonies may change in the absence of visible bleaching (Thornhill et al. 2006a). These authors also hypothesized that certain members of *Symbiodinium* clade D, such as the Caribbean D1a (also known as *S. trenchi*) and other types within clade D, are tolerant of high levels of light or of high variability in stressors in general (Toller et al. 2001a; Jones et al. 2008; LaJeunesse et al. 2009; Cooper et al. 2011; Oliver and Palumbi 2011; Cunning et al. 2015d; Bay et al. 2016; Ziegler et al. 2017). Symbionts within clade D were found to be dominant in corals on a reef flat in Thailand (Brown et al. 2002), in shallow *O. faveolata* colonies in Mexico and the Bahamas (Kemp et al. 2015), in the *O. annularis* species complex in shallow depths in Panama (Toller et al. 2001a) and in shallow corals leading up to and during bleaching (LaJeunesse et al. 2009), also suggesting that at least some *Symbiodinium* D are relatively tolerant of high light. In contrast, intra-colony zonation patterns of symbionts within clades C and D in *Acropora valida* (Ulstrup and van Oppen 2003), and the high abundance of symbionts within clade D in corals from turbid reef environments on the GBR (van Oppen et al. 2001, 2005b; LaJeunesse et al. 2010a) led to the hypothesis that some symbionts within clade D are better adapted to low-light environments (but see Tonk et al. 2014, 2017). The subtropical coral, *Oulastrea crispata*, forms a stable symbiosis with two species of *Symbiodinium* in clade D (LaJeunesse et al. 2014) in turbid non-reefal environments of Taiwan, where annual water temperatures vary between 12 °C and 35 °C (Chen et al. 2003), as does the zoanthid *Palythoa tuberculosa* on turbid Singapore reefs (Reimer and Todd 2009). Populations of the coral *Acropora palifera* in southern Taiwan harbour a mix of C and D symbionts, with clade D species decreasing in relative abundance with seasonally increasing seawater temperatures (Chen et al. 2005). Such distribution patterns suggest that some Indo-Pacific *Symbiodinium* species in clade D may be shade-adapted and that members of this clade can withstand low temperatures (see also Silverstein et al. 2017 for Caribbean D1a/*S. trenchi*).

In contrast to evidence that many *Symbiodinium* types within clade D are tolerant of both low and widely varying temperatures, clade D types are commonly associated with corals inhabiting warm waters in a variety of regions globally (Baker et al. 2004b; Fabricius et al. 2004; Ulstrup et al. 2006; Oliver and Palumbi 2009; DeBoer et al. 2012; Ziegler et al. 2017). The observation that C-dominated corals bleached while D-dominated corals were healthy during the 1997/1998 bleaching event in the far eastern Pacific (Glynn et al. 2001; Baker et al. 2004a) confirms that at least some types within *Symbiodinium* D are tolerant of higher than normal water temperatures. An apparent increase in the abundance of clade D following major bleaching events also provides evidence that D-dominated corals are more thermally tolerant compared with other coral–algal associations (Baker et al. 2004a; van Oppen et al. 2005b; Jones et al. 2008; Cunning et al. 2015a; Bay et al. 2016). In summary, corals harbouring types within clade D appear to be tolerant of a range of stressors, for example, both high- and low-temperature extremes, or have thermal tolerances that vary among potentially different types within the clade (Swain et al. 2016; Silverstein et al. 2017).

Experimentally based studies of the physiological characteristics of distinct symbiont clades and types are burgeoning (Table 6.1). Cultured symbiont strains

Table 6.1 Physiological performance of in hospite/cultured/freshly isolated *Symbiodinium* of known genetic identity

Host	*Symbiodinium* type	Symbiotic state of *Symbiodinium*	Physiological characteristic	References
Stylophora pistillata	C	In hospite	Heat sensitive	Tchernov et al. (2004)
Montipora samarensis	C	In hospite	Heat tolerant	Tchernov et al. (2004)
Pocillopora verrucosa, *Pavona gigantea*	D1, C1c, respectively	In hospite	D1 is “sun-loving”, C1c is “shade-adapted”	Iglesias-Prieto et al. (2004)
Pocillopora damicornis, *P. verrucosa*	C, D	In hospite	D-corals not photosynthetically impaired by experimental high temperature, while C-corals are	Rowan (2004)
Acropora millepora, *A. tenuis*	**C1, D**	In hospite	C1-juveniles grow 2–3 times faster than D-juveniles	Little et al. (2004)
Exaiptasia pallida	A, B	In hospite	A-anemones have higher photosynthetic rates than B-anemones at elevated temperatures	Goulet et al. (2005)
Orbicella faveolata	*B1*, *C7*	In hospite	C7 loses capacity for diel recovery of photoinhibition earlier than B1 during short-term elevated temperature stress	Warner et al. (2006)
Porites astreoides	*A4a*	In hospite	High resistance to photodamage during non-stress and high-temperature stress conditions	Warner et al. (2006)
Palythoa caribaeorum	*C1*, *D1a*	In hospite	No differences in thermal tolerance	Kemp et al. (2006)
A. millepora	**C2, D**	In hospite	D-corals have higher thermal tolerance than C2-corals	Berkelmans and van Oppen (2006)

(continued)

Table 6.1 (continued)

Host	*Symbiodinium* type	Symbiotic state of *Symbiodinium*	Physiological characteristic	References
A. cytherea	A, C	*In hospite*	Fixation and transfer of carbon to host lower in A compared to C	Stat et al. (2008)
A. tenuis	**C1, D**	*In hospite*	C1-juveniles have higher thermal tolerance than D-juveniles	Abrego et al. (2008)
A. millepora	**C1, D**	*In hospite*	Photosynthetic capacity ($rETR_{MAX}$) 45% greater in C1 than in D, resulting in doubling of 14C photosynthate (energy) incorporation into juvenile coral tissue	Cantin et al. (2009)
A. tenuis, A. millepora	**D, C1**	*In hospite*	Infection and/or propagation of D faster at higher temperatures compared to C1, and increased under high light	Abrego et al. (2012)
O. faveolata	*B1, D1a*	*In hospite*	Photochemical efficiency (Fv/Fm) of B1 higher than D1 under nonstressful conditions, but lower under heat stress	Cunning et al. (2015b)
Montastraea cavernosa	*C3, D1a*	*In hospite*	Photochemical efficiency (Fv/Fm) of C3 lower than D1a under heat stress. Higher tolerance (1–2 °C) in corals of same genotype containing D1a	Silverstein et al. (2015)
O. faveolata	A1, B1	*In hospite*	Growth higher but photophysiological measures lower for juveniles infected with A1, with no difference in survival pre- or post-thermal stress treatment	McIlroy et al. (2016)
Siderastrea siderea	*C, D1a*	*In hospite*	Photochemical efficiency (Fv/Fm) of C lower than D1a under heat stress	Cunning et al. (2017a, b)

M. cavernosa	*C3*, *D1a*	In hospite	Photochemical efficiency (Fv/Fm) of D1a lower than C3 under cold stress, but D1a symbionts still retained more than C3	Silverstein et al. (2017)
Cassiopea xamachana, *M. verrucosa*, *Zoanthus sociatus*	*S. microadriaticum* (A1), *S. kawagutii* (F1), *S. pilosum* (A2), respectively (*Symbiodinium* genetic identity from LaJeunesse 2001)	Cultured	Three types have different photosynthetic characteristics and differ in their photo-acclimatory capabilities which correlates with their ecological distribution when endosymbiotic	Iglesias-Prieto and Trench (1994)
54 species, including hydrozoans, anemones, gorgonians, scleractinians	A, B, C, D	Cultured, in hospite	In culture, clade A symbionts produce mycosporine-like amino acids (MAAs) whereas clade B, C, D and E symbionts do not. In hospite, all *Symbiodinium* types examined (members of clades A–D) produce MAAs	Banaszak et al. (2000, 2006)
Cassiopea xamachana, *M. verrucosa*, *P. damicornis*, *Cassiopea* KB8, *Zoanthus sociatus*, *Tridacna gigas*, *E. pulchella*	A, C, B, A, A, A, B, respectively	Cultured	Different growth rates at three temperatures among types within clade A; some types show increased growth, some decreased growth and some no change in growth with increasing temperature	Kinzie et al. (2001)
Agaricia spp., *M. cavernosa*, *M. franksi*, *Oculina diffusa*, *P. astreoides*, *E. pallida*, *Bartholomea annulata*, *Condylactis gigantea*, *Cassiopea xamachana*	A, B, C	Freshly isolated	PI curves are not uniform for all members of one clade and not consistently different between clades	Savage et al. (2002)
Unknown	B, A	Cultured	Both heat-sensitive and heat-tolerant strains within clades A and B	Tchernov et al. (2004)
Cassiopea xamachana, *Condylactis gigantea*, *E. pallida*, *Meandrina meandrites*	*A1*, *A1.1*, *B1*, *F2*, respectively	Cultured	A1.1 has most limited capacity to acclimate to high light levels. B1 and A1.1 have greatest thermal sensitivity, F2 and A1 show high degree of thermal tolerance	Robison and Warner (2006)

(continued)

Table 6.1 (continued)

Host	*Symbiodinium* type	Symbiotic state of *Symbiodinium*	Physiological characteristic	References
A. millepora	**A/C2* mixture, C1, C2, C·, D**	Cultured	D conferred greatest thermal tolerance to juveniles of *A. millepora*	Mieog et al. (2009)
Cassiopea xamachana, *Pseudopterogorgia bipinnata*, *Oculina diffusa*, *Discosoma sanctithomae*, *O. faveolata*, *Anthopleura elegantissima*, *M. meandrites*	A1, B1, B2, C1, D, E1, F2	Cultured	Production of reactive oxygen species (ROS) varied by type with increased sublethal temperature	McGinty et al. (2012)
Multiple hosts	110 types compared	Combinations	D_{Ber06} had greatest thermotolerance compared to 109 other types	Swain et al. (2016)
Cassiopea spp., *O. faveolata*, *Hippopus hippopus*, *Plexaura kuna*	A3, B1, B2, C2, D1a, F	Cultured	PI curves and growth rates differed among strains exposed to temperatures ranging from 20 to 33°C	Grégoire et al. (2017)

Symbiodinium types in italics refer to LaJeunesse's nomenclature (LaJeunesse 2001, 2002) and those in bold to van Oppen's nomenclature (van Oppen et al. 2001), noting that van Oppen ITS1 C2 is equivalent to LaJeunesse ITS2 type C3. Asterisks are the authors' notation

can differ considerably in physiological characteristics, such as growth rates or photo-acclimatory responses to changes in irradiance (Iglesias-Prieto and Trench 1994, 1997; Kinzie et al. 2001; McGinty et al. 2012; Karim et al. 2015; Klueter et al. 2015; Suggett et al. 2015; Swain et al. 2016; Grégoire et al. 2017). For example, *S. thermophilum* (distinct from another C3 type) and symbiont types A1, A20, D_{Ber06} and F2 have high thermal tolerances in culture or in hospite, but types A_{mie09} and D_{Abr08} do not (Sawall et al. 2014; Hume et al. 2015; Swain et al. 2016). Infectivity of early life-history stages of corals varies among types, with types A3, B1 and D1a exhibiting high infectivity compared to types such as A2 (Coffroth et al. 2001; Abrego et al. 2009b; Kuniya et al. 2015; Poland and Coffroth 2017). Further evidence of distinct physiological characteristics conferred on coral hosts by different *Symbiodinium* taxa includes variation in transcriptional profiles (Parkinson et al. 2016; Gierz et al. 2017) and bacterial communities established for different host–symbiont associations (Littman et al. 2009). However, photophysiological responses are known to differ significantly between in hospite and expelled, freshly isolated or cultured *Symbiodinium* (Ralph et al. 2001; Bhagooli and Hidaka 2003; Howells et al. 2012; Chakravarti et al. 2017); thus, it is not completely clear how data from symbiont cultures relate to the natural, symbiotic situation. Importantly, in hospite variation at the inter- and intra-type level significantly impacts host physiology (Parkinson et al. 2015b). For example, experimental in hospite studies of ITS-type D1 suggest that it is more thermally tolerant than *Symbiodinium* ITS1-type C2 (equivalent to ITS2-type C3) or C1 in the Indo-West Pacific (Rowan 2004; Berkelmans and van Oppen 2006; Abrego et al. 2008, 2012; Mieog et al. 2009), and the same is true of D1a and C1b–c in the Caribbean (Silverstein et al. 2015). D1 may also be more light-loving compared to other *Symbiodinium* types (Iglesias-Prieto et al. 2004; Abrego et al. 2012). Given that taxonomic resolution was at the clade level in many of these studies, differences among them may well be due to physiological variation of D symbionts at the type or genotype level, where, for example, D_{Ber06}, D1–4 (D1a), D1 and D1–11 all have different thermotolerances (Swain et al. 2016). Therefore, physiological diversity exists among *Symbiodinium* clades, types and genotypes, for example, among different *S. fitti*-A3 strains, among others (Kinzie et al. 2001; Tchernov et al. 2004; Baums et al. 2014; Parkinson et al. 2015a, b, 2016; Swain et al. 2016). Furthermore, either host factors or the interaction between host and symbiont may cause the same *Symbiodinium* types to function very differently in different host species (Goulet et al. 2005; Abrego et al. 2008, 2012; Mieog et al. 2009). Thus, the physiological responses of corals to environmental change may not be due to plasticity of the coral or the algal symbiont within the coral but due to plasticity in the holobiont itself, i.e. the same host may manifest different physiologies and responses depending on the host–symbiont pairing (Baker 2003; Little et al. 2004; Goulet et al. 2005; Stat et al. 2006; Parkinson and Baums 2014).

6.5 Shifts in Symbiont Communities as a Mechanism to Cope with Environmental Change

Initial uptake of symbionts by juvenile colonies is selective relative to the full diversity of available free-living *Symbiodinium* (Poland et al. 2013; Quigley et al. 2017b) but non-selective relative to adult diversity. This relative flexibility may allow for a mix of symbiont types to establish symbioses with a single host colony in early ontogeny (Coffroth et al. 2001; Little et al. 2004; Abrego et al. 2009a, b; Cumbo et al. 2013; Poland et al. 2013; Quigley et al. 2016, 2017a, b; Poland and Coffroth 2017). Subsequently, one or a few symbionts typically increase in abundance relative to others (Kinzie 1974; Coffroth et al. 2001; Little et al. 2004; Abrego et al. 2009a; Poland et al. 2013; Poland and Coffroth 2017). Although the host influences and potentially constrains the earliest *Symbiodinium* communities to varying degrees (Quigley et al. 2016, 2017a, b; Poland and Coffroth 2017), the greater diversity of symbiont types during early ontogeny compared to adults may be an adaptive trait, as it permits changes in the relative abundance of symbionts with distinct physiological characteristics within a single coral host (Little et al. 2004). For example, the initial acquisition of generalist, hardy or stress-tolerant *Symbiodinium* types/communities may provide a mechanism for rapid acclimatization or adaptation to environmental change (Baker 2003), either during range expansions of generalist hosts (Grupstra et al. 2017) or by providing options for enhanced host survival or growth (McIlroy et al. 2016; Quigley et al. 2016), similar to patterns documented in adults pre- (LaJeunesse et al. 2009) and post-bleaching (Silverstein et al. 2015). This may be a mechanism for the holobiont to acclimate to changes in the environment (Buddemeier and Fautin 1993; Ware et al. 1996; Buddemeier and Smith 1999). Superficially, this hypothesis seems to be contradicted by the observation that some corals exhibiting strong symbiont specificity and temporal stability at adulthood also take up a range of symbiont types and clades shortly after settlement (Coffroth et al. 2001; Goulet and Coffroth 2003a). However, dissimilarity in symbiont communities between juvenile and adult corals may be driven by contrasting selective pressures, which necessitate different communities during early ontogeny (Gómez-Cabrera et al. 2008; Abrego et al. 2009a) to cope with microhabitat changes in irradiance with colony growth, or the need to increase in size rapidly to minimize mortality.

It is also feasible that such differences in *Symbiodinium* communities through ontogeny are selectively neutral and, therefore, have not been lost through evolution. Furthermore, this phenomenon may be related to the fact that immunity (including allorecognition) is suppressed during the first ~2–13 months of a coral's life (Frank et al. 1997; Nozawa and Loya 2005; Puill-Stephan et al. 2009, 2012) and the suppression of immunity itself may have selective advantages during early ontogeny in corals both with and without symbiont specificity. It is also possible that specificity does not result in the total exclusion of heterologous symbionts at this early developmental stage but is expressed as lower densities and/or differing patterns of localization of heterologous compared with homologous symbionts inside host tissues (Rodriguez-Lanetty et al. 2004, 2006). Finally, even though multiple types are initially acquired, some selectivity is exhibited, as not all symbiont types are taken up while others are hosted only temporarily (LaJeunesse 2001; Weis et al. 2001; Rodriguez-Lanetty et al. 2006; Dunn and Weis 2009; Coffroth et al. 2010; Poland et al. 2013). Such selectivity may be governed by host genetic architectures that are heritable and transmitted maternally (Quigley et al. 2016, 2017a, b; Poland and Coffroth 2017). For example, a range of mechanisms associated with both innate and adaptive immune responses has been implicated in shaping invertebrate symbiont communities, including those involving T cells, Nod2, and defensins (as reviewed in Franzenburg et al. 2013), some of which have been associated with the establishment of coral–*Symbiodinium* symbioses (Wood-Charlson et al. 2006; Bay et al. 2011; Davy et al. 2012). Depending on the level to which such mechanisms are genetically determined, these mechanisms may potentially allow hosts to exclude some symbiont species, as in *B. asbestinum* (Poland and Coffroth 2017).

Preliminary findings suggested that most coral colonies are dominated by a single *Symbiodinium* type (Goulet 2006), with one or more additional types (from the same or a different clade) maintained at such low densities that they have generally gone undetected by the earlier genetic methods applied (LaJeunesse 2001; Santos et al. 2001; Goulet and Coffroth 2003a; Ulstrup and van Oppen 2003; Mieog et al. 2007). It is now clear that the majority of coral species host *Symbiodinium* from multiple clades and types (Baker and Romanski 2007; Silverstein et al. 2012). NGS estimates suggest that corals typically host at least 3–17 different symbiont types across their distributions (Kenkel et al. 2013; Green et al. 2014; Pochon et al. 2014; Thomas et al. 2014; Kennedy et al. 2016; Cunning et al. 2017b; Ziegler et al. 2017), with individual colonies harbouring between 2 and 17 different symbiont types (Arif et al. 2014; Boulotte et al. 2016; Kennedy et al. 2016; Quigley et al. 2017a) and eggs and juveniles typically hosting greater diversity (13–31 symbionts) than adults (Quigley et al. 2017a). These more sensitive methods have confirmed the almost ubiquitous presence of background *Symbiodinium* (Silverstein et al. 2012; Arif et al. 2014; Quigley et al. 2014; Boulotte et al. 2016), although the role of these low-abundance symbionts is controversial (Lee et al. 2016). Scleractinian corals also exhibit varying degrees of symbiont specificity, with some species establishing symbioses with a wider range of symbionts but others being relatively specific (Silverstein et al. 2012). Some corals are able to establish and maintain stable symbioses with a range of evolutionarily divergent *Symbiodinium* types (simultaneously within a single colony or separately in individual colonies) (Fabina et al. 2012, 2013). Key examples are *Orbicella annularis*, *O. faveolata* and *O. franksi* in the Caribbean (Rowan and Knowlton 1995; Rowan et al. 1997; Toller et al. 2001a) and *Acropora millepora* on the GBR (van Oppen et al. 2001; Berkelmans and van Oppen 2006). Other corals exhibit high specificity/fidelity to one or a few closely related *Symbiodinium* types, at least within a geographic region (populations in different geographic regions may evolve specificity for different symbiont types).

Changes in the dominant symbiont type harboured by a single coral colony may occur through changes in the relative abundance of symbiont types that are already present in host tissues (a process termed "shuffling"; Baker 2003) or by uptake of new symbiont types from the environment (a process called "switching"; Baker 2003). For scleractinian corals, recent evidence suggests that the temporal window for symbiont uptake persists into adulthood, even for vertically transmitting coral species (Boulotte et al. 2016), although uptake after about 4–6 months may be greatly reduced (Abrego et al. 2009a; McIlroy and Coffroth 2017). Evidence for exogenous uptake of new symbiont types by corals is also available for a Caribbean soft coral and a scleractinian (Lewis and Coffroth 2004; Coffroth et al. 2010), although in both cases the new symbiont was lost over time.

Despite not knowing the exact mechanism(s) responsible, it has been shown that the types of symbionts dominating coral tissues can change over time, most commonly in response to disturbance. Bleaching was shown to promote symbiont community changes following transplantation of Caribbean corals from deep to shallow water, and these changes were found to reduce mortality of corals compared to reciprocal transplants from shallow to deep, which did not bleach and did not change their symbionts (Baker 2001). Indeed, changes post-stress may be relatively routine, although complete turnover in symbiont communities may be rare and dependent on genetic constraints and the magnitude of the bleaching disturbance and the recovery environment (Cunning et al. 2015a). For example, some colonies of *Orbicella annularis*, *O. franksi* and, to a lesser extent, *O. faveolata* displayed changes in the *Symbiodinium* types dominating their tissues over a 6-year survey period in the Florida Keys. As no changes were observed in populations from the Bahamas, it was suggested that changes in Florida were linked to the greater environmental fluctuations that occur there (Thornhill et al. 2006b). Similarly, the longevity of symbiont changes following disturbance may be short-lived. A considerable increase in the occurrence of *Symbiodinium trenchi* (D1–4 = D1a) after the 1997–1998 and 2005 bleaching events was also detected in populations of *O. annularis*, but a reversion to pre-bleaching symbiont community composition took place over a number of years (Thornhill et al. 2006a; LaJeunesse et al. 2009).

Transplantation of colonies of the GBR coral, *Acropora millepora*, from two cooler reefs (a southern GBR and a mid-shelf reef in the central GBR) to a warm inshore reef was followed by severe bleaching of all transplanted colonies during the following austral summer ~6–11 months after transplantation (Berkelmans and van Oppen 2006). The transplants from the mid-shelf reef recovered with the *Symbiodinium* type they originally harboured (type C2*), but the transplants from the southern location recovered with a different symbiont type (clade D) than the one previously dominating their tissues (van Oppen ITS1 C2 is equivalent to LaJeunesse ITS2 type C3). With this change from *Symbiodinium* C2 to a type within clade D, the thermal tolerance of these corals increased by 1.0–1.5 °C, while transplants from the mid-shelf reef, which did not change symbiont type, maintained the same thermal stress response as before. The authors hypothesized that shuffling rather than switching was the most likely mechanism responsible for the change, as clade D *Symbiodinium* occurred at low abundance in the corals from the southern reef. Indeed, there may be a threshold abundance for symbiont shuffling before different types can rise to dominance following bleaching or disturbance (perhaps as low as

0.3%, Bay et al. 2016). In addition, differences in the extent of bleaching between the two transplanted populations (the transplants from the southern reef, which changed symbiont type, bleached more severely than the ones from the mid-shelf reef) may have triggered a change in the one population but not in the other (Toller et al. 2001b).

Differences in the photochemical dynamics of symbionts can also be excellent predictors of the tendency of corals to shuffle symbionts in favour of more thermotolerant symbionts following bleaching. In an experimental study of three species of Caribbean coral, 92.9% of variation in the degree of symbiont shuffling could be explained by the time-integrated photochemical advantage of *S. trenchi* (D1a) under heat stress (Cunning et al. 2017a, b). This robust empirical relationship between the performance of different symbionts and their fate in corals following heat stress shows that the likelihood of symbiont displacement by a thermotolerant symbiont is not just a function of symbionts that are increasing in dominance but also of the symbionts that are being displaced.

Similarly, in the Caribbean, shuffling from B1 to D1a was only brought on after severe bleaching and warmer recovery temperatures in corals with less than 5% abundance of D1a (Cunning et al. 2015a). Even vertically transmitting species from the genus *Pocillopora* shuffle background populations of *Symbiodinium* in response to bleaching (McGinley et al. 2012; Boulotte et al. 2016). Populations of the coral *Acropora palifera* in southern Taiwan show seasonal variation in the relative abundance of *Symbiodinium* species within clades C and D (Chen et al. 2005), while *O. annularis* across the Caribbean vary in symbionts hosted from clades B, C and D (Kennedy et al. 2016), suggesting that the symbiont shuffling response to environmental changes such as temperature, nutrients and turbidity routinely destabilizes symbiont communities and creates extreme variability. Shifts in *Symbiodinium* communities during bleaching may also be attributed to greater susceptibility of infection by opportunistic *Symbiodinium* in health-compromised corals due to stress (Toller et al. 2001b; LaJeunesse et al. 2010b). For example, dysbiosis may be caused by bleaching, facilitating the proliferation of particular *Symbiodinium* (e.g. $\mathrm{B1}_{Exaiptasia}$) that are later replaced by the original dominant taxa (C1b–c) once conditions return to normal (LaJeunesse et al. 2010b). While a change in community composition to favour a particular *Symbiodinium* type (e.g. D1a) is key to coral thermal tolerance (Silverstein et al. 2015), it is important to note that thermal tolerance can also be independent of the *Symbiodinium* community harboured (Kenkel et al. 2013), and instead reflects the host's prior thermal history (Barshis et al. 2013), or the host's physiological condition post temperature stress, as mediated by increased plasticity of host gene expression (Kenkel and Matz 2016).

In contrast to patterns of change in response to environmental conditions described above, temporally stable coral–algal associations have been documented for both spawning and brooding species. Thornhill et al. (2006b) observed no significant changes in symbiont communities within individual colonies of the broadcast spawners *Acropora palmata*, *A. cervicornis* and *Siderastrea siderea* over a period of 6 years, which included a bleaching event, despite diversity in symbiont types. The brooding species *Agaricia agaricites*, *Porites astreoides* and

Siderastrea radians showed complete spatial and temporal symbiont fidelity to one *Symbiodinium* type over a 3–4 year period, despite significant visual bleaching of *A. agaricites* colonies in 2005 (Thornhill et al. 2006b). *Symbiodinium* types harboured by colonies of the gorgonian coral, *Plexaura kuna*, remained unchanged over a 10-year period, both across different habitats and after transplantation to deeper water (Goulet and Coffroth 2003b) as did *Fungia scutaria* colonies transplanted from the Indo-Pacific to Jamaica 35 years prior (LaJeunesse et al. 2005). The same results were obtained for the scleractinian corals, *Pocillopora verrucosa* and *Pavona gigantea*, 1 year after transplantation to a different depth (Iglesias-Prieto et al. 2004), for either dominant or background types (Ziegler et al. 2014). The brooding coral, *Oulastrea crispata*, also showed no change in its symbiont community over a 1-year survey period (Chen et al. 2003). Reduced bleaching prevalence can also occur in temporally stable coral–algal associations, for example, in Caribbean octocorals harbouring various clones within *Symbiodinium* type B1 (Lasker et al. 1984; Lasker 2003; Goulet and Coffroth 2004). However, as most shuffling events may be related predominantly to low-abundance types (Boulotte et al. 2016), it may be important to reassess the stability of these symbioses using qPCR or NGS methods that are more sensitive to the detection of low-abundance types that may have been initially overlooked in previous studies using DNA fingerprinting, DGGE and/or RFLP. It is also important to assess how stable these partnerships may remain if subjected to repeated episodes of disturbance, such as predicted under climate change scenarios (Chap. 13).

6.6 Prediction of Changes in *Symbiodinium* Diversity and Distribution over the Next Several Decades

The findings summarized above show that changes in the dominant and background symbiont types within individual coral colonies can occur and that the triggers to shuffle include the strength of the disturbance, recovery temperature, initial abundances of the constituent symbiont community and differences in their photosynthetic performance. However, there are also clear differences in the ability of different coral species to readily change their symbiont community and the speed with which these changes can occur. Moderate to high heritability of the *Symbiodinium* community in broadcast-spawning species with both horizontal and vertical transmission (Poland and Coffroth 2017; Quigley et al. 2017a), as well as in a planulating, vertically transmitting species (Quigley et al. 2018), demonstrates that there is ample material for selection. However, brooding, vertically transmitting species may be more limited in their capacity to change, given larger heritability estimates (and, therefore, larger host genetic influence) (Quigley et al. 2017a). A growing body of evidence supports the notion that a change in the dominant symbiont type is accompanied by a change in the physiological response of the holobiont (Table 6.1). Goulet et al. (2005) showed that the same clone of *Exaiptasia pallida* (formerly *Aiptasia pallida*, Grajales and Rodriguez 2014) responds differently to elevated temperatures with different clades of *Symbiodinium*. Those harbouring *Symbiodinium* ITS-type A4 had higher photosynthetic rates than those with *Symbiodinium minutum* (B1) at elevated temperatures. Similarly, McIlroy and Coffroth (2017) demonstrated that *Orbicella faveolata* juveniles with *S. microadriaticum* increased surface area faster than those harbouring *S. minutum*. Berkelmans and van Oppen (2006) showed that shuffling symbionts from *Symbiodinium* clade C to clade D in *Acropora millepora* colonies increased thermal tolerance by 1.0–1.5 °C. An increase in thermal tolerance by 1.5 °C could slow the onset of thermal stress by 50–80 years (Chap. 13); however, severe bleaching conditions are expected to manifest annually in over 75% of reefs worldwide by 2070, giving coral reefs little time for adaptation (van Hooidonk et al. 2016). Moreover, shuffling from C symbiont types to D in this coral species comes at a cost of reduced growth (Little et al. 2004; Jones and Berkelmans 2010; but see Cunning et al. 2015a), lipid stores and reproduction (Jones and Berkelmans 2011) and is linked to variable electron transport rates and light absorption capacity (Jones and Berkelmans 2012) in D and C2 *Symbiodinium* in both laboratory and field settings. Indeed, modelling of the ecological trade-offs between growth and thermal tolerance reveals that the increased abundance of thermally tolerant D1a may detrimentally affect reef recovery in the Caribbean through significant reductions in growth rate (Ortiz et al. 2013).

At this stage, it is still difficult to make reliable predictions about future changes in the composition of symbiont types harboured by reef corals due to the uncertainties involved. In the short term, corals with flexible symbioses may shuffle or switch symbionts, and an increase in the abundance of thermally tolerant symbiont types (such as some types within clades C and D) is expected with an increasing frequency of bleaching conditions. However, coral hosts that are less capable of switching or shuffling may become (locally) extinct, leading to a loss in the diversity of both corals and symbionts on many reefs [including changes in community composition (Ruzicka et al. 2013; Lenz et al. 2015)] unless these holobionts are able to adapt by other means to the changed environmental conditions. Indeed, the preferential survival of host and symbiont generalists over specialists accompanied by high background diversity of symbionts maximizes the probability of community stability (Fabina et al. 2013, but see Putnam et al. 2012 who found that flexibility correlated with increased sensitivity). The potential to adapt to increasing sea surface temperatures depends mainly on the extent of heritable genetic variation for heat tolerance, the generation times of the coral host and algal endosymbionts and the strength of selection. Generation times are likely to be long for corals, yet heat tolerance has been found to be heritable in the coral host, with offspring produced from parents from warmer regions providing a tenfold increase in thermal tolerance to offspring (Dixon et al. 2015). However, the coral host exhibited reduced heritability of multiple traits related to thermal tolerance compared to *Symbiodinium* C2 and D (Császár et al. 2010). Thus, due to the enormous population sizes and rapid turnover of symbiont populations in hospite, the frequency of somatic mutations in these algal symbionts may be significant in terms of adaptive evolution (van Oppen et al. 2011; Levin et al. 2016; Chakravarti et al. 2017). Selection on genotypes that differ physiologically (e.g. those that have undergone an advantageous mutation) has been shown to lead to extremely fast changes in relative abundances of those genotypes in *Chlorella* and *Emiliania huxleyi* (Meyer et al. 2006; Lohbeck et al. 2012). Finally, virtually nothing is known about the population structure and reproductive dynamics of free-living *Symbiodinium*, although sexual recombination in hospite is probable (Baums et al. 2014; Wilkinson et al. 2015). To examine whether bleaching events represent bottlenecks for symbiont populations and whether thermally tolerant strains can spread across reef systems, population genetic approaches should be employed in future research.

6.7 Conclusions, Overview of Recent Progress and New Questions

- *Symbiodinium* play an important role in determining the physiological performance of corals. Although substantial progress has been made in untangling the relative contributions of host versus symbiont to the bleaching response, the molecular mechanisms and feedbacks between host and symbiont that lead to cell expulsion remain poorly understood. Furthermore, the potential for bacterial and viral partners to influence the bleaching process is still largely unknown.
- *Symbiodinium* exhibit significant diversity within phylogenetic clades. Many of these intra-cladal variants ("types", and in some cases species) exhibit different physiological tolerances. Significant progress has been made in understanding the role of key symbionts (e.g. *S. trenchi*-D1a) in bleaching response and recovery. While the development of qPCR, microsatellite and NGS techniques has accelerated our understanding of intra-cladal diversity within *Symbiodinium* and the ability to link taxonomy to symbiont and holobiont physiology, little information exists as to the roles that the majority of *Symbiodinium* types may have in holobiont physiology and the bleaching response. Current and expanding genome sequencing efforts will enable the discovery of single-copy markers, further facilitating the discovery and resolution of *Symbiodinium* taxonomic diversity.
- Within our range of detection, it is now known that corals typically host many different symbiont types and that *Symbiodinium* communities are predominately made up of a single genotype per *Symbiodinium* type in the Caribbean but may comprise multiple genotypes per type in the Indo-Pacific. However, genetically distinct symbionts have been detected at lower abundance within coral colonies, and advances in sequencing have led to significant improvements in the detection and characterization of background types, including their prevalence and roles in bleaching recovery, as moderated by shuffling and/or uptake of exogenous symbionts (switching). However, controversy remains as to the role and importance of the many hundreds of background types and variants that have been detected to date.

- Symbiont shuffling is an important mechanism by which some (but likely not all) coral species can increase their thermal tolerance. On its own, however, this is insufficient to cope with increasingly higher sea surface temperatures as a result of global warming, as evidenced by the high thermal bleaching-related coral mortality seen across the globe in 2014–2016 and on the GBR again in 2017. However, more work is needed to determine which coral species can shuffle or switch symbionts and the ecological trade-offs between growth and reproduction associated with hosting altered symbionts and assess if these changes can keep pace with the current rate and severity of ocean warming.
- Significant improvements have been made in understanding the adaptive potential and limits of corals. Although *Symbiodinium* demonstrate both positive acclimatory and adaptive potential at the type and community level, further work is needed to assess if this translates to improvements in holobiont survival over long timescales. Finally, efforts should also be directed at integrating these results into initiatives aimed at assisted evolution, including assisted gene flow.

Acknowledgements This work was funded by NSF grants to A.C.B. (OCE-1358699 and OCE-1447306) and M.A.C. (OCE-04-24996 and OCE-09-26822) and funding through the ARC CoE for Coral Reef Studies to B.L.W. (CE140100020).

References

Abrego D, Ulstrup KE, Willis BL, van Oppen MJH (2008) Species–specific interactions between algal endosymbionts and coral hosts define their bleaching response to heat and light stress. Proc R Soc B 275:2273–2282. https://doi.org/10.1098/rspb.2008.0180

Abrego D, van Oppen MJH, Willis BL (2009a) Onset of algal endosymbiont specificity varies among closely related species of Acropora corals during early ontogeny. Mol Ecol 18:3532–3543. https://doi.org/10.1111/j.1365-294X.2009.04276.x

Abrego D, van Oppen MJH, Willis BL (2009b) Highly infectious symbiont dominates initial uptake in coral juveniles. Mol Ecol 18:3518–3531. https://doi.org/10.1111/j.1365-294X.2009.04275.x

Abrego D, Willis BL, van Oppen MJH (2012) Impact of light and temperature on the uptake of algal symbionts by coral juveniles. PLoS One 7. https://doi.org/10.1371/journal.pone.0050311

Andras JP, Kirk NL, Harvell CD (2011) Range-wide population genetic structure of *Symbiodinium* associated with the Caribbean sea fan coral, *Gorgonia ventalina*. Mol Ecol 20:2525–2542. https://doi.org/10.1111/j.1365-294X.2011.05115.x

Anthony KRN, Kline DI, Diaz-Pulido G, Dove S, Hoegh-Guldberg O (2008) Ocean acidification causes bleaching and productivity loss in coral reef builders. Proc Natl Acad Sci U S A 105:17442–17446. https://doi.org/10.1073/pnas.0804478105

Anthony K, Hoogenboom MO, Maynard JA, Grottoli AG, Middlebrook R (2009) Energetics approach to predicting mortality risk from environmental stress: a case study of coral bleaching. Funct Ecol 23:539–550. https://doi.org/10.1111/j.1365-2435.2008.01531.x

Aranda M, Li Y, Liew YJ, Baumgarten S, Simakov O, Wilson MC et al (2016) Genomes of coral dinoflagellate symbionts highlight evolutionary adaptations conducive to a symbiotic lifestyle. Sci Rep 6. https://doi.org/10.1038/srep39734

Arif C, Daniels C, Bayer T, Hinestroza E, Barbrook A, Howe CJ et al (2014) Assessing *Symbiodinium* diversity in scleractinian corals via Next Generation Sequencing based genotyping of the ITS2 rDNA region. Mol Ecol 23:4418–4433. https://doi.org/10.1111/mec.12869

Babcock RC, Heyward AJ (1986) Larval development of certain gamete-spawning scleractinian corals. Coral Reefs 5:111–116. https://doi.org/10.1007/BF00298178

Baillie BK, Belda-Baillie CA, Maruyama T (2000a) Conspecificity and Indo-Pacific distribution of *Symbiodinium* genotypes (Dinophyceae) from giant clams. J Phycol 36:1153–1161. https://doi.org/10.1046/j.1529-8817.2000.00010.x

Baillie BK, Belda-Baillie CA, Silvestre V, Sison M, Gomez A, Gomez E et al (2000b) Genetic variation in *Symbiodinium* isolates from giant clams based on random-amplified-polymorphic DNA (RAPD) patterns. Mar Biol 136:829–836. https://doi.org/10.1007/s002270000290

Baird AH, Bhagooli R, Ralph PJ, Takahashi S (2009a) Coral bleaching: the role of the host. Trends Ecol Evol 24:16–20. https://doi.org/10.1016/j.tree.2008.09.005

Baird AH, Guest JR, Willis BL (2009b) Systematic and biogeographical patterns in the reproductive biology of scleractinian corals. Annu Rev Ecol Evol Syst 40:551–571. https://doi.org/10.1146/annurev.ecolsys.110308.120220

Baker AC (2001) Reef corals bleach to survive change. Nature 411:765–766. https://doi.org/10.1038/35081151

Baker AC (2003) Flexibility and specificity in coral-algal symbiosis: diversity, ecology, and biogeography of *Symbiodinium*. Annu Rev Ecol Evol Syst 34:661–689. https://doi.org/10.1146/annurev.ecolsys.34.011802.132417

Baker AC, Cunning R (2016) Coral "bleaching" as a generalized stress response to environmental disturbance. In: Woodley CM, Downs CA, Bruckner AW, Porter JW, Galloway SB (eds) Diseases of coral. Wiley pp 396–409

Baker AC, Romanski AM (2007) Multiple symbiotic partnerships are common in scleractinian corals, but not in octocorals: comment on Goulet (2006). Mar Ecol Prog Ser 335:237–242. https://doi.org/10.3354/meps335237

Baker AC, Rowan R (1997) Diversity of symbiotic dinoflagellates (zooxanthellae) in scleractinian corals of the Caribbean and eastern Pacific. In: Proceedings of the 8th international coral reef symposium, Panama, pp 1301–1305

Baker AC, Rowan R, Knowlton N (1997) Symbiosis ecology of two Caribbean acroporid corals. In: Proceedings of the 8th international coral reef symposium, Panama, pp 1295–1300

Baker AC, Starger CJ, McClanahan TR, Glynn PW (2004a) Coral reefs: corals' adaptive response to climate change. Nature 430:741. https://doi.org/10.1038/430741a

Baker AC, Jones SH, Lee TS, Abuzinada AH, Joubert E, Krupp F (2004b) Symbiont diversity in Arabian corals and its relation to patterns of contemporary and historical environmental stress. In: Abuzinada AH, Jouberte E, Krupp F (eds) Proceedings of the international symposium on the extent and impact of coral bleaching in the Arabian Region. National Commission for Wildlife Conservation and Development, Riyadh, pp 24–36

Banaszak AT, LaJeunesse TC, Trench RK (2000) The synthesis of mycosporine-like amino acids (MAAs) by cultured, symbiotic dinoflagellates. J Exp Mar Biol Ecol 249:219–233. https://doi.org/10.1016/s0022-0981(00)00192-1

Banaszak AT, Santos MB, LaJeunesse TC, Lesser MP (2006) The distribution of mycosporine-like amino acids (MAAs) and the phylogenetic identity of symbiotic dinoflagellates in cnidarian hosts from the Mexican Caribbean. J Exp Mar Biol Ecol 337:131–146. https://doi.org/10.1016/j.jembe.2006.06.014

Barbrook AC, Visram S, Douglas AE, Howe CJ (2006) Molecular diversity of dinoflagellate symbionts of Cnidaria: the psbA minicircle of *Symbiodinium*. Protist 157:159–171. https://doi.org/10.1016/j.protis.2005.12.002

Barbrook AC, Voolstra CR, Howe CJ (2014) The chloroplast genome of a *Symbiodinium* sp. clade C3 isolate. Protist 165:1–13. https://doi.org/10.1016/j.protis.2013.09.006

Barneah O, Weis VM, Perez S, Benayahu Y (2004) Diversity of dinoflagellate symbionts in Red Sea soft corals: mode of symbiont acquisition matters. Mar Ecol Prog Ser 275:89–95. https://doi.org/10.3354/meps275089

Barnes DJ, Chalker BE (1990) Calcification and photosynthesis in reef-building corals and algae. In: Dubinsky Z (ed) Coral Reefs. Ecosystems of the World 25. Elsevier, pp 109–131
Barshis DJ, Ladner JT, Oliver TA, Seneca FO, Traylor-Knowles N, Palumbi SR (2013) Genomic basis for coral resilience to climate change. Proc Natl Acad Sci U S A 110:1387–1392. https://doi.org/10.1073/pnas.1210224110
Baums IB, Devlin-Durante MK, LaJeunesse TC (2014) New insights into the dynamics between reef corals and their associated dinoflagellate endosymbionts from population genetic studies. Mol Ecol 23:4203–4215. https://doi.org/10.1111/mec.12788
Bay LK, Cumbo VR, Abrego D, Kool JT, Ainsworth TD, Willis BL (2011) Infection dynamics vary between *Symbiodinium* types and cell surface treatments during establishment of endosymbiosis with coral larvae. Diversity 3:356–374. https://doi.org/10.3390/d3030356
Bay LK, Doyle J, Logan M, Berkelmans R (2016) Recovery from bleaching is mediated by threshold densities of background thermo-tolerant symbiont types in a reef-building coral. Roy Soc Open Sci 3. https://doi.org/10.1098/rsos.160322
Berkelmans R, van Oppen MJH (2006) The role of zooxanthellae in the thermal tolerance of corals: a "nugget of hope" for coral reefs in an era of climate change. Proc R Soc B 273:2305–2312. https://doi.org/10.1098/prsb.2006.3567
Bhagooli R, Hidaka M (2003) Comparison of stress susceptibility of in hospite and isolated zooxanthellae among five coral species. J Exp Mar Biol Ecol 291:181–197. https://doi.org/10.1016/S0022-0981(03)00121-7
Bieri T, Onishi M, Xiang T, Grossman AR, Pringle JR (2016) Relative contributions of various cellular mechanisms to loss of algae during cnidarian bleaching. PLoS One 11. https://doi.org/10.1371/journal.pone.0152693
Blank RJ, Huss VA (1989) DNA divergency and speciation in *Symbiodinium* (*Dinophyceae*). Plant Syst Evol 163:153–163. https://doi.org/10.1007/BF00936511
Blank RJ, Trench RK (1985) Speciation and symbiotic dinoflagellates. Science 229:656–659. https://doi.org/10.1126/science.229.4714.656
Blank RJ, Trench RK (1986) Nomenclature of endosymbiotic dinoflagellates. Taxon 35:286–294. https://doi.org/10.2307/1221270
Bo M, Baker AC, Gaino E, Wirshing HH, Scoccia F, Bavestrello G (2011) First description of algal mutualistic endosymbiosis in a black coral (*Anthozoa: Antipatharia*). Mar Ecol Prog Ser 435:1–11. https://doi.org/10.3354/meps09228
Boldt L, Yellowlees D, Leggat W (2008) Measuring *Symbiodinium* sp. gene expression patterns with quantitative real-time PCR. Proceedings of the 11th international coral reef symposium, Fort Lauderdale, pp 118–122
Bongaerts P, Riginos C, Ridgway T, Sampayo EM, van Oppen MJH, Englebert N et al (2010) Genetic divergence across habitats in the widespread coral *Seriatopora hystrix* and its associated *Symbiodinium*. PLoS One 5. https://doi.org/10.1371/journal.pone.0010871
Bongaerts P, Frade PR, Hay KB, Englebert N, Latijnhouwers KRW, Bak RPM et al (2015a) Deep down on a Caribbean reef: lower mesophotic depths harbor a specialized coral-endosymbiont community. Sci Rep 5. https://doi.org/10.1038/srep07652
Bongaerts P, Carmichael M, Hay KB, Tonk L, Frade PR, Hoegh-Guldberg O (2015b) Prevalent endosymbiont zonation shapes the depth distributions of scleractinian coral species. Roy Soc Open Sci 2. https://doi.org/10.1098/rsos.140297
Boulotte NM, Dalton SJ, Carroll AG, Harrison PL, Putnam HM, Peplow LM et al (2016) Exploring the *Symbiodinium* rare biosphere provides evidence for symbiont switching in reef-building corals. ISME J 10:2693–2701. https://doi.org/10.1038/ismej.2016.54
Brazeau DA, Lasker HR (1990) Sexual reproduction and external brooding by the Caribbean gorgonian *Briareum asbestinum*. Mar Biol 104:465–474. https://doi.org/10.1007/BF01314351
Brown BB, Dunne RD, Goodson MG, Douglas AD (2002) Experience shapes the susceptibility of a reef coral to bleaching. Coral Reefs 21:119–126. https://doi.org/10.1007/s00338-002-0215-z
Buddemeier RW, Fautin DG (1993) Coral bleaching as an adaptive mechanism. Bioscience 43:320–326. https://doi.org/10.2307/1312064

Buddemeier RW, Smith SV (1999) Coral adaptation and acclimatization: a most ingenious paradox. Am Zool 39:1–9

Buddemeier RW, Baker AC, Fautin DG, Jacobs JR (2004) The adaptive hypothesis of bleaching. In: Rosenberg E, Loya Y (eds) Coral health and disease. Springer, Berlin, pp 427–444

Burnett WJ (2002) Longitudinal variation in algal symbionts (zooxanthellae) from the Indian Ocean zoanthid *Palythoa caesia*. Mar Ecol Prog Ser 234:105–109. https://doi.org/10.3354/meps234105

Byler KA, Carmi-Veal M, Fine M, Goulet TL (2013) Multiple symbiont acquisition strategies as an adaptive mechanism in the coral *Stylophora pistillata*. PLoS One 8. https://doi.org/10.1371/journal.pone.0059596

Cantin N, van Oppen MJH, Willis B, Mieog J, Negri A (2009) Juvenile corals can acquire more carbon from high-performance algal symbionts. Coral Reefs 28:405–414. https://doi.org/10.1007/s00338-009-0478-8

Carlos AA, Baillie BK, Kawachi M, Maruyama T (1999) Phylogenetic position of *Symbiodinium* (*Dinophyceae*) isolates from Tridacnids (*Bivalvia*), Cardiids (*Bivalvia*), a sponge (*Porifera*), a soft coral (*Anthozoa*), and a free-living strain. J Phycol 35:1054–1062. https://doi.org/10.1046/j.1529-8817.1999.3551054.x

Chakravarti LJ, Beltran VH, Oppen MJH (2017) Rapid thermal adaptation in photosymbionts of reef-building corals. Glob Chang Biol 23:4675–4688. https://doi.org/10.1111/gcb.13702

Chen CA, Lam KK, Nakano Y, Tsai W-S (2003) A stable association of the stress-tolerant zooxanthellae, *Symbiodinium* clade D, with the low-temperature-tolerant coral, *Oulastrea crispata* (Scleractinia: Faviidae) in subtropical non-reefal coral communities. Zool Stud 42:540–550

Chen CA, Wang AT, Fang LS, Yang YW (2005) Fluctuating algal symbiont communities in *Acropora palifera* (Scleractinia: Acroporidae) from Taiwan. Mar Ecol Prog Ser 295:113–121. https://doi.org/10.3354/meps295113

Coffroth MA, Santos SR (2005) Genetic diversity of symbiotic dinoflagellates in the genus *Symbiodinium*. Protist 156:19–34. https://doi.org/10.1016/j.protis.2005.02.004

Coffroth MA, Santos S, Goulet T (2001) Early ontogenetic expression of specificity in a cnidarian-algal symbiosis. Mar Ecol Prog Ser 222:85–96. https://doi.org/10.3354/meps222085

Coffroth MA, Lewis CF, Santos SR, Weaver JL (2006) Environmental populations of symbiotic dinoflagellates in the genus *Symbiodinium* can initiate symbioses with reef cnidarians. Curr Biol 16:R985–R987. https://doi.org/10.1016/j.cub.2006.10.049

Coffroth MA, Poland DM, Petrou EL, Brazeau DA, Holmberg JC (2010) Environmental symbiont acquisition may not be the solution to warming seas for reef-building corals. PLoS One 5. https://doi.org/10.1371/journal.pone.0013258

Cooper TF, Berkelmans R, Ulstrup KE, Weeks S, Radford B, Jones AM et al (2011) Environmental factors controlling the distribution of *Symbiodinium* harboured by the coral *Acropora millepora* on the Great Barrier Reef. PLoS One 6. https://doi.org/10.1371/jpurnal.pone.0025536

Correa A, McDonald M, Baker A (2009) Development of clade-specific *Symbiodinium* primers for quantitative PCR (qPCR) and their application to detecting clade symbionts in Caribbean corals. Mar Biol 156:2403–2411. https://doi.org/10.1007/s00227-009-1263-5

Császár NBM, Ralph PJ, Frankham R, Berkelmans R, van Oppen MJH (2010) Estimating the potential for adaptation of corals to climate warming. PLoS One 5. https://doi.org/10.1371/journal.pone.0009751

Cumbo VR, Baird AH, van Oppen MJH (2013) The promiscuous larvae: flexibility in the establishment of symbiosis in corals. Coral Reefs 32:111–120. https://doi.org/10.1007/s00338-012-0951-7

Cunning R, Baker AC (2013) Excess algal symbionts increase the susceptibility of reef corals to bleaching. Nat Clim Chang 3:259–262. https://doi.org/10.1038/nclimate1711

Cunning R, Gillette P, Capo T, Galvez K, Baker AC (2015a) Growth tradeoffs associated with thermotolerant symbionts in the coral *Pocillopora damicornis* are lost in warmer oceans. Coral Reefs 34:155–160. https://doi.org/10.1007/s00338-014-1216-4

Cunning R, Silverstein RN, Baker AC (2015b) Investigating the causes and consequences of symbiont shuffling in a multi-partner reef coral symbiosis under environmental change. Proc R Soc B 282. https://doi.org/10.1098/rspb.2014.1725

Cunning R, Yost DM, Guarinello ML, Putnam HM, Gates RD (2015c) Variability of *Symbiodinium* communities in waters, sediments, and corals of thermally distinct reef pools in American Samoa. PLoS One 10. https://doi.org/10.1371/journal.pone.0145099

Cunning R, Silverstein RN, Baker AC (2017a) Symbiont shuffling linked to differential photochemical dynamics of *Symbiodinium* in three Caribbean reef corals. Coral Reefs. https://doi.org/10.1007/s00338-017-1640-3

Cunning R, Gates RD, Edmunds PJ (2017b) Using high-throughput sequencing of ITS2 to describe *Symbiodinium* metacommunities in St. John, US Virgin Islands. PeerJ 5. https://doi.org/10.7717/peerj.3472

Davies SW, Wham DC, Kanke MR, Matz MV (2016) Ecological factors rather than physical barriers to dispersal shape genetic structure of algal symbionts in horizontally-transmitting corals. bioRxiv. https://doi.org/10.1101/037994

Davy SK, Allemand D, Weis VM (2012) Cell biology of cnidarian-dinoflagellate symbiosis. Microbiol Mol Biol Rev 76:229–261. https://doi.org/10.1128/MMBR.05014-11

DeBoer TS, Baker AC, Erdmann MV, Ambariyanto, Jones PR, Barber PH (2012) Patterns of *Symbiodinium* distribution in three giant clam species across the biodiverse Bird's Head region of Indonesia. Mar Ecol Prog Ser 444:117–132. https://doi.org/10.3354/meps09413

del C Gómez-Cabrera M, Ortiz JC, WKW L, Ward S, Hoegh-Guldberg O (2008) Acquisition of symbiotic dinoflagellates (*Symbiodinium*) by juveniles of the coral *Acropora longicyathus*. Coral Reefs 27:219–226. https://doi.org/10.1007/s00338-007-0315-x

Dixon GB, Davies SW, Aglyamova GV, Meyer E, Bay LK, Matz MV (2015) Genomic determinants of coral heat tolerance across latitudes. Science 348:1460–1462. https://doi.org/10.1126/science.1261224

Dunn SR, Weis VM (2009) Apoptosis as a post-phagocytic winnowing mechanism in a coral–dinoflagellate mutualism. Environ Microbiol 11:268–276. https://doi.org/10.1111/j.1462-2920.2008.01774.x

Fabina NS, Putnam HM, Franklin EC, Stat M, Gates RD (2012) Transmission mode predicts specificity and interaction patterns in coral-*Symbiodinium* networks. PLoS One 7. https://doi.org/10.1371/journal.pone.0044970

Fabina NS, Putnam HM, Franklin EC, Stat M, Gates RD (2013) Symbiotic specificity, association patterns, and function determine community responses to global changes: defining critical research areas for coral *Symbiodinium* symbioses. Glob Chang Biol 19:3306–3316. https://doi.org/10.1111/gcb.12320

Fabricius KE, Klumpp DW (1995) Widespread mixotrophy in reef-inhabiting soft corals: the influence of depth, and colony expansion and contraction on photosynthesis. Mar Ecol Prog Ser 125:195–204. https://doi.org/10.3354/meps125195

Fabricius KE, Mieog JC, Colin PL, Idip D, van Oppen MJH (2004) Identity and diversity of coral endosymbionts (zooxanthellae) from three Palauan reefs with contrasting bleaching, temperature and shading histories. Mol Ecol 13:2445–2458. https://doi.org/10.1111/j.1365-294X.2004.02230.x

Fadlallah YH (1983) Sexual reproduction, development and larval biology in scleractinian corals. Coral Reefs 2:129–150. https://doi.org/10.1007/BF00336720

Finney JC, Pettay DT, Sampayo EM, Warner ME, Oxenford HA, LaJeunesse TC (2010) The relative significance of host-habitat, depth, and geography on the ecology, endemism, and speciation of coral endosymbionts in the genus *Symbiodinium*. Microb Ecol 60:250–263. https://doi.org/10.1007/s00248-010-9681-y

FitzPatrick SK, Liberatore KL, Garcia JR, Burghardt I, Colman DR, Moquin SA et al (2012) *Symbiodinium* diversity in the soft coral Heteroxenia sp. and its nudibranch predator *Phyllodesmium lizardensis*. Coral Reefs 31:895–905. https://doi.org/10.1007/s00338-012-0913-0

Frade PR, De Jongh F, Vermeulen F, van Bleijswijk J, Bak RPM (2008) Variation in symbiont distribution between closely related coral species over large depth ranges. Mol Ecol 17:691–703. https://doi.org/10.1111/j.1365-294X.2007.03612.x

Frank U, Oren U, Loya Y, Rinkevich B (1997) Alloimmune maturation in the coral *Stylophora pistillata* is achieved through three distinctive stages, 4 months post–metamorphosis. Proc R Soc B 264:99–104. https://doi.org/10.1098/prsb.1997.0015

Franklin EC, Stat M, Pochon X, Putnam HM, Gates RD (2012) GeoSymbio: a hybrid, cloud-based web application of global geospatial bioinformatics and ecoinformatics for *Symbiodinium*-host symbioses. Mol Ecol Resour 12:369–373. https://doi.org/10.1111/j.1755-0998.2011.03081.x

Franzenburg S, Walter J, Künzel S, Wang J, Baines JF, Bosch TCG et al (2013) Distinct antimicrobial peptide expression determines host species-specific bacterial associations. Proc Natl Acad Sci U S A 110:E3730–E3738. https://doi.org/10.1073/pnas.1304960110

Freudenthal H (1962) *Symbiodinium* gen. nov. and *Symbiodinium microadriaticum* sp. nov., a zooxanthella: taxonomy, life cycle, and morphology. J Eukaryot Microbiol 9:45–52. https://doi.org/10.1111/j.1550-7408.1962.tb02579.x

Gierz SL, Forêt S, Leggat W (2017) Transcriptomic analysis of thermally stressed *Symbiodinium* reveals differential expression of stress and metabolism genes. Front Plant Sci 8. https://doi.org/10.3389/fpls.2017.00271

Glynn PW, Maté JL, Baker AC, Calderon MO (2001) Coral bleaching and mortality in Panama and Ecuador during the 1997-1998 El Niño Southern Oscillation event: spatial/temporal patterns and comparisons with the 1982-1983 event. Bull Mar Sci 69:79–109

Gou W, Sun J, Li X, Zhen Y, Xin Z, Yu Z et al (2003) Phylogenetic analysis of a free-living strain of *Symbiodinium* isolated from Jiaozhou Bay, P.R. China. J Exp Mar Biol Ecol 296:135–144. https://doi.org/10.1016/S0022-098(03)00242-9

Goulet TL (2006) Most corals may not change their symbionts. Mar Ecol Prog Ser 321:1–7. https://doi.org/10.3354/meps321001

Goulet TL, Coffroth MA (2003a) Genetic composition of zooxanthellae between and within colonies of the octocoral *Plexaura kuna*, based on small subunit rDNA and multilocus DNA fingerprinting. Mar Biol 142:233–239. https://doi.org/10.1007/s00227-002-0936-0

Goulet TL, Coffroth MA (2003b) Stability of an octocoral-algal symbiosis over time and space. Mar Ecol Prog Ser 250:117–124. https://doi.org/10.3354/meps250117

Goulet TL, Coffroth MA (2004) The genetic identity of dinoflagellate symbionts in Caribbean octocorals. Coral Reefs 23:465–472. https://doi.org/10.1007/s00338-004-0408-8

Goulet TL, Cook CB, Goulet D (2005) Effect of short-term exposure to elevated temperatures and light levels on photosynthesis of different host-symbiont combinations in the *Aiptasia pallida*/*Symbiodinium* symbiosis. Limnol Oceanogr 50:1490–1498. https://doi.org/10.4319/lo.2005.50.5.1490

Goulet TL, Simmons C, Goulet D (2008) Worldwide biogeography of *Symbiodinium* in tropical octocorals. Mar Ecol Prog Ser 355:45–58. https://doi.org/10.3354/meps07214

Grajales A, Rodriguez E (2014) Morphological revision of the genus *Aiptasia* and the family Aiptasiidae (Cnidaria, Actiniaria, Metridioidea). Zootaxa 3826:55–100. https://doi.org/10.11646/zootaxa.3826.1.2

Granados-Cifuentes C, Neigel J, Leberg P, Rodriguez-Lanetty M (2015) Genetic diversity of free-living *Symbiodinium* in the Caribbean: the importance of habitats and seasons. Coral Reefs 34:927–939. https://doi.org/10.1007/s00338-015-1291-1

Green EA, Davies SW, Matz MV, Medina M (2014) Quantifying cryptic *Symbiodinium* diversity within *Orbicella faveolata* and *Orbicella franksi* at the Flower Garden Banks, Gulf of Mexico. PeerJ 2. https://doi.org/10.7717/peerj.386

Grégoire V, Schmacka F, Coffroth MA, Karsten U (2017) Photophysiological and thermal tolerance of various genotypes of the coral endosymbiont *Symbiodinium* sp. (Dinophyceae). J Appl Phycol 29:1893–1905. https://doi.org/10.1007/s10811-017-1127-1

Grottoli AG, Rodrigues LJ, Palardy JE (2006) Heterotrophic plasticity and resilience in bleached corals. Nature 440:1186–1189. https://doi.org/10.1038/nature04565

Grupstra CGB, Coma R, Ribes M, Leydet KP, Parkinson JE, McDonald K et al (2017) Evidence for coral range expansion accompanied by reduced diversity of *Symbiodinium* genotypes. Coral Reefs 36:981–985. https://doi.org/10.1007/s00338-017-1589-2

Gutiérrez-Rodríguez C, Lasker HR (2004) Reproductive biology, development, and planula behavior in the Caribbean gorgonian *Pseudopterogorgia elisabethae*. Invertebr Biol 123:54–67. https://doi.org/10.1111/j.1744-7410.2004.tb00141.x

Hansen G, Daugbjerg N (2009) *Symbiodinium natans* sp. nov.: a "free-living" dinoflagellate from Tenerife (Northeast-Atlantic Ocean). J Phycol 45:251–263. https://doi.org/10.1111/j.1529-8817.2008.00621.x

Harrison PL, Wallace CC (1990) Reproduction, dispersal and recruitment of scleractinian corals. In: Dubinsky Z (ed) Coral Reefs. Ecosystems of the World 25. Elsevier, pp 133–207

Harvell D, Jordán-Dahlgren E, Merkel S, Rosenberg E, Raymundo L, Smith G et al (2007) Coral disease, environmental drivers, and the balance between coral and microbial associates. Oceanography 20:172–195. https://doi.org/10.5670/oceanog.2007.91

Hirose M, Kinzie R, Hidaka M (2001) Timing and process of entry of zooxanthellae into oocytes of hermatypic corals. Coral Reefs 20:273–280. https://doi.org/10.1007/s003380100171

Hirose M, Reimer JD, Hidaka M, Suda S (2008) Phylogenetic analyses of potentially free-living *Symbiodinium* spp. isolated from coral reef sand in Okinawa, Japan. Mar Biol 155:105–112. https://doi.org/10.1007/s00227-008-1011-2

Hoegh-Guldberg O (1999) Climate change, coral bleaching and the future of the world's coral reefs. Mar Freshw Res 50:839–866. https://doi.org/10.1071/MF99078

Hoogenboom MO, Campbell DA, Beraud E, DeZeeuw K, Ferrier-Pagès C (2012) Effects of light, food availability and temperature stress on the function of photosystem II and photosystem I of coral symbionts. PLoS One 7. https://doi.org/10.1371/journal.pone.0030167

Howells EJ, van Oppen MJH, Willis BL (2009) High genetic differentiation and cross-shelf patterns of genetic diversity among Great Barrier Reef populations of *Symbiodinium*. Coral Reefs 28:215–225. https://doi.org/10.1007/s00338-008-0450-z

Howells EJ, Beltran VH, Larsen NW, Bay LK, Willis BL, van Oppen MJH (2012) Coral thermal tolerance shaped by local adaptation of photosymbionts. Nature. Clim Chang 2:116–120. https://doi.org/10.1038/nclimate1330

Howells EJ, Willis BL, Bay LK, van Oppen MJH (2013) Spatial and temporal genetic structure of *Symbiodinium* populations within a common reef-building coral on the Great Barrier Reef. Mol Ecol 22:3693–3708. https://doi.org/10.1111/mec.12342

Howells EJ, Willis BL, Bay LK, van Oppen MJH (2016a) Microsatellite allele sizes alone are insufficient to delineate species boundaries in *Symbiodinium*. Mol Ecol 25:2719–2723. https://doi.org/10.1111/mec.13631

Howells EJ, Abrego D, Meyer E, Kirk NL, Burt JA (2016b) Host adaptation and unexpected symbiont partners enable reef-building corals to tolerate extreme temperatures. Glob Chang Biol 22:2702–2714. https://doi.org/10.1111/gcb.13250

Huang H, Zhou G, Yang J, Liu S, You F, Lei X (2013) Diversity of free-living and symbiotic *Symbiodinium* in the coral reefs of Sanya, South China Sea. Mar Biol Res 9:117–128. https://doi.org/10.1080/17451000.2012.708045

Hume BCC, D'Angelo C, Smith EG, Stevens JR, Burt J, Wiedenmann J (2015) *Symbiodinium thermophilum* sp. nov., a thermotolerant symbiotic alga prevalent in corals of the world's hottest sea, the Persian/Arabian Gulf. Sci Rep 5. https://doi.org/10.1038/srep08562

Hunter CL, Morden CW, Smith CM (1997) The utility of ITS sequences in assessing relationships among zooxanthellae and corals. In: Proceedings of the 8th international coral reef symposium, Panama, pp 1599–1602

Iglesias-Prieto R, Trench RK (1994) Acclimation and adaptation to irradiance in symbiotic dinoflagellates. I. Responses of the photosynthetic unit to changes in photon flux density. Mar Ecol Prog Ser 113:163–175. https://doi.org/10.3354/meps113163

Iglesias-Prieto R, Trench RK (1997) Acclimation and adaptation to irradiance in symbiotic dinoflagellates. II. Response of chlorophyll–protein complexes to different photon-flux densities. Mar Biol 130:23–33. https://doi.org/10.1007/s002270050221

Iglesias-Prieto R, Beltran VH, LaJeunesse TC, Reyes-Bonilla H, Thome PE (2004) Different algal symbionts explain the vertical distribution of dominant reef corals in the eastern Pacific. Proc R Soc B 271:1757–1763. https://doi.org/10.1098/prsb.2004.2757

Jeong HJ, Lee SY, Kang NS, Yoo YD, Lim AS, Lee MJ et al (2014) Genetics and morphology characterize the dinoflagellate *Symbiodinium voratum*, n. sp., (Dinophyceae) as the sole representative of *Symbiodinium* clade E. J Eukaryot Microbiol 61:75–94. https://doi.org/10.1111/jeu.12088

Jones A, Berkelmans R (2010) Potential costs of acclimatization to a warmer climate: growth of a reef coral with heat tolerant vs. sensitive symbiont types. PLoS One 5. https://doi.org/10.1371/journal.pone.0010437

Jones AM, Berkelmans R (2011) Tradeoffs to thermal acclimation: energetics and reproduction of a reef coral with heat tolerant *Symbiodinium* type-D. J Mar Biol 2011. https://doi.org/10.1155/2011/185890

Jones A, Berkelmans R (2012) The photokinetics of thermo-tolerance in *Symbiodinium*. Mar Ecol 33:490–498. https://doi.org/10.1111/j.1439-0485.2012.00514.x

Jones AM, Berkelmans R, van Oppen MJH, Mieog JC, Sinclair W (2008) A community change in the algal endosymbionts of a scleractinian coral following a natural bleaching event: field evidence of acclimatization. Proc R Soc B 275:1359–1365. https://doi.org/10.1098/prsb.2008.0069

Kamezaki M, Higa M, Hirose M, Suda S, Reimer JD (2013) Different zooxanthellae types in populations of the zoanthid *Zoanthus sansibaricus* along depth gradients in Okinawa, Japan. Mar Biodivers 43:61–70. https://doi.org/10.1007/s12526-012-0119-2

Karim W, Nakaema S, Hidaka M (2015) Temperature effects on the growth rates and photosynthetic activities of *Symbiodinium* cells. J Mar Sci Eng 3:368–381. https://doi.org/10.3390/jmse3020368

Kemp DW, Cook CB, LaJeunesse TC, Brooks WR (2006) A comparison of the thermal bleaching responses of the zoanthid *Palythoa caribaeorum* from three geographically different regions in south Florida. J Exp Mar Biol Ecol 335:266–276. https://doi.org/10.1016/j.jemb.2006.03.017

Kemp DW, Fitt WK, Schmidt GW (2008) A microsampling method for genotyping coral symbionts. Coral Reefs 27:289–293. https://doi.org/10.1007/s00338-007-0333-8

Kemp DW, Hernandez-Pech X, Iglesias-Prieto R, Fitt WK, Schmidt GW (2014) Community dynamics and physiology of *Symbiodinium* spp. before, during, and after a coral bleaching event. Limnol Oceanogr 59:788–797. https://doi.org/10.4319/lo.2014.59.3.0788

Kemp DW, Thornhill DJ, Rotjan RD, Iglesias-Prieto R, Fitt WK, Schmidt GW (2015) Spatially distinct and regionally endemic *Symbiodinium* assemblages in the threatened Caribbean reef-building coral *Orbicella faveolata*. Coral Reefs 34:535–547. https://doi.org/10.1007/s00338-015-1277-z

Kenkel C, Matz MV (2016) Enhanced gene expression plasticity as a mechanism of adaptation to a variable environment in a reef-building coral. bioRxiv. https://doi.org/10.1101/059667

Kenkel CD, Goodbody-Gringley G, Caillaud D, Davies SW, Bartels E, Matz MV (2013) Evidence for a host role in thermotolerance divergence between populations of the mustard hill coral *(Porites astreoides)* from different reef environments. Mol Ecol 22:4335–4348. https://doi.org/10.1111/mec.12391

Kennedy EV, Foster NL, Mumby PJ, Stevens JR (2015) Widespread prevalence of cryptic *Symbiodinium* D in the key Caribbean reef builder, *Orbicella annularis*. Coral Reefs 34:519–531. https://doi.org/10.1007/s00338-015-1264-4

Kennedy EV, Tonk L, Foster NL, Chollett I, Ortiz J-C, Dove S et al (2016) *Symbiodinium* biogeography tracks environmental patterns rather than host genetics in a key Caribbean reef-builder, *Orbicella annularis*. Proc R Soc B 283. https://doi.org/10.1098/prsb.2016.1938

Kinzie RA (1974) Experimental infection of aposymbiotic gorgonian polyps with zooxanthellae. J Exp Mar Biol Ecol 15:335–342. https://doi.org/10.1016/0022-0981(74)90054-9

Kinzie RA, Takayama M, Santos SR, Coffroth MA (2001) The adaptive bleaching hypothesis: experimental tests of critical assumptions. Biol Bull 200:51–58. https://doi.org/10.2307/1543084

Kirk NL, Andras JP, Harvell CD, Santos SR, Coffroth MA (2009) Population structure of *Symbiodinium* sp. associated with the common sea fan, *Gorgonia ventalina*, in the Florida Keys across distance, depth, and time. Mar Biol 156:1609–1623. https://doi.org/10.1007/s00227-009-1196-z

Klueter A, Crandall JB, Archer FI, Teece MA, Coffroth MA (2015) Taxonomic and environmental variation of metabolite profiles in marine dinoflagellates of the genus *Symbiodinium*. Metabolites 5:74–99. https://doi.org/10.3390/metabo5010074

Kuniya N, Jimbo M, Tanimoto F, Yamashita H, Koike K, Harii S et al (2015) Possible involvement of Tachylectin-2-like lectin from *Acropora tenuis* in the process of *Symbiodinium* acquisition. Fish Sci 81:473–483. https://doi.org/10.1007/s12562-015-0862-y

LaJeunesse TC (2001) Investigating the biodiversity, ecology, and phylogeny of endosymbiont dinoflagellates in the genus *Symbiodinium* using the ITS region: in search of a "species" level marker. J Phycol 37:866–880. https://doi.org/10.1046/j.1529-8817.2001.01031.x

LaJeunesse TL (2002) Diversity and community structure of symbiotic dinoflagellates from Caribbean coral reefs. Mar Biol 141:387–400. https://doi.org/10.1007/s00227-002-0829-2

LaJeunesse TC (2005) "Species" radiations of symbiotic dinoflagellates in the Atlantic and Indo-Pacific since the Miocene-Pliocene transition. Mol Biol Evol 22:570–581. https://doi.org/10.1093/molbev/msi042

LaJeunesse TC, Loh WKW, van Woesik R, Hoegh-Guldberg O, Schmidt GW, Fitt WK (2003) Low symbiont diversity in southern Great Barrier Reef corals, relative to those of the Caribbean. Limnol Oceanogr 48:2046–2054. https://doi.org/10.4319/lo.2003.48.5.2046

LaJeunesse T, Bhagooli R, Hidaka M, Devantier L, Done T, Schmidt G et al (2004) Closely related *Symbiodinium* spp. differ in relative dominance in coral reef host communities across environmental, latitudinal and biogeographic gradients. Mar Ecol Prog Ser 284:147–161. https://doi.org/10.3354/meps284147

LaJeunesse TC, Lee S, Bush S, Bruno JF (2005) Persistence of non-Caribbean algal symbionts in Indo-Pacific mushroom corals released to Jamaica 35 years ago. Coral Reefs 24:157–159. https://doi.org/10.1007/s00338-004-0436-4

LaJeunesse TC, Smith RT, Finney J, Oxenford H (2009) Outbreak and persistence of opportunistic symbiotic dinoflagellates during the 2005 Caribbean mass coral "bleaching" event. Proc R Soc B 276:4139–4148. https://doi.org/10.1098/prsb.2009.1405

LaJeunesse TC, Pettay DT, Sampayo EM, Phongsuwan N, Brown B, Obura DO et al (2010a) Long-standing environmental conditions, geographic isolation and host–symbiont specificity influence the relative ecological dominance and genetic diversification of coral endosymbionts in the genus *Symbiodinium*. J Biogeogr 37:785–800. https://doi.org/10.1111/j.1365-2699.2010.02273.x

LaJeunesse TC, Smith R, Walther M, Pinzón J, Pettay DT, McGinley M et al (2010b) Host-symbiont recombination versus natural selection in response of coral-dinoflagellate symbioses to environmental disturbance. Proc R Soc B 277:2925–2934. https://doi.org/10.1098/rspb.2010.0385

Lajeunesse TC, Parkinson JE, Reimer JD (2012) A genetics-based description of *Symbiodinium minutum* sp. nov. and *S. psygmophilum* sp. nov. (Dinophyceae), two dinoflagellates symbiotic with cnidaria. J Phycol:1380–1391. https://doi.org/10.1111/j.1529-8817.2012.01217.x

LaJeunesse TC, Wham DC, Pettay DT, Parkinson JE, Keshavmurthy S, Chen CA (2014) Ecologically differentiated stress-tolerant endosymbionts in the dinoflagellate genus *Symbiodinium* (Dinophyceae) Clade D are different species. Phycologia 53:305–319. https://doi.org/10.2216/13-186.1

Lasker HR (2003) Zooxanthella densities within a Caribbean octocoral during bleaching and non-bleaching years. Coral Reefs 22:23–26. https://doi.org/10.1007/s00338-003-0276-7

Lasker HR, Peters EC, Coffroth MA (1984) Bleaching of reef coelenterates in the San Blas Islands, Panama. Coral Reefs 3:183–190. https://doi.org/10.1007/BF00288253

Lee SY, Jeong HJ, Kang NS, Jang TY, Jang SH, Lim AS (2014) Morphological characterization of *Symbiodinium minutum* and *S. psygmophilum* belonging to clade B. Algae 29:299–310. https://doi.org/10.4490/algae.2014.29.4.299

Lee SY, Jeong HJ, Kang NS, Jang TY, Jang SH, Lajeunesse TC (2015) *Symbiodinium tridacnidorum* sp. nov., a dinoflagellate common to Indo-Pacific giant clams, and a revised morphological description of *Symbiodinium microadriaticum* Freudenthal, emended Trench & Blank. Eur J Phycol 50:155–172. https://doi.org/10.1080/09670262.2015.1018336

Lee MJ, Jeong HJ, Jang SH, Lee SY, Kang NS, Lee KH et al (2016) Most low-abundance "background" *Symbiodinium* spp. are transitory and have minimal functional significance for symbiotic corals. Microb Ecol 71:771–783. https://doi.org/10.1007/s00248-015-0724-2

Lenz EA, Bramanti L, Lasker HR, Edmunds PJ (2015) Long-term variation of octocoral populations in St. John, US Virgin Islands. Coral Reefs 34:1099–1109. https://doi.org/10.1007/s00338-015-1315-x

Levin RA, Beltran VH, Hill R, Kjelleberg S, McDougald D, Steinberg PD et al (2016) Sex, scavengers, and chaperones: transcriptome secrets of divergent *Symbiodinium* thermal tolerances. Mol Biol Evol 33:2201–2215. https://doi.org/10.1093/molbev/msw119

Levin RA, Voolstra CR, Weynberg KD, van Oppen MJH (2017a) Evidence for a role of viruses in the thermal sensitivity of coral photosymbionts. ISME J 11:808–812. https://doi.org/10.1038/ismej.2016.154

Levin RA, Voolstra CR, Agrawal S, Steinberg PD, Suggett DJ, van Oppen MJH (2017b) Engineering strategies to decode and enhance the genomes of coral symbionts. Front Microbiol 8. https://doi.org/10.3389/fmicb.2017.01220

Lewis CL, Coffroth MA (2004) The acquisition of exogenous algal symbionts by an octocoral after bleaching. Science 304:1490–1492. https://doi.org/10.1126/science.1097323

Lin S, Cheng S, Song B, Zhong X, Lin X, Li W et al (2015) The *Symbiodinium kawagutii* genome illuminates dinoflagellate gene expression and coral symbiosis. Science 350:691–694. https://doi.org/10.1126/science.aad0408

Little AF, van Oppen MJH, Willis BL (2004) Flexibility in algal endosymbioses shapes growth in reef corals. Science 304:1492–1494. https://doi.org/10.1126/science.1095733

Littman RA, Willis BL, Bourne DG (2009) Bacterial communities of juvenile corals infected with different *Symbiodinium* (dinoflagellate) clades. Mar Ecol Prog Ser 389:45–59. https://doi.org/10.3354/meps08180

Loh W, Loi T, Carter D, Hoegh-Guldberg O (2001) Genetic variability of the symbiotic dinoflagellates from the wide ranging coral species *Seriatopora hystrix* and *Acropora longicyathus* in the Indo-West Pacific. Mar Ecol Prog Ser 222:97–107. https://doi.org/10.3354/meps222097

Lohbeck KT, Riebesell U, Reusch TBH (2012) Adaptive evolution of a key phytoplankton species to ocean acidification. Nat Geosci 5:346–351. https://doi.org/10.1038/ngeo1441

Lohr J, Munn CB, Wilson WH (2007) Characterization of a latent virus-like infection of symbiotic zooxanthellae. Appl Environ Microbiol 73:2976–2981. https://doi.org/10.1128/AEM.02449-06

Loram JE, Boonham N, O'Toole P, Trapido-Rosenthal HG, Douglas AE (2007) Molecular quantification of symbiotic dinoflagellate algae of the genus *Symbiodinium*. Biol Bull 212:259–268. https://doi.org/10.2307/25066608

Lucas MQ, Stat M, Smith MC, Weil E, Schizas NV (2016) *Symbiodinium* (internal transcribed spacer 2) diversity in the coral host *Agaricia lamarcki* (Cnidaria: Scleractinia) between shallow and mesophotic reefs in the Northern Caribbean (20–70 m). Mar Ecol 37:1079–1087. https://doi.org/10.1111/maec.12367

Magalon H, Baudry E, Husté A, Adjeroud M, Veuille M (2006) High genetic diversity of the symbiotic dinoflagellates in the coral *Pocillopora meandrina* from the South Pacific. Mar Biol 148:913–922. https://doi.org/10.1007/s00227-005-0133-z

Manning MM, Gates RD (2008) Diversity in populations of free-living *Symbiodinium* from a Caribbean and Pacific reef. Limnol Oceanogr 53:1853–1861. https://doi.org/10.4319/lo.2008.53.5.1853

Marhaver KL, Edwards RA, Rohwer F (2008) Viral communities associated with healthy and bleaching corals. Environ Microbiol 10:2277–2286. https://doi.org/10.1111/j.1462-2920.2008.01652.x

McGinley MP, Aschaffenburg MD, Pettay DT, Smith RT, LaJeunesse TC, Warner ME (2012) *Symbiodinium* spp. in colonies of eastern Pacific *Pocillopora* spp. are highly stable despite the prevalence of low-abundance background populations. Mar Ecol Prog Ser 462:1–7. https://doi.org/10.3354/meps09914

McGinty ES, Pieczonka J, Mydlarz LD (2012) Variations in reactive oxygen release and antioxidant activity in multiple *Symbiodinium* types in response to elevated temperature. Microb Ecol 64:1000–1007. https://doi.org/10.1007/s00248-012-0085-z

McIlroy SE, Coffroth MA (2017) Coral ontogeny affects early symbiont acquisition in laboratory-reared recruits. Coral Reefs 36:927–932. https://doi.org/10.1007/s00338-017-1584-7

McIlroy SE, Gillette P, Cunning R, Klueter A, Capo T, Baker AC et al (2016) The effects of *Symbiodinium (Pyrrhophyta)* identity on growth, survivorship, and thermal tolerance of newly settled coral recruits. J Phycol 52:1114–1124. https://doi.org/10.1111/jpy.12471

Meyer JR, Ellner SP, Hairston NG, Jones LE, Yoshida T (2006) Prey evolution on the time scale of predator–prey dynamics revealed by allele-specific quantitative PCR. Proc Natl Acad Sci U S A 103:10690–10695. https://doi.org/10.1073/pnas.0600434103

Mieog JC, van Oppen MJH, Cantin NE, Stam WT, Olsen JL (2007) Real-time PCR reveals a high incidence of *Symbiodinium*; clade D at low levels in four scleractinian corals across the Great Barrier Reef: implications for symbiont shuffling. Coral Reefs 26:449–457. https://doi.org/10.1007/s00338-007-0244-8

Mieog JC, Olsen JL, Berkelmans R, Bleuler-Martinez SA, Willis BL, van Oppen MJH (2009) The roles and interactions of symbiont, host and environment in defining coral fitness. PLoS One 4. https://doi.org/10.1371/journal.pone.0006364

Moore RB, Ferguson KM, Loh WKW, Hoegh-Guldberg O, Carter DA (2003) Highly organized structure in the non-coding region of the psbA minicircle from clade C *Symbiodinium*. Int J Syst Evol Microbiol 53:1725–1734. https://doi.org/10.1099/ijs.0.02504-0

Moya A, Tambutté S, Tambutté E, Zoccola D, Caminiti N, Allemand D (2006) Study of calcification during a daily cycle of the coral *Stylophora pistillata*: implications for 'light-enhanced calcification'. J Exp Biol 209:3413–3419. https://doi.org/10.1242/jeb.02382

Muscatine L, Porter JW (1977) Reef corals: mutualistic symbioses adapted to nutrient-poor environments. Bioscience 27:454–460. https://doi.org/10.2307/1297526

Nozawa Y, Loya Y (2005) Genetic relationship and maturity state of the allorecognition system affect contact reactions in juvenile Seriatopora corals. Mar Ecol Prog Ser 286:115–123. https://doi.org/10.3354/meps286115

Oliver TA, Palumbi SR (2009) Distributions of stress-resistant coral symbionts match environmental patterns at local but not regional scales. Mar Ecol Prog Ser 378:93–103. https://doi.org/10.3354/meps07871

Oliver T, Palumbi S (2011) Do fluctuating temperature environments elevate coral thermal tolerance? Coral Reefs 30:429–440. https://doi.org/10.1007/s00338-011-0721-y

Ortiz JC, González-Rivero M, Mumby PJ (2013) Can a thermally tolerant symbiont improve the future of Caribbean coral reefs? Glob Chang Biol 19:273–281. https://doi.org/10.1111/gcb.12027

Padilla-Gamiño JL, Pochon X, Bird C, Concepcion GT, Gates RD (2012) From parent to gamete: vertical transmission of *Symbiodinium* (Dinophyceae) ITS2 sequence assemblages in the reef building coral *Montipora capitata*. PLoS One 7. https://doi.org/10.1371/journal.pone.0038440

Parkinson JE, Baums IB (2014) The extended phenotypes of marine symbioses: ecological and evolutionary consequences of intraspecific genetic diversity in coral-algal associations. Front Microbiol 5. https://doi.org/10.3389/fmicb.2014.00445

Parkinson JE, Banaszak AT, Altman NS, LaJeunesse TC, Baums IB (2015a) Intraspecific diversity among partners drives functional variation in coral symbioses. Sci Rep 5. https://doi.org/10.1038/srep15667

Parkinson JE, Coffroth MA, LaJeunesse TC (2015b) New species of Clade B *Symbiodinium* (Dinophyceae) from the greater Caribbean belong to different functional guilds: *S. aenigmaticum* sp. nov., *S. antillogorgium* sp. nov., *S. endomadracis* sp. nov., and *S. pseudominutum* sp. nov. J Phycol 51:850–858. https://doi.org/10.1111/jpy.12340

Parkinson J, Baumgarten S, Michell CT, Baums IB, LaJeunesse TC, Voolstra CR (2016) Gene expression variation resolves species and individual strains among coral-associated dinoflagellates within the genus *Symbiodinium*. Genome Biol Evol 8:665–680. https://doi.org/10.1093/gbe/evw019

Pettay DT, Lajeunesse TC (2009) Microsatellite loci for assessing genetic diversity, dispersal and clonality of coral symbionts in "stress-tolerant" clade D *Symbiodinium*. Mol Ecol Resour 9:1022–1025. https://doi.org/10.1111/j.1755-0998.2009.02561.x

Pettay DT, Wham DC, Pinzón JH, Lajeunesse TC (2011) Genotypic diversity and spatial–temporal distribution of *Symbiodinium* clones in an abundant reef coral. Mol Ecol 20:5197–5212. https://doi.org/10.1111/j.1365-294X.2011.05357.x

Pinzón JH, LaJeunesse TC (2011) Species delimitation of common reef corals in the genus *Pocillopora* using nucleotide sequence phylogenies, population genetics and symbiosis ecology. Mol Ecol 20:311–325. https://doi.org/10.1111/j.1365-294X.2010.04939.x

Pinzón J, Devlin-Durante M, Weber M, Baums I, LaJeunesse T (2011) Microsatellite loci for *Symbiodinium* A3 (*S. fitti*) a common algal symbiont among Caribbean *Acropora* (stony corals) and Indo-Pacific giant clams (*Tridacna*). Conserv Genet Resour 3:45–47. https://doi.org/10.1007/s12686-010-9283-5

Pochon X, Gates RD (2010) A new *Symbiodinium* clade (Dinophyceae) from soritid foraminifera in Hawai'i. Mol Phylogenet Evol 56:492–497. https://doi.org/10.1016/j.ympev.2010.03.040

Pochon X, Montoya-Burgos JI, Stadelmann B, Pawlowski J (2006) Molecular phylogeny, evolutionary rates, and divergence timing of the symbiotic dinoflagellate genus *Symbiodinium*. Mol Phylogenet Evol 38:20–30. https://doi.org/10.1016/j.ympev.2005.04.028

Pochon X, Stat M, Takabayashi M, Chasqui L, Chauka LJ, Logan DDK et al (2010) Comparison of endosymbiotic and free-living *Symbiodinium* (Dinophyceae) diversity in a Hawaiian reef environment. J Phycol 46:53–65. https://doi.org/10.1111/j.1529-8817.2009.00797.x

Pochon X, Putnam HM, Burki F, Gates RD (2012) Identifying and characterizing alternative molecular markers for the symbiotic and free-living dinoflagellate genus *Symbiodinium*. PLoS One 7. https://doi.org/10.1371/journal.pone.0029816

Pochon X, Gates RD, Vik D, Edmunds PJ (2014) Molecular characterization of symbiotic algae (*Symbiodinium spp.*) in soritid foraminifera (*Sorites orbiculus*) and a scleractinian coral (*Orbicella annularis*) from St John, US Virgin Islands. Mar Biol 161:2307–2318. https://doi.org/10.1007/s00227-014-2507-6

Poland DM, Coffroth MA (2017) Trans-generational specificity within a cnidarian–algal symbiosis. Coral Reefs 36:119–129. https://doi.org/10.1007/00338-016-1514-0

Poland DM, Mansfield JM, Hannes AR, Lewis CLF, Shearer TL, Connelly SJ et al (2013) Variation in *Symbiodinium* communities in juvenile *Briareum asbestinum (Cnidaria: Octocorallia)* over temporal and spatial scales. Mar Ecol Prog Ser 476:23–37. https://doi.org/10.3354/meps10165

Porto I, Granados C, Restrepo JC, Sanchez JA (2008) Macroalgal-associated dinoflagellates belonging to the genus *Symbiodinium* in Caribbean reefs. PLoS One 3. https://doi.org/10.1371/journal.pone.0002160

Prada C, McIlroy SE, Beltrán DM, Valint DJ, Ford SA, Hellberg ME et al (2014) Cryptic diversity hides host and habitat specialization in a gorgonian-algal symbiosis. Mol Ecol 23:3330–3340. https://doi.org/10.1111/mec.12808

Puill-Stephan E, Willis BL, van Herwerden L, van Oppen MJH (2009) Chimerism in wild adult populations of the broadcast spawning coral *Acropora millepora* on the Great Barrier Reef. PLoS One 4. https://doi.org/10.1371/journal.pone.0007751

Puill-Stephan E, Willis BL, Abrego D, Raina JB, van Oppen MJH (2012) Allorecognition maturation in the broadcast-spawning coral *Acropora millepora*. Coral Reefs 31:1019–1028. https://doi.org/10.1007/00338-012-0912-1

Putnam HM, Stat M, Pochon X, Gates RD (2012) Endosymbiotic flexibility associates with environmental sensitivity in scleractinian corals. Proc Roy Soc Biol 279:4352–4361. https://doi.org/10.1098/prsb.2012.1454

Quigley KM, Davies SW, Kenkel CD, Willis BL, Matz MV, Bay LK (2014) Deep-sequencing method for quantifying background abundances of *Symbiodinium* types: exploring the rare *Symbiodinium* biosphere in reef-building corals. PLoS One 9. https://doi.org/10.1371/journal.pone.0094297

Quigley KM, Willis BL, Bay LK (2016) Maternal effects and *Symbiodinium* community composition drive differential patterns in juvenile survival in the coral *Acropora tenuis*. Roy Soc Open Sci 3. https://doi.org/10.1098/rsos.160471

Quigley KM, Willis BL, Bay LK (2017a) Heritability of the *Symbiodinium* community in vertically- and horizontally-transmitting broadcast spawning corals. Sci Rep 7. https://doi.org/10.1038/s41598-017-08179-4

Quigley KM, Bay LK, Willis BL (2017b) Temperature and water quality-related patterns in sediment-associated *Symbiodinium* communities impact symbiont uptake and fitness of juveniles in the genus Acropora. Front Mar Sci. https://doi.org/10.3389/fmars.2017.00401

Quigley KM, Warner PA, Bay LK, Willis BL (2018) Unexpected mixed-mode transmission and moderate genetic regulation of *Symbiodinium* communities in a brooding coral. Heredity. https://doi.org/10.1038/s41437-018-0059-0

Ralph PJ, Gademann R, Larkum AWD (2001) Zooxanthellae expelled from bleached corals at 33°C are photosynthetically competent. Mar Ecol Prog Ser 220:163–168. https://doi.org/10.3354/meps220163

Reich HG, Robertson DL, Goodbody-Gringley G (2017) Do the shuffle: changes in *Symbiodinium* consortia throughout juvenile coral development. PLoS One 12. https://doi.org/10.1371/journal.pone.0171768

Reimer JD, Todd PA (2009) Preliminary molecular examination of zooxanthellate zoanthids (Hexacorallia: Zoantharia) and associated zooxanthellae (*Symbiodinium* spp.) diversity in Singapore. Raffles Bull Zool 22:103–120

Reimer JD, Shah MMR, Sinniger F, Yanagi K, Suda S (2010) Preliminary analyses of cultured *Symbiodinium* isolated from sand in the oceanic Ogasawara Islands, Japan. Mar Biodivers 40:237–247. https://doi.org/10.1007/s12526-010-0044-1

Reimer JD, Herrera M, Gatins R, Roberts MB, Parkinson JE, Berumen ML (2016) Latitudinal variation in the symbiotic dinoflagellate *Symbiodinium* of the common reef zoantharian *Palythoa tuberculosa* on the Saudi Arabian coast of the Red Sea. J Biogeogr 44:661–673. https://doi.org/10.1111/jbi.12795

Robison JD, Warner ME (2006) Differential impacts of photoacclimation and thermal stress on the photobiology of four different phylotypes of *Symbiodinium* (Pyrrhophyta). J Phycol 42:568–579. https://doi.org/10.1111/j.1529-8817.2006.00232.x

Rodriguez-Lanetty M, Krupp DA, Weis VM (2004) Distinct ITS types of *Symbiodinium* in Clade C correlate with cnidarian/dinoflagellate specificity during onset of symbiosis. Mar Ecol Prog Ser 275:97–102. https://doi.org/10.3354/meps275097

Rodriguez-Lanetty M, Wood-Charlson EM, Hollingsworth LL, Krupp DA, Weis VM (2006) Temporal and spatial infection dynamics indicate recognition events in the early hours of a dinoflagellate/coral symbiosis. Mar Biol 149:713–719. https://doi.org/10.1007/s00227-006-0272-x

Rowan R (2004) Coral bleaching: Thermal adaptation in reef coral symbionts. Nature 430:742. https://doi.org/10.1038/430742a

Rowan R, Knowlton N (1995) Intraspecific diversity and ecological zonation in coral-algal symbiosis. Proc Natl Acad Sci U S A 92:2850–2853. https://doi.org/10.1073/pnas.92.7.2850

Rowan R, Powers DA (1991) A molecular genetic classification of zooxanthellae and the evolution of animal-algal symbioses. Science 251:1348–1351. https://doi.org/10.1126/science.251.4999.1348

Rowan R, Knowlton N, Baker A, Jara J (1997) Landscape ecology of algal symbionts creates variation in episodes of coral bleaching. Nature 388:265–269. https://doi.org/10.1038/40843

Ruzicka RR, Colella MA, Porter JW, Morrison JM, Kidney JA, Brinkhuis V et al (2013) Temporal changes in benthic assemblages on Florida Keys reefs 11 years after the 1997/1998 El Niño. Mar Ecol Prog Ser 489:125–141. https://doi.org/10.3354/meps10427

Sampayo EM, Dove S, Lajeunesse TC (2009) Cohesive molecular genetic data delineate species diversity in the dinoflagellate genus *Symbiodinium*. Mol Ecol 18:500–519. https://doi.org/10.1111/j.1365-294X.2008.04037.x

Santos SR, Taylor DJ, Coffroth MA (2001) Genetic comparisons of freshly isolated versus cultured symbiotic dinoflagelaltes: implications for extrapolating to the intact symbiosis. J Phycol 37:900–912. https://doi.org/10.1046/j.1529-8817.2001.00194.x

Santos SR, Taylor DJ, Kinzie RA, Sakai K, Coffroth MA (2002) Evolution of length variation and heteroplasmy in the chloroplast rDNA of symbiotic dinoflagellates (*Symbiodinium*, Dinophyta) and a novel insertion in the universal core region of the large subunit rDNA. Phycologia 41:311–318. https://doi.org/10.2216/i0031-8884-41-4-311.1

Santos SR, Gutierrez-Rodriguez C, Coffroth MA (2003) Phylogenetic identification of symbiotic dinoflagellates via length heteroplasmy in domain V of chloroplast large subunit (cp23S)-ribosomal DNA sequences. Mar Biotechnol 5:130–140. https://doi.org/10.1007/s10126-002-0076-9

Santos SR, Shearer TL, Hannes AR, Coffroth MA (2004) Fine-scale diversity and specificity in the most prevalent lineage of symbiotic dinoflagellates (*Symbiodinium*, Dinophyceae) of the Caribbean. Mol Ecol 13:459–469. https://doi.org/10.1046/j.1365-294X.2003.02058.x

Savage AM, Trapido-Rosenthal H, Douglas AE (2002) On the functional significance of molecular variation in *Symbiodinium*, the symbiotic algae of Cnidaria: photosynthetic response to irradiance. Mar Ecol Prog Ser 244:27–37. https://doi.org/10.3354/meps244027

Sawall Y, Al-Sofyani A, Banguera-Hinestroza E, Voolstra CR (2014) Spatio-temporal analyses of *Symbiodinium* physiology of the coral *Pocillopora verrucosa* along large-scale nutrient and temperature gradients in the Red Sea. PLoS One 9. https://doi.org/10.1371/journal.pone.0103179

Schoenberg DA, Trench RK (1980) Genetic variation in *Symbiodinium* (= *Gymnodinium*) *microadriaticum* Freudenthal, and specificity in its symbiosis with marine invertebrates. I. Isoenzyme and soluble protein patterns of axenic cultures of *Symbiodinium microadriaticum*. Proc R Soc B 207:405–427. https://doi.org/10.1098/prsb.1980.0031

Serrano X, Baums IB, O'Reilly K, Smith TB, Jones RJ, Shearer TL et al (2014) Geographic differences in vertical connectivity in the Caribbean coral *Montastraea cavernosa* despite high levels of horizontal connectivity at shallow depths. Mol Ecol 23:4226–4240. https://doi.org/10.1111/mec.12861

Shoguchi E, Shinzato C, Kawashima T, Gyoja F, Mungpakdee S, Koyanagi R et al (2013) Draft assembly of the *Symbiodinium minutum* nuclear genome reveals dinoflagellate gene structure. Curr Biol 23:1399–1408. https://doi.org/10.1016/j.cub.2013.05.062

Shoguchi E, Shinzato C, Hisata K, Satoh N, Mungpakdee S (2015) The large mitochondrial genome of *Symbiodinium minutum* reveals conserved noncoding sequences between dinoflagellates and apicomplexans. Genome Biol Evol 7:2237–2244. https://doi.org/10.1093/gbe/evv137

Silverstein RN, Correa AMS, LaJeunesse TC, Baker AC (2011) Novel algal symbiont (*Symbiodinium* spp.) diversity in reef corals of Western Australia. Mar Ecol Prog Ser 422:63–75. https://doi.org/10.3354/meps08934

Silverstein RN, Correa AMS, Baker AC (2012) Specificity is rarely absolute in coral–algal symbiosis: implications for coral response to climate change. Proc R Soc B (1738):2609–2618. https://doi.org/10.1098/rspb.2012.0055

Silverstein RN, Cunning R, Baker AC (2015) Change in algal symbiont communities after bleaching, not prior heat exposure, increases heat tolerance of reef corals. Glob Chang Biol 21:236–249. https://doi.org/10.1111/gcb.12706

Silverstein RN, Cunning R, Baker AC (2017) Tenacious D: *Symbiodinium* in clade D remain in reef corals at both high and low temperature extremes despite impairment. J Exp Biol 220:1192–1196. https://doi.org/10.1242/jeb.148239

Smith EG, Ketchum RN, Burt JA (2017) Host specificity of *Symbiodinium* variants revealed by an ITS2 metahaplotype approach. ISME J 11:1500–1503. https://doi.org/10.1038/ismej.2016.206

Stat M, Carter D, Hoegh-Guldberg O (2006) The evolutionary history of *Symbiodinium* and scleractinian hosts—symbiosis, diversity, and the effect of climate change. Perspect Plant Ecol Evol Syst 8:23–43. https://doi.org/10.1016/j.ppees.2006.04.001

Stat M, Morris E, Gates RD (2008) Functional diversity in the coral-dinoflagellate symbiosis. Proc Natl Acad Sci U S A 105:9256–9261. https://doi.org/10.1073/pnas.0801328105

Stat M, Pochon X, Cowie RO, Gates RD (2009) Specificity in communities of *Symbiodinium* in corals from Johnston Atoll. Mar Ecol Prog Ser 386:83–96. https://doi.org/10.3354/meps08080

Stat M, Yost D, Gates R (2015) Geographic structure and host specificity shape the community composition of symbiotic dinoflagellates in corals from the Northwestern Hawaiian Islands. Coral Reefs 34:1075–1086. https://doi.org/10.1007/s00338-015-1320-0

Suggett DJ, Goyen S, Evenhuis C, Szabó M, Pettay DT, Warner ME et al (2015) Functional diversity of photobiological traits within the genus *Symbiodinium* appears to be governed by the interaction of cell size with cladal designation. New Phytol 208:370–381. https://doi.org/10.1111/nph.13483

Swain TD, Chandler J, Backman V, Marcelino L (2016) Consensus thermotolerance ranking for 110 *Symbiodinium* phylotypes: an exemplar utilization of a novel iterative partial rank aggregation tool with broad application potential. Funct Ecol 31:172–183. https://doi.org/10.1111/1365-2435.12694

Sweet MJ (2013) *Symbiodinium* diversity within *Acropora muricata* and the surrounding environment. Mar Ecol 35:343–353. https://doi.org/10.1111/maec.12092

Takabayashi M, Santos SR, Cook CB (2004) Mitochondrial DNA phylogeny of the symbiotic dinoflagellates (*Symbiodinium*, Dinophyta). J Phycol 40:160–164. https://doi.org/10.1111/j.0022-3646.2003.03-097.x

Takabayashi M, Adams LM, Pochon X, Gates RD (2012) Genetic diversity of free-living *Symbiodinium* in surface water and sediment of Hawai'i and Florida. Coral Reefs 31:157–167. https://doi.org/10.1007/s00338-011-0832-5

Taylor DL (1974) Symbiotic marine algae: taxonomy and biological fitness. In: Vernberg WB (ed) Symbiosis in the Sea. University of South Carolina Press, Columbia, pp 245–262

Tchernov D, Gorbunov MY, de Vargas C, Yadav SN, Milligan AJ, Haggblom M et al (2004) Membrane lipids of symbiotic algae are diagnostic of sensitivity to thermal bleaching in corals. Proc Natl Acad Sci U S A 101:13531–13535. https://doi.org/10.1073/pnas.0402907101

Thomas L, Kendrick GA, Kennington WJ, Richards ZT, Stat M (2014) Exploring *Symbiodinium* diversity and host specificity in *Acropora* corals from geographical extremes of Western Australia with 454 amplicon pyrosequencing. Mol Ecol 23:3113–3126. https://doi.org/10.1111/mec.12801

Thornhill DJ, Fitt WK, Schmidt GW (2006a) Highly stable symbioses among western Atlantic brooding corals. Coral Reefs 25:515–519. https://doi.org/10.1007/s00338-006-0157-y

Thornhill DJ, LaJeunesse TC, Kemp DW, Fitt WK, Schmidt GW (2006b) Multi-year, seasonal genotypic surveys of coral-algal symbioses reveal prevalent stability or post-bleaching reversion. Mar Biol 148:711–722. https://doi.org/10.1007/s00227-005-0114-2

Thornhill DJ, Lajeunesse TC, Santos SR (2007) Measuring rDNA diversity in eukaryotic microbial systems: how intragenomic variation, pseudogenes, and PCR artifacts confound biodiversity estimates. Mol Ecol 16:5326–5340. https://doi.org/10.1111/j.1365-294X.2007.03576.x

Thornhill DJ, Kemp DW, Bruns BU, Fitt WK, Schmidt GW (2008) Correspondence between cold tolerance and temperate biogeography in a western Atlantic *Symbiodinium* (Dinophyta) lineage. J Phycol 44:1126–1135. https://doi.org/10.1111/j.1529-8817.2008.00567.x

Thornhill DJ, Xiang Y, Fitt WK, Santos SR (2009) Reef endemism, host specificity and temporal stability in populations of symbiotic dinoflagellates from two ecologically dominant caribbean corals. PLoS One 4. https://doi.org/10.1371/journal.pone.0006262

Thornhill DJ, Kemp DW, Sampayo EM, Schmidt GW (2010) Comparative analyses of amplicon migration behavior in differing denaturing gradient gel electrophoresis (DGGE) systems. Coral Reefs 29:83–91. https://doi.org/10.1007/s00338-009-0550-4

Thornhill DJ, Lewis AM, Wham DC, LaJeunesse TC (2014) Host-specialist lineages dominate the adaptive radiation of reef coral endosymbionts. Evolution 68:352–367. https://doi.org/10.1111/evo.12270

Thornhill DJ, Howells EJ, Wham DC, Steury TD, Santos SR (2017) Population genetics of reef coral endosymbionts (*Symbiodinium*, Dinophyceae). Mol Ecol 26:2640–2659. https://doi.org/10.1111/mec.14055

Toller WW, Rowan R, Knowlton N (2001a) Zooxanthellae of the *Montastraea annularis* species complex: patterns of distribution of four taxa of *Symbiodinium* on different reefs and across depths. Biol Bull 201:348–359. https://doi.org/10.2307/1543613

Toller WW, Rowan R, Knowlton N (2001b) Repopulation of zooxanthellae in the Caribbean corals *Montastraea annularis* and *M. faveolata* following experimental and disease-associated bleaching. Biol Bull 201:360–373. https://doi.org/10.2307/1543614

Tonk L, Bongaerts P, Sampayo EM, Hoegh-Guldberg O (2013) SymbioGBR: a web-based database of *Symbiodinium* associated with cnidarian hosts on the Great Barrier Reef. BMC Ecol 13. https://doi.org/10.1186/1472-6785-13-7

Tonk L, Sampayo EM, LaJeunesse TC, Schrameyer V, Hoegh-Guldberg O (2014) *Symbiodinium* (Dinophyceae) diversity in reef-invertebrates along an offshore to inshore reef gradient near Lizard Island, Great Barrier Reef. J Phycol 50:552–563. https://doi.org/10.1111/jpy.12185

Tonk L, Sampayo EM, Chai A, Schrameyer V, Hoegh-Guldberg O (2017) *Symbiodinium* (Dinophyceae) community patterns in invertebrate hosts from inshore marginal reefs of the southern Great Barrier Reef, Australia. J Phycol 26:2640–2659. https://doi.org/10.1111/jpy.12523

Trench RK, Blank RJ (1987) *Symbiodinium microadriaticum* Freudenthal, *S. goreauii* sp. nov., *S. kawagutii* sp. nov. and *S. pilosum* sp. nov.: Gymnodinioid Dinoflagellate symbionts of marine invertebrates. J Phycol 23:469–481. https://doi.org/10.1111/j.1529-8817.1987.tb02534.x

Ulstrup KE, van Oppen MJH (2003) Geographic and habitat partitioning of genetically distinct zooxanthellae (*Symbiodinium*) in Acropora corals on the Great Barrier Reef. Mol Ecol 12:3477–3484. https://doi.org/10.1046/j.1365-294X.2003.01988.x

Ulstrup KE, Berkelmans R, Ralph PJ, van Oppen MJH (2006) Variation in bleaching sensitivity of two coral species across a latitudinal gradient on the Great Barrier Reef: the role of zooxanthellae. Mar Ecol Prog Ser 314:135–148. https://doi.org/10.3354/meps314135

van Hooidonk R, Maynard J, Tamelander J, Gove J, Ahmadia G, Raymundo L et al (2016) Local-scale projections of coral reef futures and implications of the Paris Agreement. Sci Rep 6. https://doi.org/10.1038/srep39666

van Oppen MJH, Palstra FP, Piquet AM-T, Miller DJ (2001) Patterns of coral–dinoflagellate associations in Acropora: significance of local availability and physiology of *Symbiodinium* strains and host–symbiont selectivity. Proc R Soc B 268:1759–1767. https://doi.org/10.1098/rspb.2001.1733

van Oppen MJH, Mahiny A, Done T (2005a) Geographic distribution of zooxanthella types in three coral species on the Great Barrier Reef sampled after the 2002 bleaching event. Coral Reefs 24:482–487. https://doi.org/10.1007/s00338-005-0487-1

van Oppen MJH, Mieog JC, Sanchez CA, Fabricius KE (2005b) Diversity of algal endosymbionts (zooxanthellae) in octocorals: the roles of geography and host relationships. Mol Ecol 14:2403–2417. https://doi.org/10.1111/j.1365-294X.2005.02545.x

van Oppen MJH, Leong J-A, Gates RD (2009) Coral-virus interactions: a double-edged sword? Symbiosis 47:1–8. https://doi.org/10.1007/BF03179964
van Oppen MJH, Souter P, Howells EJ, Heyward A, Berkelmans R (2011) Novel genetic diversity through somatic mutations: fuel for adaptation of reef corals? Diversity 3:405–423. https://doi.org/10.3390/d3030405
Venera-Ponton DE, Diaz-Pulido G, Rodriguez-Lanetty M, Hoegh-Guldberg O (2010) Presence of *Symbiodinium* spp. in macroalgal microhabitats from the southern Great Barrier Reef. Coral Reefs 29:1049–1060. https://doi.org/10.1007/s00338-010-0666-6
Ware JR, Fautin DG, Buddemeier RW (1996) Patterns of coral bleaching: modeling the adaptive bleaching hypothesis. Ecol Model 84:199–214. https://doi.org/10.1016/0304-3800(94)00132-4
Warner ME, LaJeunesse TC, Robison JD, Thur RM (2006) The ecological distribution and comparative photobiology of symbiotic dinoflagellates from reef corals in Belize: potential implications for coral bleaching. Limnol Oceanogr 51:1887–1897. https://doi.org/10.4319/lo.2006.51.4.1887
Weis VM (2008) Cellular mechanisms of Cnidarian bleaching: stress causes the collapse of symbiosis. J Exp Biol 211:3059–3066. https://doi.org/10.1242/jeb.009597
Weis VM, Reynolds WS, deBoer MD, Krupp DA (2001) Host-symbiont specificity during onset of symbiosis between the dinoflagellates *Symbiodinium* spp. and planula larvae of the scleractinian coral *Fungia scutaria*. Coral Reefs 20:301–308. https://doi.org/10.1007/s003380100179
Weynberg KD, Neave M, Clode PL, Voolstra CR, Brownlee C, Laffy P et al (2017) Prevalent and persistent viral infection in cultures of the coral algal endosymbiont *Symbiodinium*. Coral Reefs 36:773–784. https://doi.org/10.1007/s00338-017-1568-7
Wham DC, LaJeunesse TC (2016) *Symbiodinium* population genetics: testing for species boundaries and analysing samples with mixed genotypes. Mol Ecol 25:2699–2712. https://doi.org/10.1111/mec.13623
Wham DC, Pettay DT, LaJeunesse TC (2011) Microsatellite loci for the host-generalist "zooxanthella" & *Symbiodinium trenchi* and other Clade D *Symbiodinium*. Conserv Genet Resour 3:541–544. https://doi.org/10.1007/s12686-011-9399-2
Wilkinson SP, Fisher PL, van Oppen MJH, Davy SK (2015) Intra-genomic variation in symbiotic dinoflagellates: recent divergence or recombination between lineages? BMC Evol Biol 15. https://doi.org/10.1186/s12862-015-0325-1
Wirshing HH, Baker A (2016) On the difficulty of recognizing distinct *Symbiodinium* species in mixed communities of algal symbionts. Mol Ecol 25:2724–2726. https://doi.org/10.1111/mec.13676
Wood-Charlson EM, Hollingsworth LL, Krupp DA, Weis VM (2006) Lectin/glycan interactions play a role in recognition in a coral/dinoflagellate symbiosis. Cell Microbiol 8:1985–1993. https://doi.org/10.1111/j.1462-5822.2006.00765.x
Yamashita H, Koike K (2013) Genetic identity of free-living *Symbiodinium* obtained over a broad latitudinal range in the Japanese coast. Phycol Res 61:68–80. https://doi.org/10.1111/pre.12004
Yamashita H, Suzuki G, Hayashibara T, Koike K (2013) Acropora recruits harbor "rare" *Symbiodinium* in the environmental pool. Coral Reefs 32:355–366. https://doi.org/10.1007/s00338-012-0980-2
Ziegler M, Roder CM, Büchel C, Voolstra CR (2014) Limits to physiological plasticity of the coral *Pocillopora verrucosa* from the central Red Sea. Coral Reefs 33:1115–1129. https://doi.org/10.1007/s00338-014-1192-8
Ziegler M, Arif C, Burt JA, Dobretsov S, Roder C, LaJeunesse TC et al (2017) Biogeography and molecular diversity of coral symbionts in the genus *Symbiodinium* around the Arabian Peninsula. J Biogeogr 44:674–686. https://doi.org/10.1111/jbi.12913

Chapter 7
How Does the Coral Microbiome Cause, Respond to, or Modulate the Bleaching Process?

K. M. Morrow, E. Muller, and M. P. Lesser

7.1 Introduction

Coral holobionts are formed by a dynamic multipartite symbiosis with intracellular photoautotrophs in the genus *Symbiodinium*, as well as a consortium of microorganisms that include bacteria, archaea, viruses, fungi, and protists. The coral holobiont functions as a unit to provide flexible stability in the face of constant environmental stressors. Coral bleaching, or the loss of *Symbiodinium* and their pigments, has caused significant global declines in the percentage cover of reef-building corals, particularly in recent decades as global sea surface temperatures continue to rise (Lesser 2004; Hoegh-Guldberg et al. 2007). Bleaching can be caused by a number of environmental stressors including extreme fluctuations in seawater temperature (increase or decrease), high solar irradiance, sedimentation, pollution, herbicides, and reduced salinity. Hyperoxic conditions have been shown to act synergistically with solar radiation and thermal stress to produce significantly greater fluxes of reactive oxygen species (ROS) in both host tissues and *Symbiodinium* spp. that leads to photosynthetic dysfunction, apoptosis, and bleaching (Lesser 2006, 2011; Oakley et al. 2017). We now understand that different phylotypes of *Symbiodinium* sp. within each clade represent multiple phenotypes and potentially species (Thornhill et al. 2014), and display variable rates of photosynthesis, capacity

K. M. Morrow (✉)
Department of Molecular, Cellular and Biomedical Sciences, University of New Hampshire, Durham, NH, USA

Department of Environmental Science and Policy, George Mason University, Fairfax, VA, USA

E. Muller
Mote Marine Laboratory, Saratosa, FL, USA

M. P. Lesser
Department of Molecular, Cellular and Biomedical Sciences, University of New Hampshire, Durham, NH, USA

M. J. H. van Oppen, J. M. Lough (eds.), *Coral Bleaching*, Ecological Studies 233,
https://doi.org/10.1007/978-3-319-75393-5_7

to photoacclimate, stress tolerance, ROS production, superoxide dismutase (SOD) activity, and metabolic interchange with their hosts (Banaszak et al. 2006; Robison and Warner 2006; Reynolds et al. 2008; Suggett et al. 2008; Hennige et al. 2009; Brading et al. 2011; Lesser 2011; Buxton et al. 2012; McGinty et al. 2012; Parkinson and Baums 2014; Roberty et al. 2014; Krueger et al. 2015; Warner and Suggett 2016; Grégoire et al. 2017).

However, much less attention has been paid to the coral microbiome, which includes the *Symbiodinium*, prokaryotes and viruses, and their role in coral bleaching and the bleaching response. Microbes can exhibit extensive diversity and abilities to respond to and withstand environmental pressures, and, as with other multipartite mutualisms (Hussa and Goodrich-Blair 2013), corals may be able to take advantage of these microbial partnerships by recruiting taxa with distinct environmental tolerances that provide a means to adapt and/or acclimate to environmental change. Here, we provide some context for the importance of the coral microbiome to holobiont function. We survey what is currently known about the relationship among *Symbiodinium*, temperature stress, and the associated prokaryotes and viruses. We also examine the impact of the coral bleaching response and environmental stressors conducive to bleaching (e.g., temperature and irradiance) on coral-associated microorganisms with particular emphasis on diazotrophic (nitrogen-fixing) bacteria. In conclusion, we summarize how a shifting and potentially dysbiotic microbiome may impact the coral host in the context of bleaching.

7.2 The Coral Holobiont

Coral-associated microorganisms form an intimate and often species-specific relationship with their hosts, aiding in a number of beneficial functions (reviewed in Bourne et al. 2016) and potentially provide assisted acclimatization in the face of a changing climate (Webster and Reusch 2017). Recent reviews have proposed the term "Beneficial Microorganisms for Corals" (BMC), which defines core microbiota or microbial consortia that maintain coral health and resilience and potentially act as bioindicators of environmental stress (Peixoto et al. 2017). Coral-associated microorganisms are known to provide pathogen resistance through the production of antimicrobial compounds (Ritchie 2006; Rypien et al. 2010), catabolism of dimethylsulfoniopropionate (DMSP), and the production of sulfur-based antimicrobial compounds and antioxidants (e.g., Raina et al. 2010, 2013; Todd et al. 2010; Howard et al. 2011), as well as the acquisition and cycling of critical nutrients (i.e., carbon, nitrogen, phosphorus, metals, vitamins; Wegley et al. 2007; Raina et al. 2009; Zhang et al. 2015; reviewed in Bourne et al. 2016). Microorganisms also appear to have evolved with corals at every life history stage (Apprill et al. 2009; Sharp et al. 2012) and may be critical to their early settlement success and metamorphosis onto reef substratum (Negri et al. 2001; Webster et al. 2004; Sneed et al. 2014; Sharp et al. 2015).

Although the coral microbiome is generally found to be diverse and species specific (Rohwer et al. 2002), recent literature has demonstrated its flexibility as a result of host physiology and morphology (Thompson et al. 2015), life history stage (Sharp et al. 2012), and microhabitat within the coral substructure (i.e., mucus, tissues, gastric cavity, skeleton; Sweet et al. 2011; Ainsworth et al. 2015). The composition of the coral microbiome also varies with season, geography, and environmental influences (e.g., temperature, pollution, macroalgae; Hong et al. 2009; Littman et al. 2009; Zhang et al. 2015; Glasl et al. 2016). Thus, as with many other multipartite symbioses (reviewed in Hussa and Goodrich-Blair 2013), we are finding that the composition and function of the community are not static and appear to be influenced by host physiology, health, and the surrounding environment, likely fluctuating in time and space. There may be low- or high-abundance bacterial or archaeal phylotypes that play a core role, as well as consortia of microorganisms working together to perform critical functions, but many of the ancillary commensal microbes may simply be filling a niche or providing functional redundancy.

In comparison to the host, symbiotic microbes divide and evolve more rapidly, potentially influencing the ability of the holobiont to adapt and/or acclimate to changing environmental conditions. In this regard, van Oppen et al. (2015) hypothesized that modification of the microbiome may be one way to facilitate transgenerational acclimatization of coral reef organisms. This concept was further developed in a commentary by Damjanovic et al. (2017), in which they proposed a series of manipulative experiments to examine whether microbial mediation could aid in long-term coral stability. Subsequent perspectives have also examined the importance of microorganisms to corals as an adaptive mechanism when faced with climate change stressors such as elevated seawater temperature and pCO_2 levels (Torda et al. 2017; Webster and Reusch 2017). By shuffling community composition of both the microbiome and *Symbiodinium* populations, in addition to acquiring new genetic material through mutation and/or horizontal gene transfer, while interacting with the surrounding seawater microbiota, the holobiont may or may not transfer advantageous microbial alterations to new generations that could help them avoid or withstand future bleaching events (Fig. 7.1; Webster and Reusch 2017). This theory also feeds into the founder effect, which occurs when a new population (e.g., coral planula microbiome) is established by a small number of individuals from a larger population (parent colony microbiome), leading to speciation and subsequent evolution in extreme cases (Barton and Charlesworth 1984). The founder effect originates in population genetics but has been applied to the study of microbiomes for some time. For example, founder populations from the human mother may be key to a more natural ecological succession of the infant gut, leading to stability within adult immune and metabolic responses (Mueller et al. 2015). In a similar manner, vertically and horizontally transmitted founder populations from acclimated coral holobionts may be key to the stability of future generations (Fig. 7.1). Although we still have much to learn about the function and flexibility of the dynamic relationship between coral host and microbiome, the theory that holobiont resilience may be positively influenced by a flexible, and potentially manipulable, multipartite symbiosis is met with hesitant optimism.

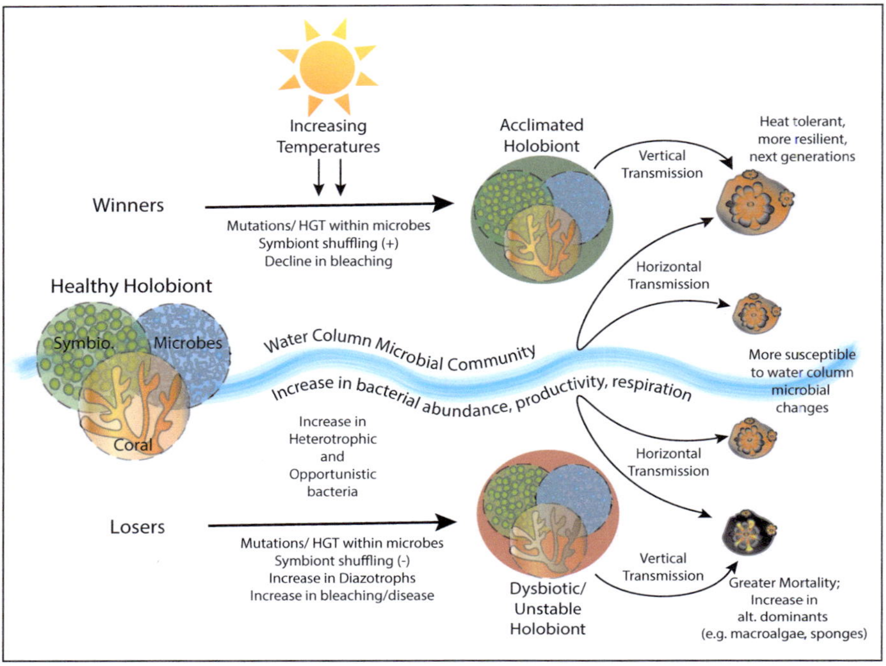

Fig. 7.1 Illustration depicts how shifts in coral microbiome and *Symbiodinium* populations, coupled with acquiring new genetic material through mutation and/or horizontal gene transfer, may be advantageous (e.g., winners) or disadvantageous (e.g., losers), to future generations of coral holobionts. Corals may be further impacted by seawater microbial community composition as this too will shift in response to changing environmental conditions and ecosystem structure resulting from repeated bleaching events [Concept adapted from Bourne et al. (2016) and Webster and Reusch (2017)]

7.3 Holobiont Response to Thermal Stress and Bleaching

For over two decades, we have known that coral bleaching events can have significant and detrimental impacts on microbial community composition (Ritchie et al. 1994), although only a handful of studies have specifically examined the impact of bleaching and the bleaching response on the coral microbiome ($n = 9$; Table 7.1). Most studies have instead focused on the impact of elevated temperature stress rather than irradiance, sedimentation, or pollution as a bleaching catalyst ($n = 15$; Table 7.1). It should be noted that although we do not include studies associated with the *Vibrio*-induced bleaching model system, we describe the concepts in Sect. 7.6.

In general, bleaching-related microbial community shifts are often characterized by a higher proportion of taxa affiliated with opportunistic bacteria (Mouchka et al. 2010), an increase in genes associated with virulence factors (Littman et al. 2011), and a shift toward *Vibrio* dominance (Bourne et al. 2008; Frydenborg et al. 2014;

Table 7.1 Summary of studies evaluating the impact of bleaching (*B*) and temperature stress (*T*) on the coral microbiome (References in chronological order)

Stressor	Impact on microbiome	Specific bacterial response	Coral spp.	Experiment type	Profiling technique	Primers	References
B	Shift in community structure	Increase in *Vibrio*/*Aeromonas* spp.	*Orbicella annularis*	In situ	Biolog plates	n/a	Ritchie et al. (1994)
B	Reduced antibacterial activity in mucus bacteria	Increase in *Vibrio* spp. dominance	*Acropora palmata*	Cultures	Assays for antibiotic activity and symbiont selection	16S (R1n/U2) BigDye direct sequencing	Ritchie (2006)
B	Shift and then recovery after stressor removed	Increase in *Vibrio* spp.	*Acropora millepora*	In situ	DGGE, clone libraries	16S DGGE (1055F/1392R) 16S clone lib. (27F/1492R)	Bourne et al. (2008)
B	Shift in bacterial community structure	Increase in *Acinetobacter* spp.	*Oculina patagonica*	In situ	Clone libraries	16S (8F/1492R)	Koren and Rosenberg (2008)
B	Shift in viral community structure	Herpes-like virus detected	*Diploria strigosa*	In situ	Metagenomics	Shotgun metagenome	Marhaver et al. (2008)
B	Healthy and bleached corals have similar dominant taxa	Higher proportion of *Vibrio* and *Acidobacteria* in bleached corals	Multiple	Both	Multiple	Multiple	Mouchka et al. (2010)
B (+4 °C)	Increase in bacteria, viruses, and shift in community structure	Increase in bacteria, viruses, and viral lytic production	*Fungia repanda*	Aquaria	Enumeration, respiration, lytic production, DGGE	16S (341F/519R)	Nguyen-Kim et al. (2015)
B	Large viral loads with VLPs found in *Symbiodinium*	Megavirus described that might infect and kill *Symbiodinium*	*Acropora aspera* and *A. millepora*	In situ	TEM and metagenomics	Nextera XT metagenome	Correa et al. (2016)

(continued)

Table 7.1 (continued)

Stressor	Impact on microbiome	Specific bacterial response	Coral spp.	Experiment type	Profiling technique	Primers	References
B	No change in overall community composition; decrease in diversity	Increase in diazotrophs	*Pocillopora verrucosa*	Aquaria	Illumina MiSeq and qPCR	nifH (F2/R6) & 16S (16SMiSeq-Andersson)	Pogoreutz et al. (2017)
T (+TOC, pH, nutrients)	Microbiome shift to more disease-associated state (i.e., virulence genes, disease-associated sequences)	Selection of virulence gene containing *Vibrio* spp.	*Porites compressa*	Aquaria	GenomiPhi amp + 454 metagenomes	n/a	Vega Thurber et al. (2009)
T (32 °C)	Bacterial shifts associated with juv. corals hosting *Symbiodinium* clade D but not clade C1	Increase in *Vibrio* spp. unless corals previously exposed to elevated temperatures in field	Juvenile *Acropora tenuis*	Aquaria	DGGE, clone libraries	16S clone lib. (63F/1387R) 16S DGGE (1055F/1392R)	Littman et al. (2010)
T	Diversity unchanged	n/a	*Acropora millepora*	Aquaria	DGGE	16S (341F/907R)	Bellantuono et al. (2011)
T (31 °C)	Antibiotic-treated corals had tissue loss	n/a	*Pocillopora damicornis*	Aquaria	Biolog EcoPlates	n/a	Gilbert et al. (2012)
T (30 °C)	Commensal bacteria suppressed by opportunistic pathogens	Increase in *Vibrio* spp. due to better substrate use	*Acropora palmate* mucus	Cultures	*chi*A qPCR, glycosidase, and protease assays	*chi*A and housekeeping genes	Frydenborg et al. (2014)
T (+2.5 and +4 °C)	Threefold increase in diazotroph diversity and richness	Increase in *Alphaproteobacteria*, including *Azospirillum*, *Rhizobium*, and *Rhodobacter*-like sequences	*Mussismilia harttii*	Aquaria	qPCR, DGGE, clone libraries	nifH clone lib. (FGPH19/PolR)	Santos et al. (2014)

T (31 °C)	Increase in microbiome diversity	Increase in *Vibrio* spp.	*Pocillopora damicornis*	Aquaria	Illumina MiSeq	16S (27F/1392R +27F/519R 16S *Vibrio* (VF169/ Vib2_R)	Tout et al. (2015)
T	Diversity unchanged in *OF* but *GV* microbiome highly variable but diversity declined	Within *GV*, shifts in *Clostridia*, *Oceanospirillales*, and *Pseudomonadaceae*	*Orbicella faveolata (OF) Gorgonia ventalina (GV)*	In situ	Roche 454	16S (27F/338R)	Tracy et al. (2015)
T (26–33 °C)	Bacterial community shift between coral mucus and tissues	Increase in *Alphaproteobacteria*, *Verrucomicrobiae*, *Cyanobacteria*, *Flavobacteriia*, and *Sphingobacteria*	*Acropora muricata*	Aquaria	Roche 454	16S (968F/ 1391R)	Lee et al. (2015)
T (26–31 °C)	Increase in diversity and richness; change in mucus composition	Increase in *Alphaproteobacteria* and *Verrucomicrobiae*; decrease in *Gammaproteobacteria*	*Acropora muricata*	Aquaria	Roche 454	16S (968F/ 11391R)	Lee et al. (2016)
T (+ algae, nutrients)	Algal contact + T-destabilized coral microbiome; increased variability	Increase in *Vibrionales and Oscillatoriales*; decline in *Synechococcus*	*Porites*, *Siderastrea*, and *Agaricia*	In situ	Roche 454	16S (515F/ 806R)	Zaneveld et al. (2016)
T (32 °C)	Stable first 10 days; alpha diversity decreased and richness unchanged	Decline in *Rhodobacterales*; increase in *Alteromonadales*, *Vibrionales*, and *Flavobacteriales*	*Acropora digitifera*	Aquaria	Illumina MiSeq	16S (515F/ 806R)	Gajigan et al. (2017)

(continued)

Table 7.1 (continued)

Stressor	Impact on microbiome	Specific bacterial response	Coral spp.	Experiment type	Profiling technique	Primers	References
T	Stress-induced upregulation of viruslike gene expression in *Symbiodinium*	Nucleocytoplasmic large DNA viruses found in *Symbiodinium* cultures	*Symbiodinium* cultures	Aquaria	Transcriptomics (no replication)	Whole transcriptome	Lawrence et al. (2017)
T (+4 °C)	Shifts in gene expression levels of viruses infecting heat-stressed symbiont cultures	Nucleocytoplasmic large DNA viruses and novel + ssRNAV detected	*Symbiodinium* cultures	Aquaria	Illumina HiSeq	Whole transcriptome	Levin et al. (2017)
T (29–35 °C)	Microbiome rapidly shifts (<20 h); microbiome response predicted by prior exposure to T stress	Increase in *Alphaproteobacteria* and *Cyanobacteria*	*Acropora hyacinthus*	In situ, transplant, and aquaria	Illumina MiSeq	16S (784F/1061R)	Ziegler et al. (2017)

Tout et al. 2015; Table 7.1), although some corals, such as *Porites lobata* (Hadaidi et al. 2017) and *Orbicella faveolata* (Tracy et al. 2015), are still able to maintain stable microbiomes through bleaching events. More recently, research has specifically focused on the impact of elevated temperatures on the coral microbiome, finding that even a single stressor can cause significant shifts in the microbiome, often increasing diversity (McDevitt-Irwin et al. 2017; Table 7.1), and the abundance of opportunistic bacteria such as *Vibrio* spp. (Littman et al. 2010; Frydenborg et al. 2014; Tout et al. 2015; Gajigan et al. 2017).

The coral microbiome itself may have a temperature tolerance threshold, as it was found that bacterial community changes occurred at temperature elevations >1 °C, with no evidence of community shifts at lower-temperature changes (Salerno et al. 2011). Other studies have demonstrated a link between temperature stress and bacteria associated with nitrogen cycling (reviewed in Sect. 7.3.1; Santos et al. 2014; Ziegler et al. 2017). Similar patterns emerge when corals are exposed to the compounding effects of multiple stressors (i.e., elevated temperatures, dissolved organic carbon, pH, and/or nutrients), causing shifts in the microbiome community toward a dysbiotic state, or disturbance of the "normal flora," which may be associated with a disease state leading to bleaching and/or mortality (Rosenberg et al. 2007; Vega Thurber et al. 2009; Zaneveld et al. 2016). Zaneveld et al. (2016), in a 3-year field study, showed that chronic exposure to eutrophication, algal overgrowth (e.g., simulated overfishing), and temperature stress destabilized coral microbiomes making them more susceptible to the effects of future exposures. Webster et al. (2016) also demonstrated a significant interactive effect of thermal stress and ocean acidification on the microbial communities of corals and other important calcifying species on coral reefs. These studies, and others outlined below, demonstrate that microbiomes can change rapidly in response to moderate to severe environmental stress, potentially aiding in the adaptability of their host (proposed in Webster and Reusch 2017; Peixoto et al. 2017) and/or leading to their destabilization and loss of critical functions (Zaneveld et al. 2016; McDevitt-Irwin et al. 2017; Fig. 7.1).

A recent meta-analysis found that stress, particularly during climate anomalies, is implicated in an increase in community diversity and a decline in the relative abundance of potentially key coral endosymbionts in the genus *Endozoicomonas* (class *Gammaproteobacteria*, order *Oceanospirillales*; McDevitt-Irwin et al. 2017). Members of the *Endozoicomonas* genus are often dominant members of coral microbiomes (Morrow et al. 2012; Bayer et al. 2013; Rodriguez-Lanetty et al. 2013) and can be found deep within coral tissues (Bayer et al. 2013; Neave et al. 2016a). They are hypothesized to prevent mitochondrial dysfunction and promote gluconeogenesis (Ding et al. 2016), aid in sulfur cycling (Neave et al. 2016b), and protect the coral from bleaching pathogens (Pantos et al. 2015), potentially through the production of quorum-sensing metabolites (Mohamed et al. 2008) or antimicrobial compounds (Rua et al. 2014). McDevitt-Irwin et al. (2017) and others (Zaneveld et al. 2016) provide evidence that suggests the coral microbiome becomes more diverse (increased richness), more variable (reduced evenness), and less stable when under stress. Interestingly, a similar pattern has emerged in human microbiome studies, where once it was believed that a dysbiotic state was characterized by low diversity and overabundant opportunistic or pathogenic taxa (Lozupone et al. 2012); more recent studies show that human microbiomes can also become more variable under stress (reviewed in Zaneveld et al. 2017). Perhaps this discrepancy has to do with the time frame in which we are sampling microbiomes, capturing the community as it transitions from a healthy equilibrium through a variable dysbiotic state to an alternate stable state characterized by pathogens and disease symptoms (reviewed in Bourne et al. 2016). In fact, communities are more often in a transient state of disturbance than stability, and some have suggested we shift our focus to studying alternate transient states, rather than alternate stable states, in order to make more relevant predictions about community assembly (Fukami and Nakajima 2011). Nevertheless, it would likely be difficult for the coral to return to a healthy equilibrium after a stressor is alleviated, leading to a reduction in overall ecosystem resilience, unless the ability of the coral microbiome to shift and acclimate to changing environmental conditions confers some adaptive advantage to the coral host.

One of the best examples that the coral microbiome may actually preadapt a coral to survive thermal stress is a recent study by Ziegler et al. (2017), which presents experimental data demonstrating that the microbiome of heat-sensitive and heat-tolerant corals is significantly different and that heat-tolerant corals exposed to bleaching temperatures showed no changes in their microbiomes and bleached less often. In these experiments, corals (*Acropora hyacinthus*) were exposed to both a long-term (17-month) reciprocal transplant experiment between two thermally distinct environments and a short-term heat-stress experiment. Coral microbiomes were shown to rapidly adjust to new environmental conditions (~20 h), and the thermal environment from which the corals originated (17-month experiment) predicted their microbial response to heat stress. These results suggest that long-term exposure to environmental stress such as thermal variability allows the coral microbiome to acclimatize, which in turn may play a role in a coral's resistance to thermal stress (Fig. 7.1). In an earlier laboratory-based study (Bellantuono et al. 2011), *Acropora millepora* corals were exposed to temperatures 3 °C below the bleaching threshold (generally defined as ~1 °C above summer maximum) in a short-term (10-day) experiment, which resulted in a significant reduction in bleaching susceptibility. However, no changes were detected in the *Symbiodinium* populations based on internal transcribed spacer region 2 (ITS2) sequencing or bacterial populations based on denaturing gradient gel electrophoresis (DGGE) results from a single time point 6-days into the experiment. The authors concluded that rapid temperature acclimation may be a function of host physiological plasticity rather than shifts in the symbiont community. In another study, antibiotics were applied every day over the course of a heat-stress experiment to reduce the bacterial activity associated with *Pocillopora damicornis* corals while monitoring the coral holobiont response to thermal stress (Gilbert et al. 2012). Microbial viability and activity were monitored with 96-well Biolog EcoPlates™, an assay panel that measures bacterial carbon metabolism. The heat-stressed and antibiotic-treated holobiont displayed significantly depleted host protein levels, chlorophyll *a* concentrations, and tissue loss in comparison to corals with an intact (i.e., unmanipulated) microbiome, again suggesting that the microbial consortium may provide some resilience against thermal stress. Thus, although thermal acclimatization has been widely demonstrated in reef corals (Brown et al. 2002a, b; Middlebrook et al. 2008; Barshis et al. 2010; Palumbi et al. 2014 among others), these recent studies are the first to demonstrate the potential role of the microbiome in host stability and thermal tolerance.

7.3.1 Diazotrophy and the Bleaching Response

Nitrogen is a critical and limiting nutrient that corals generally acquire through the uptake of dissolved inorganic nitrogen or heterotrophic feeding. Several studies have demonstrated the presence of nitrogen-fixing bacteria (diazotrophs) in the coral microbiome (e.g., Lesser et al. 2004) and that "new" nitrogen can be obtained from these members of the coral microbiome (Lesser et al. 2007a). More broadly, there is evidence suggesting that there is a critical link between coral bleaching and the availability of environmental nitrogen (Wooldridge 2013; Vega Thurber et al. 2014; Shantz et al. 2016; Pogoreutz et al. 2017). In fact, an increase in the acquisition of nitrogen through heterotrophy has been shown to reduce post-bleaching photoinhibition and recovery times (Grottoli et al. 2006; Ferrier-Pagès et al. 2010; Hoogenboom et al. 2012). To this end, Godinot et al. (2011) demonstrated that elevated temperatures (33 °C) caused a severe decrease in nitrate and ammonium uptake rates, even leading to release of nitrogen into seawater. Furthermore, the combination of high temperature (33 °C) and low pH (7.5) resulted in a significant decline in phosphate and nitrate uptake rates, although these would be the ecological extremes for tolerable temperatures and pH (Godinot et al. 2011).

From the perspective of diazotrophy (i.e., bacteria and archaea that fix atmospheric nitrogen gas into a more usable form such as ammonium), daylight dinitrogen (N_2) fixation was shown to significantly increase in corals exposed to a 6 °C temperature increase in comparison to controls (Cardini et al. 2016). Corals exposed to nitrogen-enriched seawater also demonstrated an increase in photoprotective pigment concentrations while maintaining rates of photosynthesis and calcifications at ca. 60% and 100% of rates for unenriched controls that experienced a significant decrease in photosynthesis and calcification (Beraud et al. 2013). The authors suggest that inorganic nitrogen availability may be akin to heterotrophic feeding in maintaining coral metabolism under stressful conditions.

Based on the above experiments, it is, therefore, not surprising that recent research has also documented an increase in the abundance and diversity of coral-associated diazotrophic bacteria during thermal stress events. Diazotroph diversity and richness within *Mussismilia harttii* corals increased by threefold during both a 2.5 and 4 °C experimental temperature increase (Santos et al. 2014). Pogoreutz et al. (2017) also documented a significant increase in diazotrophic activity during a 28-day, sugar-induced bleaching experiment with *Pocillopora verrucosa* corals. Although the Pogoreutz et al. (2017) study may not be ecologically relevant, they demonstrated the impact of bleaching on the diazotrophic community without the confounding effects of heat and light stress. High temperature and light also increase rates of organic matter release by corals into seawater which could further enhance pelagic N_2 fixation rates (Böttjer et al. 2016). Ainsworth et al. (2015) identified two potential N_2-fixing bacteria as intracellular within *Symbiodinium* and within *Symbiodinium*-containing coral cells (*Actinomycetales* and *Ralstonia*, respectively), using fluorescent in situ hybridization (FISH) and 16S rRNA gene-targeted sequencing (454 tag sequencing). Both identified phylotypes can also form diazotrophic symbiotic associations in other photosynthetic systems (Chen et al. 2003; Sellstedt and Richau 2013). Finally, although diazotrophs were not specifically examined, Ziegler et al. (2017) identified bacteria affiliated with the class *Alphaproteobacteria* (e.g., *Rhodospirillaceae*, *Rhizobia*) as responsible for a large fraction of the functional enrichment within heat-tolerant corals, as opposed to *Gammaproteobacteria* (e.g., *Hahellaceae*, *Alteromonadaceae*, *Vibrionaceae*), within heat-sensitive corals. Similarly, Lee et al. (2016) documented an increase in *Alphaproteobacteria*, *Verrucomicrobiae*, and *Cyanobacteria* in the mucus of thermally stressed corals preceding a bleaching event. Members of the *Alphaproteobacteria* and *Cyanobacteria* are often associated with nitrogen cycling in host-associated microbiomes (Lesser et al. 2004, 2007b; Tsoy et al. 2016).

As noted above, an increase in available nitrogen may benefit corals under thermal stress; however, too much may interfere with the internal equilibrium between host and *Symbiodinium*, potentially contributing to the bleaching response. Diazotrophy in corals has been shown to increase the *in hospite* growth rates of *Symbiodinium* under normal environmental conditions without an increase in biomass (Lesser et al. 2007a), presumably from a host-controlled increase in symbiont losses. This was hypothesized to be offset by an increase in daily turnover rates of *Symbiodinium* (Lesser et al. 2007a). Higher levels of nitrogen supplied by diazotrophic bacteria would likely release *Symbiodinium* completely from N-limited growth and cause high rates of cell division and reduced translocation of photosynthates to the coral (Falkowski et al. 1993; Dubinsky and Jokiel 1994; Suescún-Bolívar et al. 2016). Pogoreutz et al. (2017) theorize that retaining photosynthates could result in the energy limitation of coral carbon-concentrating mechanisms (CCMs), which would result in carbon (i.e., CO_2) limitation within photosynthetic dark reactions, causing a heightened susceptibility to photodamage and subsequent bleaching (i.e., sink limitation). Stimulated nitrogenase activity due to elevated temperatures leading to a further increase in coral-associated diazotrophic activity is believed to be another mechanism by which the internal nutrient equilibrium within the coral holobiont becomes imbalanced, disrupting the N-limited state of *Symbiodinium* and potentially inducing or prolonging bleaching events (Rädecker et al. 2015; Fig. 7.2). Pernice et al. (2012) clearly demonstrated the importance of ammonium (NH_4^+) uptake from the surrounding seawater, while Cardini et al. (2015) demonstrated that on a seasonal basis some corals could become more dependent on the contribution of fixed N_2 for their nitrogen budgets. These studies suggest that coral-associated nitrogen fixers may be more intimately tied with *Symbiodinium* and the coral bleaching response than we currently understand (Fig. 7.2).

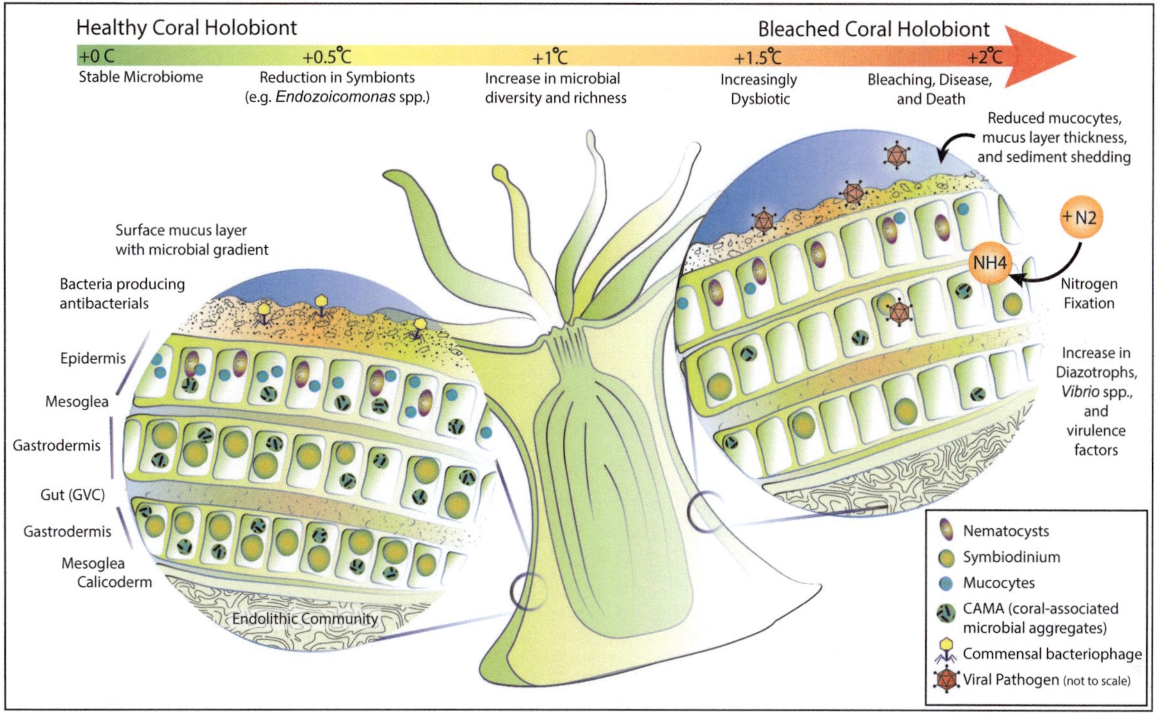

Fig. 7.2 Illustration depicts hypothesized physiological changes to the coral host and shifts in the microbial communities as a coral is exposed to increasing temperature stress. Coral mucocytes are reduced, mucus composition changes, and surface mucus layer thickness is reduced in corals experiencing bleaching temperatures, which may disrupt the microbiome composition and bacterial/viral gradient, impeding the protective benefits of high lytic activity within intermediary mucus layers. The relative abundance of putative symbionts (e.g., *Endozoicomonas* spp.), within coral-associated microbial aggregates (CAMA) decline, and bacteria affiliated with pathogens (e.g., *Vibrionales*) are stimulated. Diazotrophy (nitrogen fixation) increases as temperatures increase, which may release *Symbiodinium* from N-limited growth, causing higher rates of cell division, reduced translocation of photosynthates, and ultimately leading to greater susceptibility to photodamage and subsequent bleaching

7.3.2 Coral Mucus and the Mucus Microbiome in Response to Bleaching

Thermal stress and bleaching can also cause a compositional change in coral mucus (Wooldridge and Done 2009), which is a protective boundary layer between the coral tissues and external environment where the majority of the microbiome takes up residence, in a similar manner to the mucosal community of the human gut (Smillie et al. 2011). The surface mucus layer is loosely defined as a polysaccharide-protein-lipid complex secreted by epithelial mucocytes onto the coral surface (reviewed in Brown and Bythell 2005). Mucus production by corals not only aids in feeding and self-cleaning but also provides protection against pathogens, desiccation, UV radiation, pollutants, and other physical damage (Brown and Bythell 2005; Bythell and Wild 2011; Barr et al. 2013a). Much of the fixed carbon within the mucus layer originates from the *Symbiodinium* and serves as a rich food source for bacteria (Ritchie and Smith 2004). By providing a stable growth medium for both beneficial and detrimental bacteria, mucus plays a key role in a coral's innate immune function, enhancing susceptibility or providing protection from pathogens (Ducklow and Mitchell 1979; Banin et al. 2001; Lipp and Griffin 2004). However, as the composition of coral mucus is altered by thermal stress events, so will the stability of the mucus- and tissue-associated microbiome.

Coral bleaching and thermal stress increase organic matter and mucus production in some corals (Niggl et al. 2009) while decreasing mucus thickness in others (i.e., *Diploria* sp.; Pratte and Richardson 2014). For example, the chemical composition of *Acropora muricata* coral mucus was altered in experimentally thermally stressed corals (26–33 °C), causing a change in the proportion of simple sugars (e.g., fucose, glucose, and mannose) and a reduction in the proportion of *N*-acetyl glucosamine and C6 sugars, which also correlated with a shift in the coral-associated microbial community (Lee et al. 2016). A drop in the relative abundance of *Gammaproteobacteria* was associated with a change in the content of fucose and mannose sugars, and an increase in *Cyanobacteria* was correlated with shifts in arabinose and xylose (Lee et al. 2016). Furthermore, these changes in the composition of the mucus and microbiome began at 29 °C, prior to visual signs of bleaching, which occurred at 31 °C, suggesting that mucus composition could be used as a bioindicator of pre-bleaching conditions.

In a seminal paper describing the coral mucus microbiome, Ritchie (2006) demonstrated that healthy *Acropora palmata* mucus selected for bacteria that produced antibiotics active against a putative coral pathogen (*Serratia marcescens* PDL100). Not only was antibiotic activity lost during a summer bleaching event (i.e., increased temperatures), but coral tissues were dominated by bacteria affiliated with the genus *Vibrio*. Since then, several studies have also shown a reduction in antibacterial activity of mucus-associated bacteria under elevated temperature stress (Shnit-Orland and Kushmaro 2009; Rypien et al. 2010). Recent work has suggested additional mechanisms for the suppression of opportunistic pathogens by coral-associated commensal bacteria. For example, Frydenborg et al. (2014) demonstrated that *Vibrio* spp. were better able to use key substrates found in coral mucus (e.g., *α*-D-glucopyranosidase), allowing them to outcompete commensal coral bacteria under elevated temperatures. Lee et al. (2015) also demonstrated that the relative abundance of *Vibrio* spp. increased and the putative symbiont, *Endozoicomonas* spp., decreased at pre-bleaching temperatures (29 °C). Interestingly, this declining trend continued for *Endozoicomonas* spp., but the abundance of *Vibrio*-related OTUs also declined in the tissues but increased in the mucus layer as the coral reached bleaching temperatures (31 °C; Lee et al. 2015), perhaps avoiding the toxic environment developing within the coral tissues.

Thermally bleached corals have also demonstrated a reduced capacity to remove sediments (Bessell-Browne et al. 2017), which are laden with bacteria and viruses (Breitbart et al. 2004). Sedimentation and water turbidity can result from both anthropogenic activities (e.g., dredging, coastal runoff) and weather events. The removal of sediments is not only energetically costly (Peters and Pilson 1985; Riegl and Branch 1995) but has also been associated with the transmission of pathogens (Hodgson 1990) and increased disease prevalence (Haapkylä et al. 2011; Pollock et al. 2016). Bleaching may further interfere with sediment removal by reducing the number of epithelial mucocytes (Fitt et al. 2009; Piggot et al. 2009), as well as reduced mucus within the deeper gastrodermal layers (Fitt et al. 2009). Increased accumulation of sediments on bleached corals can lead to mucus sheet formation (Bessell-Browne et al. 2017), necrosis (Weber et al. 2012), and ultimately mortality. These studies indicate that the structure and composition of coral mucus can change under thermal stress and bleaching events (see also Krediet et al. 2009; Mao-Jones et al. 2010), which may lead to an environment that is less stable and more attractive to opportunistic microorganisms and pathogens than beneficial symbionts (Fig. 7.2).

Bacteriophages (i.e., a virus that infects and replicates within a bacterium) also play a dynamic and little recognized role in the development and maintenance of coral mucosal communities and have been shown to shift their community composition during environmental changes, including disease states (Columpsi et al. 2016; Soffer et al. 2015). Bacteriophage adherence to mucus (BAM) describes how phage can directly attach and interact with coral mucins (Barr et al. 2013a), displaying specific bacterial hunting behavior and potentially aiding in phage-mediated immunity (Barr et al. 2013b). The coral mucus layer is thought to be spatially structured in a similar manner to other metazoans (Johansson et al. 2011), forming a gradient from the seawater interface where microbial abundance is highest to the coral epithelium where microbial abundance is lowest and mucin production is greatest (Fig. 7.2). Silveira and Rohwer (2016) proposed that this gradient supports greater viral lysogeny at the mucosal-seawater interface, protecting coral commensals from superinfection and increasing their fitness, while deeper layers are protected from invading pathogens via higher levels of phage infection and cell lysis. High lytic activity within the intermediary mucus layers likely facilitates diversification of the bacterial community, aiding in the maintenance and assembly of a healthy microbial community by providing spatial refuges and coexistence stabilization (Schrag and Mittler 1996; Klimenko et al. 2016). Therefore, shifts in the structure and thickness of mucus resulting from environmental stress or disease may increase the proximity of microbes to the coral epithelium (Earle et al. 2015), shifting the phage-mediated immunity mechanisms and potentially allowing for invasion by lysogenic pathogens leading to the onset of infection (Fig. 7.2).

7.4 Microbiome-*Symbiodinium* Interactions

The control of the composition of the coral microbiome may be dependent on multiple factors, including host age, genotype and clade, *Symbiodinium* phylotype, and/or the environment (Hernandez-Agreda et al. 2016). In particular, the presence of photosymbionts is believed to influence the composition of host microbiomes through the release of a complex array of organic exudates (e.g., DMSP, amino acids, and polysaccharides; Bourne et al. 2013). Habitat-specific environmental differences have also been shown to drive the composition of the microbiome in sponges (Morrow et al. 2016) and in corals (Pantos et al. 2015). This may be why Hernandez-Agreda et al. (2016) identified a distinct group of bacteria associated with one coral species across multiple habitat types. Other studies have identified additional features of the host that are deterministic for the community structure of the coral microbiome (Williams et al. 2015; Apprill et al. 2016).

The growth and density of *in hospite Symbiodinium* populations are highly dependent on nitrogen availability; therefore, microorganisms that mediate nitrogen cycling within the holobiont may have consequences for the stability of these critical dinoflagellate populations (see Sect. 7.3.1). For example, diazotrophs (N_2-fixing bacteria and archaea) have been estimated to provide as much as 11% of the *Symbiodinium* nitrogen requirements (Cardini et al. 2015). Recent research has also confirmed earlier observations (Lesser et al. 2007a) that the majority of fixed nitrogen is translocated to the *Symbiodinium* compartment and is largely driven by bacteria affiliated with the *Alpha-* and *Gammaproteobacteria*, including the orders *Rhizobiales* and *Rhodobacterales* (Olson and Lesser 2013; Lesser et al. 2017). Members of the *Cyanobacteria* (Lesser et al. 2007a), fungi (Wegley et al. 2007), and archaea (Siboni et al. 2012) are also implicated in nitrogen cycling within corals. Also, common diazotrophs such as *Rhizobia* are found in early life stages, suggesting that these relationships develop early and provide critical photosymbiont stability within the holobiont (Lema et al. 2014).

Additional interactions between *Symbiodinium* and bacteria have been identified; for example, bacteria affiliated with the order *Roseobacterales* (within the class *Alphaproteobacteria*) were shown to form obligate associations with *Symbiodinium* in laboratory cultures and may increase *Symbiodinium* growth rates (Ritchie 2012). The *Roseobacterales*, along with other coral-associated bacteria (e.g., *Endozoicomonas*, *Halomonas*), are also affiliated with sulfur cycling in the coral holobiont (Raina et al. 2009; Todd et al. 2010). Sulfur compounds such as dimethylsulfoniopropionate (i.e., DMSP) and its breakdown products can act as antioxidants that may protect *Symbiodinium* from photosynthesis-derived oxidative stress (Sunda et al. 2002; Deschaseaux et al. 2014). Furthermore, a strong negative correlation has been demonstrated between the abundance of bacterial pathogens and the abundance of the proposed symbiont *Endozoicomonas* in bleaching corals (Pantos et al. 2015). This protective benefit was further alluded to in a study by Meyer et al. (2014), in which disease lesions on the coral *Porites astreoides* correlated with a low relative abundance of bacteria affiliated with *Endozoicomonas* spp. rather than with a specific pathogen. Although multiple studies indicate destabilization of the bacterial community is associated with stress (Vega Thurber et al. 2009), pathogen exposure (Welsh et al. 2017), or disease symptoms (Sunagawa et al. 2009), ultimately it is difficult to separate correlation from causation to determine whether *Endozoicomonas* spp. are responsible for protecting the coral from pathogens (Glasl et al. 2016) or are simply declining in response to stress and/or disease (Morrow et al. 2015, 2017).

It is clear that very little is known about the specific interactions between the microeukaryote *Symbiodinium* and prokaryotic components of the coral microbiome. The emerging story of the coral microbiome was preceded by theory and technical approaches pioneered by studies on the human microbiome. Similarly, the oceanographic community has been intensely interested in metabolic exchanges between phytoplankton, including dinoflagellates, and bacterioplankton in what has been called the "phycosphere" (Tang et al. 2010; Hu et al. 2015; Bolch et al. 2017). These studies have demonstrated that unique metabolic interdependencies exist within many pelagic eukaryotic-prokaryotic associations that could act as model systems to guide studies on similar interactions in symbiotic systems such as corals. One of the most well-studied examples is the *Roseobacter*-algae interaction, whereby *Roseobacter* bacteria supply vitamins, phytohormones, and antibacterial compounds to the alga (Sharifah and Eguchi 2011). However, when algal populations decline and release cell wall degradation products (i.e., *p*-coumaric acid), the *Roseobacter* shift from mutualistic partner to opportunistic pathogen, releasing 11 types of troponoids that eventually kill the alga and switch the *Roseobacter* from a sessile lifestyle to a motile, free-living cell (Sule and Belas 2013; reviewed in Ramanan et al. 2016). Thus, there is certainly potential for bacteria to play a critical role in the growth, stability, and perhaps demise of the coral-*Symbiodinium* relationship as was previously suggested in the "bacteria-induced bleaching" hypothesis (see Sect. 7.6).

7.5 Coral Bleaching in Relation to Coral Disease

While overfishing, pollution, and coastal development have long been drivers of coral reef degradation, the dominant cause of reef decline is climate change, which primarily manifests itself as coral bleaching. However, diseases with etiological agents, either primary or secondary in nature, are largely responsible for a 30% decline in worldwide coral cover over the past 30 years (Hughes et al. 2003). More recent research has predicted that coral diseases may cause as much mortality as bleaching within future decades (Maynard et al. 2015). While the assumptions of the model of Maynard et al. (2015) have been questioned (Lesser and van Woesik 2015), several other models have revealed that many coral diseases are not infectious (i.e., transmissible) and do not fit into a contagious disease model (Yee et al. 2011). Most studies are more consistent with disease prevalence being secondary to environmental stress, which leads to opportunistic coral disease outbreaks (Lesser et al. 2007b; Muller and van Woesik 2012; Randall and van Woesik 2015). A disease is defined as any impairment that interferes with or modifies the performance of normal functions, including responses to environmental factors, infectious agents, inherent or congenital defects, or combinations of these factors (Wobeser 2006). Identifying disease within corals is surprisingly difficult as there are very few macroscopic signs produced by corals to indicate poor physiological functioning. There is still little agreement on whether coral diseases are infectious or are opportunistic in nature, and many of the etiological agents still remain elusive (Lesser et al. 2007b; Apprill et al. 2013; Lesser and Jarett 2014).

Environmental stress causes coral bleaching (Fitt et al. 2001; Lesser 2011) and is also a prerequisite for the occurrence of many coral diseases (Burge et al. 2014). Indeed, environmental stress mediates the occurrence of coral disease by lowering host resistance, by increasing pathogen abundance and virulence, or through combinations of these responses (Bruno et al. 2007; Brandt and McManus 2009; Sokolow 2009; Reed et al. 2010; Muller and van Woesik 2014; Randall et al. 2014; Zvuloni et al. 2015). Even apparently healthy corals are often infected with microbes that may negatively impact health and host physiology (Miller et al. 2014). Despite improved efforts to identify the primary pathogens responsible for infection, most studies remain dependent on the macroscopic disease appearance for diagnosis (Ainsworth et al. 2007), with its associated limitations. Corals may show visual signs of disease such as bleached tissue, discoloration (darkening or abnormal pigmentation), abnormal growth, or tissue loss. But, the lack of baseline and basic epizootiological information has hindered our understanding of the relative importance of specific pathogens and environmental factors in the spread of disease epizootics. Interestingly, coral bleaching has also been associated with bacterial pathogens (see Sect. 7.6) and possibly viruses (Lawrence et al. 2015; Levin et al. 2017; see Sect. 7.7). Environmentally induced bleaching events can also increase the prevalence of infectious disease outbreaks on reefs through immune system suppression (Mydlarz et al. 2009), which could lead to an increase in disease susceptibility and longevity once infected. Furthermore, thermal stress conditions that often accompany coral bleaching can increase pathogen growth rates (Alker et al. 2001; Ward et al. 2007) and virulence (Ben-Haim et al. 2003; Kimes et al. 2012), changing commensal or mutualistic bacteria into pathogens.

7.5.1 Bleaching-Induced Disease

Coral bleaching events are often followed by infectious disease outbreaks (Guzman and Guevara 1998; Harvell et al. 2002; Bruno et al. 2007; Muller et al. 2008; Brandt and McManus 2009; Cróquer and Weil 2009; McClanahan et al. 2009; Miller et al. 2009), which may be due to a reduction in immune system function (e.g., Mydlarz et al. 2009). As discussed above most of the accumulating evidence supports the compromised host hypothesis, which suggests that the condition of the coral host, not necessarily the presence of a novel pathogen, is the primary reason for higher disease prevalence on many contemporary reefs (Lesser et al. 2007b; Muller and van Woesik 2014; Randall and van Woesik 2015). For example, during the 2005 coral bleaching event in the US Virgin Islands, Muller et al. (2008) showed that corals that had previously bleached also exhibited higher disease mortality than those that had not bleached. Corals that were resistant to high water temperatures were also more resistant to infectious diseases, which may have been because the immune system of temperature-tolerant corals was less impacted. Ritchie (2006) showed a significant reduction in antimicrobial properties within the mucus of bleached *Acropora palmata*, which allowed the proliferation of potential pathogens. Pinzón et al. (2015) also showed that genes that regulate the immune system within corals are downregulated up to a year after bleaching occurs. Therefore, as corals allocate resources in order to persist through a bleaching event, the immune system is suppressed and remains so for many months, likely increasing the probability of successful disease infection.

Although the host condition may play a critical role in disease dynamics, other research suggests thermal stress conditions may influence the behavior and pathogenicity of putative pathogens. Garren et al. (2014) showed that a bacterial pathogen, *Vibrio coralliilyticus*, uses chemotaxis and chemokinesis to target heat-stressed corals, using DMSP solely as a chemical cue to locate potential hosts. General behavior of coral pathogens also changes under different temperature conditions. For example, Garren et al. (2016) showed that *V. coralliilyticus* also increased chemotactic ability toward coral mucus when temperatures exceeded 23 °C, denoting an enhanced capability to track host-derived chemical cues. Further elevated temperatures (>30 °C) increased the pathogen's chemokinetic abilities, denoted by accelerated cell movement under favorable, mucus-rich, chemical conditions. Thermal stress has also been associated with an increase in virulence genes (Banin et al. 2003), lysis of coral cells (Ben-Haim et al. 2003), and infection by coral pathogens (Kushmaro et al. 1998; Ben-Haim and Rosenberg 2002). Thus, coral bleaching and the environmental parameters that are conducive to bleaching impact both host physiology and microbial community structure and function, setting the stage for subsequent and prolonged infections.

7.6 Bacteria-Induced Bleaching

Although region-wide mass coral bleaching events are most commonly attributed to unusually warm seawater temperatures, bleaching of the coral *Oculina patagonica* during the summer in the Mediterranean Sea was previously identified as the result of an infection with the bacterial pathogen *Vibrio shilonii* (previously referred to as

V. shiloi; Kushmaro et al. 1997). This particular *Vibrio* bacterium produces an extracellular superoxide dismutase (SOD) at 30 °C, but not at 16 °C, indicating a temperature-related virulence (Banin et al. 2003). The extracellular production of SOD protects the pathogen within the coral, allowing it to infect and persist within host tissue, producing an extracellular peptide toxin that inhibits algal photosynthesis ultimately leading to coral bleaching (Banin et al. 2000, 2003; Rosenberg et al. 2009). Extractions of the proline-rich toxin from *V. shilonii* also caused a reduction in the quantum yield of photosystem II (i.e., a sign of impending bleaching) of *O. patagonica* and *Acropora eurystoma*, but not several other Caribbean coral species, suggesting a regional or species-specific response to this potential phenomenon (Gil-Agudelo et al. 2017).

Although bacterial bleaching was documented a number of times prior to 2002, no additional cases of bacterial bleaching in *O. patagonica* have been found. This phenomenon led to the creation of the "probiotic hypothesis" (Reshef et al. 2006), whereby corals develop resistance to bacterial infection via an innate immune response or beneficial shift in their microbiome. Thus, the probiotic hypothesis was proposed as the mechanism of resistance to bacterial bleaching (Rosenberg et al. 2007), but this has been challenged based on the absence of bacteria-induced bleaching in *O. patagonica* after 2002 (Ainsworth et al. 2008). To date, corals have only been shown to possess a very basic innate immune system, although adaptive-like properties, reminiscent of higher organisms, have been documented in some coral species (Reed et al. 2010). For example, both soft and hard corals have shown a type of immunological memory and specificity for self-/nonself-recognition, with faster immunological responses after initial exposure (Hildemann et al. 1977; Salter-Cid and Bigger 1991; Jokiel and Bigger 1994). One study directly tested the probiotic hypothesis by treating colonies of *O. patagonica* with a broad-spectrum antibiotic and then exposing the coral to *V. shilonii* to determine whether infection and subsequent bleaching were influenced by the resident microbiome. As hypothesized, antibiotic-treated corals became sensitive to *V. shilonii* infection and bleached after 14 days, but non-treated corals remained healthy, presumably because beneficial members of the microbiome inhibited *V. shilonii* growth (Mills et al. 2013).

7.7 Virus-Induced Bleaching

Virus-like particles (VLPs) are present in all tissue layers of apparently healthy and diseased corals including the gastrodermis, mesoglea, and epidermis, as well as in the coral surface mucus layer (Patten et al. 2008; Leruste et al. 2012; Bettarel et al. 2013; Nguyen-Kim et al. 2014; Pollock et al. 2014; Wood-Charlson et al. 2015). It has been hypothesized that elevated temperatures and other stress events may trigger viral infections that contribute to coral bleaching and disease (Vega Thurber et al. 2008; Vega Thurber and Correa 2011; Wilson 2011; Lawrence et al. 2015; Levin et al. 2017). Wilson et al. (2001) documented VLPs associated with heat-stressed anemones, *Anemonia viridis*, and were the first to suggest that viruses may be involved in the cellular pathogenesis of bleaching. There is now growing evidence of specific and dynamic interactions between *Symbiodinium* cells and viruses or VLPs. For example, type C1 *Symbiodinium* cultures have been found to host multiple intracellular viral infections (Weynberg et al. 2017). In another study, *Symbiodinium* cultures exposed to UV demonstrated rapid cellular lysis, postulated to be from latent viral infections, although this response could also be a result of photochemical damage (Lawrence et al. 2015). Furthermore, all bleached corals exhibited large viral loads in a naturally occurring bleaching event, caused by low tide and heavy rainfall, with some VLPs specifically associated with the *Symbiodinium* cells (Correa et al. 2016).

Although our understanding of viruses and their role in host-associated microbial communities is currently limited, evidence is beginning to suggest that they play a significant role in bacterial bleaching events. Virulence factors located on chromosomal pathogenicity islands exist in some strains of *V. coralliilyticus*. The presence of these genetic signatures suggests that *V. coralliilyticus* virulence is driven by prophages and other horizontally acquired elements (Weynberg et al. 2015). During a bleaching event in the Caribbean, Marhaver et al. (2008) documented explicit changes in the absolute abundance and relative composition of viruses that infect the coral, *Symbiodinium*, and bacteria within partially bleached colonies of the massive coral, *Diploria strigosa*. More recent work has tested the theory that *Symbiodinium* are more susceptible to viral infections when exposed to heat stress. Lawrence et al. (2017) demonstrated upregulation of virus-like gene expression in cultured *Symbiodinium* cells following temperature stress experiments. Therefore, *Symbiodinium* cells may host latent or persistent viral infections that are induced via stress. These results were supported by analysis of host gene expression, which also showed changes consistent with viral infection after exposure to stress. Further to these experiments, transcriptomics were used to compare viral gene expression within thermosensitive and thermotolerant populations of *Symbiodinium* type C1 cells at ambient and elevated temperatures (+4 °C; Levin et al. 2017). This was the first study to indicate that the viruses infecting *Symbiodinium* may also be adversely affected by heat stress, further contributing to the endosymbiont's thermal sensitivity. If these results can be replicated in *Symbiodinium* cells *in hospite*, then we may begin to explain alternative hypotheses for the breakdown of the coral-*Symbiodinium* symbiosis that ultimately leads to bleaching.

7.8 Conclusions and Future Directions

Global climate change is currently ongoing and has already had a broad impact across every ecosystem on Earth (Scheffers et al. 2016). Global average temperatures have increased by 1 °C since preindustrial levels (Chap. 1). Thus, we are seeing the impact of temperature stress on the physiology and diversity of marine and terrestrial organisms around the world. Increasing global sea surface temperatures, specifically, are having a significant impact on both the macroscopic and microscopic composition of the coral reef benthos (Hughes et al. 2017), carving out space through bleaching and disease, whereby competitive dominants such as macroalgae and sponges, as well as opportunistic and heterotrophic microorganisms, will take over (reviewed in Rohwer and Youle 2010; Garren 2016). As the tropics continue to experience an increase in both the frequency and severity of high sea surface temperature events, research efforts aimed at understanding the impact of such events on both the seawater microbiome and coral-associated microorganisms will be critical to coral conservation efforts.

Coral bleaching correlates with a number of other triggers, but the combination of high temperatures and irradiance is thought to be primarily responsible (Lesser 2006, 2011). As discussed in this chapter, the impact of elevated temperature stress has been the subject of a number of coral microbiome studies, but the impact of irradiance (i.e., UV radiation; UV-R) independently or coupled with temperature stress has yet to be examined. UV-R has been demonstrated to suppress the immune system and immune response to pathogenic microorganisms in humans (Chapman et al. 1995; reviewed in Patra et al. 2016) and has been known for some time to trigger and/or exacerbate herpes simplex virus infections (Norval 2006). Studies conducted with seawater bacterioplankton communities also demonstrate that solar UV-R can differentially impact microbial activity, with greater sensitivity demonstrated by *Alphaproteobacteria* and greater resistance demonstrated by *Gammaproteobacteria* and *Bacteroidetes* (Alonso-Sáez et al. 2006). In one of the only studies to examine the impact of UV-R on the coral microbiome, solar radiation was shown to rapidly kill intracellular *Vibrio shiloi* and prevent bacteria-induced bleaching in the coral *Oculina patagonica* (Fine et al. 2002). These studies highlight the important role UV-R may have in structuring shallow-water microbial communities, particularly on coral reefs where solar irradiance has had an important influence on community structure over their evolutionary history (reviewed in Banaszak and Lesser 2009). However, because of the attenuation of UV-R with depth, and the differential effects of UV-B (290–320 nm) and UV-A (320–400 nm), most of the significant effects of UV-R may occur in shallow-reef environments (< 30 m depth; Shick et al. 1996; Banaszak and Lesser 2009). There is, therefore, a need for comparative studies along depth gradients from 3 to 30 m as well as across the latitudinal extent of coral reefs, particularly in areas where temperature may remain elevated while irradiance is low.

A particularly interesting new area of research focuses on the extent to which the coral microbiome (prokaryotic and eukaryotic partners) can increase or decrease coral tolerance to specific environmental disturbances. This concept is based on defining the coral as a polygenomic metaorganism (i.e., hologenome) whereby the coral phenotype is a product of the transcriptomic, proteomic, and metabolic responses of all symbiotic partners (Putnam et al. 2017). Thus, a highly flexible microbiome that confers mechanisms for rapid holobiont acclimatization to environmental stressors associated with global climate change is met with hesitant optimism (van Oppen et al. 2015; Torda et al. 2017; Webster and Reusch 2017). Ideally, molecular-based omic techniques would be combined with microscopy approaches to allow for identification of fundamental symbionts within specific compartments (Sweet et al. 2011) and cellular microniches within the coral, for example, identifying whether microbial relationships exist with either coral or *Symbiodinium* cells and/or between specific bacterial strains. Mathematical models and network-based analytical approaches can also aid in identifying these specific microbial relationships (Sweet and Bulling 2017), followed by validation with molecular methods. Robust controls should be established for experimental studies examining the impact of environmental stressors on the coral hologenome, with the realization that experimental manipulation (e.g., transplantation, aquaria rearing) may have consequences for the stability of the microbiome leading to ecologically unrealistic conclusions (Morrow et al. 2017). Corals have demonstrated natural resilience in the face of extreme bleaching events, whereby some corals retain their symbionts and others recover their symbionts over time (Cunning et al. 2016; Hughes et al. 2017). Future research needs to take a *holistic* perspective, identifying the mechanisms driving resilience as a function of the coral metaorganism (Boulotte et al. 2016; Putnam et al. 2017; Torda et al. 2017).

References

Ainsworth TD, Kramasky-Winter E, Loya Y, Hoegh-Guldberg O, Fine M (2007) Coral disease diagnostics: What's between a plague and a band? Appl Environ Microbiol 73:981–992. https://doi.org/10.1128/AEM.02172-06

Ainsworth TD, Fine M, Roff G, Hoegh-Guldberg O (2008) Bacteria are not the primary cause of bleaching in the Mediterranean coral *Oculina patagonica*. ISME J 2:67–73. https://doi.org/10.1038/ismej.2007.88

Ainsworth TD, Krause L, Bridge T, Torda G, Raina J-P, Zakrzewski M et al (2015) The coral core microbiome identifies rare bacterial taxa as ubiquitous endosymbionts. ISME J 9:2261–2274. https://doi.org/10.1038/ismej.2015.39

Alker AP, Smith GW, Kim K (2001) Characterization of *Aspergillus sydowii* (Thom et Church), a fungal pathogen of Caribbean sea fan corals. Hydrobiologia 460:105–111. https://doi.org/10.1023/A:1013145524136

Alonso-Sáez L, Gasol JM, Lefort T, Hofer J, Sommaruga R (2006) Effect of natural sunlight on bacterial activity and differential sensitivity of natural bacterioplankton groups in Northwestern Mediterranean coastal waters. Appl Environ Microbiol 72:5806–5813. https://doi.org/10.1128/AEM.00597-06

Apprill A, Marlow HQ, Martindale MQ, Rappé MS (2009) The onset of microbial associations in the coral *Pocillopora meandrina*. ISME J 3:685–699. https://doi.org/10.1038/ismej.2009.3

Apprill A, Hughen K, Mincer T (2013) Major similarities in the bacterial communities associated with lesioned and healthy Fungiidae corals. Environ Microbiol 15:2063–2072. https://doi.org/10.1111/1462-2920.12107

Apprill A, Weber LG, Santoro AE (2016) Distinguishing between microbial habitats unravels ecological complexity in coral microbiomes. mSystems 1. https://doi.org/10.1128/mSystems.00143-16

Banaszak AT, Lesser MP (2009) Effects of solar ultraviolet radiation on coral reef organisms. Photochem Photobiol Sci 8:1276–1294. https://doi.org/10.1039/b902763g

Banaszak AT, Barba Santos MG, LaJeunesse TC, Lesser MP (2006) The distribution of mycosporine-like amino acids (MAAs) and the phylogenetic identity of symbiotic dinoflagellates in cnidarian hosts from the Mexican Caribbean. J Exp Mar Biol Ecol 337:131–146. https://doi.org/10.1016/j.jembe.2006.06.014

Banin E, Israely T, Kushmaro A, Loya Y, Orr E, Rosenberg E (2000) Penetration of the coral-bleaching bacterium *Vibrio Shiloi* into *Oculina patagonica*. Appl Environ Microbiol 66:3031–3036. https://doi.org/10.1128/AEM.66.7.3031-3036.2000

Banin E, Israely T, Fine M, Loya Y, Rosenberg E (2001) Role of endosymbiotic zooxanthellae and coral mucus in the adhesion of the coral-bleaching pathogen *Vibrio shiloi* to its host. FEMS Microbiol Lett 199:33–37. https://doi.org/10.1016/S0378-1097(01)00162-8

Banin E, Vassilakos D, Orr E, Martinez RJ, Rosenberg E (2003) Superoxide dismutase is a virulence factor produced by the coral bleaching pathogen *Vibrio shiloi*. Curr Microbiol 46:418–422. https://doi.org/10.1007/s00284-002-3912-5

Barr JJ, Auro R, Furlan M, Whiteson KL, Erb ML, Pogliano J et al (2013a) Bacteriophage adhering to mucus provide a non-host-derived immunity. Proc Natl Acad Sci U S A 110:10771–10776. https://doi.org/10.1073/pnas.1305923110

Barr JJ, Youle M, Rohwer F (2013b) Innate and acquired bacteriophage-mediated immunity. Bacteriophage 3:e25857. https://doi.org/10.4161/bact.25857

Barshis DJ, Stillman JH, Gates RD, Toonen RJ, Smith LW, Birkeland C (2010) Protein expression and genetic structure of the coral *Porites lobata* in an environmentally extreme Samoan back reef: Does host genotype limit phenotypic plasticity? Mol Ecol 19:1705–1720. https://doi.org/10.1111/j.1365-294X.2010.04574.x

Barton NH, Charlesworth B (1984) Genetic revolutions, founder effects, and speciation. Annu Rev Ecol Syst 15:133–164. https://doi.org/10.1146/annurev.es.15.110184.001025

Bayer T, Neave MJ, Alsheikh-Hussain A, Aranda M, Yum LK, Mincer T et al (2013) The microbiome of the red sea coral *Stylophora pistillata* is dominated by tissue-associated *Endozoicomonas* bacteria. Appl Environ Microbiol 79:4759–4762. https://doi.org/10.1128/AEM.00695-13

Bellantuono AJ, Hoegh-Guldberg O, Rodriguez-Lanetty M (2011) Resistance to thermal stress in corals without changes in symbiont composition. Proc R Soc B 279:1100–1107. https://doi.org/10.1098/rspb.2011.1780

Ben-Haim Y, Rosenberg E (2002) A novel *Vibrio* sp. pathogen of the coral *Pocillopora damicornis*. Mar Biol 141:47–55. https://doi.org/10.1007/s00227-002-0797-6

Ben-Haim Y, Zicherman-Keren M, Rosenberg E (2003) Temperature-regulated bleaching and lysis of the coral *Pocillopora damicornis* by the novel pathogen *Vibrio coralliilyticus*. Appl Environ Microbiol 69:4236–4242. https://doi.org/10.1128/AEM.69.7.4236-4242.2003

Beraud E, Gevaert F, Rottier C, Ferrier-Pagès C (2013) The response of the scleractinian coral *Turbinaria reniformis* to thermal stress depends on the nitrogen status of the coral holobiont. J Exp Biol 216:2665–2674. https://doi.org/10.1242/jeb.085183

Bessell-Browne P, Negri AP, Fisher R, Clode PL, Duckworth A, Jones R (2017) Impacts of turbidity on corals: the relative importance of light limitation and suspended sediments. Mar Pollut Bull 117:161–170. https://doi.org/10.1016/j.marpolbul.2017.01.050

Bettarel Y, Thuy NT, Huy TQ, Hoang PK (2013) Observation of virus-like particles in thin sections of the bleaching scleractinian coral *Acropora cytherea*. J Mar Biol Assoc UK 93:909–912. https://doi.org/10.1017/S0025315411002062

Bolch CJS, Bejoy TA, Green DH (2017) Bacterial associates modify growth dynamics of the dinoflagellate *Gymnodinium catenatum*. Front Microbiol 8. https://doi.org/10.3389/fmicb.2017.00670

Böttjer D, Dore JE, Karl DM, Letelier RM, Mahaffey C, Wilson ST et al (2016) Temporal variability of nitrogen fixation and particulate nitrogen export at Station ALOHA. Limnol Oceanogr 62:200–216. https://doi.org/10.1002/lno.10386

Boulotte NM, Dalton SJ, Carroll AG, Harrison PL, Putnam HM, Peplow LM et al (2016) Exploring the *Symbiodinium* rare biosphere provides evidence for symbiont switching in reef-building corals. ISME J 10:2693–2701. https://doi.org/10.1038/ismej.2016.54

Bourne D, Iida Y, Uthicke S, Smith-Keune C (2008) Changes in coral-associated microbial communities during a bleaching event. ISME J 2112:350–363. https://doi.org/10.1038/ismej.2007.112

Bourne DG, Dennis PG, Uthicke S, Soo RM, Tyson GW, Webster N (2013) Coral reef invertebrate microbiomes correlate with the presence of photosymbionts. ISME J 7:1452–1458. https://doi.org/10.1038/ismej.2012.172

Bourne DG, Morrow KM, Webster NS (2016) Insights into the coral microbiome: underpinning the health and resilience of reef ecosystems. Annu Rev Microbiol 70:317–340. https://doi.org/10.1146/annurev-micro-102215-095440

Brading P, Warner ME, Davey P, Smith DJ, Achterberg EP, Suggett DJ (2011) Differential effects of ocean acidification on growth and photosynthesis among phylotypes of *Symbiodinium* (Dinophyceae). Limnol Oceanogr 56:927–938. https://doi.org/10.4319/lo.2011.56.3.0927

Brandt ME, McManus JW (2009) Dynamics and impact of the coral disease white plague: insights from a simulation model. Dis Aquat Org 87:117–133. https://doi.org/10.3354/dao02137

Breitbart M, Felts B, Kelley S, Mahaffy JM, Nulton J, Salamon P et al (2004) Diversity and population structure of a near-shore marine-sediment viral community. Proc R Soc B 271:565–574. https://doi.org/10.1098/rspb.2003.2628

Brown BE, Bythell JC (2005) Perspectives on mucus secretion in reef corals. Mar Ecol Prog Ser 296:291–309. https://doi.org/10.3354/meps296291

Brown BE, Downs CA, Dunne RP, Gibb SW (2002a) Exploring the basis of thermotolerance in the reef coral *Goniastrea aspera*. Mar Ecol Prog Ser 242:119–129. https://doi.org/10.3354/meps242119

Brown BE, Dunne RP, Goodson MS, Douglas A (2002b) Experience shapes the susceptibility of a reef coral to bleaching. Coral Reefs 21:119–126. https://doi.org/10.1007/s00338-002-0215-z

Bruno JF, Selig ER, Casey KS, Page CA, Willis BL, Harvell CD et al (2007) Thermal stress and coral cover as drivers of coral disease outbreaks. PLoS Biol 5:1220–1227. https://doi.org/10.1371/journal.pbio.0050124

Burge CA, Eakin CM, Friedman CS, Froelich B, Hershberger PK, Hofmann EE et al (2014) Climate change influences on marine infectious diseases: implications for management and society. Annu Rev Mar Sci 6:249–277. https://doi.org/10.1146/annurev-marine-010213-135029

Buxton L, Takahashi S, Hill R, Ralph PJ (2012) Variability in the primary site of photosynthetic damage in *Symbiodinium* sp. (Dinophyceae) exposed to thermal stress. J Phycol 48:117–126. https://doi.org/10.1111/j.1529-8817.2011.01099.x

Bythell JC, Wild C (2011) Biology and ecology of coral mucus release. J Exp Mar Biol Ecol 408:88–93. https://doi.org/10.1016/j.jembe.2011.07.028

Cardini U, Bednarz VN, Naumann MS, van Hoytema N, Rix L, Foster RA et al (2015) Functional significance of dinitrogen fixation in sustaining coral productivity under oligotrophic conditions. Proc R Soc B 282. https://doi.org/10.1098/rspb.2015.2257

Cardini U, van Hoytema N, Bednarz VN, Rix L, Foster RA, Al-Rshaidat MMD et al (2016) Microbial dinitrogen fixation in coral holobionts exposed to thermal stress and bleaching. Environ Microbiol 18:2620–2633. https://doi.org/10.1111/1462-2920.13385

Chapman RS, Cooper KD, De Faro EC (1995) Solar ultraviolet radiation and the risk of infectious disease: summary of a workshop. Photochem Photobiol 61:223–247. https://doi.org/10.1111/j.1751-1097.1995.tb03966.x

Chen W-M, Moulin L, Bontemps C, Vandamme P, Bèna G, Boivon-Masson C (2003) Legume symbiotic nitrogen fixation by beta-proteobacteria is widespread in nature. J Bacteriol 185:7266–7272. https://doi.org/10.1128/JB.185.24.7266

Columpsi P, Sacchi P, Zuccaro V, Cima S, Sarda C, Mariani M et al (2016) Beyond the gut bacterial microbiota: the gut virome. J Med Virol 88:1467–1472. https://doi.org/10.1002/jmv.24508

Correa AMS, Ainsworth TD, Rosales SM, Thurber AR, Butler CR, Vega Thurber RL (2016) Viral outbreak in corals associated with an *in situ* bleaching event: atypical herpes-like viruses and a new megavirus infecting *Symbiodinium*. Front Microbiol. https://doi.org/10.3389/fmicb.2016.00127

Cróquer A, Weil E (2009) Changes in Caribbean coral disease prevalence after the 2005 bleaching event. Dis Aquat Org 87:33–43. https://doi.org/10.3354/dao02164

Cunning R, Ritson-Williams R, Gates RD (2016) Patterns of bleaching and recovery of *Montipora capitata* in Kane'ohe Bay, Hawai'i, USA. Mar Ecol Prog Ser 551:131–139. https://doi.org/10.3354/meps11733

Damjanovic K, Blackall LL, Webster NS, van Oppen MJH (2017) The contribution of microbial biotechnology to mitigating coral reef degradation. Microb Biotechnol 10:1236–1243. https://doi.org/10.1111/1751-7915.12769

Deschaseaux ESM, Jones GB, Deseo MA, Shepherd KM, Kiene RP, Swan HB et al (2014) Effects of environmental factors on dimethylated sulfur compounds and their potential role in the antioxidant system of the coral holobiont. Limnol Oceanogr 59:758–768. https://doi.org/10.4319/lo.2014.59.3.0758

Ding JY, Shiu JH, Chen WM, Chiang YR, Tang SL (2016) Genomic insight into the host-endosymbiont relationship of *Endozoicomonas montiporae* CL-33T with its coral host. Front Microbiol 7. https://doi.org/10.3389/fmicb.2016.00251

Dubinsky Z, Jokiel PL (1994) Ratio of energy and nutrient fluxes regulates symbiosis between zooxanthellae and corals. Pac Sci 48:313–324

Ducklow HW, Mitchell R (1979) Bacterial populations and adaptations in the mucus layers on living corals. Limnol Oceanogr 24:715–725. https://doi.org/10.4319/lo.1979.24.4.0715

Earle KA, Billings G, Sigal M, Lichtman JS, Hansson GC, Elias JE et al (2015) Quantitative imaging of gut microbiota spatial organization. Cell Host Microbe 18:478–488. https://doi.org/10.1016/j.chom.2015.09.002

Falkowski PG, Dubinsky Z, Muscatine L, McCloskey L (1993) Population control in symbiotic corals. Bioscience 43:606–611. https://doi.org/10.2307/1312147

Ferrier-Pagès C, Rottier C, Beraud E, Levy O (2010) Experimental assessment of the feeding effort of three scleractinian coral species during a thermal stress: effect on the rates of photosynthesis. J Exp Mar Biol Ecol 390:118–124. https://doi.org/10.1016/j.jembe.2010.05.007

Fine M, Banin E, Israely T, Rosenberg E, Loya Y (2002) Ultraviolet radiation prevents bleaching in the mediterranean coral *Oculina patagonica*. Mar Ecol Prog Ser 226:249–254. https://doi.org/10.3354/meps226249

Fitt WK, Brown BE, Warner ME, Dunne RP (2001) Coral bleaching: Interpretation of thermal tolerance limits and thermal thresholds in tropical corals. Coral Reefs 20:51–65. https://doi.org/10.1007/s003380100146

Fitt WK, Gates RD, Hoegh-Guldberg O, Bythell JC, Jatkar AA, Grottoli A et al (2009) Response of two species of Indo-Pacific corals, *Porites cylindrica* and *Stylophora pistillata*, to short-term thermal stress: the host does matter in determining the tolerance of corals to bleaching. J Exp Mar Biol Ecol 373:102–110. https://doi.org/10.1016/j.jembe.2009.03.011

Frydenborg BR, Krediet CJ, Teplitski M, Ritchie KB (2014) Temperature-dependent inhibition of opportunistic *Vibrio* pathogens by native coral commensal bacteria. Microb Ecol 67:392–401. https://doi.org/10.1007/s00248-013-0334-9

Fukami T, Nakajima M (2011) Community assembly: alternative stable states or alternative transient states? Ecol Lett 14:973–984. https://doi.org/10.1111/j.1461-0248.2011.01663.x

Gajigan AP, Diaz LA, Conaco C (2017) Resilience of the prokaryotic microbial community of *Acropora digitifera* to elevated temperature. MicrobiologyOpen 6. https://doi.org/10.1002/mbo3.478

Garren M (2016) Algae feed a shift on coral reefs. Nat Microbiol 1. https://doi.org/10.1038/nmicrobiol.2016.61

Garren M, Son K, Raina J-B, Menolascina F, Shapiro OH, Tout J et al (2014) A bacterial pathogen uses dimethylsulfoniopropionate as a cue to target heat-stressed corals. ISME J 8:999–1007. https://doi.org/10.1038/ismej.2013.210

Garren M, Son K, Tout J, Seymour JR, Stocker R (2016) Temperature-induced behavioral switches in a bacterial coral pathogen. ISME J 10:1363–1372. https://doi.org/10.1038/ismej.2015.216

Gil-Agudelo DL, Ban-Haim Y, Rosenberg E, Smith GW (2017) Differential response of coral symbiotic dinoflagellates to bacterial toxins that produce bleaching in stony corals. J South Carolina Acad Sci 15(1):6

Gilbert JA, Hill R, Doblin MA, Ralph PJ (2012) Microbial consortia increase thermal tolerance of corals. Mar Biol 159:1763–1771. https://doi.org/10.1007/s00227-012-1967-9

Glasl B, Herndl GJ, Frade PR (2016) The microbiome of coral surface mucus has a key role in mediating holobiont health and survival upon disturbance. ISME J 10:2280–2292. https://doi.org/10.1038/ismej.2016.9

Godinot C, Houlbrèque F, Grover R, Ferrier-Pagès C (2011) Coral uptake of inorganic phosphorus and nitrogen negatively affected by simultaneous changes in temperature and pH. PLoS One 6. https://doi.org/10.1371/journal.pone.0025024

Grégoire V, Schmacka F, Coffroth MA, Karsten U (2017) Photophysiological and thermal tolerance of various genotypes of the coral endosymbiont *Symbiodinium* sp. (Dinophyceae). J Appl Phycol:1–13. https://doi.org/10.1007/s10811-017-1127-1

Grottoli AG, Rodrigues LJ, Palardy JE (2006) Heterotrophic plasticity and resilience in bleached corals. Nature 440:1186–1189. https://doi.org/10.1038/nature04565

Guzman HM, Guevara C (1998) Massive mortality of zooxanthelate reef organisms during the 1995 bleaching Cayos Cochinos, Honduras. Rev Biol Trop 46:165–173

Haapkylä J, Unsworth RKF, Flavell M, Bourne DG, Schaffelke B, Willis BL (2011) Seasonal rainfall and runoff promote coral disease on an inshore reef. PLoS One 6. https://doi.org/10.1371/journal.pone.0016893

Hadaidi G, Röthig T, Yum LK, Ziegler M, Arif C, Roder C et al (2017) Stable mucus-associated bacterial communities in bleached and healthy corals of *Porites lobata* from the Arabian Seas. Sci Rep 7. https://doi.org/10.1038/srep45362

Harvell CD, Mitchell CE, Ward JR, Altizer S, Dobson AP, Ostfeld RS et al (2002) Climate warming and disease risks for terrestrial and marine biota. Science 296:2158–2162. https://doi.org/10.1126/science.1063699

Hennige SJ, Suggett DJ, Warner ME, McDougall KE, Smith DJ (2009) Photobiology of *Symbiodinium* revisited: bio-physical and bio-optical signatures. Coral Reefs 28:179–195. https://doi.org/10.1007/s00338-008-0444-x

Hernandez-Agreda A, Leggat W, Bongaerts P, Ainsworth TD (2016) The microbial signature provides insight into the mechanistic basis of coral success across reef habitats. MBio 7. https://doi.org/10.1128/mBio.00560-16

Hildemann WH, Raison RL, Cheung G, Hull CJ, Akaka L, Okamoto J (1977) Immunological specificity and memory in a scleractinian coral. Nature 270:219–223. https://doi.org/10.1038/270219a0

Hodgson G (1990) Tetracycline reduces sedimentation damage to corals. Mar Biol 104:493–496. https://doi.org/10.1007/BF01314355

Hoegh-Guldberg O, Mumby PJ, Hooten AJ, Steneck RS, Greenfield P, Gomez E et al (2007) Coral reefs under rapid climate change and ocean acidification. Science 318:1737–1742. https://doi.org/10.1126/science.1152509

Hong MJ, Yu YT, Chen CA, Chiang PW, Tang SL (2009) Influence of species specificity and other factors on bacteria associated with the coral *Stylophora pistillata* in Taiwan. Appl Environ Microbiol 75:7797–7806. https://doi.org/10.1128/AEM.01418-09

Hoogenboom MO, Campbell DA, Beraud E, DeZeeuw K, Ferrier-Pagès C (2012) Effects of light, food availability and temperature stress on the function of photosystem II and photosystem I of coral symbionts. PLoS One 7. https://doi.org/10.1371/journal.pone.0030167

Howard EC, Sun S, Reisch CR, del Valle DA, Bürgmann H, Liene RP et al (2011) Changes in dimethylsulfoniopropionate demethylase gene assemblages in response to an induced phytoplankton bloom. Appl Environ Microbiol 77:524–531. https://doi.org/10.1128/AEM.01457-10

Hu L, Peng X, Zhou J, Zhang Y, Xu S, Mao X et al (2015) Characterizing the interactions among a dinoflagellate, flagellate and bacteria in the phycosphere of *Alexandrium tamarense* (Dinophyta). Front Mar Sci 2. https://doi.org/10.3389/fmars.2015.00100

Hughes TP, Baird AH, Bellwood DR, Card M, Connolly SR, Folke C et al (2003) Climate change, human impacts, and the resilience of coral reefs. Science 301:929–933. https://doi.org/10.1126/science.1085046

Hughes TP, Kerry JT, Álvarez-Noriega M, Álvarez- Romero JG, Anderson KD, Baird AH et al (2017) Global warming and recurrent mass bleaching of corals. Nature 543:373–377. https://doi.org/10.1038/nature21707

Hussa EA, Goodrich-Blair H (2013) It takes a village: ecological and fitness impacts of multipartite mutualism. Annu Rev Microbiol 67:161–178. https://doi.org/10.1146/annurev-micro-092412-155723

Johansson MEV, Larsson JMH, Hansson GC (2011) The two mucus layers of colon are organized by the MUC2 mucin, whereas the outer layer is a legislator of host-microbial interactions. Proc Natl Acad Sci U S A 108:4659–4665. https://doi.org/10.1073/pnas.1006451107

Jokiel PL, Bigger CH (1994) Aspects of histocompatibility and regeneration in the solitary reef coral *Fungia scutaria*. Biol Bull 186:72–80. https://doi.org/10.2307/1542037

Kimes NE, Grim CJ, Johnson WR, Hasan NA, Tall BD, Kothary MH et al (2012) Temperature regulation of virulence factors in the pathogen *Vibrio coralliilyticus*. ISME J 6:835–846. https://doi.org/10.1038/ismej.2011.154

Klimenko AI, Matushkin YG, Kolchanov NA, Lashin SA (2016) Bacteriophages affect evolution of bacterial communities in spatially distributed habitats: a simulation study. BMC Microbiol 16. https://doi.org/10.1186/s12866-015-0620-4

Koren O, Rosenberg E (2008) Bacteria associated with the bleached and cave coral *Oculina patagonica*. Microb Ecol 55:523–529. https://doi.org/10.1007/s00248-007-9297-z

Krediet CJ, Ritchie KB, Cohen M, Lipp EK, Sutherland KP, Teplitski M (2009) Utilization of mucus from the coral *Acropora palmata* by the pathogen *Serratia marcescens* and by environmental and coral commensal bacteria. Appl Environ Microbiol 75:3851–3858. https://doi.org/10.1128/AEM.00457-09

Krueger T, Fisher PL, Becker S, Pontasch S, Dove S, Hoegh-Guldberg O et al (2015) Transcriptomic characterization of the enzymatic antioxidants FeSOD, MnSOD, APX and KatG in the dinoflagellate genus *Symbiodinium*. BMC Evol Biol 15:48. https://doi.org/10.1186/s12862-015-0326-0

Kushmaro A, Rosenberg E, Fine M, Loya Y (1997) Bleaching of the coral *Oculina patagonica* by *Vibrio* AK-1. Mar Ecol Prog Ser 147:159–165. https://doi.org/10.3354/meps147159

Kushmaro A, Rosenberg E, Fine M, Ben Haim Y, Loya Y (1998) Effect of temperature on bleaching of the coral *Oculina patagonica* by *Vibrio* AK-1. Mar Ecol Prog Ser 171:131–137. https://doi.org/10.3354/meps147159

Lawrence SA, Davy JE, Wilson WH, Hoegh-Guldberg O, Davy SK (2015) *Porites* white patch syndrome: associated viruses and disease physiology. Coral Reefs 34:249–257. https://doi.org/10.1007/s00338-014-1218-2

Lawrence SA, Floge SA, Davy JE, Davy SK, Wilson WH (2017) Exploratory analysis of *Symbiodinium* transcriptomes reveals potential latent infection by large dsDNA viruses. Environ Microbiol 19. https://doi.org/10.1111/1462-2920.13782

Lee STM, Davy SK, Tang SL, Fan TY, Kench PS (2015) Successive shifts in the microbial community of the surface mucus layer and tissues of the coral *Acropora muricata* under thermal stress. FEMS Microbiol Ecol 91. https://doi.org/10.1093/femsec/fiv142

Lee STM, Davy SK, Tang SL, Kench PS (2016) Mucus sugar content shapes the bacterial community structure in thermally stressed *Acropora muricata*. Front Microbiol 7. https://doi.org/10.3389/fmicb.2016.00371

Lema KA, Willis BL, Bourne DG (2014) Amplicon pyrosequencing reveals spatial and temporal consistency in diazotroph assemblages of the *Acropora millepora* microbiome. Environ Microbiol 16:3345–3359. https://doi.org/10.1111/1462-2920.12366

Leruste A, Bouvier T, Bettarel Y (2012) Enumerating viruses in coral mucus. Appl Environ Microbiol 78:6377–6379. https://doi.org/10.1128/aem.01141-12

Lesser MP (2004) Experimental biology of coral reef ecosystems. J Exp Mar Biol Ecol 300:217–252. https://doi.org/10.1016/j.jembe.2003.12.027

Lesser MP (2006) Oxidative stress in marine environments: biochemistry and physiological ecology early history of oxygen. Annu Rev Physiol 68:253–278. https://doi.org/10.1146/annurev.physiol.68.040104.110001

Lesser MP (2011) Coral bleaching: causes and mechanisms. In: Dubinsky Z, Stambler N (eds) Coral reefs: an ecosystem in transition. Springer, Dordrecht, pp 405–419

Lesser MP, Jarett JK (2014) Culture-dependent and culture-independent analyses reveal no prokaryotic community shifts or recovery of *Serratia marcescens* in *Acropora palmata* with white pox disease. FEMS Microbiol Ecol 88:457–467. https://doi.org/10.1111/1574-6941.12311

Lesser MP, Van Woesik R (2015) Modelling climate-change effects on coral disease. Reef Encounter 30:31–35

Lesser MP, Mazel C, Gorbunov M, Falkowski PG (2004) Discovery of symbiotic nitrogen-fixing cyanobacteria in corals. Science 305:997–1000. https://doi.org/10.1126/science.1099128

Lesser MP, Falcón LI, Rodríguez-Román A, Enriquez S, Hoegh-Guldberg O, Iglesias-Prieto R (2007a) Nitrogen fixation by symbiotic cyanobacteria provides a source of nitrogen for the scleractinian coral *Montastraea cavernosa*. Mar Ecol Prog Ser 346:143–152. https://doi.org/10.3354/meps07008

Lesser MP, Bythell JC, Gates RD, Johnstone RW, Hoegh-Guldberg O (2007b) Are infectious diseases really killing corals? Alternative interpretations of the experimental and ecological data. J Exp Mar Biol Ecol 346:36–44. https://doi.org/10.1016/j.jembe.2007.02.015

Lesser MP, Morrow KM, Pankey SM, Noonan SHC (2017) Diazotroph diversity and nitrogen fixation in the coral *Stylophora pistillata* from the Great Barrier Reef. ISME J. https://doi.org/10.1038/s41396-017-008-6

Levin RA, Voolstra CR, Weynberg KD, van Oppen MJH (2017) Evidence for a role of viruses in thermal sensitivity of coral photosymbionts. ISME J 11:808–812. https://doi.org/10.1038/ismej.2016.154

Lipp EK, Griffin DW (2004) Analysis of coral mucus as an improved medium for detection of enteric microbes and for determining patterns of sewage contamination in reef environments. EcoHealth 1:317–323. https://doi.org/10.1007/s10393-004-0132-4

Littman RA, Willis BL, Pfeffer C, Bourne DG (2009) Diversities of coral-associated bacteria differ with location, but not species, for three acroporid corals on the Great Barrier Reef. FEMS Microbiol Ecol 68:152–163. https://doi.org/10.1111/j.1574-6941.2009.00666.x

Littman RA, Bourne DG, Willis BL (2010) Responses of coral-associated bacterial communities to heat stress differ with Symbiodinium type on the same coral host. Mol Ecol 19:1978–1990. https://doi.org/10.1111/j.1365-294X.2010.04620.x

Littman R, Willis BL, Bourne DG (2011) Metagenomic analysis of the coral holobiont during a natural bleaching event on the Great Barrier Reef. Environ Microbiol Rep 3:651–660. https://doi.org/10.1111/j.1758-2229.2010.00234.x

Lozupone CA, Stombaugh JI, Gordon JI, Jansson JK, Knoght R (2012) Diversity, stability and resilience of the human gut microbiota. Nature 489:220–230. https://doi.org/10.1038/nature11550

Mao-Jones J, Ritchie KB, Jones LE, Ellner SP (2010) How microbial community composition regulates coral disease development. PLoS Biol 8. https://doi.org/10.1371/journal.pbio.1000345

Marhaver KL, Edwards RA, Rohwer F (2008) Viral communities associated with healthy and bleaching corals. Environ Microbiol 10:2277–2286. https://doi.org/10.1111/j.1462-2920.2008.01652.x

Maynard J, van Hooidonk R, Eakin CM, Puotinen M, Garren M, Williams G et al (2015) Projections of climate conditions that increase coral disease susceptibility and pathogen abundance and virulence. Nat Clim Chang 5:688–694. https://doi.org/10.1038/nclimate2625

McClanahan TR, Weil E, Maina J (2009) Strong relationship between coral bleaching and growth anomalies in massive *Porites*. Glob Chang Biol 15:1804–1816. https://doi.org/10.1111/j.1365-2486.2008.01799.x

McDevitt-Irwin JM, Baum JK, Garren M, Vega Thurber RL (2017) Responses of coral-associated bacterial communities to local and global stressors. Front Mar Sci. https://doi.org/10.3389/fmars.2017.00262

McGinty ES, Pieczonka J, Mydlarz LD (2012) Variations in reactive oxygen release and antioxidant activity in multiple *Symbiodinium* types in response to elevated temperature. Microb Ecol 64:1000–1007. https://doi.org/10.1007/s00248-012-0085-z

Meyer JL, Paul VJ, Teplitski M (2014) Community shifts in the surface microbiomes of the coral *Porites astreoides* with unusual lesions. PLoS One 9. https://doi.org/10.1371/journal.pone.0100316

Middlebrook R, Hoegh-Guldberg O, Leggat W (2008) The effect of thermal history on the susceptibility of reef-building corals to thermal stress. J Exp Biol 211:1050–1056. https://doi.org/10.1242/jeb.013284

Miller J, Muller E, Rogers C, Waara R, Atkinson A, Whelan KRT et al (2009) Coral disease following massive bleaching in 2005 causes 60% decline in coral cover on reefs in the US Virgin Islands. Coral Reefs 28:925–937. https://doi.org/10.1007/s00338-009-0531-7

Miller MW, Lohr KE, Cameron CM, Williams DE, Peters EC (2014) Disease dynamics and potential mitigation among restored and wild staghorn coral, *Acropora cervicornis*. PeerJ 2. https://doi.org/10.7717/peerj.541

Mills E, Shechtman K, Loya Y, Rosenberg E (2013) Bacteria appear to play important roles in both causing and preventing the bleaching of the coral *Oculina patagonica*. Mar Ecol Prog Ser 489:155–162. https://doi.org/10.3354/meps10391

Mohamed NM, Colman AS, Tal Y, Hill RT (2008) Diversity and expression of nitrogen fixation genes in bacterial symbionts of marine sponges. Environ Microbiol 10:2910–2921. https://doi.org/10.1111/j.1462-2920.2008.01704.x

Morrow KM, Moss AG, Chadwick NE, Liles MR (2012) Bacterial associates of two caribbean coral species reveal species-specific distribution and geographic variability. Appl Environ Microbiol 78:6438–6449. https://doi.org/10.1128/AEM.01162-12

Morrow KM, Bourne DG, Humphrey C, Botté E, Laffy P, Zaneveld J et al (2015) Natural volcanic CO_2 seeps reveal future trajectories for host-microbial associations in corals and sponges. ISME J 9:894–908. https://doi.org/10.1038/ismej.2014.188

Morrow KM, Fiore CL, Lesser MP (2016) Environmental drivers of microbial community shifts in the giant barrel sponge, *Xestospongia muta*, over a shallow to mesophotic depth gradient. Environ Microbiol 18:2025–2038. https://doi.org/10.1111/1462-2920.13226

Morrow KM, Bromhall K, Motti CA, Munn CB, Bourne DG (2017) Allelochemicals produced by brown macroalgae of the *Lobophora* genus are active against coral larvae and associated bacteria, supporting pathogenic shifts to *Vibrio* dominance. Appl Environ Microbiol 83. https://doi.org/10.1128/AEM.02391-16

Mouchka ME, Hewson I, Harvell CD (2010) Coral-associated bacterial assemblages: current knowledge and the potential for climate-driven impacts. Integr Comp Biol 50:662–674. https://doi.org/10.1093/icb/icq061

Mueller NT, Bakacs E, Combellick J, Grigoryan Z, Dominiguez-Bello MG (2015) The infant microbiome development: mom matters. Trends Mol Med 21:109–117. https://doi.org/10.1016/j.molmed.2014.12.002

Muller EM, van Woesik R (2012) Caribbean coral diseases: primary transmission or secondary infection? Glob Chang Biol 18:3529–3535. https://doi.org/10.1111/gcb.12019

Muller EM, van Woesik R (2014) Genetic susceptibility, colony size, and water temperature drive white-pox disease on the coral *Acropora palmata*. PLoS One 9. https://doi.org/10.1371/journal.pone.0110759

Muller EM, Rogers CS, Spitzack AS, van Woesik R (2008) Bleaching increases likelihood of disease on *Acropora palmata* (Lamarck) in Hawksnest Bay, St John, US Virgin Islands. Coral Reefs 27:191–195. https://doi.org/10.1007/s00338-007-0310-2

Mydlarz LD, Couch CS, Weil E, Smith G, Harvell CD (2009) Immune defenses of healthy, bleached and diseased *Montastraea faveolata* during a natural bleaching event. Dis Aquat Org 87:67–78. https://doi.org/10.3354/dao02088

Neave MJ, Rachmawati R, Xun L, Mitchell CT, Bourne DG, Apprill A et al (2016a) Differential specificity between closely related corals and abundant *Endozoicomonas* endosymbionts across global scales. ISME J 11:186–200. https://doi.org/10.1038/ismej.2016.95

Neave MJ, Apprill A, Ferrier-Pagès C, Voolstra CR (2016b) Diversity and function of prevalent symbiotic marine bacteria in the genus *Endozoicomonas*. Appl Microbiol Biotechnol 100:8315–8324. https://doi.org/10.1007/s00253-016-7777-0

Negri AP, Webster NS, Hill RT, Heyward AJ (2001) Metamorphosis of broadcast spawning corals in response to bacteria isolated from crustose algae. Mar Ecol Prog Ser 223:121–131. https://doi.org/10.3354/meps223121

Nguyen-Kim H, Bouvier T, Bouvier C, Doan-Nhu H, Nguyen-Ngoc L, Rochelle-Newall E et al (2014) High occurrence of viruses in the mucus layer of scleractinian corals. Environ Microbiol Rep 6:675–682. https://doi.org/10.1111/1758-2229.12185

Nguyen-Kim H, Bouvier T, Bouvier C, Bui VN, Le-Lan H, Bettarel Y (2015) Viral and bacterial epibionts in thermally-stressed corals. J Mar Sci Eng 3:1272–1286. https://doi.org/10.3390/jmse3041272

Niggl W, Glas M, Laforsch C, Mayr C, Wild C (2009) First evidence of coral bleaching stimulating organic matter release by reef corals. In: Proceedings of the 11th international coral reef symposium, Ft. Lauderdale. Session 19, pp 905–910

Norval M (2006) The effect of ultraviolet radiation on human viral infections. Photochem Photobiol 82:1495–1504. https://doi.org/10.1562/2006-07-28-IR-987

Oakley CA, Durand E, Wilkinson SP, Peng L, Weis VM, Grossman AR et al (2017) Thermal shock induces host proteostasis disruption and endoplasmic reticulum stress in the model symbiotic cnidarian *Aiptasia*. J Proteome Res 16:2121–2134. https://doi.org/10.1021/acs.jproteome.6b00797

Olson ND, Lesser MP (2013) Diazotrophic diversity in the Caribbean coral, *Montastraea cavernosa*. Arch Microbiol 195:853–859. https://doi.org/10.1007/s00203-013-0937-z

Palumbi SR, Barshis DJ, Traylor-Knowles N, Bay RA (2014) Mechanisms of reef coral resistance to future climate change. Science 344:895–898. https://doi.org/10.1126/science.1251336

Pantos O, Bongaerts P, Dennis PG, Tyson GW, Hoegh-Guldberg O (2015) Habitat-specific environmental conditions primarily control the microbiomes of the coral *Seriatopora hystrix*. ISME J 9:1916–1927. https://doi.org/10.1038/ismej.2015.3

Parkinson JE, Baums IB (2014) The extended phenotypes of marine symbioses: ecological and evolutionary consequences of intraspecific genetic diversity in coral-algal associations. Front Microbiol. https://doi.org/10.3389/fmicb.2014.00445

Patra VK, Byrne SN, Wolf P (2016) The skin microbiome: Is it affected by UV-induced immune suppression? Front Microbiol 7. https://doi.org/10.3389/fmicb.2016.01235

Patten NL, Harrison PL, Mitchell JG (2008) Prevalence of virus-like particles within a staghorn scleractinian coral (*Acropora muricata*) from the Great Barrier Reef. Coral Reefs 27:569–580. https://doi.org/10.1007/s00338-008-0356-9

Peixoto RS, Rosado PM, Leite DC de A, Rosado AS, Bourne DG (2017) Beneficial microorganisms for corals (BMC): proposed mechanisms for coral health and resilience. Front Microbiol 8:1–16. https://doi.org/10.3389/fmicb.2017.00341

Pernice M, Meibom A, Van Den Heuvel A, Kopp C, Domart-Coulon I, Hoegh-Guldberg O et al (2012) A single-cell view of ammonium assimilation in coral–dinoflagellate symbiosis. ISME J 6:1314–1324. https://doi.org/10.1038/ismej.2011.196

Peters EC, Pilson MEQ (1985) A comparative study of the effects of sedimentation on symbiotic and asymbiotic colonies of the coral *Astrangia danae* Milne Edwards and Haime 1849. J Exp Mar Biol Ecol 92:215–230. https://doi.org/10.1016/0022-0981(85)90096-6

Piggot AM, Fouke BW, Sivaguru M, Sanford RA, Gaskins HR (2009) Change in zooxanthellae and mucocyte tissue density as an adaptive response to environmental stress by the coral, *Montastraea annularis*. Mar Biol 156:2379–2389. https://doi.org/10.1007/s00227-009-1267-1

Pinzón JH, Kamel B, Burge CA, Harvell CD, Medina M, Weil E et al (2015) Whole transcriptome analysis reveals changes in expression of immune-related genes during and after bleaching in a reef-building coral. Roy Soc Open Sci 2. https://doi.org/10.1098/rsos.140214

Pogoreutz C, Rädecker N, Cárdenas A, Gärdes A, Voolstra CR, Wild C (2017) Sugar enrichment provides evidence for a role of nitrogen fixation in coral bleaching. Glob Chang Biol 23:3838–3848. https://doi.org/10.1111/gcb.13695

Pollock FJ, M. Wood-Charlson E, van Oppen MJH, Bourne DG, Willis BL, Weynberg KD (2014) Abundance and morphology of virus-like particles associated with the coral *Acropora hyacinthus* differ between healthy and white syndrome-infected states. Mar Ecol Prog Ser 510:39–43. doi:https://doi.org/10.3354/meps10927

Pollock FJ, Lamb JB, Field SN, Heron SF, Schaffelke B, Shedrawi G et al (2016) Correction: sediment and turbidity associated with offshore dredging increase coral disease prevalence on nearby reefs. PLoS One 11. https://doi.org/10.1371/journal.pone.0165541

Pratte ZA, Richardson LL (2014) Impacts of temperature increase and acidification on thickness of the surface mucopolysaccharide layer of the Caribbean coral *Diploria* spp. Coral Reefs 33:487–496. https://doi.org/10.1007/s00338-013-1115-0

Putnam HM, Barott KL, Ainsworth TD, Gates RD (2017) The vulnerability and resilience of reef-building corals. Curr Biol 27:R528–R540. https://doi.org/10.1016/j.cub.2017.04.047

Rädecker N, Pogoreutz C, Voolstra CR, Wiedenmann J, Wild C (2015) Nitrogen cycling in corals: The key to understanding holobiont functioning? Trends Microbiol 23:490–497. https://doi.org/10.1016/j.tim.2015.03.008

Raina JB, Tapiolas D, Willis BL, Bourne DG (2009) Coral-associated bacteria and their role in the biogeochemical cycling of sulfur. Appl Environ Microbiol 75:3492–3501. https://doi.org/10.1128/AEM.02567-08

Raina JB, Dinsdale EA, Willis BL, Bourne DG (2010) Do the organic sulfur compounds DMSP and DMS drive coral microbial associations? Trends Microbiol 18:101–108. https://doi.org/10.1016/j.tim.2009.12.002

Raina J-B, Tapiolas DM, Forêt S, Lutz A, Abrego D, Ceh J et al (2013) DMSP biosynthesis by an animal and its role in coral thermal stress response. Nature 502:677–680. https://doi.org/10.1038/nature12677

Ramanan R, Kim BH, Cho DH, Oh HM, Kim HS (2016) Algae-bacteria interactions: evolution, ecology and emerging applications. Biotechnol Adv 34:14–29. https://doi.org/10.1016/j.biotechadv.2015.12.003

Randall CJ, van Woesik R (2015) Contemporary white-band disease in Caribbean corals driven by climate change. Nat Clim Chang 5:375–379. https://doi.org/10.1038/nclimate2530

Randall CJ, Jordan-Garza AG, Muller EM, van Woesik R (2014) Relationships between the history of thermal stress and the relative risk of diseases of Caribbean corals. Ecology 95:1981–1994. https://doi.org/10.1890/13-0774.1

Reed KC, Muller EM, van Woesik R (2010) Coral immunology and resistance to disease. Dis Aquat Org 90:85–92. https://doi.org/10.3354/dao02213

Reshef L, Koren O, Loya Y, Zilber-Rosenberg I, Rosenberg E (2006) The coral probiotic hypothesis. Environ Microbiol 8:2068–2073. https://doi.org/10.1111/j.1462-2920.2006.01148.x

Reynolds JM, Bruns BU, Fitt WK, Schmidt GW (2008) Enhanced photoprotection pathways in symbiotic dinoflagellates of shallow-water corals and other cnidarians. Proc Natl Acad Sci U S A 105:13674–13678. https://doi.org/10.1073/pnas.0805187105

Riegl B, Branch GM (1995) Effects of sediment on the energy budgets of four scleractinian (Bourne 1900) and five alcyonacean (Lamouroux 1816) corals. J Exp Mar Biol Ecol 186:259–275. https://doi.org/10.1016/0022-0981(94)00164-9

Ritchie KB (2006) Regulation of microbial populations by coral surface mucus and mucus-associated bacteria. Mar Ecol Prog Ser 322:1–14. https://doi.org/10.3354/meps322001

Ritchie KB (2012) Bacterial symbionts of corals and *Symbiodinium*. In: Rosenberg E, Gophina U (eds) Beneficial microorganisms in multicellular life forms. Springer, Berlin, pp 139–150. https://doi.org/10.1007/978-3-642-21680-0

Ritchie KB, Smith GW (2004) Microbial communities of coral surface mucopolysaccharide layers. In: Rosenberg E, Loya Y (eds) Coral health and disease. Springer, Berlin, pp 259–264

Ritchie KB, Dennis J, McGrath T, Smith GW (1994) Bacteria associated with bleached and nonbleached areas of *Montastrea annularis*. In: Proceedings of the 5th symposium on the natural history of the Bahamas, pp 75–80

Roberty S, Bailleul B, Berne N, Franck F, Cardol P (2014) PSI Mehler reaction is the main alternative photosynthetic electron pathway in *Symbiodinium* sp., symbiotic dinoflagellates of cnidarians. New Phytol 204:81–91. https://doi.org/10.1111/nph.12903

Robison JD, Warner ME (2006) Differential impacts of photoacclimation and thermal stress on the photobiology of four different phylotypes of *Symbiodinium* (Pyrrhophyta). J Phycol 42:568–579. https://doi.org/10.1111/j.1529-8817.2006.00232.x

Rodriguez-Lanetty M, Granados-Cifuentes C, Barberan A, Bellantuono AJ, Bastidas C (2013) Ecological inferences from a deep screening of the complex bacterial consortia associated with the coral, *Porites astreoides*. Mol Ecol 22:4349–4362. https://doi.org/10.1111/mec.12392

Rohwer F, Youle M (2010) Coral Reefs in the Microbial Seas. Plaid Press, Granada Hills

Rohwer F, Seguritan V, Azam F, Knowlton N (2002) Diversity and distribution of coral-associated bacteria. Mar Ecol Prog Ser 243:1–10. https://doi.org/10.3354/meps243001

Rosenberg E, Koren O, Reshef L, Efrony R, Zilber-Rosenberg I (2007) The role of microorganisms in coral health, disease and evolution. Nat Rev Microbiol 5:355–362. https://doi.org/10.1038/nrmicro1635

Rosenberg E, Kushmaro A, Kramarsky-Winter E, Banin E, Loya Y (2009) The role of microorganisms in coral bleaching. ISME J 3:139–146. https://doi.org/10.1038/ismej.2008.104

Rua CPJ, Trindade-Silva AE, Appolinario LR, Venas TM, Garcia GD, Carvalho LS et al (2014) Diversity and antimicrobial potential of culturable heterotrophic bacteria associated with the endemic marine sponge *Arenosclera brasiliensis*. PeerJ 2. https://doi.org/10.7717/peerj.419

Rypien KL, Ward JR, Azam F (2010) Antagonistic interactions among coral-associated bacteria. Environ Microbiol 12:28–39. https://doi.org/10.1111/j.1462-2920.2009.02027.x

Salerno JL, Reineman DR, Gates RD, Rappé MS (2011) The effect of a sublethal temperature elevation on the structure of bacterial communities associated with the coral *Porites compressa*. J Mar Biol 2011:1–9. https://doi.org/10.1155/2011/969173

Salter-Cid L, Bigger CH (1991) Alloimmunity in the gorgonian coral *Swiftia exserta*. Biol Bull 181:127–134. https://doi.org/10.2307/1542495

Santos HF, Carmo FL, Duarte G, Dini-Andreote F, Castro CB, Rosado AS et al (2014) Climate change affects key nitrogen-fixing bacterial populations on coral reefs. ISME J 8:2272–2279. https://doi.org/10.1038/ismej.2014.70

Scheffers BR, De Meester L, Bridge TCL, Hoffmann AA, Pandolfi JM, Corlett RT et al (2016) The broad footprint of climate change from genes to biomes to people. Science 354. https://doi.org/10.1126/science.aaf7671

Schrag SJ, Mittler JE (1996) Host-parasite coexistence: the role of spatial refuges in stabilizing bacteria-phage interactions. Am Nat 148:348–376. https://doi.org/10.1086/285929

Sellstedt A, Richau KH (2013) Aspects of nitrogen-fixing actinobacteria, in particular free-living and symbiotic *Frankia*. FEMS Microbiol Lett 342:179–186. https://doi.org/10.1111/1574-6968.12116

Shantz AA, Lemoine NP, Burkepile DE (2016) Nutrient loading alters the performance of key nutrient exchange mutualisms. Ecol Lett 19:20–28. https://doi.org/10.1111/ele.12538

Sharifah EN, Eguchi M (2011) The phytoplankton *Nannochloropsis oculata* enhances the ability of Roseobacter clade bacteria to inhibit the growth of fish pathogen *Vibrio anguillarum*. PLoS One 6. https://doi.org/10.1371/journal.pone.0026756

Sharp KH, Distel D, Paul VJ (2012) Diversity and dynamics of bacterial communities in early life stages of the Caribbean coral *Porites astreoides*. ISME J 6:790–801. https://doi.org/10.1038/ismej.2011.144

Sharp KH, Sneed JM, Ritchie KB, Mcdaniel LD, Paul VJ (2015) Induction of larval settlement in the reef coral *Porites astreoides* by a cultivated marine *Roseobacter* strain. Biol Bull 228:98–107. https://doi.org/10.1086/BBLv228n2p98

Shick JM, Lesser MP, Jokiel PL (1996) Effects of ultraviolet radiation on corals and other coral reef organisms. Glob Chang Biol 2:527–545. https://doi.org/10.1111/j.1365-2486.1996.tb00065.x

Shnit-Orland M, Kushmaro A (2009) Coral mucus-associated bacteria: a possible first line of defense. FEMS Microbiol Ecol 67:371–380. https://doi.org/10.1111/j.1574-6941.2008.00644.x

Siboni N, Ben-Dov E, Sivan A, Kushmaro A (2012) Geographic specific coral-associated ammonia-oxidizing Archaea in the Northern Gulf of Eilat (Red Sea). Microb Ecol 64:18–24. https://doi.org/10.1007/s00248-011-0006-6

Silveira CB, Rohwer FL (2016) Piggyback-the-Winner in host-associated microbial communities. npj Biofilms Microbiomes 2. https://doi.org/10.1038/npjbiofilms.2016.10

Smillie CS, Smith MB, Friedman J, Cordero OX, David LA, Alm EJ (2011) Ecology drives a global network of gene exchange connecting the human microbiome. Nature 480:241–244. https://doi.org/10.1038/nature10571

Sneed JM, Sharp KH, Ritchie KB, Paul VJ (2014) The chemical cue tetrabromopyrrole from a biofilm bacterium induces settlement of multiple Caribbean corals. Proc R Soc B 281. https://doi.org/10.1098/rspb.2013.3086

Soffer N, Zaneveld J, Vega Thurber R (2015) Phage-bacteria network analysis and its implication for the understanding of coral disease. Environ Microbiol 17:1203–1218. https://doi.org/10.1111/1462-2920.12553

Sokolow S (2009) Effects of a changing climate on the dynamics of coral infectious disease: a review of the evidence. Dis Aquat Org 87:5–18. https://doi.org/10.3354/dao02099

Suescún-Bolívar LP, Traverse GMI, Thomé PE (2016) Glycerol outflow in *Symbiodinium* under osmotic and nitrogen stress. Mar Biol 163. https://doi.org/10.1007/s00227-016-2899-6

Suggett DJ, Warner ME, Smith DJ, Davey P, Henninge S, Baker NR (2008) Photosynthesis and production of hydrogen peroxide by *Symbiodinium* (Pyrrhophyta) phylotypes with different thermal tolerances. J Phycol 44:948–956. https://doi.org/10.1111/j.1529-8817.2008.00537.x

Sule P, Belas R (2013) A novel inducer of *Roseobacter* motility is also a disruptor of algal symbiosis. J Bacteriol 195:637–646. https://doi.org/10.1128/JB.01777-12

Sunagawa S, DeSantis TZ, Piceno YM, Brodie EL, DeSalvo MK, Voolstra CR et al (2009) Bacterial diversity and white plague disease-associated community changes in the Caribbean coral *Montastraea faveolata*. ISME J 3:512–521. https://doi.org/10.1038/ismej.2008.131

Sunda W, Kieber DJ, Kiene RP, Huntsman S (2002) An antioxidant function for DMSP and DMS in marine algae. Nature 418:317–320. https://doi.org/10.1038/nature00851

Sweet MJ, Bulling MT (2017) On the importance of the microbiome and pathobiome in coral health and disease. Front Mar Sci. https://doi.org/10.3389/fmars.2017.00009

Sweet MJ, Croquer A, Bythell JC (2011) Bacterial assemblages differ between compartments within the coral holobiont. Coral Reefs 30:39–52. https://doi.org/10.1007/s00338-010-0695-1

Tang X, He LY, Tao XQ, Dang Z, Guo CL, Lu GN et al (2010) Construction of an artificial microalgal-bacterial consortium that efficiently degrades crude oil. J Hazard Mater 181:1158–1162. https://doi.org/10.1016/j.jhazmat.2010.05.033

Thompson JR, Rivera HE, Closek CJ, Medina M (2015) Microbes in the coral holobiont: partners through evolution, development, and ecological interactions. Front Cell Infect Microbiol 4. https://doi.org/10.3389/fcimb.2014.00176

Thornhill DJ, Lewis AM, Wham DC, LaJeunesse TC (2014) Host-specialist lineages dominate the adaptive radiation of reef coral endosymbionts. Evolution 68:352–367. https://doi.org/10.1111/evo.12270

Todd JD, Curson ARJ, Nikolaidou-Katsaraidou N, Brealey CA, Watmough NJ, Chan Y et al (2010) Molecular dissection of bacterial acrylate catabolism – unexpected links with dimethylsulfoniopropionate catabolism and dimethyl sulfide production. Environ Microbiol 12:327–343. https://doi.org/10.1111/j.1462-2920.2009.02071.x

Torda G, Donelson JM, Aranda M, Barshis DJ, Bay L, Berumen ML et al (2017) Rapid adaptive responses to climate change in corals. Nat Clim Chang 7:627–636. https://doi.org/10.1038/nclimate3374

Tout J, Siboni N, Messer LF, Garren M, Stocker R, Webster NS et al (2015) Increased seawater temperature increases the abundance and alters the structure of natural *Vibrio* populations associated with the coral *Pocillopora damicornis*. Front Microbiol. https://doi.org/10.3389/fmicb.2015.00432

Tracy AM, Koren O, Douglas N, Weil E, Harvell CD (2015) Persistent shifts in Caribbean coral microbiota are linked to the 2010 warm thermal anomaly. Environ Microbiol Rep 7:471–479. https://doi.org/10.1111/1758-2229.12274

Tsoy OV, Ravcheev DA, Čuklina J, Gelfand MS (2016) Nitrogen fixation and molecular oxygen: comparative genomic reconstruction of transcription regulation in Alphaproteobacteria. Front Microbiol 7. https://doi.org/10.3389/fmicb.2016.01343

van Oppen MJH, Oliver JK, Putnam HM, Gates RD (2015) Building coral reef resilience through assisted evolution. Proc Natl Acad Sci U S A 112:2307–2313. https://doi.org/10.1073/pnas.1422301112

Vega Thurber RL, Correa AMS (2011) Viruses of reef-building scleractinian corals. J Exp Mar Biol Ecol 408:102–113. https://doi.org/10.1016/j.jembe.2011.07.030

Vega Thurber RL, Barott KL, Hall D, Liu H, Rodriguez-Mueller B, Desnues C et al (2008) Metagenomic analysis indicates that stressors induce production of herpes-like viruses in the coral *Porites compressa*. Proc Natl Acad Sci U S A 105:18413–18418. https://doi.org/10.1073/pnas.0808985105

Vega Thurber RL, Willner-Hall D, Rodriguez-Mueller B, Desnues C, Edwards RA, Angly F et al (2009) Metagenomic analysis of stressed coral holobionts. Environ Microbiol 11:2148–2163. https://doi.org/10.1111/j.1462-2920.2009.01935.x

Vega Thurber RL, Burkepile DE, Fuchs C, Shantz AA, McMinds R, Zaneveld JR (2014) Chronic nutrient enrichment increases prevalence and severity of coral disease and bleaching. Glob Chang Biol 20:544–554. https://doi.org/10.1111/gcb.12450

Ward JR, Kim K, Harvell CD (2007) Temperature affects coral disease resistance and pathogen growth. Mar Ecol Prog Ser 329:115–121. https://doi.org/10.3354/meps329115

Warner ME, Suggett DJ (2016) The photobiology of symbiodinium spp.: linking physiological diversity to the implications of stress and resilience. In: Goffredo S, Dubinsky Z (eds) The cnidaria, past, present and future. Springer, Cham, pp 489–509

Weber M, de Beer D, Lott C, Polerecky L, Kohls K, Abed RMM et al (2012) Mechanisms of damage to corals exposed to sedimentation. Proc Natl Acad Sci U S A 109:E1558–E1567. https://doi.org/10.1073/pnas.1100715109

Webster NS, Reusch TBH (2017) Microbial contributions to the persistence of coral reefs. ISME J 11:2167–2174. https://doi.org/10.1038/ismej.2017.66

Webster NS, Smith LD, Heyward AJ, Watts JE, Webb RI, Blackall LL et al (2004) Metamorphosis of a scleractinian coral in response to microbial biofilms metamorphosis of a scleractinian coral in response to microbial biofilms. Appl Environ Microbiol 70:1213–1221. https://doi.org/10.1128/AEM.70.2.1213

Webster NS, Negri AP, Botté ES, Laffy PW, Flores F, Noonan S et al (2016) Host-associated coral reef microbes respond to the cumulative pressures of ocean warming and ocean acidification. Sci Rep 6. https://doi.org/10.1038/srep19324

Wegley L, Edwards R, Rodriguez-Brito B, Liu H, Rohwer F (2007) Metagenomic analysis of the microbial community associated with the coral *Porites astreoides*. Environ Microbiol 9:2707–2719. https://doi.org/10.1111/j.1462-2920.2007.01383.x

Welsh RM, Rosales SM, Zaneveld JR, Payet JP, McMinds R, Hubbs SL et al (2017) Alien vs. predator: bacterial challenge alters coral microbiomes unless controlled by *Halobacteriovorax* predators. PeerJ 5. https://doi.org/10.7717/peerj.3315

Weynberg KD, Voolstra CR, Neave MJ, Buerger P, van Oppen MJH (2015) From cholera to corals: viruses as drivers of virulence in a major coral bacterial pathogen. Sci Rep 5. https://doi.org/10.1038/srep17889

Weynberg KD, Neave M, Clode PL, Voolstra CR, Brownlee C, Laffy P et al (2017) Prevalent and persistent viral infection in cultures of the coral algal endosymbiont *Symbiodinium*. Coral Reefs 36:773–784. https://doi.org/10.1007/s00338-017-1568-7

Williams AD, Brown BE, Putchim L, Sweet MJ (2015) Age-related shifts in bacterial diversity in a reef coral. PLoS One 10. https://doi.org/10.1371/journal.pone.0144902

Wilson WH (2011) Coral viruses. In: Hurst CJ (ed) Studies in viral ecology: animal host systems. Wiley-Blackwell, pp 143–151

Wilson WH, Francis I, Ryan K, Davy SK (2001) Temperature induction of viruses in symbiotic dinoflagellates. Aquat Microb Ecol 25:99–102. https://doi.org/10.3354/ame025099

Wobeser GA (2006) Essentials of disease in wild animals. Blackwell, Ames, IA

Wood-Charlson EM, Weynberg KD, Suttle CA, Roux S, van Oppen MJH (2015) Metagenomic characterization of viral communities in corals: mining biological signal from methodological noise. Environ Microbiol 17:3440–3449. https://doi.org/10.1111/1462-2920.12803

Wooldridge SA (2013) Breakdown of the coral-algae symbiosis: towards formalising a linkage between warm-water bleaching thresholds and the growth rate of the intracellular zooxanthellae. Biogeosciences 10:1647–1658. https://doi.org/10.5194/bg-10-1647-2013

Wooldridge S, Done TJ (2009) Improved water quality can ameliorate effects of climate change on corals. Ecol Appl 19:1492–1499. https://doi.org/10.1890/08-0963.1

Yee SH, Santavy DL, Kern JW, Hession S (2011) Consideration of species community composition in statistical analyses of coral disease risk. Mar Ecol Prog Ser 431:83–96. https://doi.org/10.3354/meps09135

Zaneveld JR, Burkepile DE, Shantz AA, Pritchard CE, McMinds R, Payet JP et al (2016) Overfishing and nutrient pollution interact with temperature to disrupt coral reefs down to microbial scales. Nat Commun 7. https://doi.org/10.1038/ncomms11833

Zaneveld JR, McMinds R, Vega Thurber R (2017) Stress and stability: applying the Anna Karenina principle to animal microbiomes. Nat Microbiol 2. https://doi.org/10.1038/nmicrobiol.2017.121

Zhang Y, Ling J, Yang Q, Wen C, Yan Q, Sun H et al (2015) The functional gene composition and metabolic potential of coral-associated microbial communities. Sci Rep 5. https://doi.org/10.1038/srep16191

Ziegler M, Seneca FO, Yum LK, Palumbi SR, Voolstra CR (2017) Bacterial community dynamics are linked to patterns of coral heat tolerance. Nat Commun 8. https://doi.org/10.1038/ncomms14213

Zvuloni A, Artzy-Randrup Y, Katriel G, Loya Y, Stone L (2015) Modeling the impact of white-plague coral disease in climate change scenarios. PLoS Comput Biol 11. https://doi.org/10.1371/journal.pcbi.1004151

Chapter 8
Cell Biology of Coral Bleaching

C. A. Oakley and S. K. Davy

8.1 Introduction

The coral-dinoflagellate symbiosis is the energetic foundation of tropical coral reef ecosystems and depends on the cellular integration of two evolutionarily and metabolically disparate organisms. The coral host maintains photosynthetic dinoflagellates of the genus *Symbiodinium* within individual cells of its inner gastrodermal layer. These algal symbionts exist at very high densities, typically at least one million algal cells *per* square centimeter of coral tissue. The success of this association is based on complementary intercellular nutrient exchange, which requires tolerance of the symbiont's presence by the host's immune mechanisms and the coordinated growth and reproduction of both partners (Weis et al. 2008; Davy et al. 2012). Corals are highly productive, a result of cellular mechanisms of both host and symbiont that maximize photosynthetic function in the high irradiances and warm waters typical of shallow coral reefs. While this symbiosis has been successful for millions of years, the complex nutritional and immunological interactions between the animal and algal partners provide many potential points of susceptibility to external stresses.

Coral bleaching is the result of the disruption of these interactions resulting in the failure of the symbiosis, primarily due to sustained high sea surface temperatures (SST) in summer. The physiological impairment of the algal symbionts and their expulsion is primarily considered to be a result of reactive oxygen species (ROS) generation from the cnidarian host, the algal symbiont, or both, triggering a host immune response (Weis 2008). Severe and prolonged elevated SST in 2016 resulted in a global bleaching event that devastated large regions of the Great Barrier Reef, Australia (Hughes et al. 2017). Neither local water quality, fishing pressures, nor the

C. A. Oakley · S. K. Davy (✉)
School of Biological Sciences, Victoria University of Wellington, Wellington, New Zealand
e-mail: simon.davy@vuw.ac.nz

M. J. H. van Oppen, J. M. Lough (eds.), *Coral Bleaching*, Ecological Studies 233,
https://doi.org/10.1007/978-3-319-75393-5_8

presence of marine-protected areas had detectable effects on bleaching severity, and the Great Barrier Reef again suffered extensive bleaching in 2017. Therefore, the phenomenon of severe, regional-scale mass bleaching is an emergent property of cellular stress playing out at the ecosystem scale. Predicting the future of coral reefs under anthropogenic climate change scenarios requires understanding the coral cellular environment and the mechanisms that maintain the coral-algal symbiosis, as well as how these functions can be impaired.

8.2 Thermal Response of the Algal Symbiont

The coral-algal symbiosis is dependent on the productivity of the intracellular algal symbionts and requires photosynthesis to occur in the demanding microhabitat inside each host cell. This section will detail the cellular mechanisms by which the dinoflagellate endosymbionts of corals are adapted to these conditions, as well as the events that occur when these mechanisms are disrupted or overwhelmed. While the initial cellular triggers of coral bleaching are not completely understood, bleaching is highly correlated with periods of both high SST and irradiance (Hoegh-Guldberg et al. 2007; Hughes et al. 2017). As bleaching is the conspicuous loss of algae from the host, the effects of high temperature on the algal symbionts have received the most attention, with the direct generation of ROS by algal photosynthesis being an area of particular focus; see Warner and Suggett (2016) for a thorough review. The current oxidative theory of coral bleaching proposes that ROS generation increases during thermal stress and overwhelms host and symbiont mechanisms to detoxify ROS and repair oxidative damage (Weis 2008, Fig. 8.1). High temperatures and irradiances result in a large quantity and rate of energy entering and being processed by the algal photosynthetic apparatus, primarily the light-harvesting complexes and photosynthetic electron transport chain of the algal chloroplast. In addition, the surrounding host membranes and tissues constrain the efflux of photosynthetically generated oxygen (O_2) away from the algae and the import of dissolved inorganic carbon to maintain photosynthetic carbon fixation (Kuhl et al. 1995; Tansik et al. 2017). These factors impose considerable stress on the photosynthetic machinery of the algal symbionts and render them potentially susceptible to oxidative damage (Weis 2008). The failure of mechanisms that moderate these stresses and the consequences when they are overcome are likely the initial events in the coral bleaching cascade.

8.2.1 Thermal Effects on Symbiodinium *Photosynthesis*

The efficient capture of light energy and rapid flow of electrons through the photosynthetic electron transport chain are essential, not only to support carbon fixation but also to avoid photodamage, as the light absorbed can frequently surpass the capacity of the electron transport chain (Warner and Suggett 2016).

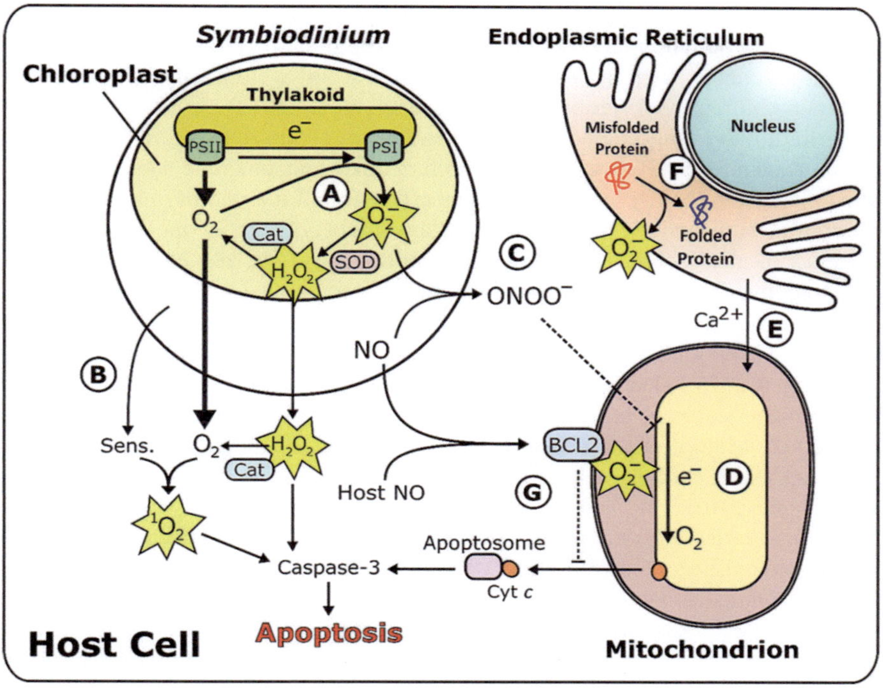

Fig. 8.1 Mechanisms of thermal stress that may initiate or participate in the bleaching cascade. (A) Inhibition of photosynthetic electron transport between photosystem II (PSII) and photosystem I (PSI) promotes the generation of superoxide (O_2^-) from photosynthetic O_2 via the Mehler reaction. O_2^- is converted to peroxide (H_2O_2) via superoxide dismutase (SOD) and then to O_2 by catalase (Cat). H_2O_2 released to the host may be decomposed by host SOD. (B) O_2 produced by photosynthesis may react with singlet oxygen sensitizers (Sens.), also released by the endosymbionts, to form the highly reactive singlet oxygen ($^1O_2$). (C) Nitric oxide (NO) reacts with O_2^- to form peroxynitrite ($ONOO^-$), which inhibits mitochondrial electron transport. (D) Mitochondrial reactive oxygen species (ROS) generation by the respiratory electron transport chain increases due to thermally induced high respiration or electron transport inhibition, independently from photosynthesis. (E) Endoplasmic reticulum (ER) stress results in the release of calcium (Ca^{2+}) to the mitochondria, promoting the release of cytochrome *c* (Cyt *c*). (F) High temperatures cause an increase in protein misfolding in the ER, and the re-folding and repair of these proteins results in ROS generation and ER stress. (G) The mitochondrial B-cell lymphoma 2 (BCL2) protein regulates apoptosis by inhibiting the release of Cyt *c*. This function can be disabled directly by NO, ROS, or by mitochondrial Ca^{2+} uptake. Cyt *c* released into the cytoplasm binds to the apoptosome, initializing caspase activity and terminating in host cell apoptosis. Caspase signaling pathways have been greatly simplified for clarity. Figure modified from Weis et al. (2008) and Oakley et al. (2017)

Symbiodinium utilizes the carotenoid peridinin, which possesses a higher absorbance of green to blue light relative to chlorophyll. Peridinin is characteristic of the peridinin-containing dinoflagellates, which include *Symbiodinium*, together with chlorophylls *a* and c_2 (Brown et al. 1999). These pigments give the algal symbionts their golden brown color and provide a dark tone to the coral, obscuring the bright white skeleton. These pigments are integrated into the intrinsic membrane-bound antenna chlorophyll *a*-chlorophyll c_2-peridinin-protein-complex, which transfers excitation energy directly to the photosystems, as well as the water-soluble peridinin-chlorophyll *a*-protein (Niedzwiedzki et al. 2014). Light absorbed by these light-harvesting complexes enters the conventional photosynthetic electron transport chain from PSII through PSI to ferredoxin-nicotinamide adenine dinucleotide phosphate ($NADP^+$) reductase. The efficiency of the photosynthetic electron transport chain, and by extension the photosynthetic state of the symbionts, is assessed by active chlorophyll induction fluorometry (Warner et al. 1996). The principal metric of interest is the maximal quantum yield of PSII chlorophyll *a* fluorescence (F_v/F_m), which reflects the proportion of absorbed light energy that enters the photosynthetic electron transport chain from PSII (Hill et al. 2004; Warner and Suggett 2016). Declines in F_v/F_m indicate nonspecific photoinactivation due to PSII damage and may precede the loss of algal symbionts from the host (Warner et al. 2002). Multiple studies have demonstrated the relationship between F_v/F_m and PSII status with cnidarian bleaching (Hill et al. 2004; Hillyer et al. 2017a; Warner and Suggett 2016).

Symbiodinium, like other photosynthetic organisms, possesses a variety of mechanisms to optimize electron flow through the photosynthetic apparatus while avoiding photodamage (Warner and Suggett 2016). PSII has been shown to be susceptible to photodamage in *Symbiodinium*, particularly the reaction center protein D1, which is potentially a site of functional differentiation between *Symbiodinium* types (Warner et al. 1999; Takahashi et al. 2004; Hennige et al. 2008; Hill et al. 2011). The relative rates of PSII repair during thermal stress have been investigated as a potential source of the variable thermal susceptibility between *Symbiodinium* genotypes (McGinley et al. 2012; Buxton et al. 2012). PSII degradation reduces the capacity of the photosynthetic apparatus to use captured light energy, resulting in increased pressure on the remaining PSII units. PSII and PSI in dinoflagellates remain in close physical proximity within the thylakoids, allowing for bidirectional energy transfer between them via the light-harvesting complexes, particularly during thermal stress (Slavov et al. 2016). By reorganizing the PSII and PSI complexes within the thylakoids, *Symbiodinium* may direct excess absorbed light energy away from PSII to PSI, where it can be converted to heat and quenched by the reduced reaction center chlorophyll $P_{700}{}^+$. This occurs with a corresponding loss of electron flow supporting carbon fixation (Slavov et al. 2016). Accessory photosynthetic pigments may also play a role in photoprotection by regulating photosynthetic electron flow and dissipating excess absorbed light energy. Nonphotochemical quenching, the dissipation of photosynthetic electron flow as heat, has been primarily attributed to the xanthophyll cycle (Brown et al. 1999; Kanazawa et al. 2014). This mechanism operates by the de-epoxidation and conversion of the light-absorbing xanthophyll pigment diadinoxanthin to the heat-dissipating diatoxanthin under high irradiance, which is sensed by increases in the pH gradient across the thylakoid membrane (Gustafsson et al. 2014).

The consequence of failure of these photoprotective mechanisms is unregulated or impaired energy flow through the photosynthetic apparatus in the high-O_2

environment of coral tissues. Hyperoxic environments greatly increase the rate of spontaneous and catalyzed ROS production by electron transport chains (Imlay 2013), and illuminated coral tissues may have internal O_2 concentrations at 250% of saturation (Kuhl et al. 1995). The role of ROS in coral bleaching has received intensive study since the mid-1990s (Lesser 1996; Warner et al. 1996). The release of ROS or ROS-sensitizing compounds from the algae is primarily a result of overexcitation of the photosynthetic apparatus (Rehman et al. 2016), and light- and temperature-stress have independent, but synergistic, effects that initiate this condition. Saturation of photosynthetic electron transport due to high temperatures or irradiance results in excitation pressure over PSI and PSII, and increased ROS production through multiple mechanisms (Roberty et al. 2015; Rehman et al. 2016), including cyclic electron flow around PSI (McCabe Reynolds et al. 2008), photo-respiration, and chlororespiration, but the increased ROS production primarily occurs via the water-water cycle/Mehler reaction (Roberty et al. 2014, 2015). The water-water cycle involves the direct reduction of O_2 by PSI to superoxide ($O_2^{\bullet-}$), which is short-lived but highly toxic and must be rapidly degraded by the symbiont's antioxidant mechanisms (Fig. 8.1A). In doing so, these mechanisms enhance the proton gradient across the thylakoid membrane, resulting in enhanced adenine triphosphate (ATP) production at the expense of NADPH (Roberty et al. 2014). O_2 reduction by the water-water cycle increases due to either high light or combined high light and temperature in cultured *Symbiodinium* (Roberty et al. 2015). In *Pocillopora damicornis*, light stress resulted in the condensation of thylakoid lamellae, indicative of oxidative damage, while thermal stress caused thylakoid membrane disorganization (Downs et al. 2013).

Antioxidant mechanisms in *Symbiodinium* include superoxide dismutase (SOD), glutathione, ascorbate peroxidase, catalases, and monodehydroascorbate reductase. SOD is an enzyme produced by both host and alga that catalyzes the conversion of $O_2^{\bullet-}$ to hydrogen peroxide (H_2O_2) (Roberty et al. 2015). If the rate of ROS production exceeds the rate of ROS detoxification, widespread cellular damage can occur, including the potential leakage of ROS from the algal cell, initiating the bleaching cascade (Fig. 8.1). SOD and ascorbate peroxidase activities have been shown to increase during extended periods of elevated temperature in cultured *Symbiodinium* coincident with oxidation of the glutathione pool (Krueger et al. 2014). A long-standing puzzle in this model is how highly reactive, short-lived $O_2^{\bullet-}$, singlet oxygen ($^1O_2$), or other ROS are able to pass across multiple membranes to interact with the host cell (Krueger et al. 2015), with evidence that leakage of ROS is possible (Saragosti et al. 2010). Moreover, symbionts may play a role in ameliorating host oxidative stress, as superoxide dismutases and multiple other ROS-mediating enzymes have been found to be upregulated in the model symbiotic anemone *Exaiptasia pallida* (Grajales and Rodríguez 2014, commonly known as *Aiptasia*) when it is rendered experimentally aposymbiotic (Nii and Muscatine 1997, Oakley et al. 2016, but see Richier et al. 2005). The release of algal-derived ROS into the host cells may vary with symbiont species, corresponding with both increased photoinhibition and expression of ROS-scavenging genes (Levin et al. 2016). One recent advance is the detection of extracellular $^1O_2$ produced by temperature- and light-stressed *Symbiodinium* cultures that has been attributed to excreted, but currently unidentified, ROS-sensitizing metabolites (Fig. 8.1B, Rehman et al. 2016). These findings provide a possible link between the redox states of the host and symbiont.

Additional evidence for the exchange of metabolites involved in oxidative balance exists for nitric oxide (NO) and dimethylsulfoniopropionate (DMSP). NO is an important cellular signaling molecule that is involved in immunity, responses to endosymbiotic microbes, and the apoptosis pathway (Weis 2008; Hawkins et al. 2013). *Symbiodinium* constitutively produces NO at a rate that is increased under elevated temperatures, coincidental with photoinhibition; however, different genotypes show varying toxicity (Hawkins and Davy 2012). Increased NO production has also been documented during bleaching (Trapido-Rosenthal et al. 2005). Importantly, NO is lipophilic and may be capable of crossing membranes into the host cell (Fig. 8.1C). NO is just one reactive nitrogen species, and reacts with $O_2^{\bullet -}$ to form peroxynitrite, a highly toxic oxidant and electron transport inhibitor (Hawkins and Davy 2013). The relative contributions of ROS and reactive nitrogen species to the bleaching cascade are currently unknown. DMSP is a ROS scavenger, among many other functions, readily crosses cell membranes, and can be produced by both corals and *Symbiodinium* (Hopkins et al. 2016). DMSP production by the holobiont increases markedly in response to aerial exposure and correspondingly high O_2 concentrations (Hopkins et al. 2016), as well as after thermal stress (Jones et al. 2014). Furthermore, DMSP is differentially produced by some coral holobionts in response to osmotic stress, and may serve as an additional antioxidant system in the symbiosis (Gardner et al. 2016).

8.2.2 Thermal Effects on the Calvin-Benson Cycle

The high demand for dissolved inorganic carbon (DIC) from seawater by the coral holobiont to support both photosynthesis and host calcification presents a physiological challenge. DIC in seawater exists overwhelmingly as bicarbonate (HCO_3^-), which is effectively incapable of crossing membranes due to its charge. As a result, inorganic carbon is frequently limiting to the algal symbionts, in contrast to phytoplankton in the water column (Tansik et al. 2017). Ensuring adequate inorganic carbon supply is essential to the algal symbionts' productivity and redox state, as carbon fixation is the ultimate result of photosynthetic electron transport. If there is insufficient inorganic carbon as carbon dioxide (CO_2) available to the chloroplast, photosynthetic electron transport may be inhibited due to the lack of an energy sink. The presence of many layers of membranes between the site of carbon fixation and the external seawater, combined with the high densities of algal symbionts in coral tissues, requires carbon-concentrating mechanisms by both partners to supply adequate amounts of DIC (Leggat et al. 1999; Tansik et al. 2015). The carbon-concentrating mechanisms of cultured *Symbiodinium* have been found to be enhanced at higher temperatures (Oakley et al. 2014a); however, whether or not these are sufficient *in hospite* under elevated temperatures is unclear. Additionally, the potential susceptibility of the host's carbon-concentrating mechanisms to elevated temperatures is currently unknown.

High temperatures have been proposed to have indirect effects on photosynthetic electron transport via inhibition of the Calvin-Benson cycle, limiting carbon fixation as an energy sink (Jones et al. 1998; Wooldridge 2009). This hypothesis, while justified in that the peridinin-containing dinoflagellates are unique among eukaryotes in their possession of a thermally labile Form II ribulose 1,5-bisphosphate carboxylase/oxygenase (Rubisco), has not been demonstrated in *Symbiodinium* (Leggat et al. 1999; Lilley et al. 2010; Oakley et al. 2014a). Form II Rubisco exhibits very low CO_2:O_2 specificity relative to the common Form I Rubisco of higher plants (Tcherkez et al. 2006). The fixation of O_2 by Rubisco, termed photorespiration, is enhanced by high O_2:CO_2 concentrations and produces the Calvin cycle inhibitor phosphoglycerate (Badger et al. 1998; Crawley et al. 2010). Photorespiration is not metabolically useful, except as an energy sink, but it may consume some of the abundant O_2 in illuminated coral tissues and, therefore, limit ROS production (Smith et al. 2005). Notably, inhibition of phosphoglycolate phosphatase, the enzyme responsible for recycling phosphoglycolate produced by photorespiration, has been documented in *Symbiodinium in hospite* under ocean acidification conditions (1100 ppm CO_2, pH 7.6) (Crawley et al. 2010).

In addition to the photosynthetic apparatus, other algal organelles also possess mechanisms to absorb and utilize excess energy (Warner and Suggett 2016). Mitochondrial respiration constitutively results in a steady rate of ROS generation, and the same high electron flows and O_2 concentrations that challenge photosynthetic electron transport are prone to increasing ROS generation by the mitochondrial respiratory electron transport chain. This is also exacerbated during thermal stress, which can elevate respiration and, potentially, mitochondrial ROS generation as H_2O_2. Mitochondrial alternative oxidase competes for electron flow with the conventional cytochrome *c* oxidase, counteracting the mitochondrial proton gradient necessary for ATP production (Oakley et al. 2014b). Alternative oxidase activity has been shown to double under thermal stress and may account for 25–50% of respiratory O_2 consumption by *Symbiodinium*. The role of the mitochondrion in oxidative stress is particularly important in the independent thermal response of the host cell, and is considered in more detail below.

8.2.3 Nutrient Availability and Susceptibility to Bleaching

Beyond the biological activities of host and symbiont, abiotic factors may also contribute to the resilience of the coral symbiosis to thermal stress, with direct impacts at the cellular level. Of particular note, recent studies have explored the relationship between reef eutrophication and bleaching, in an effort to explain the observed lower bleaching thresholds of inshore reefs exposed to terrestrial sources of dissolved inorganic nitrogen on the Great Barrier Reef, Australia (Wooldridge

2009). This ecological phenomenon may be a result of the cellular effects of imbalanced nitrogen/phosphorus on the stability of the symbiosis (see Rädecker et al. 2015 for a review). Corals are adapted to oligotrophic conditions by recycling nitrogen compounds between partners and limiting the availability of nitrogen to the symbionts to control their growth (Yellowlees et al. 2008). Elevated nitrogen, such as from anthropogenic outflows, may release this limitation and result in a relative deficiency of phosphate in the algae (Wiedenmann et al. 2013). Coral phosphate uptake increases at elevated temperatures while nitrogen uptake generally decreases, reflecting an active response by the holobiont to resolve the nutrient imbalance (Ezzat et al. 2016). Corals with high nitrate and low phosphate availabilities bleached strongly under relatively low light levels, exhibiting greatly reduced F_v/F_m values, symbiont densities, and significant tissue mortality, while those with high nitrate and phosphate availability were unharmed (Wiedenmann et al. 2013). The addition of sugar resulted in increased N_2 fixation by the *Pocillopora verrucosa* microbiome, and thus a higher N:P ratio, and induced bleaching under non-stressful irradiance and temperature (Pogoreutz et al. 2017). *Symbiodinium* cells under phosphate starvation also exhibit increases in starch, lipid, and uric acid storage bodies, indicating imbalanced growth (Rosset et al. 2017). Importantly, these effects are not seen when both nitrogen and phosphorus availability together are either high or low. The cellular basis of this response may lie in the substitution of sulfolipids for phospholipids in the photosynthetic membranes during extreme phosphate limitation (Wiedenmann et al. 2013), resulting in changes to thylakoid membrane fluidity and integrity (Tchernov et al. 2004). Thylakoid membrane fluidity has been linked to thermal resilience in *Symbiodinium*, with thermally susceptible species exhibiting lower concentrations of saturated polyunsaturated fatty acids, elevated ROS production, and greater bleaching (Tchernov et al. 2004). This line of evidence may link the observed effects of eutrophication on coral health and reef degradation with the prevailing model of bleaching as a result of algal-derived oxidative stress destabilizing the symbiosis.

8.3 Thermal Response of the Coral Host

The coral host is also subject to many of the same high-energy pressures and nutrient demands of the reef environment as its symbionts, and bleaching is not purely a light-driven process (Nii and Muscatine 1997; Tolleter et al. 2013). The direct effects of temperature on the coral host prior to bleaching, and the independent role of the host cell in the production and amelioration of oxidative stress, have been the focus of recent research, and these may be sufficient to induce coral bleaching even when the symbionts are not physiologically impaired (Ainsworth et al. 2008; Dunn et al. 2012). ROS and reactive nitrogen species are important signaling molecules in many symbioses due to their universal nature among aerobic organisms (Moné et al. 2014). The oxidative states of the host and symbiont are inescapably intertwined, resulting in considerable difficulty in determining whether the initial point of thermal damage

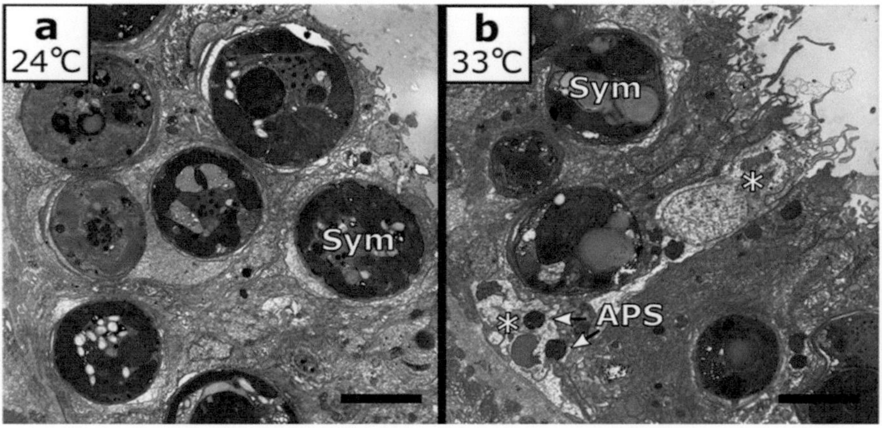

Fig. 8.2 Transmission electron micrographs of *Aiptasia* tentacles under control and elevated temperatures. (**a**) *Aiptasia* under control conditions (24 °C, 50 μmol photons m^{-2} s^{-1}). (**b**) *Aiptasia* after 48 h temperature and light stress (33 °C, 140 μmol photons m^{-2} s^{-1}). Asterisks indicate autophagic host cells. Sym = *Symbiodinium* cell. *APS* = autophagic structures. Scale bars = 5 μm. Images courtesy of Shanna D. Hanes

lies within the host or symbiont; moreover, given the physiological diversity of cnidarians and *Symbiodinium*, it is likely that the relative contribution of host and symbiont to thermal degradation varies between different pairings (Abrego et al. 2008; Hawkins et al. 2016).

The relative susceptibilities of the host and symbiont to thermal and oxidative stress are complex. There is mounting evidence that, in many holobionts, the host may be more heat-susceptible than the symbionts (Baird et al. 2009). Krueger et al. (2015) demonstrated that host antioxidant activity was elevated in two coral species during high temperatures, days before thermal impairment of the symbionts was observed. These observations are supported by measurements of the host mitochondrial coenzyme Q and algal plastoquinone pools in experimentally bleached *A. millepora*, in which the coenzyme Q pool exhibited an oxidative shift characteristic of ROS stress days before a shift in the plastoquinone pool (Lutz et al. 2015). Hillyer et al. (2017b) detected an increased production of antioxidant precursors in heat-stressed *Aiptasia*, but not in the symbionts, and Oakley et al. (2017) showed the induction of a heat shock response from the host in the absence of bleaching or photoinhibition of the algae. Thermal stress in the host has been proposed to initiate a complex suite of responses including the elevation of host endogenous oxidative stress, induction of the host innate immune response against the algal symbionts and, ultimately, host cell apoptosis (Fig. 8.2, Weis 2008). The generation of, and response to, ROS are the events that unify our understanding of the host's and alga's responses to elevated temperatures.

The primary source of ROS generation in a non-photosynthetic organism is the mitochondrion, due to its roles in central metabolism (Dunn et al. 2012). These activities, as well as its role in calcium storage and apoptosis, make the mitochondrion a potential locus of host thermal stress. The mitochondrial respiratory electron transport chain uses O_2 as the terminal electron acceptor, and ROS are constantly generated as a result of normal respiratory electron transport (Turrens 2003). In all aerobic organisms, including cnidarians, this background level of ROS generation by the mitochondria increases greatly in response to high temperatures (Nii and Muscatine 1997). Thus, highly elevated respiratory ROS generation initially impacts the mitochondria (Fig. 8.1D). The *Aiptasia* host mitochondria undergo morphological changes after thermal shock, followed by mitochondrial permeabilization and degradation, host cell apoptosis, and symbiont cell degradation (Dunn et al. 2012). Further evidence of the mitochondrial role in the host bleaching response is the reduced mRNA abundance of cytochrome *c* and ATP synthase, which are both central components of the respiratory electron transport chain (Dunn et al. 2012). Inhibited respiration hinders the ability of the host to survive or recover from sustained thermal stress.

The mitochondrion also stores calcium, a secondary messenger that regulates many cell functions and is central to sensing cell stress (Orrenius et al. 2003). Under stress conditions, the calcium pool within the mitochondrion may greatly increase, resulting in mitochondrial permeabilization and the release of calcium into the cell, subsequently triggering caspase-mediated apoptosis (Orrenius et al. 2003). The endoplasmic reticulum (ER) is the primary store of calcium in the cell, and it directly exchanges calcium ions with the mitochondrion across tight junctions (Fig. 8.1E, Rainbolt et al. 2014). Multiple studies have described a role for calcium regulation in the cnidarian symbiosis and thermal response (Fang et al. 1997; Sandeman 2006; Desalvo et al. 2008; Ganot et al. 2011; Bellantuono et al. 2012; Oakley et al. 2017). Calumenin, an ER protein that binds calcium, is upregulated in symbiosis (Ganot et al. 2011) and is further upregulated following exposure to elevated temperatures, particularly in the gastrodermal layer (Bellantuono et al. 2012; Oakley et al. 2017). Calcium-binding proteins may be more generally involved in the heat stress response, given that thermal shock induced an upregulation of multiple calcium-binding proteins in *Acropora microphthalma* host cells concurrent with the higher expression of antioxidant, heat shock, and calcium-binding proteins in the symbiont population (Weston et al. 2015). Bleaching resistance may, therefore, depend on the host's ability to maintain calcium homeostasis during thermal stress.

Beyond its primary role in protein synthesis, folding, and export, the ER is involved in stress signaling between itself and the Golgi, nucleus, and mitochondrion (Chaudhari et al. 2014). Communication between the ER and the mitochondrion allows for a coherent cellular response to thermal stress, which directly impacts the ER's role in protein synthesis and folding (Fig. 8.1F, Orrenius et al. 2003). Protein folding is temperature-sensitive and requires the activity of molecular chaperones that, due to their pronounced upregulation during elevated temperature, are known as heat shock proteins (Richter et al. 2010). Misfolded proteins are potentially toxic, and their accumulation in the ER results in the induction of the unfolded protein response (UPR; Ron and Walter 2007). The UPR is characterized by decreased overall protein synthesis, increased protein chaperone and degradation activity, and finally, if the ER stress persists, cell death (Ron and Walter 2007).

Apoptosis triggered by ER stress is thought to be induced by the leakage of calcium into the cytoplasm, where it is taken up by the mitochondria, or through caspase induction (Ron and Walter 2007). Short-term thermal stress in *Aiptasia* induced an upregulation of many proteins involved in protein folding, protein degradation, and calcium binding, and a concurrent downregulation of many cytoskeletal proteins, consistent with ER stress (Oakley et al. 2017). Roles for the ER and the UPR were also detected in the coral *Acropora hyacinthus* when it was exposed to short-term thermal stress, suggesting that the ER may be one of the initial sites of thermal stress and cellular acclimation (Ruiz-Jones and Palumbi 2017). Protein degradation and re-folding in the ER are performed by protein disulfide isomerases, which also produce ROS as a result of their activity (Tu and Weissman 2004). In this manner the ER, in addition to the mitochondria, may both generate and ameliorate oxidative stress.

NO may play a major role in the bleaching cascade, as bleaching followed the production of significant quantities of NO by the symbiotic anemone *Aiptasia* during thermal stress (Perez and Weis 2006). Hawkins et al. (2013) demonstrated elevated NO synthesis in both the anemone *Aiptasia* and in *Symbiodinium in hospite* under thermal stress co-occurring with the activation of pro-apoptotic host caspases and bleaching. Caspase activity was also induced by an NO donor and reduced by an NO scavenger, indicating a strong role for NO-dependent host apoptosis pathways in the initial stages of bleaching. NO production within symbiont cells was detected after host NO production, and this may be an alternative or additional mechanism of algal oxidative stress. These results were extended to and confirmed in multiple coral species (Hawkins et al. 2014). Importantly, the production of NO within host tissues prior to algal photoinhibition indicates a host origin of thermal dysfunction in the cnidarian-dinoflagellate symbiosis. NO also has a role in initiating apoptosis through the release of mitochondrial cytochrome *c* (Snyder et al. 2009). The role of NO in the response to parasites in other systems is particularly interesting given the intracellular nature of the cnidarian-dinoflagellate symbiosis (Perez and Weis 2006).

The ultimate result of sustained, uncontrolled oxidative stress in the host cell, whether it originates from dysfunction of the host or the symbiont, is the induction of host cell apoptosis. Apoptosis, or programmed cell death, is a highly conserved process by which extraneous or compromised cells are degraded following a caspase-mediated signaling cascade (Taylor et al. 2008; Kvitt et al. 2011). Cnidarian apoptosis pathways are complex and highly conserved, being more similar to those of vertebrates than those of arthropods or other invertebrate taxa (Moya et al. 2016). A major checkpoint in the apoptotic signaling cascade is B-cell lymphoma 2 (Bcl-2, Fig. 8.1G), an anti-apoptotic protein that mediates mitochondrial membrane permeability and has an unknown, likely indirect, antioxidant function (Susnow et al. 2009; Kvitt et al. 2011, 2016). The chemical inhibition of caspase activity is able to prevent bleaching and apoptosis (Tchernov et al. 2011). When anti-apoptotic mechanisms are insufficient, the apoptosis pathway proceeds via the permeabilization of the mitochondrial outer membrane and the release of cytochrome *c* into the cytoplasm, where it binds to the "apoptosome" containing apoptotic protease-activating factor 1 (Man and Kanneganti 2016). This signals a chain of caspases, principally caspase-3, to begin the apoptotic process by widespread protein degradation.

The role of caspases in coral bleaching has been established by multiple studies. In *Stylophora pistillata* and *Acropora millepora*, caspase expression increases rapidly in the gastrodermal tissue during thermal stress, followed by Bcl-2-mediated stabilization and recovery (Kvitt et al. 2011; Pernice et al. 2011). Moreover, the addition of the apoptosis-inducing reagent colchicine produces a similar response to thermal stress (Pernice et al. 2011). Both studies suggest a model in which the host's susceptibility to rapid thermal stress is dependent on the relative intensity and timing of caspase and Bcl-2 upregulation. In this scenario, during the initial phases of the thermal insult (24–48 hours), a subset of host cells may be lost, but thermally resistant corals are able to induce anti-apoptotic mechanisms afterward (Kvitt et al. 2016). At the organismal level, host apoptosis has been found to occur during thermal stress, notably prior to bleaching, in the gastrodermal layer coincidental with decreases in gastrodermal tissue thickness in *Acropora aspera* (Ainsworth et al. 2008). Thermal or chemical induction of host apoptosis may be followed by death of the resident *Symbiodinium* cells (Paxton et al. 2013). Bleaching is ultimately an active process on the part of the coral host, and understanding the host's role in the early stages of bleaching, prior to visible paling or photoinhibition, may be critical to understanding coral bleaching susceptibility.

8.4 Mechanisms of Symbiont Loss

The ways in which the coral and its algal symbionts are affected by high temperatures and the initial cellular responses to stress are well-studied, but the downstream mechanisms that result in the loss of symbionts from host cells once the bleaching cascade has been initiated are unclear (Weis 2008; Bieri et al. 2016). A multitude of studies have tracked the fate of the symbionts during the bleaching process, primarily using histology, in a variety of organisms and environmental conditions (Gates et al. 1992; Dunn et al. 2004, 2007; Ainsworth and Hoegh-Guldberg 2008; Tchernov et al. 2011; Hanes and Kempf 2013). These studies have attributed the loss of symbiont cells to several mechanisms (Fig. 8.3); however, the contributions of each during natural bleaching events are still unclear, and they may act in combination depending on the intensity and duration of the thermal stress (Gates et al. 1992; Brown et al. 1995; Bieri et al. 2016).

The first proposed mechanism of coral bleaching is the exocytosis of whole symbiont cells, whether alive or degraded, in which algal cells are expelled from the host cells into the gastrovascular cavity and out of the mouth of the animal. Symbiont expulsion occurs normally on a daily basis as a means to control the symbiont population in many, though not all, cnidarians (Baghdasarian and Muscatine 2000; Davy et al. 2012). Strychar et al. (2004) demonstrated that elevated temperatures (28 °C) resulted in the expulsion of viable *Symbiodinium* cells from tissues of both scleractinian corals and octocorals, but increasing temperatures

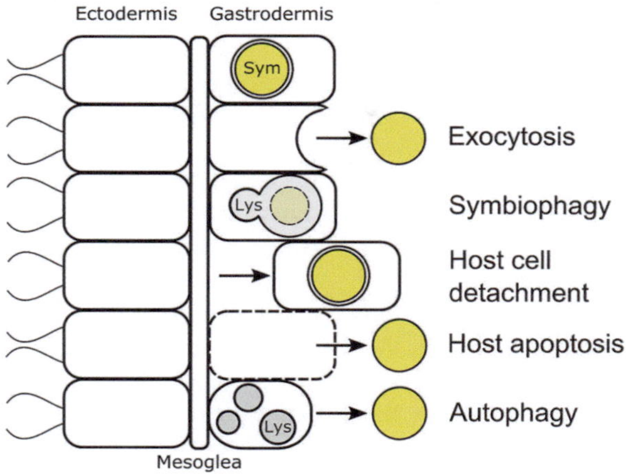

Fig. 8.3 Mechanisms of symbiont loss from cnidarian host cells after the initiation of bleaching. Sym = *Symbiodinium* cell. Lys = host lysosome. Adapted from Gates et al. (1992), Weis (2008), and Bieri et al. (2016)

beyond 28 °C resulted in an increasing proportion of the expelled cells being apoptotic or necrotic (but see Fujise et al. 2013). This contrasts with the previous study by Ralph et al. (2001), which demonstrated that expelled *Symbiodinium* cells in the coral *Cyphastrea* were expelled from the host at 33 °C but were not themselves photoinhibited, as measured by PAM fluorometry, until they were heated to 37 °C. Symbionts may even be released intact during the autophagic degradation of the host cell (Hanes and Kempf 2013). Using the model cnidarian *Aiptasia*, Bieri et al. (2016) determined that bleaching was primarily a result of expulsion of intact algae during thermal and light stress.

An alternative to symbiont expulsion is the degradation or consumption of the symbionts within the host cells. Symbiophagy involves the degradation of the symbiont cells by host autophagic pathways (Downs et al. 2009). The algal symbiosome is a host-derived late endosome that exists in a state of arrested phagocytosis, where the normal progression of consumption of foreign materials is paused indefinitely in a successful symbiosis (Davy et al. 2012). The symbiophagic process may involve the reactivation of the phagocytic pathway during thermal or oxidative stress, likely by Rab marker signaling to the symbiosome membrane (Chen et al. 2005; Downs et al. 2009). Symbiophagy was found to be primarily a function of thermal, not light, stress in the coral *Pocillopora damicornis*, suggesting that it may be a result of thermally induced disruption or activation of the host innate immune response (Downs et al. 2009). Alternately, algal cells may directly suffer from thermal or oxidative stress and degrade *in situ* via either controlled (programmed cell death) or uncontrolled (necrosis) mechanisms (Weis 2008). *In situ* degradation has been determined to be the primary means of symbiont loss in several natural bleaching events in scleractinian corals (Brown et al. 1995; Ainsworth and Hoegh-Guldberg 2008), and it has also been noted in *Aiptasia* under experimental thermal stress (Dunn et al. 2004). Bleaching via exocytosis or *in situ* degradation of symbionts would seem to be a less traumatic and metabolically costly mechanism of bleaching than host cell degradation.

In contrast to symbiont expulsion or degradation, it is possible that entire host cells containing *Symbiodinium* detach from the gastrodermal layer. Host cell release was first described by Gates et al. (1992) after both severe high- and low-temperature stress in *Aiptasia* and *Pocillopora*, and the cold response has been replicated by Bieri et al. (2016). This mechanism has also been noted in response to chemically induced stress (Sawyer and Muscatine 2001). The role of host cell detachment in typical thermal bleaching events is questionable, however, and may be a transitory, indirect outcome of more general host degradation and apoptosis mechanisms (Brown et al. 1995; Bieri et al. 2016).

Finally, *Symbiodinium* cells may be lost as a result of host cell death. Two controlled host death pathways have been proposed due to thermal or oxidative stress: apoptosis, as discussed previously, or autophagy, the controlled sequestration of and destruction of cellular structures (Dunn et al. 2007; Paxton et al. 2013). Autophagy is a highly conserved mechanism to remove compromised or extraneous organelles and cells, in which marked target structures are enveloped by phagocytic vacuoles and broken down by the fusion of the autophagosomes with lysosomes containing digestive enzymes (Hanes and Kempf 2013). In some cases, entire cells can be degraded in this way. Autophagic structures were found to greatly increase following sustained thermal shock in *Aiptasia*, during which the symbiont cells remained competent and were released into the gastrovascular cavity (Hanes and Kempf 2013). Dunn et al. (2007) found simultaneous action of both host apoptosis and autophagy in thermally stressed *Aiptasia*. Large organism-scale reductions in cytoskeletal and structural proteins provide additional suggestive evidence for the occurrence of autophagy and apoptosis in *Aiptasia* during heat shock prior to symbiont dysfunction (Oakley et al. 2016). Both autophagy and apoptosis can be initiated by ROS stress, but the sequence of each and the relative importance of autophagy remain unknown. Uncontrolled cell death, or necrosis, has been documented in cases of severe thermal stress in both host and symbiont cells (Dunn et al. 2004) and is assumed to occur when the controlled apoptotic or autophagic pathways are overcome by severe and sustained cellular damage.

8.5 Future Directions

The study of the ecosystem-wide process of coral bleaching is increasingly propelled by the use of modern "omics" techniques, aided by advances in genetic sequencing and mass spectrometry, to study the cellular mechanisms of both partners (Davy et al. 2012). The genomes of multiple coral species, including *Acropora digitifera* (Shinzato et al. 2011), are now available, as are those of several *Symbiodinium* species (Shoguchi et al. 2013; Lin et al. 2015; Aranda et al. 2016; Levin et al. 2016; Gierz et al. 2017; Wang et al. 2017) along with transcriptomic datasets of many species during bleaching (Pinzón et al. 2015). High-resolution genetic data provide insight into the ways in which symbiosis has molded the evolution and physiology of the two partners and serve as a means to predict their response to future climate change (Bay et al. 2011; Bhattacharya et al. 2016). Genomes of target species permit mass-spectrometry-based comprehensive identification and quantification of entire proteomes (Weston et al. 2015; Oakley et al. 2017), allowing for the detection of fine-scale cellular responses to thermal perturbation. Gene expression information is increasingly augmented by metabolomics, the analysis of entire classes of compounds, such as lipids, sugars, and secondary metabolites, to gain a finely-detailed description of the state of the organism under stress (Hillyer et al. 2017a). This is particularly useful in characterizing the identities and quantities of compounds exchanged between partners (Burriesci et al. 2012; Hillyer et al. 2017b; Matthews et al. 2017), which are both central features of the symbiosis and critical to assessing its health. The contributions of the bacterial consortia to the holobiont during bleaching, including their potential roles in contributing to and mediating oxidative stress, are becoming better understood (Ziegler et al. 2017). Novel imaging methods, such as nanoscale secondary ion mass spectrometry, are also moving from the wider biological literature into the coral biology field (Pernice et al. 2012).

Given the considerable difficulty in maintaining corals in a laboratory environment and their slow growth, model systems are increasingly used to study the cellular response of the symbiosis to thermal and other stresses (Weis et al. 2008; Meyer and Weis 2012). Due to its tractability and rapid growth, *Aiptasia* has become the principal model organism for studies of the cnidarian-dinoflagellate symbiosis (Lehnert et al. 2012; Baumgarten et al. 2015). Gene expression studies using both transcriptomics and proteomics of *Aiptasia* have described the effects of symbiosis on the cnidarian cell (Lehnert et al. 2014; Oakley et al. 2016). Its ability to associate with multiple *Symbiodinium* species in laboratory experiments is invaluable to its utility (Starzak et al. 2014; Wolfowicz et al. 2016). These models provide the ability to identify cellular markers and metabolic pathways involved in coral bleaching, with the goal of direct application to ecologically relevant coral species in the field (Jin et al. 2016).

Coral bleaching is an ecosystem-scale disaster arising from cellular-scale distress. "Omics" technologies and model systems are powerful tools to analyze the bleaching process and will find an increasingly prominent role in the study of coral cell biology. Our ability to describe, predict, and, perhaps, change the outcome of bleaching events depends on our ability to describe the cellular mechanisms of thermal stress and the breakdown of the symbiosis.

References

Abrego D, Ulstrup KE, Willis BL, van Oppen MJH (2008) Species-specific interactions between algal endosymbionts and coral hosts define their bleaching response to heat and light stress. Proc R Soc B 275:2273–2282. https://doi.org/10.1098/rspb.2008.0180

Ainsworth TD, Hoegh-Guldberg O (2008) Cellular processes of bleaching in the Mediterranean coral *Oculina patagonica*. Coral Reefs 27:593–597. https://doi.org/10.1007/s00338-008-0355-x

Ainsworth TD, Hoegh-Guldberg O, Heron SF, Skirving WJ, Leggat B (2008) Early cellular changes are indicators of pre-bleaching thermal stress in the coral host. J Exp Mar Biol Ecol 364:63–71. https://doi.org/10.1016/j.jembe.2008.06.032

Aranda M, Li Y, Liew YJ, Baumgarten S, Simakov WMC et al (2016) Genomes of coral dinoflagellate symbionts highlight evolutionary adaptations conducive to a symbiotic lifestyle. Sci Rep 6:39734. https://doi.org/10.1038/srep39734

Badger MR, Andrews TJ, Whitney SM, Ludwig M, Yellowlees DC, Leggat W et al (1998) The diversity and coevolution of Rubisco, plastids, pyrenoids, and chloroplast-based CO_2-concentrating mechanisms in algae. Can J Bot 76:1052–1071. https://doi.org/10.1139/b98-074

Baghdasarian G, Muscatine L (2000) Preferential expulsion of dividing algal cells as a mechanism for regulating algal-cnidarian symbiosis. Biol Bull 199:278–286. https://doi.org/10.2307/1543184

Baird AH, Bhagooli R, Ralph PJ, Takahashi S (2009) Coral bleaching: the role of the host. Trends Ecol Evol 24:16–20. https://doi.org/10.1016/j.tree.2008.09.005

Baumgarten S, Simakov O, Esherick LY, Liew YJ, Lehnert EM, Mitchell CT et al (2015) The genome of *Aiptasia*, a sea anemone model for coral biology. Proc Natl Acad Sci U S A 112:11893–11898. https://doi.org/10.1073/pnas.1513318112

Bay LK, Cumbo VR, Abrego D, Kool JT, Ainsworth TD, Willis BL (2011) Infection dynamics vary between *Symbiodinium* types and cell surface treatments during establishment of endosymbiosis with coral larvae. Diversity 3:356–374. https://doi.org/10.3390/d3030356

Bellantuono AJ, Granados-Cifuentes C, Miller DJ, Hoegh-Guldberg O, Rodriguez-Lanetty M (2012) Coral thermal tolerance: tuning gene expression to resist thermal stress. PLoS One 7. https://doi.org/10.1371/journal.pone.0050685

Bhattacharya D, Agrawal S, Aranda M, Baumgarten S, Belcaid M, Drake JL et al (2016) Comparative genomics explains the evolutionary success of reef-forming corals. eLife 5. https://doi.org/10.7554/eLife.13288

Bieri T, Onishi M, Xiang T, Grossman AR, Pringle JR (2016) Relative contributions of various cellular mechanisms to loss of algae during cnidarian bleaching. PLoS One 11:e0152693. https://doi.org/10.1371/journal.pone.0152693

Brown BE, Le Tissier MDA, Bythell JC (1995) Mechanisms of bleaching deduced from histological studies of reef corals sampled during a natural bleaching event. Mar Biol 122:655–663. https://doi.org/10.1007/BF00350687

Brown BE, Ambarsari I, Warner ME, Fitt WK, Dunne RP, Gibb SW et al (1999) Diurnal changes in photochemical efficiency and xanthophyll concentrations in shallow water reef corals: evidence for photoinhibition and photoprotection. Coral Reefs 18:99–105. https://doi.org/10.1007/s003380050163

Burriesci MS, Raab TK, Pringle JR (2012) Evidence that glucose is the major transferred metabolite in dinoflagellate-cnidarian symbiosis. J Exp Biol 215:3467–3477. https://doi.org/10.1242/jeb.070946

Buxton L, Takahashi S, Hill R, Ralph PJ (2012) Variability in the primary site of photosynthetic damage in *Symbiodinium* sp. (Dinophyceae) exposed to thermal stress. J Phycol 48:117–126. https://doi.org/10.1111/j.1529-8817.2011.01099.x

Chaudhari N, Talwar P, Parimisetty A, d'Hellencourt C, Ravana P (2014) A molecular web: endoplasmic reticulum stress, inflammation, and oxidative stress. Front Cell Neurosci 8:213. https://doi.org/10.3389/fncel.2014.00213

Chen MC, Hong MC, Huang YS, Liu MC, Cheng YM, Fang LS (2005) ApRab11, a cnidarian homologue of the recycling regulatory protein Rab11, is involved in the establishment and maintenance of the *Aiptasia-Symbiodinium* endosymbiosis. Biochem Biophys Res Commun 338:1607–1616. https://doi.org/10.1016/j.bbrc.2005.10.133

Crawley A, Kline DI, Dunn S, Anthony K, Dove S (2010) The effect of ocean acidification on symbiont photorespiration and productivity in *Acropora formosa*. Glob Chang Biol 16:851–863. https://doi.org/10.1111/j.1365-2486.2009.01943.x

Davy SK, Allemand D, Weis VM (2012) Cell biology of cnidarian-dinoflagellate symbiosis. Microbiol Mol Biol Rev 76:229–261. https://doi.org/10.1128/MMBR.05014-11
Desalvo MK, Voolstra CR, Sunagawa S, Schwarz JA, Stillman JH, Coffroth MA et al (2008) Differential gene expression during thermal stress and bleaching in the Caribbean coral *Montastraea faveolata*. Mol Ecol 17:3952–3971. https://doi.org/10.1111/j.1365-294X.2008.03879.x
Downs CA, Kramarsky-Winter E, Martinez J, Kushmaro A, Woodley CM, Loya Y et al (2009) Symbiophagy as a cellular mechanism for coral bleaching. Autophagy 5:211–216. https://doi.org/10.1016/j.ecolind.2015.07.022
Downs CA, McDougall KE, Woodley CM, Fauth JE, Richmond RH, Kushmaro A et al (2013) Heat-stress and light-stress induce different cellular pathologies in the symbiotic dinoflagellate during coral bleaching. PLoS One 8. https://doi.org/10.1371/journal.pone.0077173
Dunn SR, Thomason JC, Le Tissier MDA, Bythell JC (2004) Heat stress induces different forms of cell death in sea anemones and their endosymbiotic algae depending on temperature and duration. Cell Death Differ 11:1213–1222. https://doi.org/10.1038/sj.cdd.4401484
Dunn SR, Schnitzler CE, Weis VM (2007) Apoptosis and autophagy as mechanisms of dinoflagellate symbiont release during cnidarian bleaching: every which way you lose. Proc R Soc B 274:3079–3085. https://doi.org/10.1098/rspb.2007.0711
Dunn SR, Pernice M, Green K, Hoegh-Guldberg O, Dove SG (2012) Thermal stress promotes host mitochondrial degradation in symbiotic cnidarians: are the batteries of the reef going to run out? PLoS One 7:e39024. https://doi.org/10.1371/journal.pone.0039024
Ezzat L, Maguer J-F, Grover R, Ferrier-Pagès C (2016) Limited phosphorus availability is the Achilles heel of tropical reef corals in a warming ocean. Sci Rep 6:31768. https://doi.org/10.1038/srep31768
Fang LS, Huang SP, Lin KL (1997) High temperature induces the synthesis of heat-shock proteins and the elevation of intracellular calcium in the coral *Acropora grandis*. Coral Reefs 16:127–131. https://doi.org/10.1007/s003380050066
Fujise L, Yamashita H, Suzuki G, Koike K (2013) Expulsion of zooxanthellae (*Symbiodinium*) from several species of scleractinian corals: comparison under non-stress conditions and thermal stress conditions. Galaxea J Coral Reef Stud 15:29–36. https://doi.org/10.3755/galaxea.15.29
Ganot P, Moya A, Magnone V, Allemand D, Furla P, Sabourault C (2011) Adaptations to endosymbiosis in a cnidarian-dinoflagellate association: differential gene expression and specific gene duplications. PLoS Genet 7. https://doi.org/10.1371/journal.pgen.1002187
Gardner SG, Nielsen DA, Laczka O, Shimmon R, Beltran V, Ralph P et al (2016) Dimethylsulfoniopropionate, superoxide dismutase and glutathione as stress response indicators in three corals under short-term hyposalinity stress. Proc R Soc B 283:1–9. https://doi.org/10.1098/rspb.2015.2418
Gates RD, Baghdasarian G, Muscatine L (1992) Temperature stress causes host cell detachment in symbiotic cnidarians: implications for coral bleaching. Biol Bull 182:324–332. https://doi.org/10.2307/1542252
Gierz SL, Forêt S, Leggat W (2017) Transcriptomic analysis of thermally stressed *Symbiodinium* reveals differential expression of stress and metabolism genes. Front Plant Sci 8:1–20. https://doi.org/10.3389/fpls.2017.00271
Grajales A, Rodríguez E (2014) Morphological revision of the genus *Aiptasia* and the family Aiptasiidae (Cnidaria, Actiniaria, Etridioidea). Zootaxa 3826:55–100. https://doi.org/10.11646/zootaxa.3826.1.2
Gustafsson MSM, Baird ME, Ralph PJ (2014) Modeling photoinhibition-driven bleaching in Scleractinian coral as a function of light, temperature, and heterotrophy. Limnol Oceanogr 59:603–622. https://doi.org/10.4319/lo.2014.59.2.0603
Hanes SD, Kempf SC (2013) Host autophagic degradation and associated symbiont loss in response to heat stress in the symbiotic anemone, *Aiptasia pallida*. Invertebr Biol 132:95–107. https://doi.org/10.1111/ivb.12018

Hawkins TD, Davy SK (2012) Nitric oxide production and tolerance differ among *Symbiodinium* types exposed to heat stress. Plant Cell Physiol 53:1889–1898. https://doi.org/10.1093/pcp/pcs127

Hawkins TD, Davy SK (2013) Nitric oxide and coral bleaching: is peroxynitrite generation required for symbiosis collapse? J Exp Biol 216:3185–3185. https://doi.org/10.1242/jeb.087510

Hawkins TD, Bradley BJ, Davy SK (2013) Nitric oxide mediates coral bleaching through an apoptotic-like cell death pathway: evidence from a model sea anemone-dinoflagellate symbiosis. FASEB J 27:4790–4798. https://doi.org/10.1096/fj.13-235051

Hawkins TD, Krueger T, Becker S, Fisher PL, Davy SK (2014) Differential nitric oxide synthesis and host apoptotic events correlate with bleaching susceptibility in reef corals. Coral Reefs 33:141–153. https://doi.org/10.1007/s00338-013-1103-4

Hawkins TD, Hagemeyer JCG, Warner ME (2016) Temperature moderates the infectiousness of two conspecific *Symbiodinium* strains isolated from the same host population. Environ Microbiol 18:5204–5217. https://doi.org/10.1111/1462-2920.13535

Hennige SJ, Suggett DJ, Warner ME, McDougall KE, Smith DJ (2008) Photobiology of *Symbiodinium* revisited: bio-physical and bio-optical signatures. Coral Reefs 28:179–195. https://doi.org/10.1007/s00338-008-0444-x

Hill R, Larkum AWD, Frankart C, Kühl M, Ralph PJ (2004) Loss of functional Photosystem II reaction centres in zooxanthellae of corals exposed to bleaching conditions: Using fluorescence rise kinetics. Photosynth Res 82:59–72. https://doi.org/10.1023/B:PRES.0000040444.41179.09

Hill R, Brown CM, DeZeeuw K, Campbell DA, Ralph PJ (2011) Increased rate of D1 repair in coral symbionts during bleaching is insufficient to counter accelerated photo-inactivation. Limnol Oceanogr 56:139–146. https://doi.org/10.4319/lo.2011.56.1.0139

Hillyer KE, Dias DA, Lutz A, Wilkinson SP, Roessner U, Davy SK (2017a) Metabolite profiling of symbiont and host during thermal stress and bleaching in the coral *Acropora aspera*. Coral Reefs 36:105–118. https://doi.org/10.1007/s00338-016-1508-y

Hillyer KE, Dias DA, Lutz A, Roessner U, Davy SK (2017b) Mapping carbon fate during bleaching in a model cnidarian symbiosis: the application of 13C metabolomics. New Phytol 214:1551–1562. https://doi.org/10.1111/nph.14515

Hoegh-Guldberg O, Mumby PJ, Hooten AJ, Steneck RS, Greenfiled P, Gomez E et al (2007) Coral reefs under rapid climate change and ocean acidification. Science 318:1737–1742. https://doi.org/10.1126/science.1152509

Hopkins FE, Bell TG, Yang M, Suggett DJ, Steinke M (2016) Air exposure of coral is a significant source of dimethylsulfide (DMS) to the atmosphere. Sci Rep 6:36031. https://doi.org/10.1038/srep36031

Hughes TP, Kerry JT, Álvarez-Noriega M, Álvarez-Romero JG, Anderson KD, Baird AH et al (2017) Global warming and recurrent mass bleaching of corals. Nature 543:373–377. https://doi.org/10.1038/nature21707

Imlay JA (2013) The molecular mechanisms and physiological consequences of oxidative stress: lessons from a model bacterium. Nat Rev Microbiol 11:443–454. https://doi.org/10.1038/nrmicro3032

Jin YK, Lundgren P, Lutz A, Raina J-B, Howells EJ, Paley AS (2016) Genetic markers for antioxidant capacity in a reef-building coral. Sci Adv 2:e1500842. https://doi.org/10.1126/sciadv.1500842

Jones RJ, Hoegh-Guldberg O, Larkum AWD, Schreiber U (1998) Temperature-induced bleaching of corals begins with impairment of the CO_2 fixation mechanism in zooxanthellae. Plant Cell Environ 21:1219–1230. https://doi.org/10.1046/j.1365-3040.1998.00345.x

Jones GB, Fischer E, Deschaseaux ESM, Harrison PL (2014) The effect of coral bleaching on the cellular concentration of dimethylsulphoniopropionate in reef corals. J Exp Mar Biol Ecol 460:19–31. https://doi.org/10.1016/j.jembe.2014.06.003

Kanazawa A, Blanchard GJ, Szabó M, Ralph PJ, Kramer DM (2014) The site of regulation of light capture in *Symbiodinium*: Does the peridinin–chlorophyll *a*–protein detach to regulate light capture? BBA-Bioenergetics 1837:1227–1234. https://doi.org/10.1016/j.bbabio.2014.03.019

Krueger T, Becker S, Pontasch S, Dove S, Hoegh-Guldberg O, Leggat W et al (2014) Antioxidant plasticity and thermal sensitivity in four types of *Symbiodinium* sp. J Phycol 50:1035–1047. https://doi.org/10.1111/jpy.12232

Krueger T, Hawkins TD, Becker S, Pontasch S, Dove S, Hoegh-Guldberg O et al (2015) Differential coral bleaching – contrasting the activity and response of enzymatic antioxidants in symbiotic partners under thermal stress. Comp Biochem Physiol A Mol Integr Physiol 190:15–25. https://doi.org/10.1016/j.cbpa.2015.08.012

Kuhl M, Cohen Y, Dalsgaard T, Jørgensen BB, Revsbech NP (1995) Microenvironment and photosynthesis of zooxanthellae in scleractinian corals studied pH and light with microsensors for O_2, pH and light. Mar Ecol Prog Ser 117:159–172. https://doi.org/10.3354/meps117159

Kvitt H, Rosenfeld H, Zandbank K, Tchernov D (2011) Regulation of apoptotic pathways by *Stylophora pistillata* (Anthozoa, Pocilloporidae) to survive thermal stress and bleaching. PLoS One 6. https://doi.org/10.1371/journal.pone.0028665

Kvitt H, Rosenfeld H, Tchernov D (2016) The regulation of thermal stress induced apoptosis in corals reveals high similarities in gene expression and function to higher animals. Sci Rep 6:30359. https://doi.org/10.1038/srep30359

Leggat W, Badger MR, Yellowlees D (1999) Evidence for an inorganic carbon concentrating mechanism in the symbiotic dinoflagellate *Symbiodinium* sp. Plant Physiol 121:1247–1255. https://doi.org/10.1104/pp.121.4.1247

Lehnert EM, Burriesci MS, Pringle JR (2012) Developing the anemone *Aiptasia* as a tractable model for cnidarian-dinoflagellate symbiosis: the transcriptome of aposymbiotic *A. pallida*. BMC Genomics 13:271. https://doi.org/10.1186/1471-2164-13-271

Lehnert EM, Mouchka ME, Burriesci MS, Gallo N, Schwarz J, Pringle JR (2014) Extensive differences in gene expression between symbiotic and aposymbiotic cnidarians. Genes Genomes Genet 4:277–295. https://doi.org/10.1534/g3.113.009084

Lesser MP (1996) Elevated temperatures and ultraviolet radiation cause oxidative stress and inhibit photosynthesis in symbiotic dinoflagellates. Limnol Oceanogr 41:271–283. https://doi.org/10.4319/lo.1996.41.2.0271

Levin RA, Beltran VH, Hill R, Kjelleberg S, McDougald D, Steinberg PD et al (2016) Sex, scavengers, and chaperones: transcriptome secrets of divergent *Symbiodinium* thermal tolerances. Mol Biol Evol 33:2201–2215. https://doi.org/10.1093/molbev/msw119

Lilley RM, Ralph PJ, Larkum AWD (2010) The determination of activity of the enzyme Rubisco in cell extracts of the dinoflagellate alga *Symbiodinium* sp. by manganese chemiluminescence and its response to short-term thermal stress of the alga. Plant Cell Environ 33:995–1004. https://doi.org/10.1111/j.1365-3040.2010.02121.x

Lin S, Cheng S, Song B, Zhong X, Lin X, Li W et al (2015) The *Symbiodinium kawagutii* genome illuminates dinoflagellate gene expression and coral symbiosis. Science 350:691–694. https://doi.org/10.1126/science.aad0408

Lutz A, Raina JB, Motti CA, Miller DJ, van Oppen MJH (2015) Host coenzyme Q redox state is an early biomarker of thermal stress in the coral *Acropora millepora*. PLoS One 10. https://doi.org/10.1371/journal.pone.0139290

Man SM, Kanneganti T-D (2016) Converging roles of caspases in inflammasome activation, cell death and innate immunity. Nat Rev Immunol 16:7–21. https://doi.org/10.1038/nri.2015.7

Matthews JL, Crowder CM, Oakley CA, Lutz A, Roessner U, Meyer E, Grossman AR, Weis VM, Davy SK (2017) Optimal nutrient exchange and immune responses operate in partner specificity in the cnidarian-dinoflagellate symbiosis. Proc Natl Acad Sci 114(50):13194–13199

McCabe Reynolds J, Bruns BU, Fitt WK, Schmidt GW (2008) Enhanced photoprotection pathways in symbiotic dinoflagellates of shallow-water corals and other cnidarians. Proc Natl Acad Sci U S A 105:13674–13678. https://doi.org/10.1073/pnas.0805187105

McGinley MP, Aschaffenburg MD, Pettay DT, Smith RT, LaJeunesse TC, Warner ME (2012) Transcriptional response of two core photosystem genes in *Symbiodinium* spp. exposed to thermal stress. PLoS One 7. https://doi.org/10.1371/journal.pone.0050439

Meyer E, Weis VM (2012) Study of cnidarian-algal symbiosis in the "omics" age. Biol Bull 223:44–65. https://doi.org/10.1086/BBLv223n1p44

Moné Y, Monnin D, Kremer N (2014) The oxidative environment: a mediator of interspecies communication that drives symbiosis evolution. Proc R Soc B 281:20133112. https://doi.org/10.1098/rspb.2013.3112

Moya A, Sakamaki K, Mason BM, Huisman L, Forêt S, Weiss Y et al (2016) Functional conservation of the apoptotic machinery from coral to man: the diverse and complex Bcl-2 and caspase repertoires of *Acropora millepora*. BMC Genomics 17:62. https://doi.org/10.1186/s12864-015-2355-x

Niedzwiedzki DM, Jiang J, Lo CS, Blankenship RE (2014) Spectroscopic properties of the chlorophyll *a* – chlorophyll c_2 – peridinin-protein-complex (acpPC) from the coral symbiotic dinoflagellate *Symbiodinium*. Photosynth Res 120:125–139. https://doi.org/10.1007/s11120-013-9794-5

Nii CM, Muscatine L (1997) Oxidative stress in the symbiotic sea anemone *Aiptasia pulchella* (Carlgren, 1943): contribution of the animal to superoxide ion production at elevated temperature. Biol Bull 192:444–456. https://doi.org/10.2307/1542753

Oakley CA, Schmidt GW, Hopkinson BM (2014a) Thermal responses of *Symbiodinium* photosynthetic carbon assimilation. Coral Reefs 33:501–512. https://doi.org/10.1007/s00338-014-1130-9

Oakley CA, Hopkinson BM, Schmidt GW (2014b) Mitochondrial terminal alternative oxidase and its enhancement by thermal stress in the coral symbiont *Symbiodinium*. Coral Reefs 33:543–552. https://doi.org/10.1007/s00338-014-1147-0

Oakley CA, Ameismeier MF, Peng L, Weis VM, Grossman AR, Davy SK (2016) Symbiosis induces widespread changes in the proteome of the model cnidarian *Aiptasia*. Cell Microbiol 18:1009–1023. https://doi.org/10.1111/cmi.12564

Oakley CA, Durand E, Wilkinson SP, Peng L, Weiss VM, Grossman AR et al (2017) Thermal shock induces host proteostasis disruption and endoplasmic reticulum stress in the model symbiotic cnidarian *Aiptasia*. J Proteome Res 16:2121–2134. https://doi.org/10.1021/acs.jproteome.6b00797

Orrenius S, Zhivotovsky B, Nicotera P (2003) Regulation of cell death: the calcium-apoptosis link. Nat Rev Mol Cell Biol 4:552–565. https://doi.org/10.1038/nrm1150

Paxton CW, Davy SK, Weis VM (2013) Stress and death of cnidarian host cells play a role in cnidarian bleaching. J Exp Biol 216:2813–2820. https://doi.org/10.1242/jeb.087858

Perez S, Weis V (2006) Nitric oxide and cnidarian bleaching: an eviction notice mediates breakdown of a symbiosis. J Exp Biol 209:2804–2810. https://doi.org/10.1242/jeb.02309

Pernice M, Dunn SR, Miard T, Dufour S, Dove S, Hoegh-Guldberg O (2011) Regulation of apoptotic mediators reveals dynamic responses to thermal stress in the reef building coral *Acropora millepora*. PLoS One 6. https://doi.org/10.1371/journal.pone.0016095

Pernice M, Meibom A, Van Den Heuvel A, Kopp C, Domart-Coulon I, Hoegh-Guldberg O et al (2012) A single-cell view of ammonium assimilation in coral–dinoflagellate symbiosis. ISME J 6:1314–1324. https://doi.org/10.1038/ismej.2011.196

Pinzón JH, Kamel B, Burge CA, Harvell CD, Medina M, Weil E et al (2015) Whole transcriptome analysis reveals changes in expression of immune-related genes during and after bleaching in a reef-building coral. Roy Soc Open Sci 2:140214. https://doi.org/10.1098/rsos.140214

Pogoreutz C, Rädecker N, Cárdenas A, Gärdes A, Voolstra CR, Wild C (2017) Sugar enrichment provides evidence for a role of nitrogen fixation in coral bleaching. Glob Chang Biol 23:3838–3848. https://doi.org/10.1111/gcb.13695

Rädecker N, Pogoreutz C, Voolstra CR, Weidenmann J, Wild C (2015) Nitrogen cycling in corals: the key to understanding holobiont functioning? Trends Microbiol 23:490–497. https://doi.org/10.1016/j.tim.2015.03.008

Rainbolt TK, Saunders JM, Wiseman RL (2014) Stress-responsive regulation of mitochondria through the ER unfolded protein response. Trends Endocrinol Metab 25:528–537. https://doi.org/10.1016/j.tem.2014.06.007

Ralph PJ, Gademann R, Larkum AWD (2001) Zooxanthellae expelled from bleached corals at 33C are photosynthetically competent. Mar Ecol Prog Ser 220:163–168. https://doi.org/10.3354/meps220163

Rehman AU, Szabó M, Deak Z, Sass L, Larkum A, Ralph P et al (2016) *Symbiodinium* sp. cells produce light-induced intra- and extracellular singlet oxygen, which mediates photodamage of the photosynthetic apparatus and has the potential to interact with the animal host in coral symbiosis. New Phytol 212:472–484. https://doi.org/10.1111/nph.14056

Richier S, Furla P, Plantivaux A, Merle PL, Allemand D (2005) Symbiosis-induced adaptation to oxidative stress. J Exp Mar Biol 208:277–285. https://doi.org/10.1242/jeb.01368

Richter K, Haslbeck M, Buchner J (2010) The heat shock response: life on the verge of death. Mol Cell 40:253–266. https://doi.org/10.1016/j.molcel.2010.10.006

Roberty S, Bailleul B, Berne N, Franck F, Cardol P (2014) PSI Mehler reaction is the main alternative photosynthetic electron pathway in *Symbiodinium* sp., symbiotic dinoflagellates of cnidarians. New Phytol 204:81–91. https://doi.org/10.1111/nph.12903

Roberty S, Fransolet D, Cardol P, Plumier J-C, Franck F (2015) Imbalance between oxygen photoreduction and antioxidant capacities in *Symbiodinium* cells exposed to combined heat and high light stress. Coral Reefs 34:1063–1073. https://doi.org/10.1007/s00338-015-1328-5

Ron D, Walter P (2007) Signal integration in the endoplasmic reticulum unfolded protein response. Nat Rev Mol Cell Biol 8:519–529. https://doi.org/10.1038/nrm2199

Rosset S, Wiedenmann J, Reed AJ, D'Angelo C (2017) Phosphate deficiency promotes coral bleaching and is reflected by the ultrastructure of symbiotic dinoflagellates. Mar Pollut Bull 118:180–187. https://doi.org/10.1016/j.marpolbul.2017.02.044

Ruiz-Jones LJ, Palumbi SR (2017) Tidal heat pulses on a reef trigger a fine-tuned transcriptional response in corals to maintain homeostasis. Sci Adv 3. https://doi.org/10.1126/sciadv.1601298

Sandeman IM (2006) Fragmentation of the gastrodermis and detachment of zooxanthellae in symbiotic cnidarians: a role for hydrogen peroxide and Ca^{2+} in coral bleaching and algal density control. Rev Biol Trop 54:79–96. https://doi.org/10.15517/rbt.v54i3.26899

Saragosti E, Tchernov D, Katsir A, Shaked Y (2010) Extracellular production and degradation of superoxide in the coral *Stylophora pistillata* and cultured *Symbiodinium*. PLoS One 5. https://doi.org/10.1371/journal.pone.0012508

Sawyer SJ, Muscatine L (2001) Cellular mechanisms underlying temperature-induced bleaching in the tropical sea anemone *Aiptasia pulchella*. J Exp Biol 204:3443–3456

Shinzato C, Shoguchi E, Kawashima T, Hamada M, Hisata K, Tanaka M et al (2011) Using the *Acropora digitifera* genome to understand coral responses to environmental change. Nature 476:320–323. https://doi.org/10.1038/nature10249

Shoguchi E, Shinzato C, Kawashima T, Gyoja F, Munqpakdee S, Takeuchi T et al (2013) Draft assembly of the *Symbiodinium minutum* nuclear genome reveals dinoflagellate gene structure. Curr Biol 23:1399–1408. https://doi.org/10.1016/j.cub.2013.05.062

Slavov C, Schrameyer V, Reus M, Ralph PJ, Hill R, Büchel C et al (2016) "Super-quenching" state protects *Symbiodinium* from thermal stress – implications for coral bleaching. Biochim Biophys Acta (BBA) Bioenerg 1857:840–847. https://doi.org/10.1016/j.bbabio.2016.02.002

Smith DJ, Suggett DJ, Baker NR (2005) Is photoinhibition of zooxanthellae photosynthesis the primary cause of thermal bleaching in corals? Glob Chang Biol 11:1–11. https://doi.org/10.1111/j.1365-2486.2004.00895.x

Snyder CM, Shroff EH, Liu J, Chandel NS (2009) Nitric oxide induces cell death by regulating anti-apoptotic BCL-2 family members. PLoS One 4. https://doi.org/10.1371/journal.pone.0007059

Starzak DE, Quinnell RG, Nitschke MR, Davy SK (2014) The influence of symbiont type on photosynthetic carbon flux in a model cnidarian – dinoflagellate symbiosis. Mar Biol 16:711–724. https://doi.org/10.1007/s00227-013-2372-8

Strychar KB, Coates M, Sammarco PW, Piva TJ (2004) Bleaching as a pathogenic response in scleractinian corals, evidenced by high concentrations of apoptotic and necrotic zooxanthellae. J Exp Mar Biol Ecol 304:99–121. https://doi.org/10.1016/j.jembe.2003.11.023

Susnow N, Zhang L, Margineantu D, Hockenbery DM (2009) Bcl-2 family proteins as regulators of oxidative stress. Semin Cancer Biol 19:42–49. https://doi.org/10.1016/j.semcancer.2008.12.002

Takahashi S, Nakamura T, Sakamizu M, van Woesik R, Yamasaki H (2004) Repair machinery of symbiotic photosynthesis as the primary target of heat stress for reef-building corals. Plant Cell Physiol 45:251–255. https://doi.org/10.1093/pcp/pch208

Tansik AL, Fitt WK, Hopkinson BM (2015) External carbonic anhydrase in three Caribbean corals: quantification of activity and role in CO_2 uptake. Coral Reefs 34:703–713. https://doi.org/10.1007/s00338-015-1289-8

Tansik AL, Fitt WK, Hopkinson BM (2017) Inorganic carbon is scarce for symbionts in scleractinian corals. Limnol Oceanogr 62:2045–2055. https://doi.org/10.1002/lno.10550

Taylor RC, Cullen SP, Martin SJ (2008) Apoptosis: controlled demolition at the cellular level. Nat Rev Mol Cell Biol 9:231–241. https://doi.org/10.1038/nrm2312

Tcherkez G, Farquhar G, Andrews T (2006) Despite slow catalysis and confused substrate specificity, all ribulose bisphosphate carboxylases may be nearly perfectly optimized. Proc Natl Acad Sci U S A 103:7246–7251. https://doi.org/10.1073/pnas.0600605103

Tchernov D, Gorbunov MY, de Vargas C, Yadav SN, Milligan AJ, Häggblom M et al (2004) Membrane lipids of symbiotic algae are diagnostic of sensitivity to thermal bleaching in corals. Proc Natl Acad Sci U S A 101:13531–13535. https://doi.org/10.1073/pnas.0402907101

Tchernov D, Kvitt H, Haramaty L, Bibby TS, Gorbunov MY, Rosenfeld H et al (2011) Apoptosis and the selective survival of host animals following thermal bleaching in zooxanthellate corals. Proc Natl Acad Sci U S A 108:9905–9909. https://doi.org/10.1073/pnas.1106924108

Tolleter D, Seneca O, DeNofrio JC, Krediet CJ, Palumbi SR, Pringle JR et al (2013) Coral bleaching independent of photosynthetic activity. Curr Biol 23:1782–1786. https://doi.org/10.1016/j.cub.2013.07.041

Trapido-Rosenthal H, Zielke S, Owen R, Buxton L, Boeing B, Bhagooli R et al (2005) Increased zooxanthellae nitric oxide synthase activity is associated with coral bleaching. Biol Bull 208:3–6. https://doi.org/10.2307/3593094

Tu BP, Weissman JS (2004) Oxidative protein folding in eukaryotes: mechanisms and consequences. J Cell Biol 164:341–346. https://doi.org/10.1083/jcb.200311055

Turrens JF (2003) Mitochondrial formation of reactive oxygen species. J Physiol 552:335–344. https://doi.org/10.1113/jphysiol.2003.049478

Wang X, Liew YJ, Li Y, Zoccola D, Tambutte S, Aranda M (2017) Draft genomes of the corallimorpharians *Amplexidiscus fenestrafer* and *Discosoma* sp. Mol Ecol Resour. https://doi.org/10.1111/1755-0998.12680

Warner ME, Suggett DJ (2016) The photobiology of *Symbiodinium* spp.: linking physiological diversity to the implications of stress. In: Goffredo S, Dubinsky Z (eds) The cnidaria, past, present and future: the world of Medusa and her sisters. Springer, pp 489–509

Warner ME, Fitt WK, Schmidt GW (1996) The effects of elevated temperature on the photosynthetic efficiency of zooxanthellae in hospite from four different species of reef coral: a novel approach. Plant Cell Environ 19:291–299. https://doi.org/10.1111/j.1365-3040.1996.tb00251.x

Warner M, Fitt W, Schmidt G (1999) Damage to photosystem II in symbiotic dinoflagellates: a determinant of coral bleaching. Proc Natl Acad Sci U S A 96:8007–8012. https://doi.org/10.1073/pnas.96.14.8007

Warner ME, Chilcoat GC, Mcfarland FK, Fitt WK (2002) Seasonal fluctuations in the photosynthetic capacity of photosystem II in symbiotic dinoflagellates in the Caribbean reef-building coral *Montastraea*. Mar Biol 141:31–38. https://doi.org/10.1007/s00227-002-0807-8

Weis VM (2008) Cellular mechanisms of cnidarian bleaching: stress causes the collapse of symbiosis. J Exp Biol 211:3059–3066. https://doi.org/10.1242/jeb.009597

Weis VM, Davy SK, Hoegh-Guldberg O, Rodriguez-Lanetty M, Pringle JR (2008) Cell biology in model systems as the key to understanding corals. Trends Ecol Evol 23:369–376. https://doi.org/10.1016/j.tree.2008.03.004

Weston AJ, Dunlap WC, Beltran VH, Starcevic A, Hranueli D, Ward M et al (2015) Proteomics links the redox state to calcium signaling during bleaching of the scleractinian coral *Acropora*

microphthalma on exposure to high solar irradiance and thermal stress. Mol Cell Proteomics 14:585–595. https://doi.org/10.1074/mcp.M114.043125
Wiedenmann J, D'Angelo C, Smith EG, Hunt AN, Legiret F-E, Postle AD et al (2013) Nutrient enrichment can increase the susceptibility of reef corals to bleaching. Nat Clim Chang 3:160–164. https://doi.org/10.1038/nclimate1661
Wolfowicz I, Baumgarten S, Voss PA, Hambleton EA, Voolstra CR, Hatta M et al (2016) *Aiptasia* sp. larvae as a model to reveal mechanisms of symbiont selection in cnidarians. Sci Rep 6:32366. https://doi.org/10.1038/srep32366
Wooldridge SA (2009) A new conceptual model for the warm-water breakdown of the coral–algae endosymbiosis. Mar Freshw Res 60:483–496. https://doi.org/10.1071/MF08251
Yellowlees D, Rees TAV, Leggat W (2008) Metabolic interactions between algal symbionts and invertebrate hosts. Plant Cell Environ 31:679–694. https://doi.org/10.1111/j.1365-3040.2008.01802.x
Ziegler M, Seneca FO, Yum LK, Palumbi SR, Voolstra CR (2017) Bacterial community dynamics are linked to patterns of coral heat tolerance. Nat Commun 8. https://doi.org/10.1038/ncomms14213

Chapter 9
Bleaching and Mortality Thresholds: How Much Is Too Much?

R. Berkelmans

9.1 Introduction

A considerable number of studies have investigated the link between coral bleaching events and environmental factors. Whilst localized bleaching events can have many causes (Dove and Hoegh-Guldberg 2006), widespread bleaching events have, almost without fail, been demonstrably linked to unusually warm temperatures often in conjunction with increased light. Prior to the 1990s, there was a paucity of representative instrumental temperature data to link with observations of mass coral bleaching (Chap. 3). However, temperatures beyond the envelope normally experienced by corals have always been implicated as the cause of mass bleaching. For example, as early as 1914 Alfred Mayer wrote:

> Thus on July 21–22, 1911, at Tortugas, Florida, after several hot, calm days, the shallow water over Bird Key Reef rose to 33 to 38 °C and Dr. L. R. Cary observed that large numbers of *Diadema*, *Octopus*, *Fissurella*, and other molluscs and small fishes were killed in considerable numbers over extensive areas, and the corals were injured even when not exposed to the air. (Mayer 1914)

Finding a suitable metric for temperature, light and other meteorological variables that adequately predicts bleaching and can be universally applied has proven much more challenging. In theory, a bleaching threshold for a particular coral species at a particular location is a function of absolute temperature, light and exposure time (Fitt et al. 2001). Ideally, it also incorporates possible additional stress factors such as salinity (Coles and Jokiel 1992) and water quality (Marshall and Schuttenberg 2006) or mitigating factors such as water motion (Nakamura and van Woesik 2001; van Woesik and Koksal 2006). However, constructing and applying such a multivariate model for predicting impending bleaching events is highly problematic and

R. Berkelmans (✉)
Australian Institute of Marine Science, Townsville MC, QLD, Australia
e-mail: r.berkelmans@aims.gov.au

M. J. H. van Oppen, J. M. Lough (eds.), *Coral Bleaching*, Ecological Studies 233,
https://doi.org/10.1007/978-3-319-75393-5_9

impractical. The problems include the difficulty in establishing the model parameters for each species, the applicability of laboratory-derived values to real-world field conditions and the availability of representative real-time data for variables such as underwater light, water motion and water quality parameters. As a result, a more parsimonious approach needs to be considered. Published papers on bleaching thresholds and predictors are invariably empirically derived using correlations between bleaching events and environmental factors. The developed relations often involve multispecies coral assemblages, use a single parameter and simple metrics of readily available data and are either location-specific or "best-fit" approximations applied over large spatial scales (from 10s to 1000s of kilometres). Consequently, "bleaching thresholds" have taken many different forms and are not easily applied to other locations, nor are they comparable between locations. Examples of the kinds of bleaching indices include satellite-derived sea surface temperature (SST) metrics such as monthly means (Goreau et al. 1993; Brown et al. 1996), anomalies above monthly means (e.g. "HotSpots" and "ReefTemp"; Goreau and Hayes 1994; Strong et al. 1997; Maynard et al. 2008; Chap. 4), "degree heating weeks" or "degree heating days" (Gleeson and Strong 1995; Maynard et al. 2008; Chap. 4) and maximum 3-day temperature (Berkelmans et al. 2004). For indices derived from in situ data, examples include maximum daily SST (Jones et al. 1997; Winter et al. 1998), monthly means (Podestá and Glynn 2001), weekly means and anomalies (Vargas-Ángel et al. 2001), degree-days (Podestá and Glynn 2001), days above certain temperatures (Winter et al. 1998), coefficient of variation of SST (Sammarco et al. 2006) and time–temperature curves (Berkelmans 2002a; Manzello et al. 2007). Thus, although highly diverse, each of these indices has merit for its own application driven by specific research questions, local conditions and responses, the spatial and temporal scale of interest and using the best locally available data.

The spatial and temporal scale of interest is an important factor in determining which metric to use as a bleaching threshold. "HotSpots" and "degree heating weeks" are operational web-based products produced by NOAA at a global scale (http://www.osdpd.noaa.gov/PSB/EPS/SST/climohot.html; Chap. 4). One of the best features of these products is the "global view" of bleaching risk. The downside is that globally applied algorithms inherently suffer from error (i.e. false positive or negative bleaching predictions) at local/regional scales (e.g. McClanahan et al. 2007). Regionally applied algorithms for satellite-derived SSTs such as "ReefTemp" Maynard et al. 2008, which is specific for the Great Barrier Reef (GBR), should reduce error rates, but this is as yet untested. On the other end of the spectrum, highly localized (reef-specific) bleaching indices such as time–temperature curves (Berkelmans 2002a) offer little spatial overview but potentially highly accurate bleaching predictions. Because these are "calibrated" to local conditions and responses, they can potentially also be extended to defining mortality thresholds for selected coral taxa. Used together, the satellite- and in situ-derived products can offer both locally accurate bleaching and mortality predictions and wider spatial interpolation of likely thermal stress.

Time–temperature curves were first developed for 13 reefs on the GBR following the 1998 bleaching event (Berkelmans and Oliver 1999; Berkelmans 2002a). Since then there has been another GBR-wide bleaching event in early 2002 and a localized but intense bleaching event in early 2006, which affected the southern GBR (GBRMPA 2006). The purpose of this chapter is to:

- Evaluate the appropriateness of in situ temperature in modelling bleaching thresholds and determine what improvement in predictive capacity could be gained by incorporating solar radiation and other environmental data into a model, using Magnetic Island in the central GBR as a case study.
- Review and evaluate the accuracy of the time-integrated bleaching thresholds for the GBR which were constructed after the 1998 bleaching event, with the benefit of an additional 7 years of SST data and in light of the 2002 and 2006 bleaching events.
- Construct mortality thresholds for coral taxa at specific sites which suffered high mortality (>50%) during the 1998, 2002 and 2006 bleaching events on the GBR.

9.2 Methods

9.2.1 Statistical Modelling

The physiological response of corals to bleaching is, in large part, a function of the amount of light (photosynthetically active radiation, PAR; ultra violet, UV) received after high temperatures have damaged the carbon-fixing processes of the zooxanthellae (Jones et al. 1998). To determine the relative influence of solar and UV radiation, SST and a number of other potential weather variables in explaining past coral bleaching events at Magnetic Island, a statistical exploratory model was constructed using classification trees. Classification trees examine the effects of predictor variables one at a time on a categorical response variable (e.g. bleaching) using a hierarchical system of splits, each one resulting in more homogeneous groups (De'ath and Fabricius 2000). Trees are an alternative to traditional statistical methods used with categorical response variables such as logistic regression and discriminant analysis and are often preferred because they can be used with a variety of data types and they can handle missing data and have the ability to uncover patterns and associations missed by traditional linear models (De'ath and Fabricius 2000). The categorical regression trees (CRT) exhaustive search algorithm of computing univariate splits was used since it deals with missing data better than other algorithms by using surrogates (Breiman et al. 1984). A tenfold cross-validation was performed on the final tree model with a learning data set to evaluate the robustness of the model with larger data sets. Statistical analyses were performed using SPSS software ver. 15.0.

The onset and intensification (if these occurred in separate months) of bleaching (dependent variable) were tested against 17 potential explanatory variables. These included monthly mean, minimum and maximum SST (calculated from daily averages), total monthly global radiation (measured horizontal and 19° off horizontal—the angle of maximum quantum exposure for Magnetic Island), total monthly UV-A radiation, maximum and minimum monthly air temperature, mean monthly relative humidity and dew point (9 a.m. and 9 p.m.), monthly total rainfall and evaporation and monthly total wet and sunshine hours. Data covered the summer months from December to March between 1991 and 2004 and included four bleaching events: 1992, 1994, 1998 and 2002. SST data were obtained from SeaTemps, a long-term sea temperature monitoring program on the GBR (www. aims.gov.au/pages/facili ties/adc/seatemps.html). Weather data were obtained from Allunga Exposure Laboratories (global and UV radiation, rain, wet hours), a materials testing facility ~20 km south of Magnetic Island and the Australian Bureau of Meteorology (air temperature, dew point, humidity, evaporation, sun hours) at the Townsville airport, ~8 km south of Magnetic Island. The efficacy of using weather data 20 km away from the study site was tested by comparing two related variables: monthly global radiation (measured at Allunga Exposure Laboratories) and PAR (measured at the study site) for the period December 1999 to December 2005 when overlapping data existed. The two variables were significantly correlated ($r^2 = 0.62$, $n = 73$), indicating that the Allunga radiation data are likely to be reasonably representative of Magnetic Island.

9.2.2 Bleaching Thresholds

Bleaching thresholds were previously constructed for 13 locations on the GBR using in situ water temperature records and concurrent observations of coral bleaching during the anomalously warm summer of 1998 (Berkelmans and Oliver 1999; Berkelmans 2002a). Bleaching thresholds were reconstructed using the methodology outlined in Berkelmans (2002a, b) with a temperature record extended by 7 years and observations of coral bleaching from two more bleaching events, one that was GBR-wide in 2002 (Berkelmans et al. 2004) and another in 2006 that predominantly affected reefs in the southern GBR (GBRMPA 2006). For 2 of the original 13 locations (Wallace Islet and Norman Reef), there was a gap in the temperature record during the 2002 bleaching event, and these sites were omitted from the analysis. Agincourt 3 Reef (~51 km north of Norman Reef) was substituted in place of Norman Reef since its temperature record was complete and, although it suffered no bleaching in 1998, it bleached extensively in 2002. The bleaching period for which time–temperature curves were calculated covered the warmest period during and on the shoulder of each austral summer from 1 November to 30 April.

The accuracy of the bleaching curves was assessed in two ways. First, the performance of the Magnetic Island curve was evaluated in predicting the 2002 bleaching event. Seven visits to Magnetic Island were made during the 2001/2002 summer at roughly fortnightly intervals, and for each visit a cumulative exposure curve was calculated allowing the onset and development of bleaching and the associated thermal conditions to be assessed in a stepwise fashion. During the field visits, bleaching was assessed by rapid visual surveys with the severity of bleaching estimated as a percentage of coral cover white on upper surfaces in the following bins: 0–1% (no bleaching), 1–10% (mild bleaching), 10–30% (heavy bleaching), 30–60% (very heavy bleaching) and >60% (extreme bleaching). Second, the performance of all original bleaching curves was evaluated in terms of their ability to accurately separate bleaching from non-bleaching years since 1998. Bleaching was assessed either by field or aerial surveys as detailed above and in Berkelmans et al. (2004).

9.2.3 Mortality Thresholds

Mortality thresholds were constructed in a similar manner to the bleaching threshold curves. During the 1998, 2002 and 2006 bleaching events, a number of locations suffered high mortality (>50%) of sensitive (and locally dominant) coral taxa. These were mostly members of the pocilloporid and acroporid families. Table 9.1 shows the locations and how much mortality was experienced in each species group. Mortality curves, based on 50% mortality (T_{L50}), were estimated by linear interpolation between the bleaching threshold curve and the time–temperature curve for the summer in which significant mortality took place. For cases with 100% mortality, the T_{L50} curve (for 50% mortality) was assumed to lie halfway between the bleaching threshold (which in theory should result in little or no mortality) and the curve which resulted in 100% mortality. This may overestimate the T_{L50} curve, particularly if the conditions causing 100% mortality were much warmer and sustained much longer than the bleaching threshold. In other words, 100% mortality occurred sooner than the warmest conditions experienced, but no direct observations were taken at the time. For cases where 50% mortality resulted, no interpolation was required: the T_{L50} curve was the final time–temperature curve for the summer. For cases with mortality levels between 50% and 100%, the position of the T_{L50} curve was scaled accordingly. As with the bleaching curves, the number of days' exposure does not necessarily represent consecutive days. For the sake of simplicity, it was assumed that damage to coral tissue continues from where it left off between one hot period and the next.

Table 9.1 Locations on the Great Barrier Reef which suffered high (≥50%) mortality as a result of bleaching

Location	Latitude (°S)	Longitude (°E)	Year	Depth range (m)	% mortality	Species group	References
Myrmidon Rf	18.3	147.4	2002	0–2	50	Arborescent *Acropora* spp.	GBRMPA (2002)
Orpheus Is	18.6	146.5	1998	0–5	70	Staghorn, plate and arborescent *Acropora* spp.	Berkelmans (2001)
Orpheus Is	18.6	146.5	1998	0–5	>99	*Millepora tenella* (fire coral) and *A. pulchra*	Shackeroff (1999)
Davies Rf	18.8	147.6	2002	0–2	50	Arborescent and plate *Acropora* spp.	Berkelmans, unpublished survey data
Stone Is, Bowen			2002	3–6	80	Staghorn and arborescent *Acropora* spp.	GBRMPA (2002)
Stone Is, Bowen			2002	0–4	50	*Millepora tenella* (fire coral)	GBRMPA (2002)
Daydream Is	20.3	148.8	2002	8–12	100	*Seriatopora hystrix*	GBRMPA (2002)
Halfway Is, Keppels	23.2	151.0	2006	0–2	57	*A. millepora* and staghorn *Acropora* spp.	Berkelmans and Jones, unpublished survey data

9.3 Results

9.3.1 *Is Temperature Appropriate for Modelling Bleaching Thresholds?*

The results of the classification tree show that sea temperature is clearly the most important variable explaining bleaching events. Of the 17 potential variables investigated, only 6 contributed significantly to the model, and, of these, the top 3 explanatory variables were maximum, average and minimum SST (Fig. 9.1a). Total monthly UV radiation rated less than 18% of the importance of maximum monthly SST to the model and total monthly sun hours less than 13%. Total global radiation did not contribute significantly to the model. Only one split in the decision tree was required by the final model to accurately predict the seven cases for the onset or continued development of bleaching, being January and February of 1992, January of 1994 and January and February of both 1998 and 2002 (Fig. 9.1b). In all cases this split was on the basis of a maximum monthly temperature >31.46 °C. The risk of misclassification, or the probability that these data could be classified by chance alone, was low at 1.3%, as indicated by the cross-validation error (Table 9.2). Thus, temperature is by far the most important parameter in predicting bleaching events, and hence temperature-based models remain an appropriate foundation for a bleaching alert system.

9.3.2 *Bleaching Thresholds*

A time-series evaluation of the bleaching threshold at Magnetic Island first published after the 1998 bleaching event on the GBR (Berkelmans 2002a) clearly shows that it accurately predicted the onset of the 2002 bleaching event (Fig. 9.2). On 21 December 2001, no bleaching was evident, but 2 weeks later on 7 January, the first signs of bleaching were evident with upper surfaces appearing pale to white on staghorn and plating *Acropora* spp. and margins of encrusting *Montipora* colonies. At that time, the time–temperature curve had just exceeded the predicted bleaching curve at temperatures >30.8 °C (Fig. 9.2). Over the ensuing weeks, the bleaching spread and intensified; and by 8 February ~30% of the hard corals on the reef crest were bleached white with another 50% pale (Berkelmans et al. 2004). By this time the bleaching threshold had been exceeded by a considerable margin. Maximum temperatures and extent and intensity of bleaching were experienced at the time of the visit on 1 March 2002. After this, temperatures declined, and early signs of zooxanthella population recovery were evident in individual colonies 3 weeks later. The Magnetic Island bleaching threshold, therefore, worked well in predicting bleaching at the same site 4 years after the 1998 bleaching event.

A review of the time-integrated bleaching thresholds for 12 of the 13 reefs in Berkelmans (2002a) also shows that, for the majority of these locations, the original

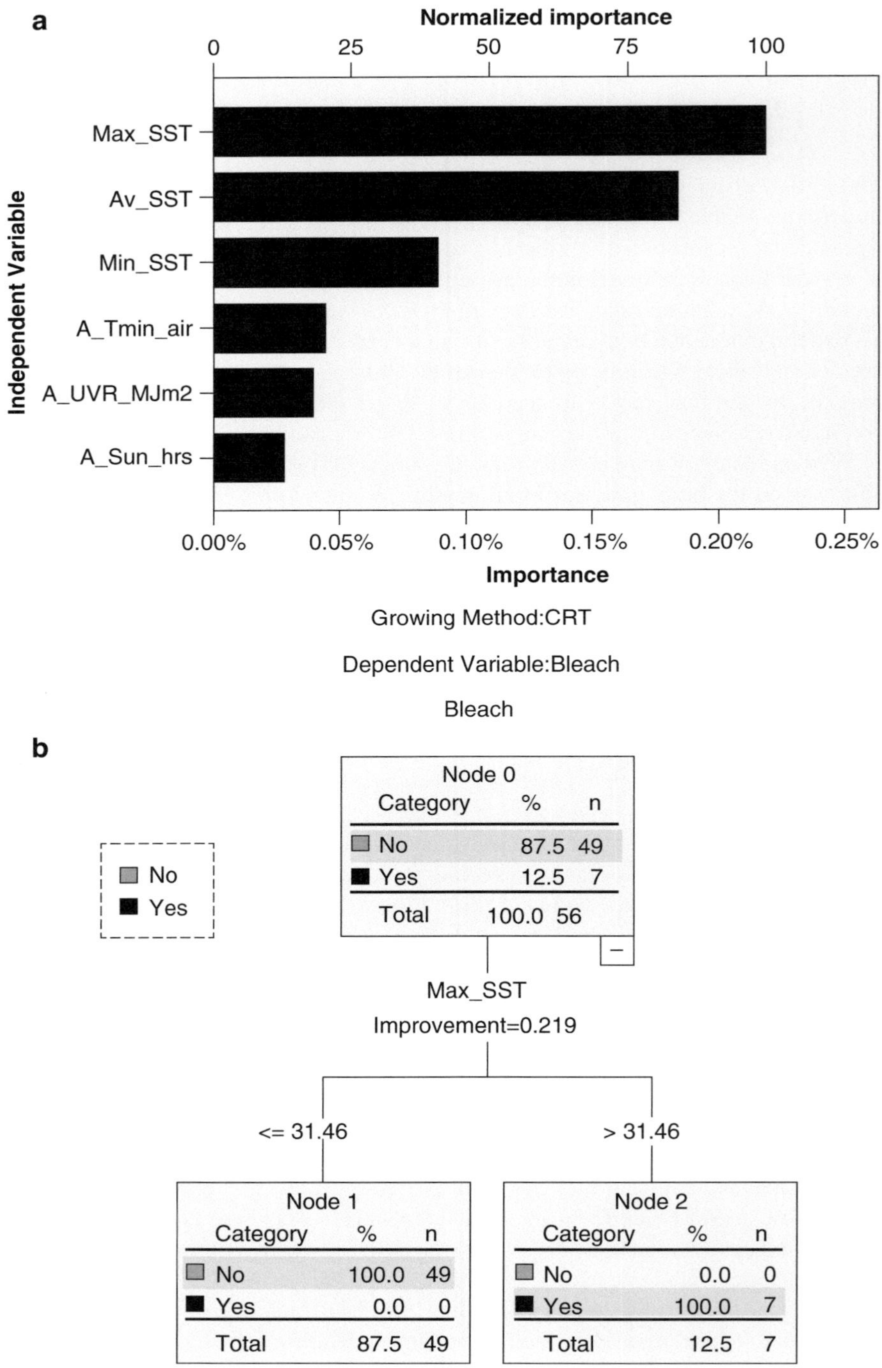

Fig. 9.1 Variables and their relative significance to a classification tree model in (**a**) predicting bleaching at Magnetic Island between 1990 and 2004 and (**b**) the results of the final model classification

Table 9.2 Specification criteria and calculated results for a classification trees model which evaluates the contribution of various environmental variables in predicting bleaching at Magnetic Island in 1992, 1994, 1998 and 2002

Model specifications	
Growing method	CRT
Dependent variable	Bleach (yes/no)
Independent variables	BoM_Sun_hrs, Av_SST, Min_SST, Max_SST, Tmax_air, Tmin_air, Dew_9am, Dew_9pm, RHumidity_9am, RHumidity_9pm, Rain, Evaporation, Wet_hrs, A_Sun_hrs, GlobalRad_hor, GlobRad_19deg, UVR
Model results	
Independent variables included in model	Max_SST, Av_SST, Min_SST, UVR, Tmin_air, A_Sun_hrs
Classification error	0.000
Estimated risk (cross-validation error)	0.013
Standard error of risk	0.025

The summary of the model results shows that only 6 of the original 17 variables contributed significantly to the model

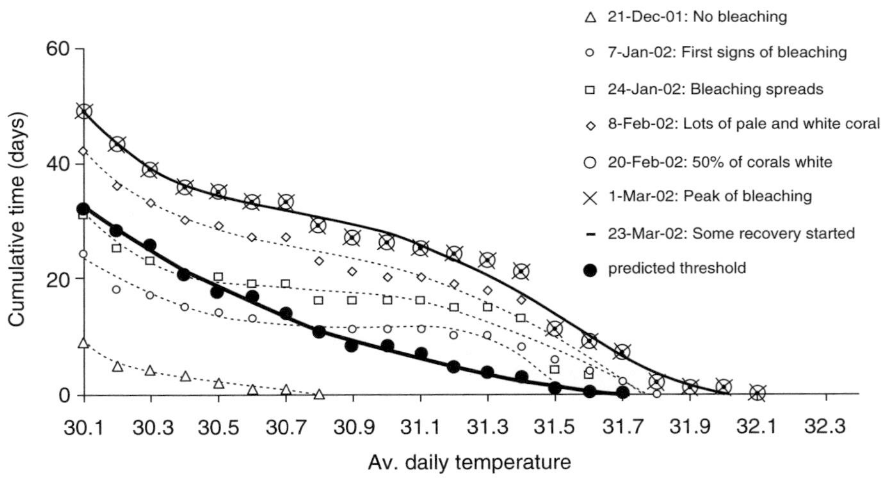

Fig. 9.2 Time-series assessment of the state of bleaching at Magnetic Island during the 2002 bleaching summer with a stepwise view of the time–temperature curve during each visit

bleaching threshold curves still accurately separated the bleaching years from the non-bleaching years (Fig. 9.3). For all but three locations (Daintree coast, Orpheus Island, Magnetic Island), the 2002 curve exceeded the curve of the 1998 bleaching year. Of the three locations that were cooler in 2002 than in 1998, both the Daintree coast and Orpheus Island did not show widespread bleaching (Berkelmans et al. 2004; T. Ayling, personal communication). The 2002 curve for Orpheus Island was

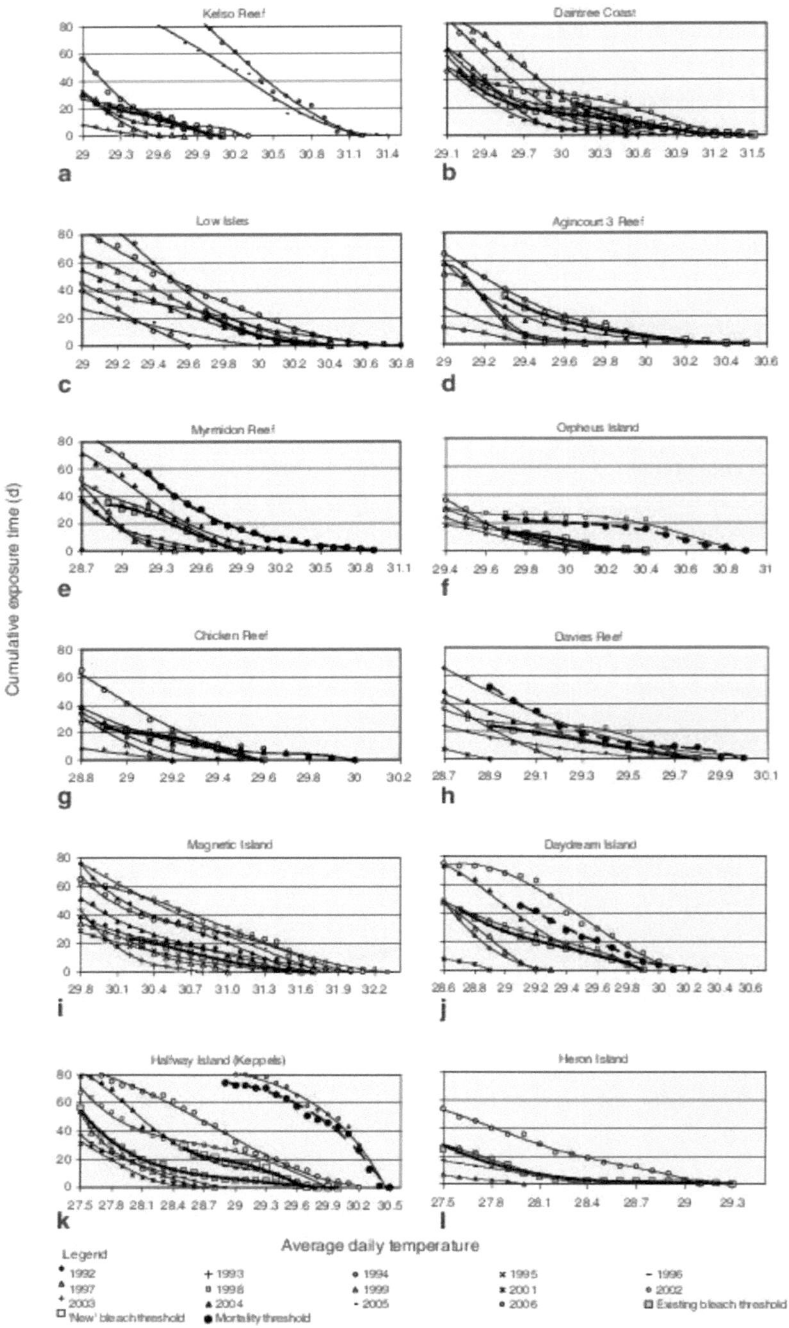

Fig. 9.3 Time–temperature curves for the bleaching summers and three warmest non-bleaching summers for 12 locations on the Great Barrier Reef (**a–l**), including the estimated position of bleaching curves and, where available, mortality curves

below the bleaching threshold as was the Daintree coast at temperatures above 30 °C (Fig. 9.3b, f). The 2002 curve at Magnetic Island was below the 1998 curve, and the bleaching intensity was also lower (Berkelmans et al. 2004).

The 2006 summer was generally warm for many reefs in the northern and central GBR, but widespread bleaching only occurred in the southern GBR, with inshore reefs most affected, especially in the Keppel Island group (Keppels). Approximately 87% of reef flat and 98% of reef slope coral bleached in this area with 61% and 78% of these habitats (respectively) bleached white (Berkelmans and Jones in prep.). The 2006 curve for the Keppels clearly shows the intensity of this event, both in terms of the record temperatures experienced and the duration of hot conditions (Fig. 9.3k).

A few locations showed anomalies with time–temperature curves exceeding the bleaching thresholds, but no bleaching reported. For example, conditions at Kelso Reef in 2005 and 2006 greatly exceeded both the predicted bleaching threshold for this location and the 2002 curve (Fig. 9.3a), but no reports of bleaching were received. Access to this location was difficult since a daily tourist operation to this reef stopped. A visit to Kelso Reef in July 2006 showed that >80% of the coral on the reef flat and upper reef slope were dead, but since this reef also had an outbreak of crown-of-thorns starfish over several years, the cause of the mortality may only be in part bleaching-related. Three more sites had their bleaching thresholds adjusted slightly in light of warmer non-bleaching years occurring above the predicted threshold but still below the coolest bleaching year. These include the Daintree coast, which had its bleaching threshold curve truncated at lower temperatures and slightly raised, as well as Orpheus Island and Halfway Island. At Halfway Island, in particular, the estimated position of the 1998 curve was a long way above the coolest non-bleaching year at that time. The non-bleaching summer of 2004 has since enabled the bleaching threshold to be better estimated. These are considered minor refinements of the estimated position of the bleaching threshold curve in light of more and better data.

More perplexing, however, are four locations where the curve for the 2004 summer exceeded the curves for the known bleaching year of 1998 and yet did not bleach. These locations included Myrmidon Reef, Chicken Reef, Magnetic Island and Daydream Island (Fig. 9.3e, h–j). Possible reasons for this are examined in Sect. 9.4. On balance, the original bleaching curves performed well during the 2002 bleaching event, but subsequent repeated warm summers without bleaching may require a reinterpretation of bleaching thresholds at some sites.

9.3.3 Mortality Thresholds

High mortality (>50%) among sensitive hard coral species at six locations in 1998, 2002 and 2006 gives rise to the possibility of estimating a T_{L50} time–temperature curve, effectively a species-specific mortality curve. These locations include Orpheus Island, which suffered high mortality in 1998; Myrmidon Reef, Davies Reef and Stone Island, which suffered high mortality in 2002; and the Keppels, which suffered high mortality in 2006 (Table 9.1). The position of these curves in relation to the warmest year and the bleaching threshold is shown in Fig. 9.3. Replotting these curves in relation to their respective bleaching thresholds revealed some interesting patterns. First, when the T_{L50} curves were recalculated as a time offset from the bleaching threshold (i.e. days above bleaching threshold), there was no consistent relationship evident (Fig. 9.4a). However, when they were recalculated as a temperature offset from the bleaching threshold (i.e. °C above bleaching threshold), there was a suggestion of a linear relationship (Fig. 9.4b). With the exception of the Keppels, the T_{L50} curves for the taxa in question were all less than 1 °C above the bleaching threshold, many only <0.5 °C above the bleaching threshold. The plating *Acropora* species on the reef flat in the Davies Reef lagoon and the pocilloporid, *Seriatopora hystrix*, at Daydream Island in particular had T_{L50} curves just 0.2–0.4 °C above their respective bleaching thresholds. In contrast, the corymbose and staghorn *Acropora* species in the Keppels were considerably more resistant to dying after bleaching. Nevertheless, their T_{L50} curve was still <2 °C higher than the bleaching threshold (Fig. 9.4b). The mortality curves for these locations and species groups were, therefore, offset from the bleaching threshold curve on the temperature axis but were still remarkably close to the bleaching thresholds.

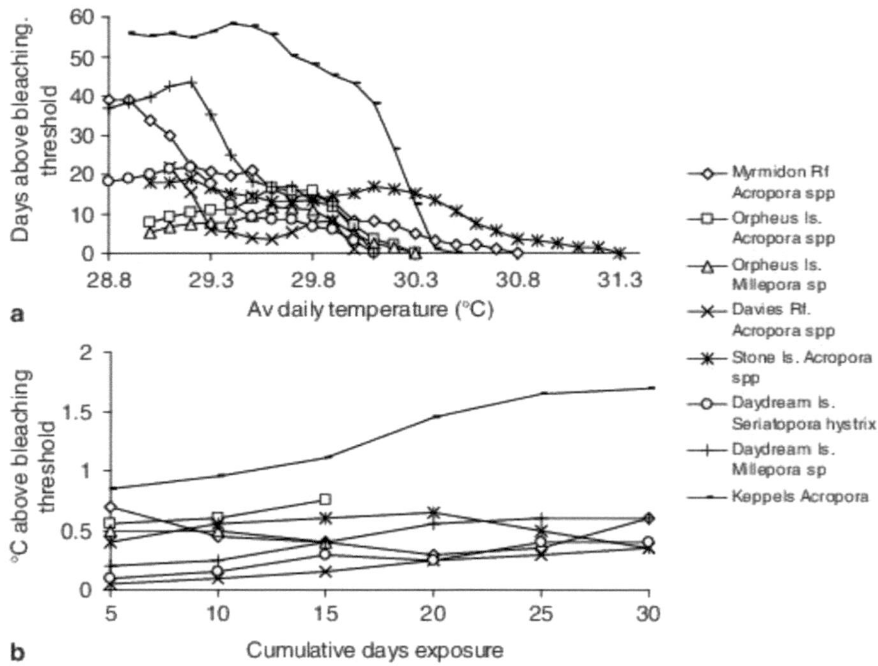

Fig. 9.4 Relationship of mortality curves to (**a**) "days above bleaching threshold" and (**b**) "temperature above bleaching threshold"

9.4 Discussion

Coral reef managers have come to rely on spatially extensive bleaching alert systems such as "HotSpots" (Goreau and Hayes 1994; Gleeson and Strong 1995; Chap. 4) and locally specific systems such as the Coral Reef Early Warning System (Hendee et al. 2001; Berkelmans et al. 2002) and time-integrated bleaching thresholds (Berkelmans 2002a) for valuable information on the build-up of stressful conditions to hard corals, the dominant builders of reefs. Reef managers value such warning systems because they allow them to be the source of timely and credible information about bleaching risk for decision-makers, stakeholders and the media (Marshall and Schuttenberg 2006). It also allows for early management responses to be put in place, including the instigation of formal monitoring programs to assess the extent and severity of bleaching and, where appropriate, take local action to ameliorate the risk of further damage to reefs from such activities as dredging, coastal development and point-source pollution. The time-integrated bleaching thresholds have been in use since 2000 as one of the monitoring tools for the Great Barrier Reef Marine Park Authority. The results of this review show they performed well in the time leading up to and including the 2002 bleaching even but also that there were some inconsistencies in recent years at some sites. The summer of 2004 in particular exceeded the bleaching threshold as well as the 1998 curve at Daydream Island, Myrmidon Reef and Chicken Reef in the central GBR, but, apart from a few pale colonies, no widespread bleaching occurred. Similarly, the 2005 summer at Magnetic Island exceeded the bleaching threshold and the conditions during the 1992 and 1994 bleaching events and was almost equivalent to the 2002 bleaching event, but, apart from pale tops on a few *Porites* bommies, no bleaching was evident. One plausible explanation is that the bleaching curves are too simplistic in modelling the bleaching thresholds because they do not take into consideration any possible recovery by the corals between intra-seasonal heat waves. This is unlikely since temperatures were well above the mean at these sites for almost the entire 2004 summer and the week-to-week fluctuations were smaller than in the 1998 summer and equivalent to the 2002 summer (data not shown). Another plausible explanation is that light levels, UV or other environmental factors were sufficiently lower in the non-bleaching 2004/2005 summers compared with the bleaching summers and effectively prevented bleaching. Again, this is unlikely because global radiation near Magnetic Island was higher during the critical 2005 summer months of January/February (total = 1403 MJ/m^2) than the same period during the bleaching summer of 2002 (total = 1301 MJ/m^2) as was total sun hours (546 vs 533 h, respectively). Total UV-A and B radiation were marginally lower (82.2 vs 86.4 MJ/m^2, respectively), but this difference was not statistically significant [$t(2) = 1.12$, $df = 116$, $P = 0.27$]. The results of the statistical analysis at Magnetic Island also show that light, UV and other environmental factors at best only correlate weakly with the bleaching events and were nowhere near as important as temperature in predicting bleaching at this site. A more likely explanation is that corals at Magnetic Island, Daydream Island, Myrmidon Reef and Chicken Reef have undergone some acclimatization after the

2002 bleaching event, possibly through shuffling zooxanthellae types within the coral tissue (Baker et al. 2004; Rowan 2004; Berkelmans and van Oppen 2006) or in part through selection of more thermally resistant coral and symbiont genotypes among surviving populations.

Since it is likely that a number of reefs have effectively achieved a "stepwise" increase in thermal tolerance since 2002, the next challenge is to adjust the bleaching threshold to a new level. For those sites which survived the extra warm summers without bleaching, this adjustment can be made on the empirical evidence. The question then becomes how far can these thresholds be moved up and how will other locations respond to the next warm summer? If the mechanism of acclimatization is by zooxanthellae shuffling (Baker et al. 2004; Buddemeier et al. 2004), then it is likely that there is a limit of around 1.0–1.5 °C by which the bleaching threshold can be increased (Berkelmans and van Oppen 2006). If there is a gradual shifting of community zooxanthella types over multiple bleaching years or warm events, there may need to be an equally gradual increment in the empirical bleaching thresholds. Early warning systems based on these bleaching thresholds may show some false positive warnings as reef communities acclimatize before their increased thermal limits are quantified. This may decrease the reliability of time-integrated bleaching thresholds and other temperature-based early warning systems for some time until any shift can be adequately taken into account in a refined model. These adjustments and improvements in early warning systems are inevitable as our understanding of thermal stress and physiological responses of corals improves and advances in technology allow relevant parameters to be better monitored and interpreted. In the meantime a few false positive warnings may be something that reef managers would rather endure than their counterpart, a false negative.

The results of the classification tree analysis clearly show that temperature is the most important variable in explaining past bleaching events at Magnetic Island and that maximum monthly SST (based on daily averages) was the most important metric. UV (A and B) radiation, sunshine hours and other variables contributed either a small amount or not at all to the final model. This is not to say that UV and light are not important in bleaching, only that the field evidence does not support these factors as explanatory variables for the purpose of bleaching predictions at this site. Unfortunately, lack of global radiation and UVR data precludes similar analyses at other GBR sites. Given the importance of light in the bleaching response of corals as demonstrated experimentally (e.g. Jones et al. 1998) and evidenced in field situations (e.g. Oliver 1985), it is important that more data sets on light regimes influencing coral reefs are collected to better understand the role of light in field conditions and further evaluate its potential in refining early warning systems.

The T_{L50} curves presented here are the first attempt to quantify mortality thresholds for corals under field conditions. A great deal of experimental work has been done to determine the upper thermal limits for a range of corals (e.g. Mayer 1914; Yonge and Nicholls 1931; Coles et al. 1976; Marcus and Thorhaug 1981; Glynn and D'Croz 1990; Berkelmans and Willis 1999; Ulstrup et al. 2006). However, it is extremely difficult to replicate field conditions in controlled laboratory experiments, and hence results are hard to apply in early warning systems or scenario modelling.

Notwithstanding the fact that there may be some variation in the bleaching thresholds due to acclimatization and, for the same reason, probably also in the mortality thresholds, these T_{L50} curves provide a useful starting point for early warning systems and modelling future effects of climate change on coral reefs (e.g. Wooldridge et al. 2006; Chap. 13). The proximity of the T_{L50} curves to the bleaching threshold curves is clear evidence that there is a fine line between bleached corals recovering or dying in the communities examined here. The fact that all bar one of the mortality thresholds were <1 °C and many only <0.5 °C above the bleaching threshold illustrates the small quantum of increased temperature required to turn the next big bleaching event into an ecological disaster, the scale of which has not been seen on the GBR before but has already been experienced in the Indian Ocean in 1998 (Wilkinson et al. 1999; Goreau et al. 2000). Keppel Island reefs in the southern GBR were remarkably resilient in comparison with other reefs which suffered high mortality with a mortality threshold 0.9–1.7 °C higher than their bleaching threshold. These communities are dominated by the same corymbose and staghorn *Acropora* species which died <0.5 °C above their respective bleaching thresholds at Davies Reef and Stone Island in the central GBR. A possible explanation for the variation in mortality thresholds is an increase in resilience by Keppel Island corals through switching to heterotrophic feeding whilst bleached (Grottoli et al. 2006). This notion is supported by long-term chlorophyll monitoring data which shows that chlorophyll *a* in the Keppel Island area is up to 3.5 times higher than in the central offshore GBR (near Davies Reef) and twice the concentration of the central inshore GBR (near Stone Island, Brodie et al. 2007). It should be remembered that these mortality curves only relate to the most sensitive coral species and are not representative of scleractinian corals generally. Nevertheless, the sensitive species including most members of the Acroporidae and Pocilloporidae families (Marshall and Baird 2000; Loya et al. 2001) on Indo-Pacific reefs generally also make up the dominant coral cover, contribute most to reef building and provide the three-dimensional complexity on which fish and invertebrates rely for habitat and shelter (e.g. Done 1982; Bell and Galzin 1984). The near-linear relationship of temperature above bleaching threshold with exposure time suggests that the mortality thresholds simply represent a lateral shift in the bleaching thresholds outwards along the temperature axis. This makes mortality thresholds an easy metric to work with and apply in scenario modelling.

In conclusion, time-integrated bleaching thresholds remain an appropriate and useful method for modelling thermal stress in corals. There is a high likelihood that bleaching thresholds have increased at a number of locations on the GBR since the major 2002 bleaching event, potentially as a result of acclimatization. Whilst these increases may be limited in terms of absolute temperature, there is no doubt that they are of great ecological significance. Mortality thresholds developed for a limited number of reefs based on time–temperature curves for 50% mortality of specific taxa indicate a very narrow margin between the bleaching and mortality thresholds for sensitive species.

Acknowledgements Allunga Exposure Laboratories are gratefully acknowledged for making their weather observations over the last 35 years available, including the most comprehensive data set on solar and UV radiation for the Townsville area. The Australian Bureau of Meteorology kindly contributed data on sun hours and other weather variables for the Townsville area.

References

Baker AC, Starger CJ, McClanahan TR, Glynn PW (2004) Corals' adaptive response to climate. Nature 430:741. https://doi.org/10.1038/430741a

Bell JD, Galzin R (1984) Influence of live coral cover on coral-reef fish communities. Mar Ecol Prog Ser 15:265–274. https://doi.org/10.3354/meps015265

Berkelmans R (2001) Bleaching, upper thermal limits and temperature adaptation in reef corals. PhD thesis, James Cook University, Townsville

Berkelmans R (2002a) Time-integrated thermal bleaching thresholds of reefs and their variation on the Great Barrier Reef. Mar Ecol Prog Ser 229:73–82. https://doi.org/10.3354/meps229073

Berkelmans R (2002b) Reply comment. 'Time-integrated bleaching thresholds of reefs and their variation on the Great Barrier Reef'. Mar Ecol Prog Ser 237:309–310

Berkelmans R, Oliver JK (1999) Large-scale bleaching of corals on the Great Barrier Reef. Coral Reefs 18:55–60. https://doi.org/10.1007/s003380050154

Berkelmans R, van Oppen MJH (2006) The role of zooxanthellae in the thermal tolerance of corals: a 'nugget of hope' for coral reefs in an era of climate change. Proc R Soc B 273:2305–2312. https://doi.org/10.1098/rspb.2006.3567

Berkelmans R, Willis BL (1999) Seasonal and local spatial patterns in the upper thermal limits of corals on the inshore central Great Barrier Reef. Coral Reefs 18:219–228. https://doi.org/10.1007/s003380050186

Berkelmans R, Hendee JC, Marshall PA, Ridd PV, Orpin AR, Irvine D (2002) Automatic weather stations: tools for managing and monitoring potential impacts to coral reefs. Mar Technol Soc J 36:29–38. https://doi.org/10.4031/002533202787914250

Berkelmans R, De'ath G, Kininmonth S, Skirving WJ (2004) A comparison of the 1998 and 2002 coral bleaching events on the Great Barrier Reef: spatial correlation, patterns and predictions. Coral Reefs 23:74–83. https://doi.org/10.1007/s00338-003-0353-y

Breiman L, Friedman JH, Olshen RA, Stone CG (1984) Classification and regression trees. Wadsworth International, Belmont

Brodie J, De'Ath G, Devlin M, Furnas M, Wright M (2007) Spatial and temporal patterns of near-surface chlorophyll *a* in the Great Barrier Reef lagoon. Mar Freshw Res 58:342–353. https://doi.org/10.1071/MF06236

Brown BE, Dunne RP, Chansang H (1996) Coral bleaching relative to elevated seawater temperature in the Andaman Sea (Indian Ocean) over the last 50 years. Coral Reefs 15:151–152. https://doi.org/10.1007/BF01145885

Buddemeier RW, Baker AC, Fautin DG, Jacobs JR (2004) The adaptive hypothesis of bleaching. In: Rosenberg E, Loya Y (eds) Coral health and disease. Springer, Berlin, pp 427–444

Coles SL, Jokiel PL (1992) Effects of salinity on coral reefs. In: Pollution in Tropical Aquatic Systems. CRC Press, Florida, pp 147–166

Coles SL, Jokiel PL, Lewis CR (1976) Thermal tolerance in tropical versus subtropical Pacific reef corals. Pac Sci 30:159–166

De'ath G, Fabricius KE (2000) Classification and regression trees: a powerful yet simple technique for ecological data analysis. Ecology 81:3178–3192. https://doi.org/10.1890/0012-9658(2000)081[3178:CARTAP]2.0.CO;2

Done TJ (1982) Patterns in the distribution of coral communities across the central Great Barrier Reef. Coral Reefs 1:95–107. https://doi.org/10.1007/BF00301691

Dove SG, Hoegh-Guldberg O (2006) The cell physiology of coral bleaching. In: Phinney JT, Hoegh-Guldberg O, Kleypas J, Skirving W, Strong A (eds) Coral reefs and climate change: science and management. (Coastal and estuarine studies 61). Am Geophys Union, Washington, DC, pp 55–71

Fitt WK, Brown BE, Warner ME, Dunne RP (2001) Coral bleaching: interpretation of thermal tolerance limits and thermal thresholds in tropical corals. Coral Reefs 20:51–65. https://doi.org/10.1007/s003380100146

GBRMPA (2002) Report on GBRMPA bleaching response strategy 2002. http://www.gbrmpa.gov.au/corp_site/info_services/science/climate_change/01 -02/final_report/index.html

GBRMPA (2006) Final bleaching summary report 2005/2006. http://www.gbrmpa.gov.au/corp_site/info_services/science/climate_change/co nditions_report.html

Gleeson MW, Strong AE (1995) Applying MCSST to coral reef bleaching. Adv Space Res 16:151–154. https://doi.org/10.1016/0273-1177(95)00396-V

Glynn PW, D'Croz L (1990) Experimental evidence for high temperature stress as the cause of El Niño-coincident coral mortality. Coral Reefs 8:181–191. https://doi.org/10.1007/BF00265009

Goreau TJ, Hayes RL (1994) Coral bleaching and ocean 'hot spots'. Ambio 3:176–180

Goreau TJ, Hayes RL, Clark JW, Basta DJ, Robertson CN (1993) Elevated sea temperatures correlate with Caribbean coral reef bleaching. In: Geyer RA (ed) A global warming forum: scientific, economic and legal overview. CRC, Boca Raton, pp 225–255

Goreau TJ, McClanahan T, Hayes RL, Strong AL (2000) Conservation of coral reefs after the 1998 global bleaching event. Conserv Biol 14:5–15. https://doi.org/10.1046/j.1523-1739.2000.00011.x

Grottoli AG, Rodrigues LJ, Palardy JE (2006) Heterotrophic plasticity and resilience in bleached corals. Nature 440:1186–1189. https://doi.org/10.1038/nature04565

Hendee JC, Mueller E, Humphrey C, Moore T (2001) A data-driven expert system for producing coral bleaching alerts at Sombrero reef. Bull Mar Sci 69:673–684

Jones RJ, Berkelmans R, Oliver JK (1997) Recurrent bleaching of corals at Magnetic Island (Australia) relative to air and seawater temperature. Mar Ecol Prog Ser 158:289–292. https://doi.org/10.3354/meps158289

Jones RJ, Hoegh-Guldberg O, Larcum AWD, Schreiber U (1998) Temperature-induced beaching of corals begins with impairment of the CO_2 fixation mechanism in zooxanthellae. Plant Cell Environ 21:1219–1230. https://doi.org/10.1046/j.1365-3040.1998.00345.x

Loya Y, Sakai K, Yamazato K, Nakano Y, Sambali H, Van Woesik R (2001) Coral bleaching: the winners and the losers. Ecol Lett 4:122–131. https://doi.org/10.1046/j.1461-0248.2001.00203.x

Manzello DP, Berkelmans R, Hendee JC (2007) Coral bleaching indices and thresholds for the Florida Reef Tract, Bahamas, and St Croix, US Virgin Islands. Mar Pollut Bull 54:1923–1931. https://doi.org/10.1016/j.marpolbul.2007.08.009

Marcus J, Thorhaug A (1981) Pacific versus Atlantic responses of the subtropical hermatypic coral *Porites* spp. to temperature and salinity effects. Proceedings of the 4th International Coral Reef Symposium, Manilla 4-2, pp 15–20

Marshall PA, Baird AH (2000) Bleaching of corals on the Great Barrier Reef: differential susceptibilities among taxa. Coral Reefs 19:155–163. https://doi.org/10.1007/s003380000086

Marshall P, Schuttenberg H (2006) A reef managers guide to coral bleaching. Great Barrier Reef Marine Park Authority, Townsville

Mayer AG (1914) The effects of temperature upon marine animals. Carnegie Inst Wash Publ Pap Mar Lab Tortugas 6:3–24

Maynard J, Turner PJ, Anthony KRN, Baird AH, Berkelmans R, Eakin CM et al (2008) *ReefTemp* an interactive monitoring system for coral bleaching using high-resolution SST and improved stress predictors. Geophys Res Lett 35:L05603. https://doi.org/10.1029/2007GL032175

McClanahan TR, Ateweberhan M, Ruiz Sebastian C, Graham NAJ, Wilson SK, Bruggeman JH et al (2007) Predictability of coral bleaching from synoptic satellite and in situ temperature observations. Coral Reefs 26:695–701. https://doi.org/10.1007/s00338-006-0193-7

Nakamura T, van Woesik R (2001) Water-flow rates and passive diffusion partially explain differential survival of corals during the 1998 bleaching event. Mar Ecol Prog Ser 212:301–304. https://doi.org/10.3354/meps212301

Oliver J (1985) Recurrent seasonal bleaching and mortality of corals on the Great Barrier Reef. Proceedings of the 5th International Coral Reef Congress, Tahiti 5-4, pp 201–206

Podestá GP, Glynn PW (2001) The 1997–98 El Niño event in Panama and Galápagos: an update of thermal stress indices relative to coral bleaching. Bull Mar Sci 69:43–59

Rowan R (2004) Thermal adaptation in reef coral symbionts. Nature 430:742. https://doi.org/10.1038/430742a

Sammarco PW, Winter A, Cody Stewart J (2006) Coefficient of variation of sea surface temperature (SST) as an indicator of coral bleaching. Mar Biol 149:1337–1344. https://doi.org/10.1007/s00227-006-0318-0

Shackeroff J (1999) The effects of the 1998 bleaching event on *Millepora tenella*: an indicator of climate change? (Unpublished report), School for International Training, Cairns

Strong AE, Barrientos CS, Duda C, Sapper J (1997) Improved satellite techniques for monitoring coral reef bleaching. Proc 8th Int Coral Reef Symp, Panama 8-2:1495–1498

Ulstrup KE, Berkelmans R, Ralph PJ, van Oppen MJH (2006) Variation in bleaching sensitivity of two coral species with contrasting bleaching thresholds across a latitudinal gradient on the Great Barrier Reef. Mar Ecol Prog Ser 314:135–148. https://doi.org/10.3354/meps314135

van Woesik R, Koksal S (2006) A coral population response (CPR) model for thermal stress. In: Phinney JT, Hoegh-Guldberg O, Kleypas J, Skirving W, Strong A (eds) Coral reefs and climate change: science and management. (Coastal and estuarine studies 61). American Geophysical Union, Washington, DC, pp 129–144

Vargas-Ángel B, Zapata FA, Hernández H, Jiménez JM (2001) Coral and coral reef responses to the 1997-98 El Niño event on the Pacific coast of Colombia. Bull Mar Sci 69:111–132

Wilkinson C, Lindén O, Cesar H, Hodgson G, Rubens J, Strong AE (1999) Ecological and socioeconomic impacts of 1998 coral mortality in the Indian Ocean: an ENZO impact and a warning of future change? Ambio 28:188–196

Winter A, Appeldoorn RS, Bruckner A, Williams EH, Goenaga C (1998) Sea surface temperatures and coral reef bleaching off La Parguera, Puerto Rico (northeastern Caribbean Sea). Coral Reefs 17:377–382. https://doi.org/10.1007/s003380050143

Wooldridge S, Done T, Berkelmans R, Jones R, Marshall P (2006) Precursors for resilience in coral communities in a warming climate: a belief network approach. Mar Ecol Prog Ser 295:157–169. https://doi.org/10.3354/meps295157

Yonge CM, Nicholls AG (1931) Studies on the physiology of corals. IV. The structure, distribution and physiology of the zooxanthellae. Sci Rep Great Barrier Reef Exped, 1928–29, 1:135–176

Chapter 10
Consequences of Coral Bleaching for Sessile Reef Organisms

T. R. McClanahan, E. Weil, and A. H. Baird

10.1 Introduction

Sessile organisms, such as corals and erect algae, provide a complex reef architecture that is important for maintaining organic and inorganic (calcium carbonate) forms of reef productivity (Enriques et al. 2005; Finelli et al. 2006). These sessile organisms also provide shelter and, in some cases, food for mobile animals such as fishes (Chap. 11). Corals and sessile algae that deposit their calcium carbonate skeletons create the geologic reefs that shelter many tropical shorelines from storm damage. Bleaching, the loss of the coral symbionts and the plant and animal pigments, often leads to a loss of energy and partial and whole-colony mortality of affected organisms. Higher than normal temperatures affect metabolic reactions that are regulated by temperature-sensitive enzymes, which can then affect physiological functions. The loss of zooxanthellae, on the other hand, causes a significant reduction in the translocation of photosynthetic products to the host cells and is, therefore, expected to have both immediate and delayed effects on individuals, which cascade to affect populations, communities, and ecosystems. Zooxanthellae have been estimated to provide 30% of the total nitrogen and 91% of the carbon needs of the coral host (Bythell 1988). Consequently, the effect of coral bleaching has major consequences for reef productivity, reef growth, and biodiversity. Despite the taxonomic extent of

T. R. McClanahan (✉)
Wildlife Conservation Society, Marine Programs, Bronx, NY, USA
e-mail: tmcclanahan@wcs.org

E. Weil
Department of Marine Sciences, University of Puerto Rico, Mayaguez, PR, USA

A. H. Baird
ARC Centre of Excellence for Coral Reef Studies, James Cook University, Townsville, QLD, Australia

M. J. H. van Oppen, J. M. Lough (eds.), *Coral Bleaching*, Ecological Studies 233,
https://doi.org/10.1007/978-3-319-75393-5_10

bleaching, scleractinian (hard corals) and octocorals (soft corals) have attracted most research attention and are, therefore, the focus of this chapter.

Mass mortality events on coral reefs due to large-scale environmental disturbances have been recorded since the 1870s. From 1876 to 1979, 63 mass coral mortalities were reported (Glynn 1993; Chap. 3). During this period, only three minor coral bleaching events were registered. Since 1979, bleaching has continued (Donner et al. 2017), and bleaching is considered the main global threat to coral reefs worldwide by many coral reef biologists, with overharvesting, costal development, and pollution being locally important (McClanahan et al. 2008a; Weil and Rogers 2011; Burge et al. 2014; Maynard et al. 2015). This perception contrasts with coral reef managers who generally perceive pollution and fishing as the main threats and manageable at the scales they work at (Jackson et al. 2014; Wear 2016). The fate of coral reefs at this global scale is dire according to many climate change-coral reef projection models (Baskett et al. 2010; Frieler et al. 2013; Logan et al. 2014; Chap. 13). Model projections remain a contentious and important debate among coral reef scientists because of the uncertain acclimation and adaptive responses of coral to rapidly warming temperatures (e.g., Buddemeier et al. 2004; van Oppen et al. 2015).

10.2 Taxa Affected by Bleaching

Almost all photo-symbiotic marine organisms bleach in response to temperature. Indeed, symbiosis seems to predispose organisms to be susceptible to a broad range of stressor (Baird et al. 2009). In general, the taxa most affected by bleaching include the hydrocorals (e.g., *Millepora* spp.), scleractinians, and octocorals. However, photo-symbiotic sponges and bivalves also bleach (Table 10.1, Fig. 10.1). Even some plants, such as the green alga *Halimeda*, lose pigments in response to thermal stress (Vicente 1990). The most detailed descriptions of the taxa affected by bleaching come from the Caribbean where numerous species bleached in response to higher than usual sea temperature in 2005 and 2010. Five species of hydrozoan (100% of the species pool), 60 species of scleractinians (90% of the species pool), and 30 octocoral species (20% of the species pool) bleached along with other cnidarians and sponges (Prada et al. 2010) (Fig. 10.1). These observations support the hypothesis that most photo-symbiotic organisms will bleach when exposed to thermal stress.

Table 10.1 List of scleractinian (61), Octocorallia (24), Hydrozoa (5), zoanthid (3), Actiniaria (5), and Corallimorpharia (2) species reported/observed with bleaching signs during bleaching events in Puerto Rico and the Caribbean

Species	1983	1987	1993	1995	1998	2003	2005	2009	2010	SU	MO
Acropora palmata	x	x	x	x	x	x	x		x	***	x
A. cervicornis	x	x	x	x	x	x	x			***	x
A. prolifera		x		x	x		x			***	x
Stephanocoenia intersepta		x	x		x	x	x	x	x	***	x
Siderastrea siderea		x	x	x	x	x	x	x	x	***	x
S. radians		x			x		x			**	x
Solenastrea bournoni		x			x		x	x	x	*	
Madracis decactis		x					x		x	*	
M. formosa							x			*	
M. auretenra[1]		x					x		x	**	x
*M. pharensis**							x			*	
M. senaria							x	x		*	
Undaria agaricites[2]	x	x		x	x		x	x	x	***	x
U. purpurea		x		x	x		x	x		**	x
U. humilis		x		x	x		x	x	x	**	x
U. danae		x		x	x	x	x		x	***	x
U. carinata		x			x	x	x		x	***	x
U. tenuifolia		x			x		x	x		***	x
Agaricia lamarcki		x			x		x	x	x	***	x
A. fragilis		x			x		x	x	x	**	x
A. pumila		x					x	x		**	x
A. grahamae		x			x		x	x		*	
A. undata		x					x	x		*	
Helioceris cucullata		x		x	x	x	x	x	x	***	x

(continued)

Table 10.1 (continued)

Species	1983	1987	1993	1995	1998	2003	2005	2009	2010	SU	MO
Orbicella annularis[3]	x	x	x	x	x	x	x	x	x	***	x
O. faveolata	x	x	x	x	x	x	x	x	x	***	x
O. franksi		x		x	x	x	x	x	x	**	x
Montastraea cavernosa		x		x	x		x	x	x	*	x
Colpophyllia natans[4]	x	x		x	x	x	x	x	x	***	x
C. amaranthus				x			x		x	***	x
Pseudodiploria strigosa		x	x		x	x	x	x	x	*	x
P. clivosa		x			x		x	x		*	
Diploria labyrinthiformis		x	x		x	x	x	x	x	**	x
Favia fragum		x	x		x	x	x	x		***	x
Manicina areolata		x			x		x			*	
Cladocora arbuscula					x		x			*	x
Meandrina meandrites		x		x	x		x	x	x	***	x
M. danae					x		x			*	
M. jacksoni[5]		x			x		x	x	x	***	x
Dichocoenia stokesi		x					x		x	**	?
D. stellaris							x			*	?
Dendrogyra cylindrus		x		x	x		x	x	x	***	x
Porites astreoides		x	x	x	x	x	x	x	x	***	x
P. colonensis							x			*	?
P. branneri		x					x		x	*	?
P. porites	x	x			x		x	x	x	***	?
P. furcata		x					x	x	x	***	x
P. divaricata		x		x		x	x		x	**	?
Mussa angulosa							x		x	*	x

Mycetophyllia ferox		x			x		x	x	x	**	x
M. aliciae		x					x	x	x	**	x
M. lamarckiana					x		x	x	x	*	x
M. danaana			x		x	x	x	x		*	x
M. resii			x	x	x	x	x	x		*	?
Scolymia lacera		x					x		x	*	x
S. cubensis		x					x		x	*	x
Isophyllia sinuosa		x			x		x		x	*	?
I. rigida					x		x			*	?
Eusmilia fastigiata		x			x	x	x	x	x	**	x
Oculina diffusa		x			x		x		x	**	x
O. varicosa							x		x	**	?
O. valenciennesi					x		x		x	*	?
Octocorals											
Gorgonia ventalina							x		x	*	?
G. flabellum							x		x	*	?
G. mariae					x		x		x	*	?
Antillogorgia americana					x	x	x		x	*	?
A. acerosa							x		x	*	?
A. bipinnata							x			*	?
Pseudopterogorgia sp.							x		x	*	?
Plexaura flexuosa							x			*	?
P. homomalla							x			*	?
Pseudoplexaura porosa							x		x	*	?
Pseudoplexaura sp.							x			*	?
Plexaurella nutans					x	x	x		x	**	x
P. anceps							x		x	*	?

(continued)

Table 10.1 (continued)

Species	1983	1987	1993	1995	1998	2003	2005	2009	2010	SU	MO
Plexaurella sp.							x		x	**	x
Briareum asbestinum				x	x	x	x		x	**	x
B. polyanthes				x	x	x	x		x	**	x
Erythropodium caribaeorum					x	x	x		x	**	x
Pterogorgia citrina							x		x	**	?
Muricea pinnata							x			**	
Muricea muricata					x		x		x	**	x
Muricea sp.							x			*	x
Muriceopsis flavida							x		x	*	?
Eunicea calyculata							x		x	*	?
Eunicea laxispica							x			*	?
Hydrozoa											
Millepora alcicornis		x	x	x	x	x	x	x	x	***	x
M. complanata		x	x	x	x	x	x	x	x	***	x
M. squarrosa		x			x		x	x		**	x
M. striata					x		x			**	?
Stylaster roseus		x			x		x		x	**	x
Actiniaria											
Condylactis gigantea		x			x		x		x	**	?
Bartholomea annulata		x			x		x		x	*	?
Lebrunia danae		x					x			*	?
Epicystis crucifer		x								*	?
Stichodactyla helianthus		x					x			*	?
Zoanthids											
Palythoa caribaeorum		x		x	x	x	x		x	***	x

Zoanthus sociatus		x			x		x			**	?
Z. pulchellus							x		x	*	?
Corallimorpharia											
Discosoma sanctithomae		x								*	?
Ricordea florida		x							x	*	?

The susceptibility index (SU) is drawn from these reports and personal observations since the 1998 bleaching event: *** = highly susceptible, bleaches frequently; ** = moderate, bleaches only during significant bleaching events; and * = resistant, only bleaches during extreme bleaching events. Mortality (MO) = any reported/observed mortality in any bleaching event (Prepared by E. Weil)

Sources: Lasker et al. (1984), Williams and Bunkley-Williams (1988, 1990), McField (1999), Weil (2002, 2004), Weil et al. (2009a, b), Cróquer and Weil (2009a, b), Weil and Cróquer (2009), Pinzon and Weil (2012), Weil and Rogers (2011), Weil et al. (2017)

[1]*M. auretenra* (Locke et al. 2007) is the former *M. mirabilis*

[2]*U. agaricites* might include all taxa in the *U. agaricites* species complex

[3]*O. annularis* includes the other two species (*O. faveolata* and *O. franksi*) in early reports

[4]*C. natans* could include *C. breviserialis* and *C. amaranthus*

[5]*Meandrina jacksoni* (Weil & Pinzon) is a new species formerly confused with *M. meandrites*

Fig. 10.1 The 2005 bleaching event in the Caribbean. View of bleached reefs in Puerto Rico (**a**, **b**) and Grenada (**c**, **d**). Acroporids were hit hard in most shallow habitats in Puerto Rico and other Caribbean localities (**e**). Side-by-side bleached and unbleached colonies of *Orbicella faveolata* raise questions about "resistant" zooxanthellae strains (**f**). Significant numbers of colonies of bleaching-resistant genera such as *Mycetophyllia* were completely white (**g**). Many *O. faveolata* colonies with on-going yellow band disease bleached completely increasing tissue mortality rates (**h**). Bleached *Stephanocoenia intersepta* with dark spots disease (**i**) did not show any mortality. Other zooxanthellate reef organisms like mileporids, crustose and branching octocorals also bleached (**j**, **k**). Significant bleaching-associated mortalities were observed in *Mycetophyllia* spp., and *Undaria* spp. in 2005 (**l**), and in acroporids in both 2005 and 2010 (**m**) (Photos E. Weil)

10.3 Spatial and Temporal Variation in Bleaching and Mortality

The scale and severity of bleaching mortality are highly correlated with the scale and severity of thermal stress (Glynn 1984; Hughes et al. 2017). Bleaching mortality is also habitat-specific. For example, mortality declines with depth and is reduced in habitats or environments where light intensity is reduced and natural background temperature variability is high (Brown et al. 1996; McClanahan et al. 2007a; Bridge et al. 2014). For example, windward habitats have less natural background temperature variability than leeward habitats, and this can result in higher bleaching in windward habitats when temperatures rise (McClanahan et al. 2005a).

Bleaching at the assemblage level can vary at many spatial scales. For example, on the Great Barrier Reef (GBR), inshore reefs were more affected than offshore reefs in both the 1998 and 2002 events (Berkelmans et al. 2004), while the opposite pattern was observed in the eastern Pacific in 1998 (Glynn et al. 2001) and Puerto Rico in 2005 (Weil et al. 2009a). Reefs on the GBR also bleached in clusters on a scale of tens or hundreds of kilometers, which was attributed to local weather patterns, oceanographic conditions, or both (Berkelmans et al. 2004). Bleaching was considerably worse in the northern than southern GBR in 2016 (Hughes et al. 2017). Spatial variation in bleaching among sites has also been linked to a habitat's coral assemblage, where assemblages dominated by susceptible species were more likely to have a higher proportion of bleached colonies (Marshall and Baird 2000; Shuail et al. 2016; Heron et al. 2016). For example, half of the bleaching response in western Indian Ocean coral assemblages was attributable to the site's initial species composition (McClanahan et al. 2007b, c).

Environmental factors that reduce the effects of stressful conditions or create conditions that promote acclimation to extreme temperatures, irradiation, and UV levels are expected to minimize the effects of bleaching. In an analysis of sea surface temperatures (SST) off East Africa, McClanahan et al. (2007a) showed that long-term temperature variability played a major role in buffering the effects of anomalous SST warming. Locations in northern Kenya and southern Somalia that had SST distributed narrowly around the mean had higher-degree heating weeks than areas where SST had wide or flat distributions (Tanzania, Comoros; Fig. 10.2). Many sites around the central Indian Ocean, such as Lakshadweep, Sri Lanka, Maldives, Chagos, and Seychelles, that suffered higher mortalities in 1998 also had narrowly distributed background SST (Ateweberhan and McClanahan 2010). Similarly, a protective, sub-bleaching stress prior to stronger bleaching-inducing temperatures can reduce the extent of bleaching in colonies in the laboratory (Ainsworth et al. 2016). In contrast, there was no evidence to suggest that prior low to moderate bleaching histories reduced the extent of bleaching on reefs on the GBR in 2016 (Hughes et al. 2017). However, repeated strong bleaching temperatures had less effect on a number of dominant coral taxa in 2016 compared to 1998 on Kenyan reefs (McClanahan 2017).

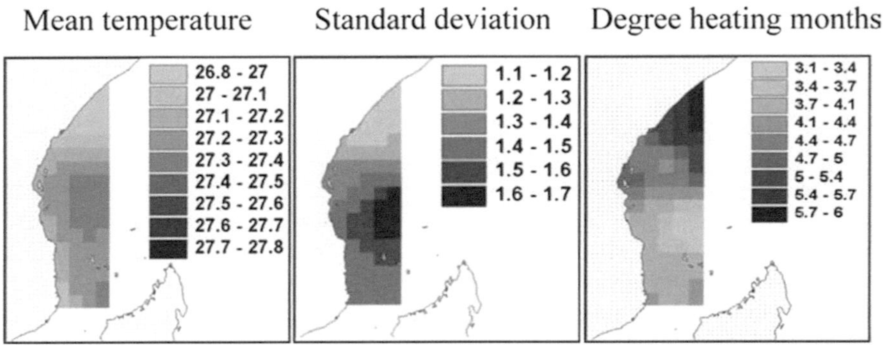

Fig. 10.2 Spatial distribution of the mean, variance (standard deviation), and degree heating months during 1998 along the East African coast, based on Hadley Centre data for 100 × 100 km grids and monthly SST (1950–2002). Degree heating months is the number of months that the temperature is 1 °C above the mean for that month

10.4 Patterns of Susceptibility to Bleaching Among Taxa

Earlier work on patterns of susceptibilities and bleaching to thermal stress is generally consistent among taxa across the Indo-Pacific, at least during the initial bouts of high temperatures (McClanahan et al. 2004a). Some more recent work suggests differences in rates of acclimation of the same taxa over time in different regions, which further complicates efforts to understand changes on the global scale (McClanahan 2017). Therefore, it is less clear how the bleaching susceptibility hierarchy changes over time and with temperature severity histories. Mortality rates are often poorly associated with bleaching intensity, which makes estimates of natural selection and genetic adaptation rates difficult to determine from bleaching observations (McClanahan 2004, 2017). It may depend on the past history and severity of the bleaching, as there are few unaffected taxa during the most severe events that have been proceeded by less severe events (Hughes et al. 2017). Colonies of some species bleach readily but recover (Baird and Marshall 2002), while fewer species may show no symptoms of stress or obvious loss of pigmentation but still suffer mortality from thermal stress and disease (McClanahan 2004; Weil and Rogers 2011).

Other patterns associated with bleaching include observations that partial and whole-colony mortality can be quite different with consequences for estimating impacts of thermal anomalies. For example, branching colonies typically have high rates of whole-colony but little partial mortality. Massive taxa, in contrast, have low rates of whole-colony mortality, and the majority of tissue is lost through partial mortality. For example, a study of individually tagged colonies of four species on the GBR following bleaching found 88% of *Acropora hyacinthus* colonies died while no whole-colony mortality occurred in *Porites lobata* (Baird and Marshall 2002). However, the mean loss of tissue per massive colony was 42%. In contrast, partial mortality was rare among *Acropora* and colonies either survived intact or died. Consequently, estimates of whole-colony mortality alone would underestimate the effect of bleaching on the massive taxa.

The time taken to respond to warmer water also varies considerably among species making the ranking of susceptibilities dependent on the time elapsed since the onset of thermal stress. Massive species take longer to respond to thermal stress, can stay bleached, and take longer to die than most branching species that bleach quickly and either quickly recover or die (Brown and Suharsono 1990; McClanahan et al. 2001; Baird and Marshall 2002). These findings are from macroscopic field observations, but studies of the *Acropora hyacinthus* transcriptome found bleaching involved approximately 20% of the host transcriptome and it was perturbed for more than 6 months after the disturbance when normal growth returned (Thomas and Palumbi 2017). In the GBR, the appearance of bleaching in individually tagged *Acropora* colonies changed markedly across a single month (Baird and Marshall 2002). In contrast, massive *Porites* have been shown to bleach slowly over a number of months but stay bleached many months beyond when warm temperatures passed (McClanahan et al. 2001). In many cases, bleached or surviving corals are more susceptible to diseases, predators, and competitors (Figs. 10.3 and 10.4).

The response of individual coral colonies can be also shaped by previous experience (Brown et al. 2000; Oliver and Palumbi 2011a; Guest et al. 2012; McClanahan 2017). Individuals can also respond to bleaching by changing the relative abundance of high-temperature-resistant symbiont strains making individuals less susceptible to subsequent bleaching events (Baker 2003; Baker et al. 2004; Oliver and Palumbi 2011b, Chap. 9). Consequently, there is increasing evidence that some corals can adjust to global warming, and, therefore, projections of the future state of coral reefs need to take adaptation and acclimation into account (Baird et al. 2007; Logan et al. 2014).

10.5 Effect of Bleaching on Individuals

Sublethal effects on individual coral reef organisms following bleaching include reduced reproductive output, reduced growth, and increased susceptibility to diseases and other disturbances (Lesser et al. 2007).

10.5.1 Effect of Bleaching on Reproduction

Coral reproductive output can be affected by bleaching in a number of ways, including changes in egg size and quality; reduced polyp fecundity, both in the number of polyps with eggs and the number of eggs per polyp; and finally in the number of whole colonies breeding in a given year (Baird and Marshall 2002; Mendes and Woodley 2002). In contrast, bleaching had no effect on reproductive output in *Montipora capitata*, possibly because this species can respond to the loss

Fig. 10.3 High-temperature-induced bleaching (**a**, **c**, **d**) and tissue mortality (**b**, **e**) in the important and endangered elkhorn scleractinian coral *A. palmata* in La Parguera during the 2005 thermal anomaly in the northeastern Caribbean (photos: E. Weil)

Fig. 10.4 High-temperature-induced bleaching and tissue mortality in the hydrocoral *Millepora alcicornis* (**a**, **b**), the foliose scleractinians *Agaricia agaricites* (**c**, **d**), *Helioceris cucullata* (**e**), *Mycetophyllia ferox* (**f**), *A. lamarcki* (**g**, **h**), and *M. lamarckiana* (**i**) during and after the intense thermal anomaly of 2005 in reefs off La Parguera Natural Reserve in Puerto Rico (Photos: E. Weil)

of symbionts by increasing heterotrophic feeding (Cox 2007). Egg size and number of eggs per polyp were both lower in experimentally bleached colonies than in unbleached colonies of the soft coral *Lobophytum compactum* (Michalek-Wagner and Willis 2001a). Egg quality was also affected, with protein, lipid, mycosporine-like amino acids, and carotenoid concentrations significantly reduced in the eggs of bleached colonies (Michalek-Wagner and Willis 2001b). The effect of bleaching on reproduction is also species-specific. For example, the proportion of colonies of *Acropora hyacinthus* (45%) breeding after bleaching on the GBR was much lower than that of *A. millepora* (88%; Baird and Marshall 2002).

On the GBR, bleaching caused a substantial decrease in reproductive output primarily as a result of a significant decline in colony abundance (Baird and Marshall 2002). Following the 1998 bleaching event, the number of gravid *A. hyacinthus* colonies at Pelorus Island was only 6% and *A. millepora* colonies 63% of that in 1997. Missing a few reproductive seasons for most long-lived coral species might not have drastic consequences for their fitness. Nevertheless, an increase in temperatures compounded with local anthropogenic disturbances is expected to reduce the capacity of these species to recovery from repeated disturbances.

10.5.2 Effect of Bleaching on Growth

Bleaching generally reduces coral growth rates (Pratchett et al. 2015). For example, the unbleached massive coral *Orbicella annularis* deposited 1.4 mm/year more aragonite than bleached colonies (Porter et al. 1989). Similarly, moderately affected colonies of *Acropora hyacinthus* and *A. millepora* grew following bleaching, whereas severely affected colonies did not. Yet, a direct association between bleaching severity and growth was not statistically significant (Baird and Marshall 2002). Measuring growth precisely in the field is difficult, and, therefore, only a few field tests are available, and these show high variation in their results.

10.5.3 Size-Specific Mortality Following Bleaching in Corals

Mortality in corals is often size-specific (Madin et al. 2014); however, the effect of bleaching-related mortality on the size structure of populations has not been fully resolved. For example, mortality rates following bleaching on the GBR in 1998 were not size-specific, but patterns were difficult to assess across the whole spectrum of sizes as only mature colonies were sampled (Baird and Marshall 2002). In contrast, experimental and theoretical work predicted that large size might actually be deleterious when corals are exposed to thermal stress (Nakamura and van Woesik 2001). Field studies backed up these predictions by showing that coral recruits with diameters of <20 mm were unbleached compared to large corals (Mumby 1999).

One predicted consequence of climate change is that it should reduce reproduction and recruitment and produce coral populations with large colonies (Bak and Meesters 1999). However, a number of bleaching field studies report high partial mortality and that small colonies survive better than larger colonies (Loya et al. 2001; Nakamura and van Woesik 2001; Bena and van Woesik 2004; Shenkar et al. 2005). Further, a long-term study of changes in coral size found that bleaching reduced mean sizes but mortality was not related to the taxa's mean size on Kenyan reefs (McClanahan et al. 2008b). This indicates that the size effect is within and mostly through partial colony mortality rather than higher mortality of taxa with large maximum sizes. Studies in the Persian Gulf confirmed these predictions in that reefs stressed by temperature and salinity had smaller colony sizes (Bauman et al. 2013).

10.5.4 Effect of Bleaching on Coral Settlement, Recruitment, and Recovery

Coral settlement is typically measured as larval recruitment to artificial substrata (Babcock et al. 2003). The few studies available indicate that coral settlement rates are reduced following bleaching. For example, Gilmour et al. (2013) recorded a 97% reduction in settlement following a bleaching event that reduced adult coral cover by between 75% and 90%. However, the studied reef, Scott Reef (Western Australia), is highly isolated (Underwood et al. 2009), which might have affected the supply of new larval recruits. Nonetheless, coral settlement in the more connected reefs of Maldives was reduced after bleaching in 1998 and continued to decline over time (McClanahan 2000; Loch et al. 2004). In the eastern Pacific, while gametes became mature in bleaching years (Glynn et al. 1991, 1996, 2000), subsequent larval recruitment to the substratum was variable (Guzmán and Cortés 2001, 2007). Similarly, while larval recruitment of *Pavona varians* in Panama was significantly correlated with high seawater temperature, recruitment failed at very high temperatures (Glynn et al. 2000). Post-recruitment survivorship is poorly studied, but evidence from East Africa and the Caribbean suggests that predation can retard the recovery of small corals recovering from bleaching (McClanahan et al. 2005b; Rotjan and Lewis 2005). The loss of coral cover after bleaching can result in more intense predation as coral predators increase their feeding focus on the surviving corals.

10.5.5 Diseases

There is a clear synergy between high water temperatures, bleaching, and disease, all possibly associated with changing environmental conditions linked to global climate change (Baird 2000; McClanahan et al. 2004b; Willis et al. 2004; Miller et al. 2006; Bruno et al. 2007; Weil et al. 2009a, 2017; Cróquer and Weil 2009a, b; Fig. 10.5). Temperature is the major driver of deadly scleractinian, octocoral, hydrocoral, and crustose coralline algae disease epizootics. As SST are projected to continue to increase, it is expected that the number of new diseases, their prevalence, and virulence will also increase (Rosenberg and Ben-Haim 2002; Lesser et al. 2007; Weil and Rogers 2011; Weil et al. 2017). There is compelling evidence that high temperatures affect immune response making corals more susceptible to infections. The destabilization of the "physiological equilibrium" between the host and resident bacteria, where the host becomes more susceptible, the vector becomes more virulent, or both promotes disease virulence (Harvell et al. 1999; Rosenberg and Ben-Haim 2002; Ritchie 2006; Ward et al. 2007; Weil et al. 2009a; Harvell et al. 2009; Weil and Rogers 2011; Burge et al. 2014; Page et al. 2016; Weil et al. 2017). Bleaching reduces the antibiotic properties of mucus, making colonies more susceptible to diseases, and also promotes an ecological imbalance among coexisting taxa in the coral holobiont, such as fungi, causing sublethal and lethal effects (Ritchie 2006; McClanahan et al. 2009; Chap. 7).

Reefs of the Caribbean have seen the greatest proliferation of coral disease, and the most devastating outbreaks have frequently been associated with high thermal stress and bleaching events. Caribbean octocorals also lose *Symbiodinium*, and this renders them more susceptible to stress and diseases (Couch et al. 2008; McClanahan et al. 2009; Burge et al. 2013; Kim 2015). The fungi *Aspergillus* spp., protozoans, and pathogenic bacteria cause most octocoral diseases. For example, *Aspergillosis* outbreaks have caused widespread mortality of sea fans associated

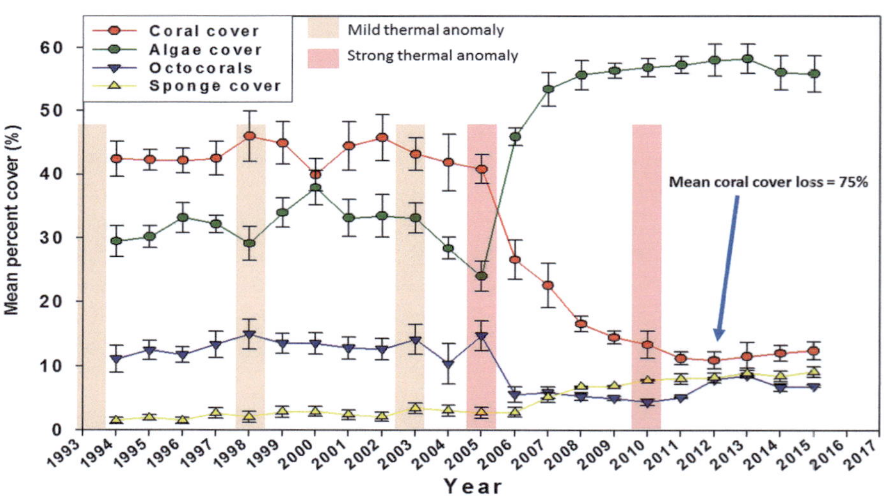

Fig. 10.5 Time series of surveys in La Parguera, Puerto Rico, over a 22-year period document the shift from coral- to an algae-dominated community after the disease and bleaching mortalities associated with the 2005 thermal anomaly. Three previous mild thermal anomalies associated with coral bleaching events but not associated disease outbreaks did not produce high mortalities

with high thermal anomalies (Harvell et al. 2009; Burge et al. 2013; Weil et al. 2017). Following the thermal anomaly and bleaching events of 1998–1999, 2005–2006, and 2010, many soft coral species of *Briareum polyanthes*, *B. asbestinum*, and *Erythropodium caribaeorum* became diseased and died in the Florida Keys, Puerto Rico, and throughout the Caribbean (Harvell et al. 2001; Weil et al. 2017) (Fig. 10.1).

The white band disease (WBD) epizootic that wiped out more than 95% of acroporids region-wide in the Caribbean in the early 1980s was associated with high water temperatures (Gladfelter 1982). Similarly, during the thermal anomalies of 2005 and 2010, outbreaks of white plague disease (WPD) and Caribbean yellow band disease (CYBD) produced coral tissue losses of 60% in northeast and 35% in the southern Caribbean of the main reef-building genera (Miller et al. 2009; Weil et al. 2006, 2009b; Bastidas et al. 2012). Therefore, since both bleaching and diseases are highly correlated with high SST, they are bound to co-occur. Co-occurrence makes it difficult to assess whether bleaching, disease, or both are responsible for coral mortalities. It is speculated that bleaching reduces energy available to resist infections that might otherwise be handled by a healthy immune system.

10.6 Population and Community Structure

Due to differential susceptibilities of taxa and populations to thermal stress (Jokiel and Coles 1990; Kayanne et al. 2002), changes in abundance almost always lead to changes in coral assemblages, including reductions in species richness (Loya et al. 2001; McClanahan and Maina 2003), diversity (Warwick et al. 1990), changing community composition (McClanahan et al. 2007b; Weil et al. 2009a; Kuo et al. 2012; Harii et al. 2014), and the dominant life history traits of the coral community (Darling et al. 2013). Some of these changes are short term (Brown 1997; McClanahan and Maina 2003), whereas others appear to be very persistent (Glynn 1994; Berumen and Pratchett 2006; McClanahan 2014). Population changes have community-level effects, such as influencing competitive interactions mediated by the different temperature responses of corals (Alino et al. 1992). Shifting species composition is expected to lead to large-scale changes in reef communities as the pace of climate change accelerates (Langmead and Sheppard 2004; van Woesik et al. 2012; Edmunds et al. 2014; McClanahan et al. 2014a).

Some of the early studies of intense bleaching in the eastern Pacific during the 1982–1983 El Niño showed large changes in the abundance of many coral populations (Glynn et al. 2001), the local or regional extinction of species (Glynn and Ault 2000; Maté 2003), and the devastation and disappearance of whole reef structures (Eakin 2001). In other cases, large switches in coral dominance were evident; for example, in Cocos Island, Panama, *Pavona clavus* became common after bleaching, and a species never recorded prior to the 2002 event, *Leptoseris scabra*, was observed (Guzmán and Cortés 2007). These early observations were alarming because they signaled the potential for major changes in reefs that seem to follow historical periods of increasing El Niño intensity events in this region (Toth et al. 2012). Over the long term, however, the main reef builders (i.e., poritids, pocilloporids) continue to be the dominant species at the study sites (Guzmán and Cortés 2001).

Similar patterns of change and dominance after bleaching disturbances have been observed in the Indo-Pacific and Caribbean (McClanahan 2008; Edmunds et al. 2014). Studies in the Indo-Pacific have shown considerable coral losses, but also recovery and responses have been variable (Ateweberhan et al. 2011; Selig et al. 2012; McClanahan et al. 2014b). For example, on the reef flats of islands in Pulau Seribu, Indonesia, up to 92% of the coral cover and most of the *Acropora* were lost following a 2–3 °C rise in temperatures in 1982 (Brown and Suharsono 1990). Similarly, the reefs of Sesoko Island, Japan, experienced a 61% reduction in species richness and an 85% reduction in coral cover following the 1998 bleaching (Loya et al. 2001). In contrast, recurrent bleaching of corals on intertidal Ko Phuket, Thailand, reefs had no discernable effect on coral community measures, such as species richness or diversity over reefs, when studied between 1991 and 1998. The lack of response is attributed to the dominance of massive corals, with other taxa acclimated to the extremes of intertidal living, and partial rather than whole-colony mortality (Brown et al. 2002).

The largest effects of the 1998 bleaching event were documented in the western Indian Ocean (Ateweberhan et al. 2011; McClanahan et al. 2015). Many high coral cover reefs, such as Maldives, Chagos, and Seychelles, were dominated by *Acropora* and *Montipora* before 1998, which was reduced to ~10% after 1998 (McClanahan 2000; Sheppard et al. 2002; Graham et al. 2006). In some places, such as the Chagos, Cocos Islands, and Tanzania, recovery of *Acropora* and other branching taxa was rapid (Sheppard et al. 2002; Ateweberhan et al. 2011; Gilmour et al. 2013). At other locations, the relative dominance of bleaching-resistant taxa, such as massive *Porites* or opportunistic taxa such as *Pocillopora*, increased and largely replaced *Acropora* and *Montipora* (McClanahan et al. 2007b; Darling et al. 2013). Large-scale surveys of the western Indian Ocean found that the relative covers of *Acropora* and *Montipora* were negatively associated with degree heating weeks in 1998 and declined over time (McClanahan et al. 2007b, 2014b). Some areas, such as the granitic islands of the Seychelles, have maintained low cover of all corals and erect algae since 1998 (Graham et al. 2006). Yet, recovery has occurred in reefs in deeper water, with high structural complexity, and when juvenile corals and herbivorous fish numbers were high and nutrient loads low (Graham et al. 2015). Maldives reefs were dominated by turf and coralline algae after 1998, and recovery has been quite variable and poorly understood (Tkachenko 2012; McClanahan and Muthiga 2014; Morri et al. 2015).

10.7 Changes Over Time

There are many studies on the immediate response of coral assemblages to thermal stress but fewer on the slower changes over time. While massive losses of coral cover have been well documented for several localities in the Caribbean (Gardner et al. 2003; Miller et al. 2009; Weil et al. 2009a; Jackson et al. 2014; Mumby et al. 2014), information from only a few sites have been published in sufficient detail to understand community change. The compilation by Edmunds et al. (2014) of mostly unpublished time series studies of common taxa is an exception. They found that of 16 genera with sufficient data, 11 genera declined while 5 increased since the early 1980s. In Puerto Rico, after the high thermal stress and bleaching of 2005, populations of *Millepora squarrosa* and *M. complanata* suffered high mortality or disappeared from local reefs. Shallow-water habitats were dominated first by algal lawns and then monopolized by zoanthids, sponges, and weedy corals, such as *Porites astreoides* and *A. agaricites* (Weil et al. 2009b). Other reported changes were associated with a variety of factors including high thermal anomalies, diseases, coastal development, and losses of herbivores. These include basin-wide loss of *Acropora* species (Greenstein et al. 1998) and localized losses of *Mycetophyllia* spp., *Dichocoenia stokesi*, *Eusmilia fastigiata*, *Meandrina jacksoni*, *Colpophyllia natans*, and *Pseudodiploria strigosa* in Florida (Dustan 1977; Richardson and Voss 2005; Precht et al. 2016); *Agaricia* in Belize (Aronson et al. 2002a); and *Orbicella* spp. in the US Virgin Islands, Puerto Rico, and Venezuela (Miller et al. 2009; Bruckner and Hill 2009; Weil et al. 2009a; Bastidas et al. 2012).

Not all demises of coral reef species are always clearly linked to thermal stress and bleaching. Shorter-term studies comparing the past few millennia have concluded that most of the change in Caribbean coral assemblages occurred very recently (Aronson et al. 2002b, 2004) but may be locally associated with diseases, fishing, and the demise of parrotfish (Jackson et al. 2014; Cramer et al. 2017). Additionally, resistance or susceptibility to high temperatures and temperature-induced diseases could result in competitive interactions between hard corals and other sessile taxa. While soft corals bleach and die from thermal stress and diseases, the overall lower effects compared to hard corals may still cause shifts from hard to soft corals and algae-dominated reefs in the near future (Ruzicka et al. 2013).

10.8 Changes in Ecological Processes

Potential effects of coral bleaching on ecosystems include a decrease in net rates of calcium carbonate accretion and changes in primary productivity. Studies of calcification rates suggest that once coral cover declines below 10%, net calcification falls below zero and reefs stop growing and keeping up with sea-level rise (Perry et al. 2013, 2015). Loss of reef growth is most likely when dominant fast-growing, thermally sensitive taxa, such as *Acropora*, decline. Increased organic carbon production is expected to follow the loss of inorganic carbon, but this change will occur at the costs of losing animal refuges and reef growth (Sebastián and McClanahan 2013a). The effect may, however, not be immediate as one study from Ishigaki, Japan, found that during a bleaching year, excess organic production was reduced by 75% when compared with a non-bleaching year (Kayanne et al. 2005). Given that coral cover changed a little, the study suggested that the change was largely due to reduced productivity of corals, possibly associated with reduced symbiont densities. Epilithic turf and encrusting coralline algae will increase after coral mortality and compensate for the loss of some coral productivity and calcification, but these taxa lack the structural complexity that corals provide (McClanahan 2008; Alvarez-Filip et al. 2011). Additionally, the recovery rate of coralline algae is considerably slower than epilithic turf algae. Consequently, after bleaching, there will be an early peak in organic production followed by a slower recovery of calcifying algae and calcification after coral losses (Sebastián and McClanahan 2013a). The full consequences of carbon and other nutrient influences associated with bleaching remain key areas for future investigations.

The coming decades will be a time of unprecedented change and reorganization of the sessile organisms on coral reefs with consequences for the ecosystem processes, including fisheries production (McClanahan 2002, 2008; Chap. 11). Bleaching may have some positive impacts on herbivores and macro-invertivores as benthic space is opened up for their feeding while having negative effects on other small-bodied species and feeding groups that rely on coral for food and refuge in the absence of heavy fishing (Graham et al. 2008, 2011; McClanahan et al. 2014a).

The bleaching influences of changing production of organic and inorganic carbon on fisheries yields are not well understood but initially may be different from the long-term responses and depend greatly on fishing pressure and selectivity (Graham et al. 2007; McClanahan and Abunge 2014). Empirical studies of fishing are few, but simulation models indicate that responses will depend on fishing pressure, catch selection, diets of the fish, and region-specific food web configurations (Sebastián and McClanahan 2013a; Bozec et al. 2016).

10.9 Interactions with Fisheries Management

Studies of fisheries closures suggest many have been preferentially established in areas with low-temperature variability that makes them susceptible to temperature anomalies, bleaching, and death after strong thermal stress events (Selig et al. 2012). Consequently, reefs affected by these closures should have lower resistance to warm temperatures; but, because recovery is an important and independent part of overall resilience, they could still be more resilient than fished reefs (McClanahan et al. 2012). Empirical evidence suggests that corals do recover faster in unfished than fished reefs in terms of coral cover but not necessarily when the metric is community composition (McClanahan 2008; Mumby and Harborne 2010; Ateweberhan et al. 2011; Darling et al. 2013). Consequently, the long-term effects of bleaching on sessile organisms may be influenced by the state of the fish and fishery, but most evidence indicates that the geographic context is very important (Ateweberhan et al. 2011).

After the 1998 coral mortality event in the western Indian Ocean, many reefs began changing based largely on geography and weakly by fisheries management (McClanahan et al. 2014b). Most reefs were quickly colonized by rapidly growing turf algae, which was eventually colonized by slower-growing coralline and erect algae (McClanahan et al. 2001; Graham et al. 2006; McClanahan 2008). Coralline algae were more prevalent in areas with high fish grazing which should increase the survival of coral recruits (O'Leary et al. 2013). However, experimental studies found that after bleaching fishes focused their predation on coral recruits growing on coralline algae (McClanahan et al. 2005b). Additionally, coral taxa changed the most in fisheries closures where weedy and stress-resistant taxa fared better than competitive fast-growing taxa such as *Acropora* (Darling et al. 2013). *Acropora* appeared to do less well in fished than unfished reefs possibly due to lower coralline algae cover required for recruitment (O'Leary et al. 2012, 2013). Additionally, a disease outbreak in 2002 swept through the region and killed many susceptible corals irrespective of the management system (McClanahan et al. 2004b). Eventually, even stress-resistant taxa began to display stress and declines after 1998, possibly associated with repeated stressful warm temperatures in 2010 and 2013 (McClanahan 2014, 2017). This generally led to declining coral cover after 2007 (Fig. 10.6). This contrasts with some remote areas where fishing levels were very low and corals were reported to recover at historically expected rates and to original composition (Sheppard et al. 2008; Gilmour et al. 2013). Consequently, there may be other factors, like coastal versus oceanic or wilderness versus small parks in fished seascapes that are poorly understood but may be influencing recovery dynamics (Humphries et al. 2014).

Reducing fishing pressure on key functional groups, such as herbivorous fishes, is commonly recommended as a way to increase reef resilience and recovery rates (Bellwood et al. 2004; Mumby et al. 2006; Hughes et al. 2007; Bozec et al. 2016). No-take areas generally have higher levels of herbivory by fish that can reduce erect algal cover and increase coral recruitment and recovery in some locations (Mumby et al. 2007; Ateweberhan et al. 2011). Yet, some scraping and excavating parrotfish and other invertebrate-feeders can eat coral and may increase their focus on the remaining corals, often small recruits, after coral mortality (McClanahan et al. 2005b). Additionally, there can be other food web effects that are not considered, such as the roles of sea urchin grazing on corals in heavily fished reefs due to a loss of sea urchin predators (McClanahan and Muthiga 2016a). Sea urchins are often associated with some negative impacts on coral cover and some suppression of the recovery of corals that may be associated with their destructive grazing on coral recruits (O'Leary et al. 2013). From a very large survey on Indian Ocean reefs, McClanahan and Muthiga (2016b) concluded that the negative effects of sea urchins grazing on reef calcifiers in many fished reefs were on the same order as stressful temperature anomalies.

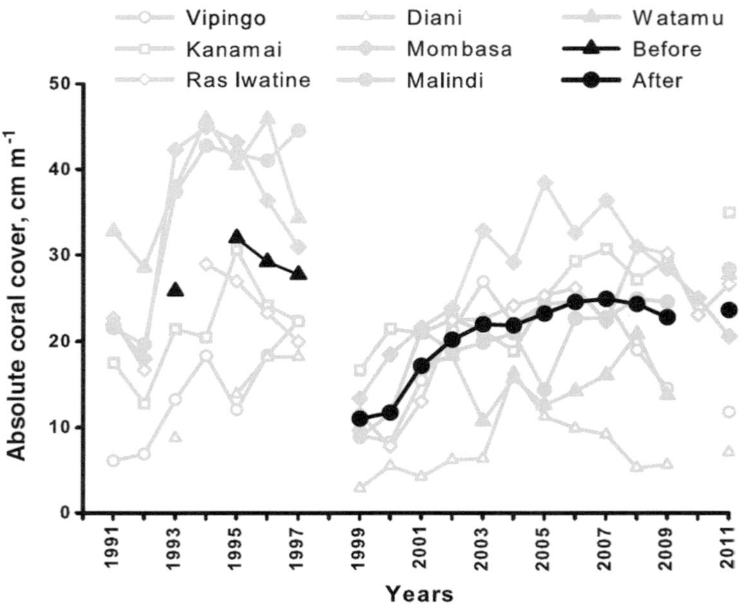

Fig. 10.6 Change in coral cover across the 1998 bleaching event and recovery on Kenyan reefs with different management. Four fished reefs (open circles), three old parks (closed circles), and one new park (diamond) were monitored. The old parks were closed in 1968 (Malindi), 1972 (Watamu), and 1973 (Kisite). The new park at Mombasa was closed in 1991. Vipingo, Kanamai, Ras Iwatine, and Diani were the fished reefs

A simulation model calibrated for East African reefs to study fishing impacts found that coral recovery after bleaching was slower when predators of sea urchins were fished using hook and line (Sebastián and McClanahan 2013b). Thus, fishing other fish groups including herbivores using nets and traps was a better option than line fishing that focused on predators of sea urchins. These recommendations are contrary to those of another simulation model calibrated for the Caribbean that suggested banning traps that capture parrotfish (Bozec et al. 2016). The protect-herbivore recommendations arise from Caribbean studies where sea urchin grazers are uncommon and parrotfish are among the main herbivores (Bellwood et al. 2004; Mumby and Harborne 2010). Consequently, these food web differences indicate that fisheries management impacts and recommendations may be context dependent. Models that promote specific gear bans also need to consider trade-offs in terms of yields and reef ecological services arising from using alternative fishing gears (McClanahan 1995; McClanahan and Cinner 2008). To emphasize the importance of context, some studies indicate that the presence of mangroves may influence the abundance, types, and diversity of herbivores, which may have consequences for coral recovery (Mumby et al. 2004; Olds et al. 2014).

Stopping fishing is not a solution to the global warming crisis for many coral reefs (Toth et al. 2012), but it does help with recovery processes in some situations

(Ateweberhan et al. 2011; McClanahan et al. 2012; Olds et al. 2014; Mellin et al. 2016). The specifics and mechanisms require more research as well as how protected areas should be planned and what they incorporate in their boundaries. One suggestion has been to not just continue establishing protected areas in high diversity reefs with low temperature variability that makes them susceptible to strong temperature deviations (Maina et al. 2008; Selig et al. 2012). Rather, protected areas should include portfolios of environments, including variable temperature and exposure environments as well as habitats such as shallow- and deep-water reefs and the inclusion of seagrass and mangrove ecosystems (Olds et al. 2014; McClanahan et al. 2015). In general, protected area designs that increase the between-site diversity are likely to play important roles in increasing stability by providing a diverse seascape where many habitats, species, and traits are present and contribute functional roles after disturbances (Mellin et al. 2014; Maina et al. 2015).

10.10 Conclusions

Reports of coral reef bleaching have been increasing in the past few decades, overtaking most other human impacts as the most significant threat that interacts with diseases to threaten reef resilience (Hughes et al. 2017; Weil and Rogers 2011; Weil et al. 2017). Reports of community-wide bleaching are more frequent during El Niño, high thermal anomalies, and other strong oceanographic events such as the Indian Ocean Dipole (Eakin et al. 2010; McClanahan 2017; Chaps. 3 and 4). Aerosols (Gill et al. 2006), local water quality (Riegl and Piller 2003; McClanahan et al. 2007c; Hughes et al. 2017), hurricanes (Manzello et al. 2007), and other large-scale oceanographic processes, such as downwelling, upwelling, and oceanographic cycles, influence the intensity and extent of bleaching (McClanahan et al. 2007a, c). These factors are all variations on the main problem, the trend in rising ocean temperature that is expected to challenge the adaptive potential of coral reefs in the coming decades (Hughes et al. 2003; Hoegh-Guldberg et al. 2007; McClanahan et al. 2012; Weil et al. 2017). Research to date suggests dire consequences for shallow-water marine sessile organisms, but the considerable patchiness in taxa, space, and time responses and the evolutionary history of cnidarians, especially scleractinians, indicate that refuges from climate change might exist and adaptation to climate change is possible. Yet, as the rate and impacts of events expand in geographic space and intensity, even what have been refuges in the past may lose this capacity in the future. Reducing carbon emissions that are causing global warming is the only truly long-term global solution for coral reefs.

References

Ainsworth TD, Heron SF, Ortiz JC, Mumby PJ, Grech A, Ogawa D et al (2016) Climate change disables coral bleaching protection on the Great Barrier Reef. Science 352:338–342. https://doi.org/10.1126/science.aac7125

Alino PM, Sammarco PW, Coll JC (1992) Competitive strategies in soft corals (Coelenterata, Octocorallia). IV. Environmentally induced reversals in competitive superiority. Mar Ecol Prog Ser 81:129–145. https://doi.org/10.3354/meps081129

Alvarez-Filip L, Gill J, Dulvy N, Perry A, Watkinson A, Cote I (2011) Drivers of region-wide declines in architectural complexity on Caribbean reefs. Coral Reefs 30:1051–1060. https://doi.org/10.1007/s00338-011-0795-6

Aronson RB, Macintyre IG, Precht WF, Murdoch TJT, Wapnick CM (2002a) The expanding scale of species turnover events on coral reefs in Belize. Ecol Monogr 72:233–249. https://doi.org/10.1890/0012-9615(2002)072[0233:TESOST]2.0.CO;2

Aronson RB, Precht W, Toscano M, Koltes KH (2002b) The 1998 bleaching event and its aftermath on a coral reef in Belize. Mar Biol 141:435–447. https://doi.org/10.1007/s00227-002-0842-5

Aronson RB, Macintyre IG, Wapnick CM, O'Neill MW (2004) Phase shifts, alternative states, and the unprecedented convergence of two reef systems. Ecology 85:1876–1891. https://doi.org/10.1890/03-0108

Ateweberhan M, McClanahan TR (2010) Relationship between historical sea-surface temperature variability and climate change-induced coral mortality in the Western Indian Ocean. Mar Poll Bull 60:964–970. https://doi.org/10.1016/j.marpolbul.2010.03.033

Ateweberhan M, McClanahan TR, Graham NAJ, Sheppard CRC (2011) Episodic heterogeneous decline and recovery of coral cover in the Indian Ocean. Coral Reefs 30:739–752. https://doi.org/10.1007/s00338-011-0775-x

Babcock RC, Baird AH, Piromvaragorn S, Thomson DP, Willis BL (2003) Identification of scleractinian coral recruits from Indo-Pacific reefs. Zool Stud 42:211–226

Baird AH (2000) Microbial menace: coral disease on the GBR. (Exploring reef science). CRC Reef Research Centre, Townsville, Australia

Baird AH, Marshall PA (2002) Mortality, growth and reproduction in scleractinian corals following bleaching on the Great Barrier Reef. Mar Ecol Prog Ser 237:133–141. https://doi.org/10.3354/meps237133

Baird AH, Cumbo VR, Leggat W, Rodriguez-Lanetty M (2007) Fidelity and flexibility in coral symbioses. Mar Ecol Prog Ser 347:307–309. https://doi.org/10.3354/meps07220

Baird AH, Bhagooli R, Ralph PJ, Takahashi S (2009) Coral bleaching: the role of the host. Trends Ecol Evol 24:16–20. https://doi.org/10.1016/j.tree.2008.09.005

Bak RPM, Meesters EH (1999) Population structure as a response of coral communities to global change. Integr Comp Biol 39:56–65. https://doi.org/10.1093/icb/39.1.56

Baker AC (2003) Flexibility and specificity in coral-algal symbiosis: diversity, ecology and biogeography of *Symbiodinium*. Annu Rev Ecol Syst 34:661–689. https://doi.org/10.1146/annurev.ecolsys.34.011802.132417

Baker AC, Starger CJ, McClanahan TR, Glynn PW (2004) Corals' adaptive response to climate change. Nature 430:741. https://doi.org/10.1038/430741a

Baskett ML, Nisbet RM, Kappel CV, Mumby PJ, Gaines SD (2010) Conservation management approaches to protecting the capacity for corals to respond to climate change: a theoretical comparison. Glob Chang Biol 16:1229–1246. https://doi.org/10.1111/j.1365-2486.2009.02062.x

Bastidas C, Bone D, Cróquer A, Debrot D, Garcia E, Humanes A, Ramos R, Rodríguez S (2012) Massive hard coral loss after a severe bleaching event in 2010 at Los Roques, Venezuela. Rev Biol Trop 60:29–37

Bauman AG, Pratchett MS, Baird AH, Riegl B, Heron SF, Feary DA (2013) Variation in the size structure of corals is related to environmental extremes in the Persian Gulf. Mar Environ Res 84:43–50. https://doi.org/10.1016/j.marenvres.2012.11.007

Bellwood DR, Hughes TP, Folke C, Nystrom M (2004) Confronting the coral reef crisis. Nature 429:827–832. https://doi.org/10.1038/nature02691

Bena C, van Woesik R (2004) The impact of two bleaching events on the survival of small coral colonies (Okinawa, Japan). Bull Mar Sci 75:115–126

Berkelmans R, De'ath G, Kininmonth S, Skirving WJ (2004) A comparison of the 1998 and 2002 coral bleaching events on the Great Barrier Reef: spatial correlation, patterns, and predictions. Coral Reefs 23:74–83. https://doi.org/10.1007/s00338-003-0353-y

Berumen ML, Pratchett MS (2006) Recovery without resilience: persistent disturbance and long-term shifts in the structure of fish and coral communities at Tiahura Reef, Moorea. Coral Reefs 25:647–653. https://doi.org/10.1007/s00338-006-0145-2

Bozec Y-M, O'Farrell S, Bruggemann JH, Luckhurst BE, Mumby PJ (2016) Tradeoffs between fisheries harvest and the resilience of coral reefs. Proc Natl Acad Sci U S A 113:4536–4541. https://doi.org/10.1073/pnas.1601529113

Bridge TC, Ferrari R, Bryson M, Hovey R, Figueira WF, Williams SB et al (2014) Variable responses of benthic communities to anomalously warm sea temperatures on a high-latitude coral reef. PLoS One 9. https://doi.org/10.1371/journal.pone.0113079

Brown BE, Suharsono (1990) Damage and recovery of coral reefs affected by El Nino related seawater warming in the Thousand Islands, Indonesia. Coral Reefs 8:163–170. https://doi.org/10.1007/BF00265007

Brown BE (1997) Coral bleaching: causes and consequences. Coral Reefs 16:S129–S138. https://doi.org/10.1007/s003380050249

Brown BE, Dunne RP, Chansang H (1996) Coral bleaching relative to elevated seawater temperature in the Andaman Sea (Indian Ocean) over the last 50 years. Coral Reefs 15:151–152. https://doi.org/10.1007/BF01145885

Brown BE, Dunne RP, Goodson MS, Douglas AE (2000) Bleaching patterns in reef corals. Nature 404:142–143. https://doi.org/10.1038/35004657

Brown BE, Dunne RP, Goodson MS, Douglas AE (2002) Experience shapes the susceptibility of a reef coral to bleaching. Coral Reefs 21:119–126

Bruckner AW, Hill RL (2009) Ten years of change to coral communities off Mona and Desecheo Islands, Puerto Rico, from disease and bleaching. Dis Aquat Organ 87:19–31. https://doi.org/10.3354/dao02120

Bruno JF, Selig ER, Casey KS, Page CA, Willis BL, Harvell CD et al (2007) Thermal stress and coral cover as drivers of coral disease outbreaks. PLoS Biol 5. https://doi.org/10.1371/journal.pbio.0050124

Buddemeier RW, Baker AC, Fautin DG, Jacobs JR (2004) The adaptive hypothesis of bleaching. In: Rosenberg E, Loya Y (eds) Coral health and disease. Springer, Berlin, pp 427–444

Burge CA, Kim CJ, Lyles JM, Harvell CD (2013) Special issue oceans and humans health: th e ecology of marine opportunists. Microb Ecol 65:869–879. https://doi.org/10.1007/s00248-013-0190-7

Burge CA, Eakin CM, Friedman CS, Froelich B, Hershberger PK, Hofmann EE et al (2014) Climate change influences on marine infectious diseases: Implications for management and society. Ann Rev Mar Sci 6:249–277. https://doi.org/10.1146/annurev-marine-010213-135029

Bythell JC (1988) A total nitrogen and carbon budget for the elkhorn coral *Acropora palmata* (Lamarck). In: Proceedings of the 6th international coral reef symposim, Townsville, vol 2, pp 535–540

Couch CS, Mydlarz LD, Harvell CD, Douglas NL (2008) Variation in measures of immunocompetence of sea fan coral, *Gorgonia ventalina*, in the Florida Keys. Mar Biol 155:281–292. https://doi.org/10.1007/s00227-008-1024-x

Cox EF (2007) Continuation of sexual reproduction in *Montipora capitata* following bleaching. Coral Reefs 26:721–724. https://doi.org/10.1007/s00338-007-0251-9

Cramer KL, O'Dea A, Clark TR, Zhao J, Norris RD (2017) Prehistorical and historical declines in Caribbean coral reef accretion rates driven by loss of parrotfish. Nat Commun 8. https://doi.org/10.1038/ncomms14160

Cróquer A, Weil E (2009a) Changes in Caribbean coral disease prevalence after the 2005 bleaching event. Dis Aquat Organ 87:33–43. https://doi.org/10.3354/dao02164

Cróquer A, Weil E (2009b) Spatial variability in distribution and prevalence of coral and octocoral diseases in the Caribbean II: genera-level analysis. Dis Aquat Organ 83:209–222. https://doi.org/10.3354/dao02012

Darling ES, McClanahan TR, Cote IM (2013) Life histories predict coral community disassembly under multiple stressors. Glob Chang Biol 19:1930–1940. https://doi.org/10.1111/gcb.12191

Donner SD, Rickbeil GJ, Heron SF (2017) A new, high-resolution global mass coral bleaching database. PLoS One 12. https://doi.org/10.1371/journal.pone.0175490

Dustan P (1977) Vitality of reef coral populations off Key Largo, Florida: recruitment and mortality. Environ Geol 2:51–58. https://doi.org/10.1007/BF02430665

Eakin CM (2001) A tale of two ENSO events: carbonate budgets and the influence of two warming disturbances and intervening variability, UVA Island, Panama. Bull Mar Sci 69:171–186

Eakin CM, Morgan JA, Heron SF, Smith TB, Liu G, Alvarez-Filip L et al (2010) Caribbean corals in crisis: record thermal stress, bleaching, and mortality in 2005. PLoS One 5. https://doi.org/10.1371/journal.pone.0013969

Edmunds PJ, Adjeroud M, Baskett ML, Baums IB, Budd AF, Carpenter RC et al (2014) Persistence and change in community composition of reef corals through present, past, and future climates. PLoS One 9. https://doi.org/10.1371/journal.pone.0107525

Enríquez S, Méndez ER, Iglesias-Prieto R (2005) Multiple scattering on coral skeletons enhances light absorption by symbiotic algae. Limnol Oceanogr 50:1025–1032. https://doi.org/10.4319/lo.2005.50.4.1025

Finelli CM, Helmuth BS, Pentcheff ND, Wethey DS (2006) Water flow influences oxygen transport and photosynthetic efficiency in corals. Coral Reefs 24:47–57. https://doi.org/10.1007/s00338-005-0055-8

Frieler K, Meinshausen M, Golly A, Mengel M, Lebek K, Donner SD et al (2013) Limiting global warming to 2°C is unlikely to save most coral reefs. Nat Clim Chang 3:165–170. https://doi.org/10.1038/nclimate1674

Gardner TA, Côté IM, Gill JA, Grant A, Watkinson AR (2003) Long-term region-wide declines in Caribbean corals. Science 301:958–960. https://doi.org/10.1126/science.1086050

Gill JA, Watkinson AR, McWilliams JP, Côté IM (2006) Opposing forces of aerosol cooling and El Niño drive coral bleaching on Caribbean reefs. Proc Natl Acad Sci U S A 103:18870–18873. https://doi.org/10.1073/pnas.0608470103

Gilmour JP, Smith LD, Heyward AJ, Baird AH, Pratchett MS (2013) Recovery of an isolated coral reef system following severe disturbance. Science 340:69–71. https://doi.org/10.1126/science.1232310

Gladfelter WB (1982) White-band disease in *Acropora palmata*: implications for the structure and growth of shallow reefs. Bull Mar Sci 32:639–645

Glynn PW (1984) Widespread coral mortality and the 1982/83 El Niño warming event. Environ Conserv 11:133–146. https://doi.org/10.1017/s0376892900013825

Glynn PW (1993) Coral reef bleaching: ecological perspectives. Coral Reefs 12:1–17. https://doi.org/10.1007/BF00303779

Glynn PW (1994) State of coral reefs in the Galapagos Islands: natural vs anthropogenic impacts. Mar Poll Bull 29:131–140. https://doi.org/10.1016/0025-326X(94)90437-5

Glynn PW, Ault JS (2000) A biogeographic analysis and review of the far eastern Pacific coral reef region. Coral Reefs 19:1–23. https://doi.org/10.1007/s003380050220

Glynn PW, Gassman NJ, Eakin CM, Cortés J, Smith DB, Guzman HM (1991) Reef coral reproduction in the eastern pacific: Costa Rica, Panama, and Galapagos Island (Ecuador). Mar Biol 109:355–368. https://doi.org/10.1007/BF01313501

Glynn PW, Colley SB, Gassman NJ, Black K, Cortés J, Maté JL (1996) Reef coral reproduction in the eastern Pacific: Costa Rica, Panama, and Galápagos Islands (Ecuador). III. Agariciidae (*Pavona gigantea* and *Gardineroseris planulata*). Mar Biol 125:579–601. https://doi.org/10.1007/BF00353270

Glynn PW, Colley SB, Ting JH, Maté JL, Guzmán HM (2000) Reef coral reproduction in the eastern Pacific: Costa Rica, Panama and Galapagos Islands (Ecuador). IV. Agariciidae, recruitment and recovery of *Pavona varians* and *Pavona* sp.a. Mar Biol 136:785–805. https://doi.org/10.1007/s002270000286

Glynn PW, Maté JL, Baker AC, Calderon MO (2001) Coral bleaching and mortality in Panama and Ecuador during the 1997-1998 El Niño-Southern Oscillation event: spatial/temporal patterns and comparisons with the 1982-1983 event. Bull Mar Sci 69:79–109

Graham NAJ, Wilson SK, Jennings S, Polunin NVC, Bijoux JP, Robinson J (2006) Dynamic fragility of oceanic coral reef ecosystems. Proc Natl Acad Sci U S A 103:8425–8429. https://doi.org/10.1073/pnas.0600693103

Graham NAJ, Wilson SK, Jennings S, Polunin NV, Robinson J, Bijoux JP et al (2007) Lag effects in the impacts of the 1998 bleaching event on coral reef fisheries in the Seychelles. Conserv Biol 21:1291–1300. https://doi.org/10.1111/j.1523-1739.2007.00754.x

Graham NAJ, McClanahan TR, MacNeil MA, Wilson SK, Polunin NV, Jennings S et al (2008) Climate warming, marine protected areas and the ocean-scale integrity of coral reef ecosystems. PLoS One 3. https://doi.org/10.1371/journal.pone.0003039

Graham NAJ, Chabanet P, Evans RD, Jennings S, Letourneur Y, MacNeil MA et al (2011) Extinction vulnerability of coral reef fishes. Ecol Lett 14:341–348. https://doi.org/10.1111/j.1461-0248.2011.01592.x

Graham NAJ, Jennings S, MacNeil MA, Mouillot D, Wilson SK (2015) Predicting climate-driven regime shifts versus rebound potential in coral reefs. Nature 518:94–97. https://doi.org/10.1038/nature14140

Greenstein BJ, Curran HA, Pandolfi JM (1998) Shifting ecological baselines and the demise of *Acropora cervicornis* in the western North Atlantic and Caribbean Province: a Pleistocene perspective. Coral Reefs 17:249–261. https://doi.org/10.1007/s003380050125

Guest JR, Baird AH, Maynard JA, Muttaqin E, Edwards AJ, Campbell SJ et al (2012) Contrasting patterns of coral bleaching susceptibility in 2010 suggest an adaptive response to thermal stress. PLoS One 7. https://doi.org/10.1371/journal.pone.0033353

Guzmán HM, Cortés J (2001) Changes in reef community structure after fifteen years of natural disturbances in the eastern Pacific (Costa Rica). Bull Mar Sci 69:133–149

Guzmán HM, Cortés J (2007) Reef recovery 20 years after the 1982-1983 El Niño massive mortality. Mar Biol 151:401–411. https://doi.org/10.1007/s00227-006-0495-x

Harii S, Hongo C, Ishihara M, Ide Y, Kayanne H (2014) Impacts of multiple disturbances on coral communities at Ishigaki Island, Okinawa, Japan, during a 15 year survey. Mar Ecol Prog Ser 509:171–180. https://doi.org/10.3354/meps10890

Harvell CD, Kim K, Burkholder JM, Colwell RR, Epstein PR, Grimes DJ et al (1999) Emerging marine diseases – climate links and anthropogenic factors. Science 285:1505–1510. https://doi.org/10.1126/science.285.5433.1505

Harvell D, Kim K, Quirolo C, Weir J, Smith G (2001) Coral bleaching and disease: contributors to 1998 mass mortality in *Briareum asbestinum* (Octocorallia, Gorgonacea). Hydrobiologia 460:97–104. https://doi.org/10.1023/A:101316933

Harvell D, Altizer S, Cattadori IM, Harrington L, Weil E (2009) Climate change and wildlife disease: when does the host matter the most? Ecology 90:912–920. https://doi.org/10.1890/08-0616.1

Heron SF, Johnston L, Liu G, Geiger EF, Maynard JA, De La Cour JL et al (2016) Validation of reef-scale thermal stress satellite products for coral bleaching monitoring. Remote Sens (Basel) 8. https://doi.org/10.3390/rs8010059

Hoegh-Guldberg O, Mumby PJ, Hooten AJ, Steneck RS, Greenfield P, Gomez E et al (2007) Coral reefs under rapid climate change and ocean acidification. Science 318:1737–1742. https://doi.org/10.1126/science.1152509

Hughes TP, Baird AH, Bellwood DR, Card M, Connolly SR, Folke C et al (2003) Climate change, human impacts, and the resilience of coral reefs. Science 301:929–933. https://doi.org/10.1126/science.1085046

Hughes TP, Rodrigues MJ, Bellwood DR, Ceccarelli D, Hoegh-Guldberg O, McCook L et al (2007) Phase shifts, herbivory, and the resilience of coral reefs to climate change. Curr Biol 17:360–365. https://doi.org/10.1016/j.cub.2006.12.049

Hughes TP, Kerry JT, Álvarez-Noriega M, Álvarez-Romero JG, Anderson KD, Baird AH et al (2017) Global warming and recurrent mass bleaching of corals. Nature 543:373–377. https://doi.org/10.1038/nature21707

Humphries AT, McClanahan TR, McQuaid CD (2014) Differential impacts of coral reef herbivores on algal succession in Kenya. Mar Ecol Prog Ser 504:119–132. https://doi.org/10.3354/meps10744

Jackson JBC, Donovan MK, Cramer KL, Lam V (2014) Status and Trends of Caribbean Coral Reefs: 1970-2012. IUCN, Gland, Switzerland

Jokiel PL, Coles SL (1990) Response of Hawaiian and other Indo-Pacific reef corals to elevated temperature. Coral Reefs 8:155–162. https://doi.org/10.1007/BF00265006

Kayanne H, Harii S, Ide Y, Akimoto F (2002) Recovery of coral populations after the 1998 bleaching on Shiraho Reef, in the southern Ryukyus, NW Pacific. Mar Ecol Prog Ser 239:93–103. https://doi.org/10.3354/meps23903

Kayanne H, Hata H, Kudo S, Yamano H, Watanabe A, Ikeda Y et al (2005) Seasonal and bleaching-induced changes in coral reef metabolism and CO_2 flux. Glob Biogeochem Cycles 19. https://doi.org/10.1029/2004GB002400

Kim K (2015) Diseases in Octocorals. In: Diseases of Coral. Wiley, Hoboken, NJ, pp 410–415

Kuo CY, Yuen YS, Meng PJ, Ho PH, Wang JT, Liu PJ et al (2012) Recurrent disturbances and the degradation of hard coral communities in Taiwan. PLoS One 7. https://doi.org/10.1371/journal.pone.0044364

Langmead O, Sheppard C (2004) Coral reef community dynamics and disturbance: a simulation model. Ecol Model 175:271–290. https://doi.org/10.1016/j.ecolmodel.2003.10.019

Lasker HR, Peters EC, Coffroth MA (1984) Bleaching of reef coelenterates in the San Blas Islands, Panama. Coral Reefs 3:183–190. https://doi.org/10.1007/BF00288253

Lesser MP, Bythell JC, Gates RD, Johnstone RW, Hoegh-Guldberg O (2007) Are infectious diseases really killing corals? Alternative interpretations of the experimental and ecological data. J Exp Mar Biol Ecol 346:36–44. https://doi.org/10.1016/j.jembe.2007.02.015

Loch K, Loch W, Schuhmacher H, See WR (2004) Coral recruitment and regeneration on a Maldivian reef four years after the coral bleaching event of 1998. Part 2: 2001-2002. Mar Ecol 25:145–154. https://doi.org/10.1111/j.1439-0485-2004.00021.x

Locke JM, Weil E, Coates KA (2007) A newly documented species of *Madracis* (Scleractinia: Pocilloporidae) from the Caribbean. Proc Biol Soc Wash 120:214–226. https://doi.org/10.2988/0006-324X(2007)120[214.ANDSOM]2.0.CO;2

Logan CA, Dunne JP, Eakin CM, Donner SD (2014) Incorporating adaptive responses into future projections of coral bleaching. Glob Chang Biol 20:125–139. https://doi.org/10.1111/gcb.12390

Loya Y, Sakai K, Yamazato K, Nakano Y, Sambali H, van Woesik R (2001) Coral bleaching: the winners and losers. Ecol Lett 4:122–131. https://doi.org/10.1046/j.1461-0248.2001.00203.x

Madin JS, Baird AH, Dornelas M, Connolly SR (2014) Mechanical vulnerability explains size-dependent mortality of reef corals. Ecol Lett 17:1008–1015. https://doi.org/10.1111/ele.12306

Maina J, Venus V, McClanahan TR, Ateweberhan M (2008) Modelling susceptibility of coral reefs to environmental stress using remote sensing data and GIS models in the western Indian Ocean. Ecol Model 212:180–199. https://doi.org/10.1016/j.ecolmodel.2007.10.033

Maina JM, Jones KR, Hicks CC, McClanahan TR, Watson JEM, Tuda AO et al (2015) Designing climate-resilient marine protected area networks by combining remotely sensed coral reef habitat with coastal multi-use maps. Remote Sens (Basel) 7:16571–16587. https://doi.org/10.3390/rs71215849

Manzello DP, Brandt M, Smith TB, Lirman D, Hendee JC, Nemeth RS (2007) Hurricanes benefit bleached corals. Proc Natl Acad Sci U S A 104:12035–12039. https://doi.org/10.1073/pnas.0701194104

Marshall PA, Baird AH (2000) Bleaching of corals on the Great Barrier Reef: differential susceptibilities among taxa. Coral Reefs 19:155–163. https://doi.org/10.1007/s003380000086

Maté JL (2003) Corals and coral reefs of the Pacific coast of Panamá. In: Cortés J (ed) Latin American Coral Reefs. Elsevier Science, Amsterdam, pp 387–418

Maynard J, van Hooidonk R, Eakin CM, Puotinen M, Garren M, Williams G et al (2015) Projections of climate conditions that increase coral disease susceptibility and pathogen abundance and virulence. Nat Clim Chang 5:688–694. https://doi.org/10.1038/nclimate2625

McClanahan T (1995) A coral reef ecosystem-fisheries model: impacts of fishing intensity and catch selection on reef structure and processes. Ecol Model 80:1–19. https://doi.org/10.1016/0304-3800(94)00042-G

McClanahan T (2000) Bleaching damage and recovery potential of Maldivian coral reefs. Mar Pollut Bull 40:587–597. https://doi.org/10.1016/S0025-326X(00)00064-3

McClanahan TR (2002) The near future of coral reefs. Environ Conserv 29:460–483. https://doi.org/10.1017/S0376892902000334

McClanahan TR (2004) The relationship between bleaching and mortality of common corals. Mar Biol 144:1239–1245. https://doi.org/10.1007/s00227-003-1271-9

McClanahan TR (2008) Response of the coral reef benthos and herbivory to fishery closure management and the 1998 ENSO disturbance. Oecologia 155:169–177. https://doi.org/10.1007/s00442-007-0890-0

McClanahan TR (2014) Decadal coral community reassembly on an African fringing reef. Coral Reefs 33:939–950. https://doi.org/10.1007/s00338-014-1178-6

McClanahan TR (2017) Changes in coral sensitivity to thermal anomalies. Mar Ecol Prog Ser 570:71–85. https://doi.org/10.3354/meps12150

McClanahan TR, Abunge CA (2014) Catch rates and income are associated with fisheries management restrictions and not an environmental disturbance, in a heavily exploited tropical fishery. Mar Ecol Prog Ser 513:201–210. https://doi.org/10.3354/meps10925

McClanahan TR, Cinner J (2008) A framework for adaptive gear and ecosystem-based management in the artisanal coral reef fishery of Papua New Guinea. Aquat Conserv: Mar Freshwat Ecosyst 18:493–507. https://doi.org/10.1002/aqc.874

McClanahan TR, Maina J (2003) Response of coral assemblages to the interaction between natural temperature variation and rare warm-water events. Ecosystems 6:551–563. https://doi.org/10.1007/s10021-002-0104-x

McClanahan TR, Muthiga NA (2014) Community change and evidence for variable warm-water temperature adaptation of corals in Northern Male Atoll, Maldives. Mar Poll Bull 80:107–113. https://doi.org/10.1016/j.marpolbul.2014.01.035

McClanahan TR, Muthiga NA (2016a) Geographic extent and variation of a coral reef trophic cascade. Ecology 97:1862–1872. https://doi.org/10.1890/15-1492.1

McClanahan TR, Muthiga NA (2016b) Similar impacts of fishing and environmental stress on calcifying organisms in Indian Ocean coral reefs. Mar Ecol Prog Ser 560:87–103. https://doi.org/10.3354/meps11921

McClanahan TR, Muthiga NA, Mangi S (2001) Coral and algal changes after the 1998 coral bleaching: interaction with reef management and herbivores on Kenyan reefs. Coral Reefs 19:380–391. https://doi.org/10.1007/s003380000133

McClanahan TR, Baird AH, Marshall PA, Toscano MA (2004a) Comparing bleaching and mortality responses of hard corals between southern Kenya and the Great Barrier Reef, Australia. Mar Poll Bull 48:327–335. https://doi.org/10.1016/j.marpolbul.2003.08.024

McClanahan TR, McLaughlin SM, Davy JE, Wilson WH, Peters CE, Price KL et al (2004b) Observations of a new source of coral mortality along the Kenyan coast. Hydrobiologia 530:469–479. https://doi.org/10.1007/s10750-004-2672-6

McClanahan TR, Maina J, Moothien-Pillay BAC (2005a) Effects of geography, taxa, water flow, and temperature variation on coral bleaching intensity in Mauritius. Mar Ecol Prog Ser 298:131–142. https://doi.org/10.3354/meps298131

McClanahan TR, Maina J, Starger CJ, Herron-Perez P, Dusek E (2005b) Detriments to post-bleaching recovery of corals. Coral Reefs 24:230–246. https://doi.org/10.1007/s00338-004-0471-1

McClanahan TR, Ateweberhan M, Muhando CA, Maina J, Mohammed SM (2007a) Effects of climate and seawater temperature variation on coral bleaching and mortality. Ecol Monogr 77:503–525. https://doi.org/10.1890/06-1182.1

McClanahan TR, Ateweberhan M, Sebastián CR, Graham NAJ, Wilson SK, Guillaume MMM et al (2007b) Western Indian Ocean coral communities: bleaching responses and susceptibility to extinction. Mar Ecol Prog Ser 337:1–13. https://doi.org/10.3354/meps337001

McClanahan TR, Ateweberhan M, Sebastián CR, Graham NAJ, Wilson SK, Bruggemann JH et al (2007c) Predictability of coral bleaching from synoptic satellite and in situ temperature observations. Coral Reefs 26:695–701. https://doi.org/10.1007/s00338-006-0193-7

McClanahan TR, Buddemeier RW, Hoegh-Guldberg O, Sammarco P (2008a) Projecting the current trajectory for coral reefs. In: Polunin NVC (ed) Aquatic ecosystems: trends and global prospects. Cambridge University Press, Cambridge, pp 242–262

McClanahan TR, Ateweberhan M, Omukoto J (2008b) Long-term changes in coral colony size distributions on Kenyan reefs under different management regimes and across the 1998 bleaching event. Mar Biol 153:755–768. https://doi.org/10.1007/s00227-007-0844-4

McClanahan TR, Weil E, Maina J (2009) Strong relationship between coral bleaching and growth anomalies in massive *Porites*. Glob Chang Biol 15:1804–1816. https://doi.org/10.1111/j.1365-2486.2008.01799.x

McClanahan TR, Donner SD, Maynard JA, MacNeil MA, Graham NA, Maina J et al (2012) Prioritizing key resilience indicators to support coral reef management in a changing climate. PLoS One 7. https://doi.org/10.1371/jounal.pone.0042884

McClanahan TR, Ateweberhan M, Darling ES, Graham NAJ, Muthiga NA (2014a) Biogeography and change among regional coral communities across the Western Indian Ocean. PLoS One 9. https://doi.org/10.1371/journal.pone.0093385

McClanahan TR, Cinner JE, Abunge C, Rabearisoa A, Mahatante P, Ramahatratra F et al (2014b) Perceived benefits of fisheries management restrictions in Madagascar. Ecol Soc:19. https://doi.org/10.5751/ES-06080-190105

McClanahan TR, Maina J, Ateweberhan M (2015) Regional coral responses to climate disturbances and warming is predicted by multivariate stress model and not temperature threshold metrics. Clim Change 131:607–620. https://doi.org/10.1007/s10584-015-1399-x

McField MD (1999) Coral response during and after mass bleaching in Belize. Bull Mar Sci 64:155–172

Mellin C, Bradshaw CJA, Fordham DA, Caley MJ (2014) Strong but opposing β-diversity–stability relationships in coral reef fish communities. Proc R Soc B 281. https://doi.org/10.1098/rspb.2013.1993

Mellin C, MacNeil MA, Cheal AJ, Emslie MJ, Caley MJ (2016) Marine protected areas increase resilience among coral reef communities. Ecol Lett 19:629–637. https://doi.org/10.1111/ele.12598

Mendes JM, Woodley JD (2002) Effect of the 1995-1996 bleaching event on polyp tissue depth, growth, reproduction and skeletal band formation in *Montastraea annularis*. Mar Ecol Prog Ser 235:93–102. https://doi.org/10.3354/meps235093

Michalek-Wagner K, Willis BL (2001a) Impacts of bleaching on the soft coral *Lobophytum compactum*. II. biochemical changes in adults and their eggs. Coral Reefs 19:240–246. https://doi.org/10.1007/PL00006959

Michalek-Wagner K, Willis BL (2001b) Impacts of bleaching on the soft coral *Lobophytum compactum*. I. fecundity, fertilization and offspring viability. Coral Reefs 19:231–239. https://doi.org/10.1007/s003380170003

Miller J, Waara R, Muller E, Rogers C (2006) Coral bleaching and disease combine to cause extensive mortality on reefs in US Virgin Islands. Coral Reefs 25:418–418. https://doi.org/10.1007/s00338-006-0125-6

Miller J, Muller E, Rogers C, Waara R, Atkinso A, Whelan KRT, Patterson M, Witcher B (2009) Coral disease following massive bleaching in 2005 causes 60% decline in coral cover on reefs in the US Virgin Island. Coral Reefs 28:925–937

Morri C, Montefalcone M, Lasagna R, Gatti G, Rovere A, Parravicini V et al (2015) Through bleaching and tsunami: coral reef recovery in the Maldives. Mar Poll Bull 98:188–200. https://doi.org/10.1016/j.marpolbul.2015.06.050

Mumby PJ (1999) Bleaching and hurricane disturbances to populations of coral recruits in Belize. Mar Ecol Prog Ser 190:27–35. https://doi.org/10.3354/meps

Mumby PJ, Harborne AR (2010) Marine reserves enhance the recovery of corals on Caribbean reefs. PLoS One 5. https://doi.org/10.1371/journal.pone.0008657

Mumby PJ, Edwards AJ, Arias-González JE, Lindeman KC, Blackwell PG, Gall A et al (2004) Mangroves enhance the biomass of coral reef fish communities in the Caribbean. Nature 427:533–536. https://doi.org/10.1038/nature02286

Mumby PJ, Dahlgren CP, Harborne AR, Kappel CV, Micheli F, Brumbaugh DR et al (2006) Fishing, trophic cascades, and the process of grazing on coral reefs. Science 311:98–101. https://doi.org/10.1126/science.1121129

Mumby PJ, Harborne AR, Williams J, Kappel CV, Brumbaugh DR, Micheli F et al (2007) Trophic cascade facilitates coral recruitment in a marine reserve. Proc Natl Acad Sci U S A 104:8362–8367. https://doi.org/10.1073/pnas.0702602104

Mumby PJ, Wolff NH, Bozec Y-M, Chollett I, Halloran P (2014) Operationalizing the resilience of coral reefs in an era of climate change. Conserv Lett 7:176–187. https://doi.org/10.1111/conl.12407

Nakamura T, van Woesik R (2001) Differential survival of corals during the 1998 bleaching event is partially explained by water-flow rates and passive diffusion. Mar Ecol Prog Ser 212:301–304. https://doi.org/10.1154/meps212301

O'Leary JK, Potts DC, Braga JC, McClanahan TR (2012) Indirect consequences of fishing: reduction of coralline algae suppresses juvenile coral abundance. Coral Reefs 31:547–559. https://doi.org/10.1007/s00338-012-0872-5

Olds AD, Pitt KA, Maxwell PS, Babcock RC, Rissik D, Connolly RM (2014) Marine reserves help coastal ecosystems cope with extreme weather. Glob Chang Biol 20:3050–3058. https://doi.org/10.1111/gcb.12606

O'Leary JK, Potts D, Schoenrock KM, McClanahan T (2013) Fish and sea urchin grazing opens settlement space equally but urchins reduce survival of coral recruits. Mar Ecol Prog Ser 493:156–177. https://doi.org/10.3354/meps10510

Oliver TA, Palumbi SR (2011a) Do fluctuating temperature environments elevate coral thermal tolerance? Coral Reefs 30:429–440. https://doi.org/10.1007/s00338-011-0721-y

Oliver TA, Palumbi SR (2011b) Many corals host thermally resistant symbionts in high-temperature habitat. Coral Reefs 30:241–250. https://doi.org/10.1007/s00338-010-0696-0

Page CA, Cróquer A, Bastidas C, Rodriguez S, Neale SJ, Weil E et al (2016) *Halofolliculina* ciliate infections on corals (skeletal eroding disease). In: Woodley CM, Downs CA, Bruckner AW, Porter JW, Galloway SB (eds) Diseases of coral. Wiley, Hoboken, New Jersey, pp 361–375

Perry CT, Murphy GN, Kench PS, Smithers SG, Edinger EN, Steneck RS et al (2013) Caribbean-wide decline in carbonate production threatens coral reef growth. Nat Commun 4. https://doi.org/10.1038/ncomms2409

Perry CT, Murphy GN, Graham NAJ, Wilson SK, Januchowski-Hartley FA, East HK (2015) Remote coral reefs can sustain high growth potential and may match future sea-level trends. Sci Rep 5. https://doi.org/10.1038/srep18289

Pinzon J, Weil E (2012) Cryptic species in the Atlantic-Caribbean genus *Meandrina* (Scleractinia): a multi-variable review of the taxonomy and description of the new species *Meandrina jacksoni*. Bull Mar Sci 87:823–853

Porter JW, Fitt WK, Spero HJ, Rogers CS, White MW (1989) Bleaching in reef corals: physiological and stable isotopic responses. Proc Natl Acad Sci U S A 86:9342–9346. https://doi.org/10.1073/pnas.86.23.9342

Prada C, Weil E, Yoshioka PM (2010) Octocoral bleaching during unusual thermal stress. Coral Reefs 29:41–45. https://doi.org/10.1007/s00338-009-0547-z

Pratchett MS, Anderson KD, Hoogenboom MO, Widman E, Baird AH, Pandolfi JM et al (2015) Spatial, temporal and taxonomic variation in coral growth: Implications for the structure and function of coral reef ecosystems. Oceanogr Mar Biol: Annu Rev 53:215–295. https://doi.org/10.1201/b18733-7

Precht WF, Gintert BE, Robbart ML, Fura R, van Woesik R (2016) Unprecedented disease-related coral mortality in Southeastern Florida. Sci Rep 6. https://doi.org/10.1038/srep31374

Richardson LL, Voss JD (2005) Changes in a coral population on reefs of the northern Florida Keys following a coral disease epizootic. Mar Ecol Prog Ser 297:147–156. https://doi.org/10.3354/meps297147

Riegl B, Piller WE (2003) Possible refugia for reefs in times of environmental stress. Int J Earth Sci 92:520–531. https://doi.org/10.1007/s00531-003-0328-9

Ritchie KB (2006) Regulation of microbial populations by coral surface mucus and mucus-associated bacteria. Mar Ecol Prog Ser 332:1–14. https://doi.org/10.3354/meps322001

Rosenberg E, Ben-Haim Y (2002) Microbial diseases of corals and global warming. Environ Microbiol 4:318–326. https://doi.org/10.1046/j.1462-2920-2002.00302.x

Rotjan RD, Lewis SM (2005) Selective predation by parrotfishes on the reef coral *Porites astreoides*. Mar Ecol Prog Ser 305:193–201. https://doi.org/10.3354/meps305193

Ruzicka RR, Colella MA, Porter JW, Morrison JM, Kidney JA, Brinkhuis V et al (2013) Temporal changes in benthic assemblages on Florida Keys reefs 11 years after the 1997/1998 El Niño. Mar Ecol Prog Ser 489:125–141. https://doi.org/10.3354/meps10427

Sebastián CR, McClanahan TR (2013a) Description and validation of production processes in the coral reef ecosystem model CAFFEE (Coral-Algae-Fish-Fisheries Ecosystem Energetics) with a fisheries closure and climatic disturbance. Ecol Model 263:326–348. https://doi.org/10.1016/j.ecolmodel.2013.05.012

Sebastián CR, McClanahan TR (2013b) Using an ecosystem model to evaluate fisheries management options to mitigate coral bleaching on Western Indian Ocean coral reefs. West Indian Ocean J Mar Sci 11:77–86

Selig ER, Casey KS, Bruno JF (2012) Temperature-driven coral decline: the role of marine protected areas. Glob Chang Biol 18:1561–1570. https://doi.org/10.1111/j.1365-2486.2012.02658.x

Shenkar N, Fine M, Loya Y (2005) Size matters: bleaching dynamics of the coral *Oculina patagonica*. Mar Ecol Prog Ser 294:181–188. https://doi.org/10.3354/meps294181

Sheppard CRC, Spalding M, Bradshaw C, Wilson S (2002) Erosion vs. recovery of coral reefs after 1998 El Niño: Chagos reefs, Indian Ocean. Ambio 31:40–48. https://doi.org/10.1639/0044-7447(2002)031[0040:EVROCR]2.0.CO;2

Sheppard CRC, Harris A, Sheppard ALS (2008) Archipelago-wide coral recovery patterns since 1998 in the Chagos Archipelago, central Indian Ocean. Mar Ecol Prog Ser 362:109–117. https://doi.org/10.3354/meps07436

Shuail D, Wiedenmann J, D'Angelo C, Baird AH, Pratchett MS, Riegl B et al (2016) Local bleaching thresholds established by remote sensing techniques vary among reefs with deviating bleaching patterns during the 2012 event in the Arabian/Persian Gulf. Mar Poll Bull 105:654–659. https://doi.org/10.1016/j.marpolbul.2016.03.001

Thomas L, Palumbi SR (2017) The genomics of recovery from coral bleaching. Proc R Soc B 284. https://doi.org/10.1098/rspb.2017.1790

Tkachenko KS (2012) The northernmost coral frontier of the Maldives: the coral reefs of Ihavandippolu Atoll under long-term environmental change. Mar Environ Res 82:40–48. https://doi.org/10.1016/j.marenvres.2012.09.004

Toth LT, Aronson RB, Vollmer SV, Hobbs JW, Urrego DH, Cheng H et al (2012) ENSO drove 2500-year collapse of Eastern Pacific coral reefs. Science 337:81–84. https://doi.org/10.1126/science.1221168

Underwood JN, Smith LD, Van Oppen MJH, Gilmour JP (2009) Ecologically relevant dispersal of corals on isolated reefs: implications for managing resilience. Ecol Appl 19:18–29. https://doi.org/10.1890/07-1461.1

van Oppen MJH, Oliver JK, Putnam HM, Gates RD (2015) Building coral reef resilience through assisted evolution. Proc Natl Acad Sci U S A 112:2307–2313. https://doi.org/10.1073/pnas.1422301112

van Woesik R, Franklin EC, O'Leary JK, McClanahan TR, Klaus JS, Budd AF (2012) Hosts of the Plio-Pleistocene past reflect modern-day coral vulnerability. Proc R Soc B 279:2448–2456. https://doi.org/10.1098/rspb.2011.2621

Vicente VP (1990) Response of sponges with autotrophic endosymbionts during the coral-bleaching episode in Puerto-Rico. Coral Reefs 8:199–202. https://doi.org/10.1007/BF00265011

Ward JR, Kim K, Harvell CD (2007) Temperature affects coral disease resistance and pathogen growth. Mar Ecol Prog Ser 329:115–121. https://doi.org/10.3354/meps329115

Warwick RM, Clarke KR, Suharsono (1990) A statistical analysis of coral community responses to the 1982-83 El Niño in the Thousand Islands, Indonesia. Coral Reefs 8:171–179. https://doi.org/10.1007/BF00265008

Wear SL (2016) Missing the boat: critical threats to coral reefs are neglected at global scale. Mar Policy 74:153–157. https://doi.org/10.1016/j.marpol.2016.09.009

Weil E (2002) Coral Disease Epizootiology: status and Research Needs. Coral Health and Disease: developing a National Research Plan. Coral Health and Disease Consortium, Charleston, South Carolina 14pp

Weil E (2004) Coral reef diseases in the wider Caribbean. In: Rosenberg E, Loya Y (eds) Coral Health and Disease. Springer, Berlin, pp 35–68

Weil E, Cróquer A (2009) Spatial variability in distribution and prevalence of Caribbean scleractinian coral and octocoral diseases. I. community-level analysis. Dis Aquat Organ 83:195–209. https://doi.org/10.3354/dao02011

Weil E, Rogers CS (2011) Coral reef disease in the Atlantic-Caribbean. In: Dubinsky Z, Stambler N (eds) Coral reefs: an ecosystem in transition. Springer, pp 465–491

Weil E, Smith G, Gil-Agudelo DL (2006) Status and progress in coral reef disease research. Dis Aquat Organ 69:1–7. https://doi.org/10.3354/dao069001

Weil E, Cróquer A, Urreiztieta I (2009a) Temporal variability and impact of coral diseases and bleaching in La Parguera, Puerto Rico from 2003-2007. Caribb J Sci 52:221–246. https://doi.org/10.18475/cjos.v45i2.a10

Weil E, Cróquer A, Urreiztieta I (2009b) Yellow band disease compromises the reproductive output of the Caribbean reef-building coral *Montastraea faveolata* (Anthozoa, Scleractinia). Dis Aquat Organ 87:45–55. https://doi.org/10.3354/dao02103

Weil E, Rogers CS, Cróquer A (2017) Octocoral diseases in a changing ocean. In: Rossi S, Bramanti L, Gori A, del Valle COS (eds) Marine animal forests: the ecology of benthic biodiversity hotspots. Springer, Cham, pp 1109–1163

Williams EH, Bunkley-Williams L (1988) Bleaching of Caribbean coral reef symbionts in 1987–1988. In: Proceedings of the 6th international coral reef symposium, Townsville, vol 3, pp 313–318

Williams EH, Bunkley-Williams L (1990) The world-wide coral reef bleaching cycle and related sources of coral mortality. Atoll Res Bull 35:1–71. https://doi.org/10.5479/si00775630.335.1

Willis B, Page CA, Dinsdale EA (2004) Coral disease on the Great Barrier Reef. In: Rosenberg E, Loya Y (eds) Coral health and disease. Springer, Berlin, pp 69–104

Chapter 11
Effects of Coral Bleaching and Coral Loss on the Structure and Function of Reef Fish Assemblages

M. S. Pratchett, C. A. Thompson, A. S. Hoey, P. F. Cowman, and S. K. Wilson

11.1 Introduction

Mass coral bleaching, caused by elevated ocean temperatures, has now emerged as a major, if not the single most important, contributor to elevated rates of coral mortality (Hughes et al. 2017; Chaps. 3, 4, and 13), greatly accelerating the degradation of coral reef ecosystems throughout the world. Coral reefs have been subject to increasing anthropogenic disturbances and threats throughout the last few decades (if not centuries), resulting in sustained declines in the cover or abundance of scleractinian corals and corresponding shifts in the structure of reef habitats (Hughes et al. 2003; Alvarez-Filip et al. 2011). Climate change (specifically resulting in coral bleaching) is almost always considered, along with a variety of other more localised anthropogenic disturbances and threats, as a key contributor to sustained and ongoing coral loss (e.g. De'ath et al. 2012). However, mass coral bleaching has previously been considered to be a relatively minor, though emerging and increasingly important, contributor to coral loss, especially relative to other major disturbances such as severe tropical storms and outbreaks of coral predators (Pratchett et al. 2011a; De'ath et al. 2012). The extent and severity of the latest (2014–2017)

M. S. Pratchett (✉) · A. S. Hoey · P. F. Cowman
ARC Centre of Excellence for Coral Reef Studies, James Cook University, Townsville, QLD, Australia
e-mail: morgan.pratchett@jcu.edu.au

C. A. Thompson
College of Science and Engineering and ARC Centre of Excellence for Coral Reef Studies, James Cook University, Townsville, QLD, Australia

S. K. Wilson
Marine Science Program, Government of Western Australia Department of Biodiversity, Conservation and Attractions, Kensington, WA, Australia

Oceans Institute, University of Western Australia, Crawley, WA, Australia

M. J. H. van Oppen, J. M. Lough (eds.), *Coral Bleaching*, Ecological Studies 233,
https://doi.org/10.1007/978-3-319-75393-5_11

global bleaching event (Hughes et al. 2017, 2018), as well as successive years of severe bleaching in many locations, have firmly heralded in an era where global climate change is the foremost threat to coral reef ecosystems.

What makes coral reef ecosystems particularly vulnerable to climate change is that reef-building (scleractinian) corals are both very sensitive to elevated temperatures (Jokiel and Coles 1990; Smith and Buddemeier 1992; Chaps. 2 and 9) and fundamental to the structure and function of coral reefs (Bellwood et al. 2004; Pratchett et al. 2015a). Scleractinian corals are the building blocks of coral reefs, not only contributing to reef accretion (Pratchett et al. 2015a) but also forming complex habitats which support a high diversity of fishes (Coker et al. 2014) and other reef-associated organisms (Stella et al. 2011). The importance of scleractinian corals is particularly apparent given marked declines in the abundance and diversity of coral reef fishes following acute and extensive coral loss (Jones et al. 2004; Wilson et al. 2006; Munday et al. 2008; Pratchett et al. 2008a; Cheal et al. 2017) caused by severe tropical storms (hurricanes, typhoons and tropical cyclones), outbreaks of coral predators and mass coral bleaching. Such effects are particularly pronounced when coral cover falls below 10% (Wilson et al. 2006; Holbrook et al. 2008), suggesting that $\geq$10% coral cover is necessary to maintain ecological functions that support diverse assemblages of coral reef fishes.

Many coral reef fishes rely on scleractinian corals for food (Cole et al. 2008), habitat (Coker et al. 2014) and/or settlement (Jones et al. 2004; Coker et al. 2012). However, the range of coral reef fishes that decline in abundance following extensive coral depletion (60–75%) far exceeds that which are known to have an explicit and direct reliance on scleractinian corals (Jones et al. 2004; Graham et al. 2007; Pratchett et al. 2011b; Cheal et al. 2017). The broadscale ecosystem consequences of extensive coral loss suggest that we have overlooked some important ecological benefits of coral-rich habitats for reef fishes (e.g. Dixson et al. 2014; Pratchett et al. 2015b). For example, extensive coral depletion may effectively remove major odour cues that are used by reef fishes and corals to orientate towards and settle within coral reef habitats (Dixson et al. 2014). The effects of coral loss on the biodiversity and abundance of reef-associated organisms may also be compounded by declines in topographic complexity (Syms and Jones 2000; Wilson et al. 2006; Graham et al. 2009; Coker et al. 2012), which occurs due to erosion and decomposition of dead coral skeletons (Sheppard et al. 2002) and disproportionate loss of key habitat-forming corals (Graham et al. 2006; Alvarez-Filip et al. 2011). Importantly, high levels of structural complexity increase habitat area and moderate key ecological interactions (e.g. competition and predation) contributing to increased species packing, as well as facilitating coexistence of large numbers of diverse species (Gratwicke and Speight 2005).

In this chapter, we revisit the effects of mass coral bleaching (and associated changes in the structure of tropical reef habitats) on the structure and function of reef fish assemblages. In particular, this chapter will investigate the ecological and economic consequences of declines in the abundance of fishes, based on the selectivity of effects within and among different functional groups, as well as testing for compensatory dynamics necessary to maintain key ecological functions following species declines or losses (Houlahan et al. 2007). It is well established that extensive coral depletion (whether caused by mass coral bleaching, tropical cyclones or outbreaks of invertebrate corallivores) results in significant declines in abundance across a broad range of reef fishes (Pratchett et al. 2011b) and overall declines in diversity of fish assemblages (Wilson et al. 2006). However, it is still not known whether far-reaching declines in the abundance of coral reef fishes will compromise ecological functions and especially those functions that are fundamental in maintaining ecosystem resilience (Bellwood et al. 2003; Hoey and Bellwood 2009). Importantly, net declines in the abundance or performance of ecologically important reef fishes may lead to feedbacks that inhibit recovery and reassembly of coral-dominated habitats (Graham et al. 2015) and/or further exacerbate the degradation of coral reef environments (Hoey and Bellwood 2011).

The extent to which declines in the abundance and diversity of fishes will cause ecological functions to be lost or compromised depends on the number of species that can perform a particular function (i.e. *functional redundancy*) as well as variation in responses to environmental perturbations among functionally equivalent fishes (i.e. *response diversity*) (Elmqvist et al. 2003). Some fishes, particularly herbivorous or generalist species, actually exhibit increases in abundance following extensive coral loss (Jones et al. 2004; Bellwood et al. 2006; Pratchett et al. 2008a; Halford and Caley 2009; Cheal et al. 2017). Intuitively, species that are ecologically equivalent would be equally or similarly affected by perturbations that lead to loss of habitat complexity or resource depletion. For example, extensive coral depletion will lead to declines in coral prey across all species of corallivores, though different species may be more or less affected depending on their degree of dietary specialisation (Pratchett et al. 2008a). Functionally equivalent species may also differ in the extent to which they are adversely affected by small-scale or patchy habitat disturbances based on differences in the scales at which they associate with reef habitats (Nash et al. 2016). To maintain ecosystem function, however, significant declines in the abundance of key ecological species must be offset by compensatory increases in the abundance of species that can perform, or contribute to, the same function, though compensatory dynamics are rarely observed in most ecosystems (Houlahan et al. 2007). To explicitly test for response diversity and compensatory dynamics among reef fishes, data were compiled from a variety of studies that have looked at species-specific declines in abundance of fishes before and then 1–7 years after distinct episodes of coral loss, following Pratchett et al. (2011b). While we were primarily interested in the effects of coral depletion caused by climate-induced coral bleaching, data were taken from all studies that have explored changes in the abundance of fishes following acute episodes of coral loss, regardless of the cause. Each species of fish was independently assigned one of four primary functional groups based on their trophic function (i.e. carnivores, omnivores, corallivores and herbivores). Species were subsequently assigned to secondary functional groups based on feeding mode, diet and behaviour to reflect their role in ecosystem processes and/or differential sources of their prey. Response diversity and compensatory dynamics were then assessed based on the distribution of responses (changes in abundance) for fishes within 19 distinct functional or trophic groups.

11.2 Coral Bleaching and Changes in the Structure of Reef Habitats

Coral reefs are among the most vulnerable ecosystems to global climate change (Walther et al. 2002), owing to the magnitude and severity of habitat loss that occurs during severe mass bleaching episodes (Hughes et al. 2017, 2018). The scale and magnitude of coral loss caused by pantropical mass bleaching events eclipse all other major acute disturbances (e.g. tropical cyclones, outbreaks of coral predators and coral disease) that have contributed to coral declines around the world. Generally, it is the cumulative effects of multiple discrete disturbances, which may or may not be increasing in incidence, that have caused coral declines in major reef regions (Gardner et al. 2003; De'ath et al. 2012). These disturbances tend to occur at the scale of individual reefs or reef clusters. However, large-scale mass bleaching is unequivocally linked to sustained increases in global sea surface temperatures (Heron et al. 2016), which may be compounded by ocean-scale climatic features (e.g. El Niño events). In 1998, for example, mass coral bleaching was reported on coral reefs throughout the Indo-Pacific and in the Caribbean (Wilkinson 2000) and killed 75–99% of corals across the worst affected regions (Goreau et al. 2000; Graham et al. 2006). This event contributed greatly to increased recognition of climate change as a significant threat to coral reef ecosystems (Hoegh-Guldberg 1999) and motivated many of the foremost studies on ecosystem effects of severe coral bleaching and coral loss (Graham et al. 2006, 2008). Graham et al. (2008) surveyed fish assemblages at 66 sites across the western Indian Ocean in the aftermath of the 1998 bleaching and assessed changes in the size structure and taxonomic composition of fish assemblages by comparing results with surveys conducted prior to the bleaching (in 1990–1998). The key finding from this study was that spatial management arrangements provided no protection against mass coral bleaching and subsequent effects of coral loss on reef fish assemblages. It was also apparent that mass coral bleaching had disproportionate impacts on small-bodied (<20 cm total length) reef fishes (Graham et al. 2008). Beyond that, the effects of the 1998 bleaching event (on both coral and fish assemblages) were highly variable (Graham et al. 2008), requiring much greater consideration of the specific changes to coral reef habitats that are caused by mass coral bleaching. It is also important to realise that mass coral bleaching was even more widespread in 2014–2017 than in 1998 (Hughes et al. 2017, 2018), though the impacts of these latest pantropical bleaching events are yet to be fully realised.

11.2.1 Bleaching Selectivity and Changes in Coral Composition

The specific effects of mass coral bleaching on coral reef fishes will depend on the magnitude (extent and severity) and selectivity of coral loss. All scleractinian corals are susceptible to bleaching at some level, but certain genera, such as *Stylophora*, *Pocillopora*, *Acropora* and *Montipora*, tend to be much more susceptible, based on the proportion of colonies and species that bleach (McClanahan et al. 2004; Hoey et al. 2016) and are also more likely to die once bleached (Baird and Marshall 2002). The selectivity of mass coral bleaching declines with increasing severity (Hughes et al. 2017), whereby there are very few corals that can withstand prolonged exposure to extreme temperatures. During moderate bleaching events, however, bleaching and mortality may be restricted to a few specific coral taxa (Baird and Marshall 2002). For the most part, it is branching corals that tend to exhibit higher rates of bleaching and mortality (Baker et al. 2008), though it is unknown whether this reflects a taxonomic bias in coral morphology (McCowan et al. 2012) or inherent physiological properties of massive or robust corals that increase resistance and resilience to coral bleaching (Loya et al. 2001). Taxonomic differences in susceptibility to bleaching can vary spatially and temporally (Guest et al. 2012; Pratchett et al. 2013), depending on depth and habitat, the recent thermal history, hydrodynamics and endosymbiont associations. Moreover, increasing incidence of coral bleaching will not necessarily favour those coral species that are most resistant to bleaching (Hughes et al. 2003; Baker et al. 2008). Rather, directional shifts in the composition of coral assemblages will depend on both rates of colony-level mortality due to bleaching (relative to normal background rates of whole colony mortality) and the differential recovery capacity of species. Importantly, corals with rapid growth and high rates of population turnover may be relatively unaffected by recurrent bleaching (Linares et al. 2011), compared to slow-growing coral species that invest significant energy in maintenance and persistence.

Declines in the species richness of coral assemblages can directly impact on diversity of fish assemblages (Messmer et al. 2011), though there are specific coral taxa that are particularly important in providing food and habitat resources and also make disproportionate contributions to topographic complexity of reef habitats (Coker et al. 2012). Corals vary in the extent to which they provide effective habitat for reef fishes mainly due to differences in gross morphology (e.g. branching versus encrusting or massive colonies), though some specialist coral-dwelling fishes only occupy very specific corals and clearly distinguish within or among coral species (Munday 2001; Messmer et al. 2011; Noonan et al. 2012). The corals that are most important in providing habitat for highly specialised coral-dwelling fishes (including digitate *Acropora*, *Stylophora* and *Pocillopora* corals; Coker et al. 2014) are particularly susceptible to coral bleaching. Some of the more bleaching-resistant corals, such as branching *Porites* spp., are also important in providing habitat for a wide range of fishes (Richardson et al. 2017), but they tend to be occupied by less-specialised species that use a wide range of different corals (Gardiner and Jones 2005; Coker et al. 2014). Coral specialists are, by definition, expected to have a much stronger reliance on live corals and are more vulnerable to any changes in coral availability (Munday 2004; Pratchett et al. 2012). Given the close association of many specialist reef fishes with *Acropora* and *Pocillopora* corals, selective depletion of these corals may be just as devastating as a wholesale loss of scleractinian corals, affecting both the availability of preferred habitats and topographic complexity.

Even if these corals are generally resilient to recurrent bleaching, temporary declines in the availability of critical habitats may have devastating effects for fishes that are directly reliant on specific coral hosts (Munday 2004).

11.2.2 Coral Loss and Increasing Predominance of Alternative Habitat-Forming Taxa

While moderate bleaching is likely to cause directional shifts in the structure of coral assemblages (described above), severe episodes of mass bleaching affect a significant proportion of scleractinian corals (Hughes et al. 2017) and are likely to lead to extensive and widespread declines in coral cover. Such reductions in the abundance or cover of corals within shallow reef environments are often accompanied by increases in abundance of other alternative habitat-forming organisms, such as macroalgae (Hughes et al. 2010), though this depends on the specific environmental settings and relevant constraints on macroalgal growth and coral dynamics (Chong-Seng et al. 2014). Accordingly, there have been documented increases in the abundance of herbivorous fishes following widespread coral loss and concomitant increases in cover and biomass of macroalgae (Jones et al. 2004; Pratchett et al. 2008a; Cheal et al. 2017). In some instances, high densities of herbivores can prevent proliferation of macroalgae even after extensive coral depletion. However, fisheries exploitation has reduced the abundance of herbivorous fishes in many reef regions, compromising their capacity to respond to increasing cover and growth of macroalgae (Rasher et al. 2013). Once established, shifts from coral to macroalgal dominance may be reinforced by reduction in grazing and increased production of algal propagules (Hoey and Bellwood 2011) as well as constraints on recovery and replenishment of coral assemblages (Hughes et al. 2007). Moreover, the estimated biomass of herbivorous fishes needed to promote recovery of coral-dominated habitats (>180 kg per hectare) is much greater than what is required to prevent the initial proliferation of macroalgae and exceeds the estimated biomass of herbivorous fishes in most reef regions (Hoey et al. 2016).

Fundamental shifts in the dominant habitat-forming biota within tropical reef environments will significantly affect the behaviour, abundance and composition of coral reef fishes. While there are some species of reef-associated fishes that will benefit from increased cover and biomass of macroalgae (Dahlgren and Eggleston 2000; Wilson et al. 2010, 2017), fishes that feed, shelter or recruit to live corals are likely to disappear from reefs dominated by macroalgae (Jones et al. 2004). Importantly, macroalgae do not provide the same level of habitat structure and complexity as scleractinian corals, and the overall abundance and diversity of fishes on algal-dominated reefs are much lower, compared with coral-dominated habitats (Sano 2001). In the Seychelles, Graham et al. (2015) showed that the biomass of herbivorous fishes was highest on reefs with high levels of underlying structural complexity and this, in turn, was a major determinant of whether reefs recovered (rather than undergoing a regime shift to macroalgae) following the 1998 mass coral bleaching. While data on overall biomass of fishes was not presented for coral- versus macroalgae-dominated reefs (Graham et al. 2015), restoration of the functional integrity of reefs that recovered will likely have significant ecological and economic benefits, such as increased contribution to local fisheries production.

11.2.3 Coral Loss Versus Declines in Topographic Complexity

Mass coral bleaching is categorised (along with outbreaks of coral predators and coral diseases) as a *biological* or *nonstructural* disturbance (Wilson et al. 2006) whereby corals are killed without directly modifying their physical structure, at least in the short term. *Physical* or *structural* disturbances (e.g. tropical cyclones), meanwhile, cause immediate reductions in both live coral cover and topographic complexity of the reef framework by dislodging and/or breaking apart coral skeletons (e.g. Madin and Connolly 2006). *Structural* disturbances are generally thought to have much more pronounced and far-reaching effects on reef fishes compared to biological disturbances, because coral loss is compounded by declines in structural complexity (Wilson et al. 2006) and it is sometimes difficult to identify the independent contributions of coral loss versus topographic collapse (Pratchett et al. 2008a; Graham and Nash 2013). In at least some cases, coral loss appears to have greater influence on abundance of fishes than declines in structural complexity (Komyakova et al. 2013), while loss of coral diversity is the foremost driver of declines in diversity of reef fishes (Messmer et al. 2011; Holbrook et al. 2015). Moreover, extensive coral depletion can lead to declines in structural complexity, regardless of whether it is caused by biological or physical disturbances; coral colonies that have died due to bleaching (or other biological disturbances) are immediately subject to physical and biological forces that cause skeletal erosion and decomposition (Glynn 1997; Sheppard et al. 2002; Ferrari et al. 2017a), leading to inevitable, though sometimes protracted, declines in structural complexity (e.g. Sano et al. 1987). There are, however, instances where habitat complexity is provided by underlying substrate rugosity and is independent of coral loss (Pratchett et al. 2008a; Emslie et al. 2014), which may actually buffer fish assemblages during periods of coral loss.

The timing and sequence of decomposition and structural collapse of dead coral skeletons have only rarely been explicitly studied, but it is increasingly apparent that this is a gradual and ongoing process (Ferrari et al. 2017a), rather than an acute and delayed effect of coral mortality. Notably, delayed declines in the abundance and diversity of fishes, which may be apparent >3 years after extensive coral depletion (Pratchett et al. 2008a), have been at least partly attributed to lags in declines in topographic complexity following extensive coral mortality (Graham et al. 2006). The extent to which coral reef habitats are actually vulnerable to structural collapse also varies, depending on the extent to which coral assemblages are dominated by branching species, variation in the structural integrity of corals (even among different branching species), localised differences in the factors that promote physical and biological erosion and the relative contribution of contemporary coral growth versus underlying reef structure to topographic complexity (Pratchett et al. 2008a; Cheal et al. 2017).

11.3 Declines in Abundance of Reef Fishes and Loss of Biodiversity and Function

Much of the current understanding regarding the effects of mass coral bleaching on coral reef fishes comes from studies that have documented species-specific changes in abundance of fishes before and after specific bleaching events (e.g. Graham et al. 2008). While the results of such studies are dependent on the specific timing of surveys and especially the time elapsed following mass coral bleaching (Pratchett et al. 2009), the magnitude of species losses is broadly comparable to the effects of coral depletion caused by tropical cyclones, outbreaks of crown-of-thorns starfish and experimentally imposed disturbances (Fig. 11.1). Where coral mortality was

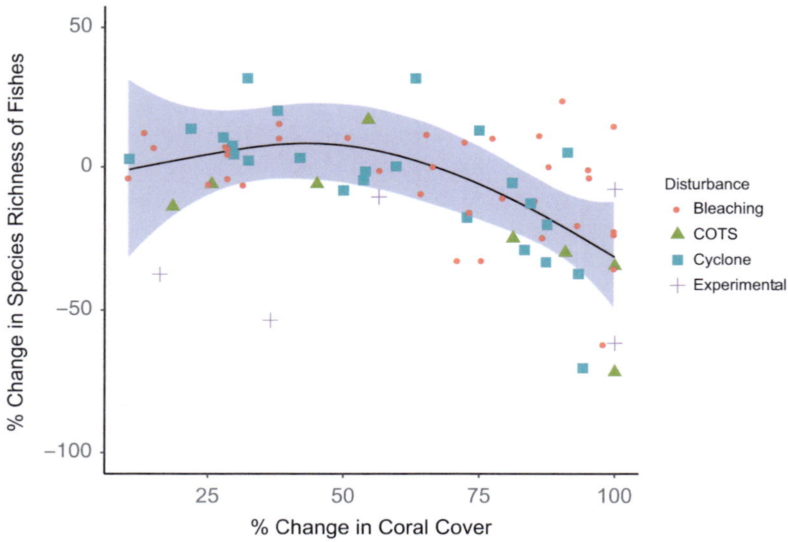

Fig. 11.1 Effects of coral loss on biodiversity (species richness) of coral reef fishes. The relationship between change in species richness (%) is plotted against increasing declines (%) in local coral cover, based on data extracted from 30 independent studies, categorised according to the major (but not exclusive) cause of localised coral loss. Confidence intervals (~95%) for the general additive model were calculated using a continuous set of predictor variables (n = 1000), with the MGCV package in *R*

>60%, the number of species recorded along belt transects or within point counts generally declined, although there was considerable variation among studies, with some documenting limited or even positive changes in diversity of fishes when coral mortality was >80%. Notably, extreme levels of species loss (60% decline in species richness) occurred following near-complete coral loss, regardless of what caused this coral mortality. However, moderate levels of coral loss had limited or positive effects on the local diversity of reef fishes (Fig. 11.1) which may reflect increases in the abundance and diversity of generalist fishes following initial loss of corals and corresponding increases in habitat heterogeneity (Wilson et al. 2006). It is important, however, to consider species-specific responses of fishes to mass coral bleaching, to understand which species are affected and potential impacts on ecosystem function and fisheries production (e.g. Brandl et al. 2016).

11.3.1 Interspecific Variation in Vulnerability to Coral Bleaching and Depletion

Fishes vary in their responses to habitat perturbations (and potential resource depletion) due to varying levels of ecological specialisation and differential reliance on specific resources (e.g. coral prey), their capacity to use alternative habitats or resources following changes in resource availability, their distribution relative to areas of major impacts and variation in the scales at which they associate with coral reef habitats (Pratchett et al. 2011b). The main group of fishes that consistently exhibit pronounced and often very rapid declines in abundance following localised coral depletion are obligate coral-feeding fishes (Wilson et al. 2006, 2014; Emslie et al. 2011; Fig. 11.2). Declines in the abundance of coral-feeding fishes are directly attributable to prey depletion and subsequent starvation, reflected in initial declines in their physiological condition (Pratchett et al. 2004). Coral-feeding fishes are initially attracted to bleached, diseased or injured corals, though they will preferentially feed on healthy (unbleached corals) in the longer term (McIlwain and Jones 1997; Cole et al. 2009). This suggests that the nutritional quality of corals declines soon after they bleach. The eventual mortality of bleached corals will then further constrain prey availability for obligate coral-feeding fishes. Accordingly, virtually all corallivorous fishes (including soft-coral feeders) exhibit significant declines in abundance, probably reflecting high rates of mortality, as opposed to movement among habitats, following localised coral depletion (Emslie et al. 2011; Wilson et al. 2014; Fig. 11.2). While localised declines in the abundance of these fishes may result from movement, the potential to find more suitable habitats is likely to be very limited, especially given the spatial extent of major mass bleaching events (e.g. Hughes et al. 2017). Even if fishes are able to find relatively undisturbed reef environments, high levels of aggression among coral specialists (e.g. coral-feeding butterflyfishes, Blowes et al. 2013) are likely to constrain the invasion of new habitats by displaced individuals. Moreover, individual fishes are likely to persist

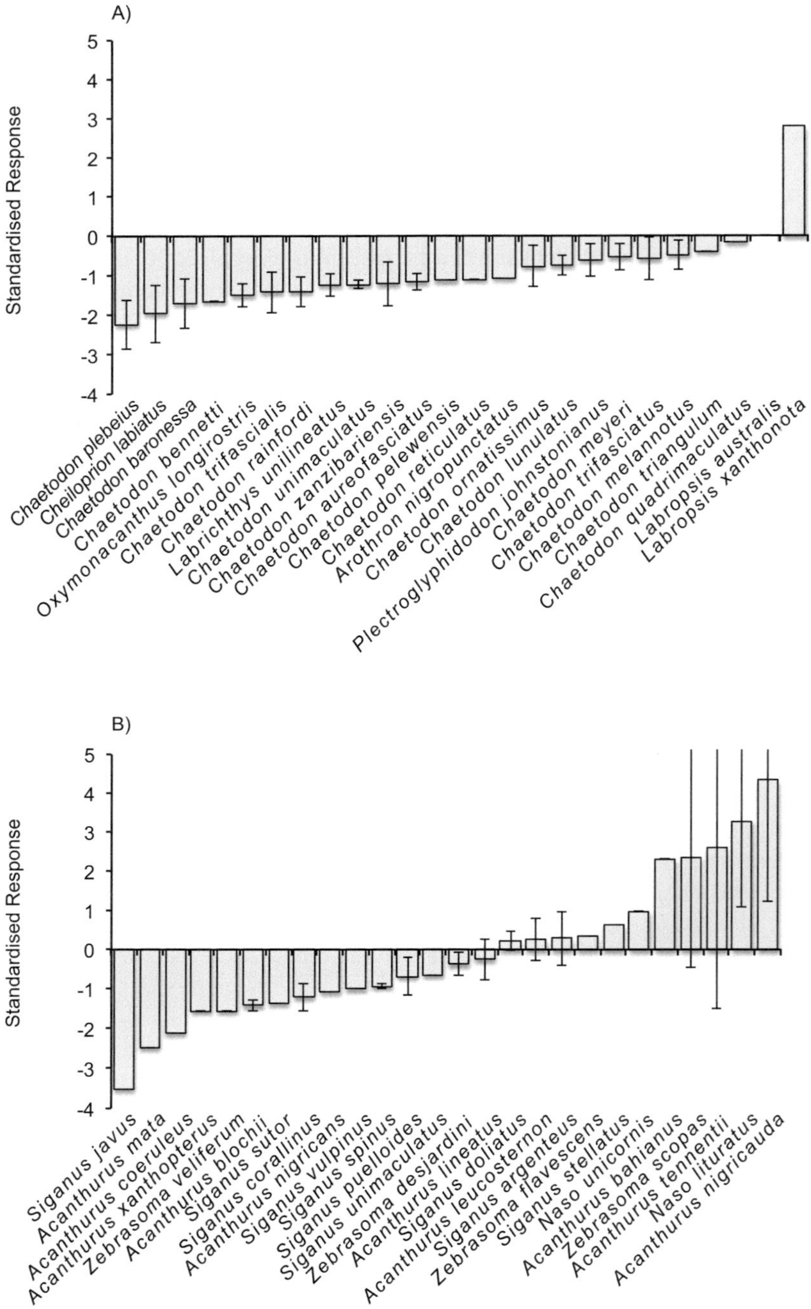

Fig. 11.2 Standardised responses of (**a**) obligate corallivores and (**b**) croppers and browsers to significant (>10%) declines in coral cover. Species-specific responses are calculated based on their proportional decline in abundance divided by proportional declines in live coral cover and averaged across multiple studies, where possible. Standard errors are calculated based on variation in responses among studies

within their established home ranges for some time after corals have bleached and died (Pratchett et al. 2004), which would further limit their ability to ultimately outcompete conspecifics and invade new habitats.

For fishes with an explicit and direct reliance on corals (e.g. for food and habitat), declines in abundance are often disproportionate to levels of local coral depletion. For example, *Chaetodon trifascialis* is often locally extirpated following relatively low (e.g. 14%) levels of coral depletion. The standardised response (which explicitly accounts for proportional coral loss) for *C. trifascialis* is, therefore, as low as −5. Such disproportionate declines in the abundance of these fishes reflect high selectivity in coral use as well as the sensitivity of preferred coral species to mass coral bleaching and other major disturbances. *Chaetodon trifascialis* is among the most highly specialised of coral reef fishes (Pratchett 2014), feeding almost exclusively on tabulate *Acropora* as well as a few other select species (e.g. *Acropora florida*). Accordingly, declines in the abundance of *C. trifascialis* relate to declines in the abundance of their major prey (tabulate *Acropora*), rather than overall declines in live coral cover. Other relatively specialised corallivores (*C. plebeius*, *C. baronessa* and *Oxymonacanthus longirostris*) also exhibit disproportionate declines in abundance following significant (>10%) reductions in local coral cover (Fig. 11.2). The only corallivorous fish that has been seen to increase in abundance following coral depletion is *Labropsis xanthonota*, for which densities of recruits actually increased (albeit from very low densities) following mass coral bleaching in Chagos (Graham et al. 2008).

Concern about the loss of key functional groups on coral reefs mostly centres around herbivorous fishes, largely due to their role in preventing and potentially reversing macroalgal overgrowth (Hughes et al. 2007; Hoey and Bellwood 2011). While all herbivorous fishes ostensibly consume and remove algae, they perform different and complimentary roles in helping reefs to resist shifts to alternate states and reassemble following disturbances. Specific differences in the feeding mode of herbivorous fishes, as well as variation in body size, group size and home ranges, suggest that each and every species of herbivorous fish may have a subtly different role in preventing or reversing proliferation of macroalgae following extensive coral depletion (Mouillot et al. 2014). Certainly, there are some individual species that contribute disproportionately to specific functions. For example, overfishing of the excavating parrotfish, *Bolbometopon muricatum*, has resulted in a near total loss of external bioerosion on several Indo-Pacific reefs (Bellwood et al. 2003, 2012) even though there are several other excavating species on these reefs. In many regions, fisheries exploitation represents the foremost threat to herbivorous fishes (Graham et al. 2011), though it is also apparent that at least some herbivorous fishes decline in abundance following extensive coral depletion (Fig. 11.2). While some herbivorous fishes do increase in abundance following coral depletion, presumably responding to increases in the areal extent and/or productivity of algae, these responses are not consistent either within or among species. Most notably, there are several species of acanthurids (e.g. *Acanthurus lineatus* and *A. tennenti*) that have exhibited contrasting responses (increases versus decreases in abundance) to coral loss in different studies conducted at different locations. Moreover, apparent declines in the abundance of several species of croppers and browsers (Fig. 11.2) are based on results from just one study location. Such vagaries in responses of herbivorous fishes to changes in resource availability may reflect the role of irregular larval supply and recruitment in facilitating population increases. However, rapid increases in the abundance of some fishes almost certainly reflect the aggregation and movement of fishes into degraded reef habitats (Hart et al. 1996), which may also be constrained by local densities and distributions of fishes.

11.3.2 Loss of Biodiversity and Function

The extent to which declines in the abundance of fishes will constrain key ecological functions depends on the functional identity of component species and the variation in responses to coral loss among species that contribute to similar functions. For fishes that are directly reliant on corals for food, it is to be expected that entire functional groups will respond similarly to the localised depletion of scleractinian corals (e.g. Pratchett et al. 2011b). However, the responses of other groups are much less clear. For herbivorous fishes, several studies have reported increases in the abundance of specific species or groups following large-scale reductions in coral cover (e.g. Adam et al. 2011; Gilmour et al. 2013), and previous meta-analyses have shown that the majority of species that respond positively to coral loss are herbivorous fishes, though such effects may be relatively short-lived (Wilson et al. 2006; Pratchett et al. 2008a). As discussed previously, increases in the abundance of herbivorous fishes are generally attributed to increased cover and availability of turfing algae, which rapidly colonise dead coral skeletons (Mumby and Steneck 2008). When large numbers of corals die, there may also be a significant, though temporary, input of nutrients that actually increase algal productivity and growth. However, beneficial effects of increased food availability following the loss of live corals may be more than offset by declines in the availability of preferred settlement habitats and increased intensity of competition and predation within degraded reef habitats.

Comparisons of the standardised responses to coral loss for each of 19 distinct functional (or trophic) groups show that while there was considerable variation in the response of individual species within each group, the net effect is a decline in abundance for 18 of the 19 functional groups examined (Fig. 11.3). The only exception was the macroalgal browsers (*Naso* spp. and *Siganus* spp.) that show a net positive response to coral loss. Interestingly, these browsing fishes do not associate with live coral or structurally complex areas at settlement, instead settling to areas of coral rubble (e.g. *Naso unicornis*, Doherty et al. 2004) or dense macroalgae, predominantly *Sargassum* (*Siganus* spp., Hoey et al. 2013; Evans et al. 2014). This, coupled with the increased availability of algal resources following coral mortality (e.g. Diaz-Pulido and McCook 2002), suggests these species are not reliant on live coral and may actually prosper in habitats with depauperate coral cover.

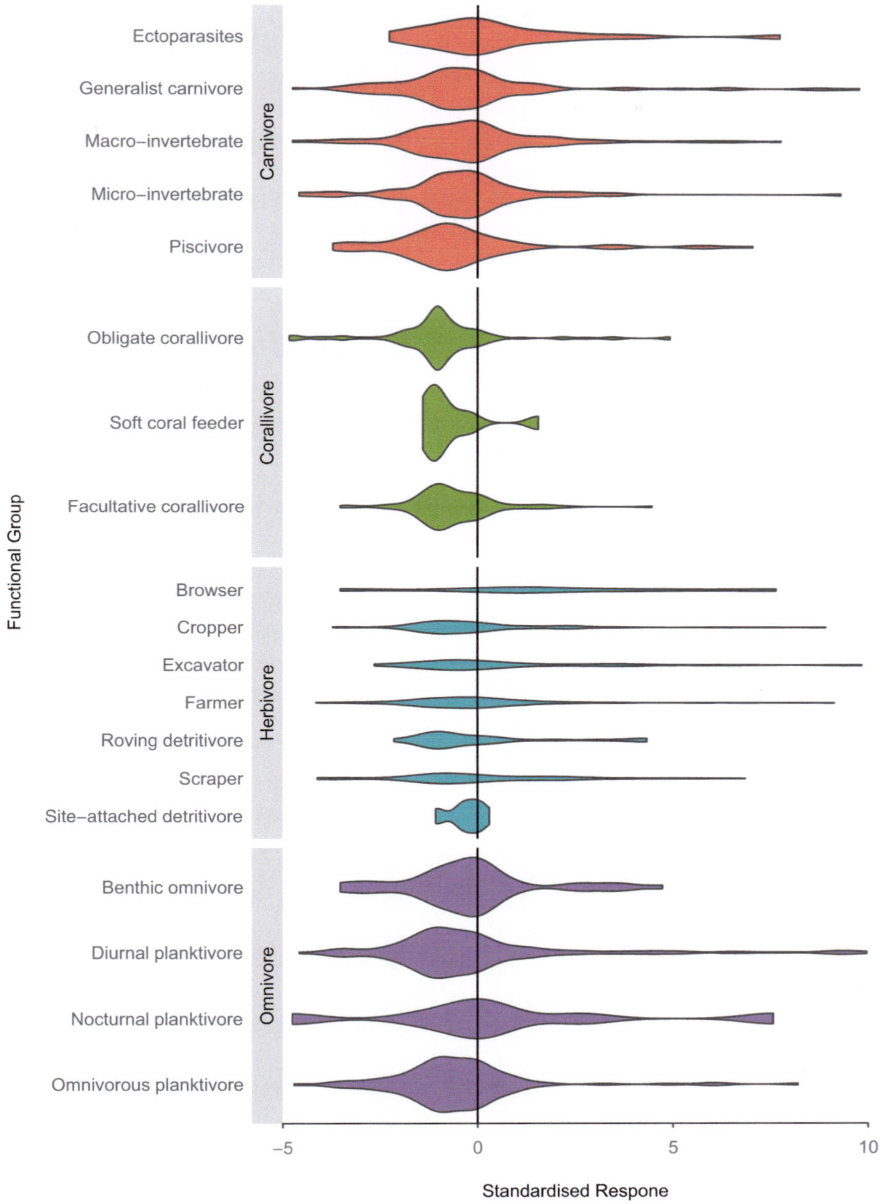

Fig. 11.3 Variation in standardised responses to coral loss (proportional change in the abundance of individual species divided by proportional declines in in local coral cover) for 19 functional (trophic) groups of reef fishes. Responses are predominantly negative showing the broad range of fishes that decline in abundance following coral loss

Although the net response of most functional groups to coral loss was negative, there was considerable variation among species within each group. This variation likely reflects differences in the intensity and/or nature of the disturbance, the temporal scale over which the changes were quantified and the longevity of the species and the reliance of individual species on live coral for settlement, food and/or shelter. For example, many scraping and excavating parrotfishes settle to macroalgal (Green 1998) or mangrove habitat (Dorenbosch et al. 2006), yet other species, such as the bumphead parrotfish *Bolbometopon muricatum*, settle to branching *Acropora* (Hamilton et al. 2017). The reliance of *B. muricatum* on live coral at settlement is of particular concern as, when present, it overwhelmingly dominates the process of external bioerosion on reefs (Bellwood et al. 2003, 2012), and as such the loss of corals could undermine this ecological process. Similarly, differences in settlement habitat are evident within other function groups. For example, the piscivorous coral trout *Plectropomus leopardus* settles to rubble habitats (Light and Jones 1997), while the congener *P. maculatus* settles to branching *Acropora* (Wen et al. 2013). Predicting the effects of coral loss on the functional composition of reef fish assemblages is complex. Despite some fishes appearing to be largely unaffected by coral loss (Emslie et al. 2017) or even thriving following coral loss (e.g. Adam et al. 2011), the net effect of extensive coral depletion on almost all functional groups is negative (Fig. 11.3). Although changes in the abundance do not account for differences in the functional contribution of individual species, the prevalence of negative impacts suggests ongoing coral loss is likely to compromise the functioning and productivity of coral reefs, especially given cumulative effects of multiple disturbances (Brandl et al. 2016).

11.3.3 Effects of Coral Reef Degradation on Tropical Fisheries Production

Coral reef ecosystems are a major contributor to tropical coastal fisheries production, with coral reef fishes accounting for up to 65% of coastal fisheries production in the tropical Pacific (Bell et al. 2013). Coral reef fisheries typically target multiple species across almost all trophic levels, including piscivores, invertivores and herbivores (Dalzell et al. 1996). Differential effects of coral bleaching and depletion across major functional groups are, therefore, expected to cause changes in catch composition, if not declines in overall fisheries production. Cheung et al. (2013) showed that there have been marked shifts in catch composition for tropical fisheries from the 1970s to 2006. These changes are related to changing thermal regimes, differential thermal sensitivities of tropical fishes and increasing redistribution of fishes to match changing climatic envelopes (Cheung et al. 2010). However, significant and widespread habitat degradation due to increasing incidence and severity of mass coral bleaching will also have consequences for coral reef fish and fisheries (Pratchett et al. 2011c; Bell et al. 2013). By 2100, the combined effects of habitat degradation and ocean warming, together with ocean acidification, are projected to cause a 20–50% decline in sustainable fisheries production of demersal fishes from tropical coastal environments across Pacific island countries and territories (Pratchett et al. 2011c). These projected declines are mostly linked to ongoing habitat loss across coral reef, seagrass and mangrove habitats, though the projected rates of habitat loss (e.g. 50% decline in mean coral cover by 2035) may have been overly conservative given the habitat loss that is likely to have occurred during recent pantropical mass coral bleaching events. Projected declines in coral reef fisheries productivity caused by climate change could equate to losses of up to US$8.4 billion per annum by the year 2100 (Speers et al. 2016). Given the increasing demand for fish with significant increases in human population across tropical island nations, the degradation of coral habitats from bleaching poses a major threat to future food security (Bell et al. 2017).

The specific effects of coral bleaching on fisheries species and production remain equivocal (Brander 2007; Cinner et al. 2013), given difficulties in discerning climatic signals against background fluctuations in catch and effort (McClanahan et al. 2002; Grandcourt and Cesar 2003). Graham et al. (2007) found declines in the size structure of fisheries target species following extensive mass bleaching in the Seychelles. It is also clear that extensive coral depletion, especially when combined with declines in topographic complexity, can impact on the abundance of large-bodied reef fishes and fisheries target species (Pratchett et al. 2011c, 2017). Most fishes that associate with live coral are small-bodied (Coker et al. 2014) and are not typically targeted by fishers. Reduced abundance of these small-bodied fish can, however, impact on prey availability for larger piscivorous species (Wen et al. 2016). For example, reduced abundance of coral-dwelling planktivores following bleaching resulted in an increased benthic, rather than pelagic, isotopic signature in the barred-cheek coral trout, *Plectropomus maculatus* (Hempson et al. 2017a). This shift in dietary composition was also associated with declines in individual condition, potentially attributable to declines in resource availability within degraded reef systems (Hempson et al. 2017b). This may explain why densities of these *P. maculatus* covary with fluctuations in live coral cover (Williamson et al. 2014), though coral habitats are also important for growth and survival of newly settled individuals (Wen et al. 2013) which feed on benthic invertebrates that are generally more prevalent in the presence of live corals.

Previous studies (and meta-analyses) on the susceptibility of reef fishes to coral depletion and changes in the structure of reef habitats emphasised the vulnerability of small-bodied species (Wilson et al. 2006; Graham et al. 2008), which often have closer association with benthic habitats and explicitly utilise complex habitats to moderate exposure to predators. However, more recent analyses show that species with large body size are equally vulnerable to coral depletion and reef degradation as their smaller-bodied counterparts (Pratchett et al. 2014), though there may be significant lags in the time required for such declines to become apparent. The mean and modal standardised responses to coral loss for all groups of carnivorous fishes, which include many large-bodied piscivores and macro-invertebrate feeders, were negative (Fig. 11.3). The mechanistic basis for such declines is likely to be complex and vary among species. However, high coral cover and topographic complexity may be critical to the feeding success of many ambush predators, such as groupers (Kerry and Bellwood 2012). Some larger-bodied fisheries species (e.g. *Bolbometopon muricatum*) may also rely on specific corals or coral-rich habitats for settlement (Rogers et al. 2014; Hamilton et al. 2017), even though adults are only loosely associated with coral habitats, which would explain why extensive coral depletion results in protracted declines in the abundance of these species. The overall importance of corals (cf. algae or other reef habitats) for inducing settlement or promoting survival of newly settled fishes is still largely unknown (but may have been greatly underappreciated, Jones et al. 2004), as habitat requirements for many juvenile coral reef fish are yet to be established (Wilson et al. 2010).

Dramatic shifts in the species composition, abundance and biomass of reef fishes are likely to occur following shifts from coral- to macroalgae-dominated systems (Chong-Seng et al. 2014; Graham et al. 2015; Ainsworth and Mumby 2015), which may become increasingly common following severe mass coral bleaching and will have obvious connotations for fisheries production. Most notably, the abundance of many traditional fisheries species is suppressed on reefs dominated by macroalgae (Ainsworth and Mumby 2015). There are, however, other fishes that are very abundant on reefs with high cover of macroalgae (Graham et al. 2014), which might still sustain high fisheries productivity, albeit based on a different suite of species (Ainsworth and Mumby 2015). Exploring new fishing opportunities and adapting to changes in resource availability may help to close the gap between productivity of coral reef fisheries and increasing fisheries demands in tropical island countries (Bell et al. 2017). Embracing such changes must, however, take account of the ecological importance of some fishes and balance the needs for fisheries production versus contributions of fishes to ecosystem function. If, for example, fisheries capitalise on the increased abundance of herbivorous fish that can occur following mass coral bleaching (Wilson et al. 2006; Pratchett et al. 2008a), this may undermine the capacity for recovery and reassembly of coral-dominated habitats (Bozec et al. 2016).

11.4 Changes in the Behaviour and Fitness of Reef Fishes

While there are many studies that have documented changes in the abundance and diversity of fishes on reefs affected by coral bleaching and coral loss (Graham et al. 2008), sublethal effects of such disturbances have often been overlooked. However, fishes may respond to coral bleaching and coral loss in ways other than absolute changes in their local abundance (Table 11.1). These differences in feeding rates and diet, physiological condition and growth may forewarn of longer-term impacts on individual survival and population viability, but understanding sublethal effects of coral depletion may also help to elucidate how and why reef fishes associate with coral-dominated habitats. For example, *Chaetodon auriga* declines in abundance following localised coral loss (Bouchon-Navaro et al. 1985; Pratchett et al. 2015b) despite feeding very little, if at all, on scleractinian corals throughout much of its

Table 11.1 Sublethal effects of mass coral bleaching and coral depletion on coral reef fishes, highlighting (A) behavioural shifts associated with declines in the local abundance, cover or diversity of corals and (B) consequences for individual fitness, such as declines in condition and growth

(A) Behavioural shifts			
Effect	Observed change	Time frame	Species and data source(s)
Feeding on bleached corals	Preferential feeding on bleached corals	Hours to days	*Labrichthys unilineatus* (McIlwain and Jones 1997; Cole et al. 2009) *Chaetodon baronessa* (Cole et al. 2009)
	Active avoidance of bleached corals	9 Days	*Chaetodon plebeius* (Pisapia et al. 2012)
Overall bite rates	Increased with increasing coral cover	Space for time comparison	*Chaetodon auriga* (Pratchett et al. 2014)
	No difference	Space for time comparison	*Chaetodon vagabundus* (Pratchett et al. 2014)
Shifts in dietary composition	Reduced intake of benthic invertebrates	1–2 Years	Juvenile *Plectropomus maculatus* (Wen et al. 2016)
	Increased intake of herbivorous fishes	3–7 Years	*Plectropomus maculatus* (Hempson et al. 2017a)
	Increased use of massive corals	2 Years	*Chaetodon lunulatus* (Pratchett et al. 2004)
	No change despite loss of preferred prey	1 Year	*Oxymonacanthus longirostris* (Brooker et al. 2014)
Territoriality and aggression	Larger territories with declining coral cover	Space for time comparison	*Labrichthys unilineatus* (McIlwain and Jones 1997)
	Larger territories and reduced aggression at low coral cover	Space for time comparison	*Chaetodon baronessa* (Berumen and Pratchett 2006)
	Increased inter- vs intraspecific aggression in degraded habitats	20 Days	*Pomacentrus moluccensis* (Kok et al. 2016)
Susceptibility and exposure to predators	Failure to respond to predator odour cues on dead corals	Immediate	*Pomacentrus amboinensis* (Lönnstedt et al. 2014)
	Failure to respond to conspecific alarm cues in degraded habitats	Immediate	*Pomacentrus moluccensis* and *Chromis viridisi* (Ferrari et al. 2017b)
	Water from dead corals impedes escape responses	Immediate	*Pomacentrus amboinensis*, *P. chrysurus*, *P. wardi*, (McCormick and Allan 2017)
	Dead coral impedes social learning and predator recognition	Immediate	*Pomacentrus amboinensis*, *P. nagasakiensis* (Chivers et al. 2016)
	Higher strike rates by predators on prey fishes against bleached corals	Immediate	*Pomacentrus moluccensis*, *Dascyllus aruanus* (Coker et al. 2009)

(continued)

Table 11.1 (continued)

(B) Consequences for individual fitness			
Effect	Observed change	Time frame	Species and data source(s)
Physiological condition	Declines in condition when feeding on suboptimal prey	2 Years	*Chaetodon lunulatus* (Pratchett et al. 2004)
	No differences between bleached versus unbleached corals	Space for time comparison	*Dascyllus aruanus* (Coker et al. 2015)
	No change, despite increased use of suboptimal habitats	1 Month	*Chrysiptera parasema*, *Dascyllus melanurus* (Feary et al. 2009)
Spawning	Lack of spawning for fishes deprived of access to preferred prey	3 Weeks	*Oxymonacanthus longirostris* (Brooker et al. 2013)
Growth rates	Limited growth following extensive depletion of coral prey	1 Year	*Oxymonacanthus longirostris* (Kokita and Nakazono 2001)
	Growth rates directly related to tissue cover of coral hosts	1 Month	*Chrysiptera parasema*, *Dascyllus melanurus* (Feary et al. 2009)
	No difference when feeding on bleached versus unbleached corals	23 Days	*Chaetodon aureofasciatus*, *C. lunulatus* (Cole et al. 2014)

geographical range. Moreover, *C. auriga* preferentially settles to near-shore patch reef habitats with limited cover of live corals (Pratchett et al. 2008b), questioning why this species would be negatively affected by coral depletion. However, Pratchett et al. (2015b) showed that feeding rates of *C. auriga* (on non-coral substrates) increase with increasing coral cover. This suggests that corals support increased abundance of prey items (such as small crustaceans and other cryptofauna living within the algal turfs) consumed by *C. auriga*. The cryptofauna associated with algal turfs is an important contributor to the trophic dynamics of shallow reef systems (Kramer et al. 2013), and though it is unknown whether high coral cover increases overall productivity of cryptobenthic assemblages or favours only specific species consumed by *C. auriga*, this is a potentially important link in understanding the far-reaching effects of mass coral bleaching and coral depletion on coral reef fishes.

11.4.1 Behavioural Changes

Mass coral bleaching and associated changes in the structure of reef habitats have important and far-reaching effects on the availability of prey resources and not only for those fishes that feed directly on live corals (Wen et al. 2013; Pratchett et al. 2015b). Where possible, fishes would be expected to respond to changes in resource availability by altering their dietary intake and feeding behaviour (e.g. Pratchett et al.

2004). However, diets of some highly specialist species are inflexible (Berumen and Pratchett 2008; Brooker et al. 2014), making them extremely susceptible to prey depletion. Even for fishes that can modify dietary intake in accordance with changing prey availability, this may have significant consequences for individual fitness and long-term survival. Following mass coral bleaching in the central Great Barrier Reef, Australia, Pratchett et al. (2004) showed that *Chaetodon lunulatus* increased its intake of bleaching-resistant coral species. This shift in diet composition enabled adult fishes to persist following the bleaching, though it did have consequences for physiological conditions (Pratchett et al. 2004), which may, in turn, affect longer-term survival and reproductive output (discussed below). Reductions in live coral cover also increase the area over which corallivores forage (e.g. McIlwain and Jones 1997; Kokita and Nakazono 2001), potentially increasing competitive encounters and exposure to predators.

One of the foremost explanations put forward to explain high abundance and diversity of fishes in coral-rich habitats is the extent to which high levels of habitat diversity and complexity will moderate outcomes of competition and predation (Almany 2004). This hypothesis is not readily testable, because the persistence and coexistence of multiple species depend on the outcomes of numerous biological interactions over extended time scales, rather than the changes in the absolute intensity of competition and predation. It would be expected, however, that resource competition would become more intense (especially among coral-dependent fishes) following local coral depletion. Similarly, predation rates might be expected to increase with coral loss and declines in fine-scale topographic complexity of reef habitats (Almany 2004). Thus far, there has been limited evidence of increased intensity of competition among reef fishes following coral loss (Table 11.1). One of the more unexpected consequences of coral loss is that fishes appear to be much more susceptible to predation when corals bleach and die (Coker et al. 2009; Chivers et al. 2016; Ferrari et al. 2017b; Table 11.1), attributable to both changes in habitat structure and apparent changes in the behaviour of fishes in degraded reef environments. Notably, Chivers et al. (2016) showed that water that has passed over dead and degraded corals impedes social learning and predator recognition among the damselfish, *Pomacentrus amboinensis*. Lönnstedt et al. (2014) suggested that dead coral masks the odours of potential predators. Accordingly, field experiments showed that damselfish (*P. amboinensis*) associated with dead coral hosts did not exhibit characteristic sheltering within corals when exposed to the odour of a predator but did so upon seeing the predator. Interestingly, *P. amboinensis* is one of the few coral-dwelling damselfishes that is commonly found on dead coral hosts, especially after major disturbances (Pratchett et al. 2012), and these effects may be even stronger for obligate coral-dwelling species.

There has been comparatively little research on the specific behavioural responses of fishes to host coral bleaching or mortality (but see Coker et al. 2009, 2013). It is implicitly assumed, for example, that declines in the abundance of coral-dwelling fishes following extensive and widespread coral loss are due to elevated rates of individual mortality, due to compromised health of the individual fishes and increased susceptibility to predation (Sano et al. 1984, 1987; Jones et al. 2004;

Pratchett et al. 2008a), which may be compounded by declines in local settlement rates (Feary et al. 2007). However, Coker et al. (2009) showed that prey fishes associated with bleached corals are much more susceptible to predation compared to conspecifics living on unbleached corals, which was attributed to both visual and chemical camouflage provided by live coral tissues. Even if predation rates are not actually any higher for fishes on bleached corals, it is possible that increased exposure to predators may provide significant motivation for coral-dwelling fishes to rapidly vacate bleached coral hosts (Sano et al. 1987; Coker et al. 2009). This might also explain the reluctance of fishes to settle on bleached corals (e.g. Feary et al. 2007).

11.4.2 Effects of Coral Bleaching Versus Coral Loss on Individual Fitness of Fishes

Where coral bleaching and/or coral mortality leads to depletion of resources for reef fishes, species losses are likely to be preceded or accompanied by declines in individual condition or fitness (Kokita and Nakazono 2001; Pratchett et al. 2006; Brooker et al. 2013). Kokita and Nakazono (2001) documented localised extirpation of *Oxymonacanthus longirostris* over 2 years following coral bleaching in Okinawa, Japan (see also Brooker et al. 2014). However, even before these fishes disappeared, growth rates were severely compromised (Kokita and Nakazono 2001), reflecting limited access to their preferred coral prey, *Acropora*, which had succumbed to the bleaching and died. Experimental studies conducted by Brooker et al. (2013) also showed that constrained access to preferred coral prey by breeding pais of *O. longirostris* resulted in reproductive failure. It is also possible that declines in the nutritional quality of coral prey may occur even during bleaching, due to rapid depletion of lipid reserves (but see Pisapia et al. 2012). Some coral-feeding fishes will initially target bleached corals over healthy coral prey, potentially due to increased mucous production or limited tissue retraction, but preferentially target healthy corals over colonies subject to prolonged bleaching (Cole et al. 2009). It is clear that corallivorous fishes will consume bleached corals, but protracted feeding on bleached corals may constrain nutritional intake. However, Cole et al. (2014) found no apparent differences in the growth rates of juvenile butterflyfishes that were restricted to feeding on bleached versus healthy corals for 23 days. The findings suggest that bleaching per se may have limited effects on corallivorous reef fishes (see also Bonin et al. 2009; McCormick et al. 2010). Rather, it is the subsequent mortality of bleached coral colonies that impacts on food availability leading to declines in the physiological condition and ultimately the survival of corallivorous fishes.

Aside from coral-feeding fishes (e.g. butterflyfishes), coral-dwelling fishes have the most direct and explicit reliance on corals (Pratchett et al. 2012) and are extremely vulnerable to local depletion of their specific coral hosts (Munday

2004). Many coral-dwelling fishes vacate their coral hosts as soon as they bleach, let alone die (Feary et al. 2007; Coker et al. 2012). Redistribution of coral-dwelling fishes among remnant coral hosts may moderate the susceptibility of these fishes to host coral depletion, though overall densities often decline in approximate accordance with the proportional loss of preferred coral habitats (Wilson et al. 2008; Pratchett et al. 2012). The proximate causes of declines in the abundance of coral-dwelling fishes following host coral mortality remain largely unknown. Explicit comparisons of the physiological condition of several different species of coral-dwelling damselfishes (Table 11.1) constrained to living on bleached versus healthy coral hosts reveal little or no differences. This suggests that declines in the abundance of coral-dwelling fishes within habitats subject to host coral depletion are due to extrinsic processes (e.g. predation) rather than intrinsic factors. However, sublethal effects of coral depletion need to be considered across a much wider range of reef fish species.

11.5 Conclusions

Highly diverse and productive assemblages of coral reef fishes rely on the combination of high abundance (cover) and diversity of scleractinian corals, as well as high levels of topographic complexity, which in itself is often provided by high coral growth. Mass coral bleaching, which can cause extensive coral loss, will, therefore, have significant effects on the structure and function of reef fish assemblages (e.g. Wilson et al. 2006; Graham et al. 2006, 2008), with potential consequences for fisheries production and coral reef resilience. The loss of entire functional groups that comprise multiple species may appear unlikely. However, biodiversity of coral reef fishes is not equally apportioned among different functional groups (Mouillot et al. 2014). There are also entire functional groups that respond similarly to coral bleaching and coral loss, such that certain ecological functions will be severely compromised, if not lost altogether (Graham et al. 2011). It is now clear that effects of coral bleaching and associated coral loss extend well beyond those species traditionally thought to have specific reliance on corals for food and shelter (e.g. butterflyfishes, damselfishes, gobies). In extreme cases, the abundance and species richness of fishes may decline >60% following extensive coral depletion and topographic collapse of reef habitats, combined with increasing dominance of non-coral biota. The spatial extent of mass bleaching events is also far greater than other major disturbances, potentially threatening widespread species with extinction. This provides significant imperative for reducing greenhouse gas emissions, to reduce the incidence and severity of future mass coral bleaching while also addressing other more localised disturbances that contribute to coral loss and reef degradation.

References

Adam TC, Schmitt RJ, Holbrook SJ, Brooks AJ, Edmunds PJ, Carpenter RC et al (2011) Herbivory, connectivity, and ecosystem resilience: response of a coral reef to a large-scale perturbation. PLoS One 6:e23717. https://doi.org/10.1371/journal.pone.0023717

Ainsworth CH, Mumby P (2015) Coral–algal phase shifts alter fish communities and reduce fisheries production. Glob Change Biol 21:165–172. https://doi.org/10.1111/gcb.12667

Almany GR (2004) Does increased habitat complexity reduce predation and competition in coral reef fish assemblages? Oikos 106:275–284. https://doi.org/10.1111/j.0030-1299.2004.13193.x

Alvarez-Filip L, Dulvy NK, Côté IM, Watkinson AR, Gill JA (2011) Coral identity underpins architectural complexity on Caribbean reefs. Ecol Appl 21:2223–2231. https://doi.org/10.1890/10-1563.1

Baird AH, Marshall PA (2002) Mortality, growth and reproduction in scleractinian corals following bleaching on the Great Barrier Reef. Mar Ecol Prog Ser 237:133–141. https://doi.org/10.3354/meps237133

Baker AC, Glynn PW, Riegl B (2008) Climate change and coral reef bleaching: an ecological assessment of long-term impacts, recovery trends and future outlook. Estuar Coast Shelf Sci 80:435–471. https://doi.org/10.1016/j.ecss.2008.09.003

Bell JB, Ganachaud A, Gehrke PC, Griffiths SP, Hobday AJ, Hoegh-Guldberg O et al (2013) Mixed responses of tropical Pacific fisheries and aquaculture to climate change. Nat Clim Change 3:591–599. https://doi.org/10.1038/nlimate1838

Bell JD, Cisneros-Montemayor A, Crowder L, Hanich Q, Johnson J, Lehodey P et al (2017) Adaptations to maintain the contributions of small-scale fisheries to food security in the Pacific Islands. Mar Policy. https://doi.org/10.1016/j.marpol.2017.05.019

Bellwood DR, Hoey AS, Choat JH (2003) Limited functional redundancy in high diversity systems: resilience and ecosystem function on coral reefs. Ecol Lett 6:281–285. https://doi.org/10.1046/j.1461-0248.2003.00432.x

Bellwood DR, Hughes TP, Folke C, Nyström M (2004) Confronting the coral reef crisis. Nature 429:827–833. https://doi.org/10.1038/nature02691

Bellwood DR, Hoey AS, Ackerman JL, Depczynski M (2006) Coral bleaching, reef fish community phase shifts and the resilience of coral reefs. Glob Change Biol 12:1587–1594. https://doi.org/10.1111/j.1365-2486.2006.01204.x

Bellwood DR, Hoey AS, Hughes TP (2012) Human activity selectively impacts the ecosystem roles of parrotfishes on coral reefs. Proc R Soc B 279:1621–1629. https://doi.org/10.1098/rspb.2011.1906

Berumen ML, Pratchett MS (2006) Effects of resource availability on the competitive behaviour of butterflyfishes (Chaetodontidae). In: Proceedings of 10th international coral reef symposium, Japan, vol 1, pp 644–650

Berumen ML, Pratchett MS (2008) Trade-offs associated with dietary specialization in corallivorous butterflyfishes (Chaetodontidae: Chaetodon). Behav Ecol Sociol 62:989–994. https://doi.org/10.1007/s00265-007-0526-8

Blowes S, Pratchett MS, Connolly S (2013) Heterospecific aggression, dominance and specialization in a guild of corallivorous reef fishes. Am Nat 182:157–168. https://doi.org/10.1086/670821

Bonin MC, Munday PL, McCormick MI, Srinivasan M, Jones GP (2009) Coral-dwelling fishes resistant to bleaching but not to mortality of host corals. Mar Ecol Prog Ser 394:215–222. https://doi.org/10.3354/meps08294

Bouchon-Navaro Y, Bouchon C, Harmelin-Vivien ML (1985) Impact of coral degradation on a chaetodontid fish assemblage (Moorea, French Polynesia). In: Proceedings of 5th coral reef symposium, Tahiti, vol 5, pp 427–432

Bozec YM, O'Farrell S, Bruggemann JH, Luckhurst BE, Mumby PJ (2016) Tradeoffs between fisheries harvest and the resilience of coral reefs. Proc Natl Acad Sci USA 113:4536–4541. https://doi.org/10.1073/pnas.1601529113

Brander KM (2007) Global fish production and climate change. Proc Natl Acad Sci USA 104:19709–19714. https://doi.org/10.1073/pnas.0702059104

Brandl SJ, Emslie MJ, Ceccarelli DM, Richards ZT (2016) Habitat degradation increases functional originality in highly diverse coral reef fish assemblages. Ecosphere 7:e01557. https://doi.org/10.1002/ecs2.1557

Brooker RM, Jones GP, Munday PL (2013) Prey selectivity affects reproductive success of a corallivorous reef fish. Oecologia 172:409–416. https://doi.org/10.1007/s00442-012-2521-7

Brooker RM, Munday PL, Brandl SJ, Jones GP (2014) Local extinction of a coral reef fish explained by inflexible prey choice. Coral Reefs 33:891–896. https://doi.org/10.1007/s00338-014-1197-3

Cheal AJ, MacNeil MA, Emslie MJ, Sweatman HPA (2017) The threat to coral reefs from more intense cyclones under climate change. Glob Change Biol 23:1511–1524. https://doi.org/10.1111/gcb.13593

Cheung WW, Lam VW, Sarmiento JL, Kearney K, Watson R, Zeller D et al (2010) Large-scale redistribution of maximum fisheries catch potential in the global ocean under climate change. Glob Change Biol 16:24–35. https://doi.org/10.1111/j.1365-2486.2009.01995.x

Cheung WW, Watson R, Pauly D (2013) Signature of ocean warming in global fisheries catch. Nature 497:365–368. https://doi.org/10.1038/nature12156

Chivers DP, McCormick MI, Allan BJ, Ferrari MC (2016) Risk assessment and predator learning in a changing world: understanding the impacts of coral reef degradation. Sci Rep 6:32542. https://doi.org/10.1038/srep32542

Chong-Seng KM, Graham NAJ, Pratchett MS (2014) Bottlenecks to coral recovery in Seychelles. Coral Reefs 33:449–461. https://doi.org/10.1007/s00338-014-1137-2

Cinner JE, Huchery C, Darling ES, Humphries AT, Graham NA, Hicks CC et al (2013) Evaluating social and ecological vulnerability of coral reef fisheries to climate change. PloS One 8:e74321. https://doi.org/10.1371/journal.pone.0074321

Coker DJ, Pratchett MS, Munday PL (2009) Coral bleaching and habitat degradation increase susceptibility to predation for coral-dwelling fishes. Behav Ecol 20:1204–1210. https://doi.org/10.1093/beheco/arp113

Coker DJ, Graham NAJ, Pratchett MS (2012) Interactive effects of live coral and structural complexity on the recruitment of reef fishes. Coral Reefs 31:919–927. https://doi.org/10.1007/s00338-012-0920-1

Coker DJ, Walker S, Munday P, Pratchett MS (2013) Social group entry rules may limit population resilience to patchy habitat disturbance. Mar Ecol Prog Ser 493:237–242. https://doi.org/10.3354/meps10493

Coker DJ, Wilson SK, Pratchett MS (2014) Importance of live coral habitat for reef fishes. Rev Fish Biol Fish 24:89–126. https://doi.org/10.1007/s11160-013-9319-5

Coker DJ, Nowicki JP, Pratchett MS (2015) Body condition of the coral-dwelling fish Dascyllus aruanus (Linnaeus 1758) following host colony bleaching. Environ Biol Fish 98:691–695. https://doi.org/10.1007/s10641-014-0277-0

Cole AJ, Pratchett MS, Jones GP (2008) Diversity and functional importance of coral-feeding fishes on tropical coral reefs. Fish Fish 9:286–307. https://doi.org/10.1111/j.1467-2979.2008.00290.x

Cole AJ, Pratchett MS, Jones GP (2009) Effects of coral bleaching on the feeding response of two species of coral-feeding fish. J Exp Mar Biol Ecol 373:11–15. https://doi.org/10.1016/j.jembe.2009.02.016

Cole AJ, Lawton RJ, Pisapia C, Pratchett MS (2014) The effects of coral bleaching on settlement preferences and growth of juvenile butterflyfishes. Mar Environ Res 98:106–110. https://doi.org/10.1016/j.marenvres.2014.03.003

Dahlgren CP, Eggleston DB (2000) Ecological processes underlying ontogenetic habitat shifts in a coral reef fish. Ecology 81:2227–2240. https://doi.org/10.1890/0012-9658(2000)081[2227,EPUOHS]2.0.CO;2

Dalzell P, Adams TJH, Polunin NVC (1996) Coastal fisheries in the Pacific Islands. Oceanogr Mar Biol Annu Rev 34:395–531

De'ath G, Fabricius KE, Sweatman H, Puotinen M (2012) The 27–year decline of coral cover on the Great Barrier Reef and its causes. Proc Natl Acad Sci USA 109:17995–17999. https://doi.org/10.1073/pnas.1208909109

Diaz-Pulido G, McCook LJ (2002) The fate of bleached corals: patterns and dynamics of algal recruitment. Mar Ecol Prog Ser 232:115–128. https://doi.org/10.3354/meps232115

Dixson DL, Abrego D, Hay ME (2014) Chemically mediated behavior of recruiting corals and fishes: a tipping point that may limit reef recovery. Science 345:892–897. https://doi.org/10.1126/science.1255057

Doherty PJ, Dufour V, Galzin R, Hixon MA, Meekan MG, Planes S (2004) High mortality during settlement is a population bottleneck for a tropical surgeonfish. Ecology 85:2422–2428. https://doi.org/10.1890/04-0366

Dorenbosch M, Grol MGG, Nagelkerken I, Van der Velde G (2006) Seagrass beds and mangroves as potential nurseries for the threatened Indo-Pacific humphead wrasse, Cheilinus undulatus and Caribbean rainbow parrotfish, Scarus guacamaia. Biol Conserv 129:277–282. https://doi.org/10.1016/j.biocon.2005.10.032

Elmqvist T, Folke C, Nyström M, Peterson M, Bengtsson J, Walker B et al (2003) Response diversity, ecosystem change, and resilience. Front Ecol Environ 1:488–494. https://doi.org/10.1890/1540-9295(2003)001[0488,RDECAR]2.0.CO;2

Emslie MJ, Pratchett MS, Cheal AJ (2011) Effects of different disturbance types on butterflyfish communities of Australia's Great Barrier Reef. Coral Reefs 30:461–471. https://doi.org/10.1007/s00338-011-0730-x

Emslie MJ, Cheal AJ, Johns KA (2014) Retention of habitat complexity minimizes disassembly of reef fish communities following disturbance: a large-scale natural experiment. PLoS One 9: e105384. https://doi.org/10.1371/journal.pone.0105384

Emslie MJ, Cheal AJ, Logan M (2017) The distribution and abundance of reef-associated predatory fishes on the Great Barrier Reef. Coral Reefs 36:829–846. https://doi.org/10.1007/s00338-017-1573-x

Evans RD, Wilson SK, Field SN, Moore JAY (2014) Importance of macroalgal fields as coral reef fish nursery habitat in north-west Australia. Mar Biol 161:599–607. https://doi.org/10.1007/s00227-013-2362-x

Feary DA, Almany GR, McCormick MI, Jones FP (2007) Habitat choice, recruitment and the response of coral reef fishes to coral degradation. Oecologia 153:727–737. https://doi.org/10.1007/s00442-007-0773-4

Feary DA, McCormick MI, Jones GP (2009) Growth of reef fishes in response to live coral cover. J Exp Mar Biol Ecol 373:45–49. https://doi.org/10.1016/j.jembe.2009.03.002

Ferrari R, Figueira WF, Pratchett MS, Boube T, Adam A, Kebelkowsky-Vidrio T et al (2017a) 3D photogrammetry quantifies growth and external erosion of individual coral colonies and skeletons. Sci Rep 7. https://doi.org/10.1038/s41598-017-16408-z

Ferrari MC, McCormick MI, Allan BJ, Chivers DP (2017b) Not equal in the face of habitat change: closely related fishes differ in their ability to use predation-related information in degraded coral. Proc R Soc B 284. https://doi.org/10.1098/rspb.2016.2758

Gardiner NM, Jones GP (2005) Habitat specialisation and overlap in a guild of coral reef cardinal fishes (Apogonidae). Mar Ecol Prog Ser 305:163–175. https://doi.org/10.3354/meps305163

Gardner TA, Côté IM, Gill JA, Grant A, Watkinson AR (2003) Long-term region-wide declines in Caribbean corals. Science 301:958–960. https://doi.org/10.1126/science.1086050

Gilmour JP, Smith LD, Heyward AJ, Baird AH, Pratchett MS (2013) Recovery of an isolated coral reef system following severe disturbance. Science 340:69–71. https://doi.org/10.1126/science.1232310

Glynn PW (1997) Bioerosion and coral reef growth: a dynamic balance. In: Birkeland C (ed) Life and death of coral reefs. Chapman & Hall, New York, pp 68–95

Goreau T, McClanahan TR, Hayes R, Strong A (2000) Conservation of coral reefs after the 1998 global bleaching event. Conserv Biol 14:5–15. https://doi.org/10.1046/j.1523-1739.2000.00011.x

Graham NAJ, Nash KL (2013) The importance of structural complexity in coral reef ecosystems. Coral Reefs 32:315–326. https://doi.org/10.1007/s00338-012-0984-y

Graham NAJ, Wilson SK, Jennings S, Polunin NVC, Bijoux JP, Robinson J (2006) Dynamic fragility of oceanic coral reef ecosystems. Proc Nat Acad Sci USA 103:8425–8429. https://doi.org/10.1073/pnas.0600693103

Graham NAJ, Wilson SK, Jennings S, Polunin NVC, Robinson J, Bijoux JP et al (2007) Lag effects in the impacts of mass coral bleaching on coral reef fish, fisheries, and ecosystems. Conserv Biol 21:1291–1300. https://doi.org/10.1111/j.1523-1739.2007.00754.x

Graham NAJ, McClanahan TR, MacNeil MA, Wilson SK, Polunin NVC, Jennings S et al (2008) Climate warming, marine protected areas and the ocean-scale integrity of coral reef ecosystems. PLoS One 3:e3039. https://doi.org/10.1371/journal.pone.0003039

Graham NAJ, Wilson SK, Pratchett MS, Polunin NVC, Spalding MD (2009) Coral mortality versus structural collapse as drivers of corallivorous butterflyfish decline. Biodivers Conserv 18:3325–3336. https://doi.org/10.1007/s10531-009-9633-3

Graham NAJ, Chabanet P, Evans RD, Jennings S, Letourneur Y, MacNeil AM et al (2011) Extinction vulnerability of coral reef fishes. Ecol Lett 14:341–348. https://doi.org/10.1111/j.1461-0248.2011.01592.x

Graham NA, Cinner JE, Norström AV, Nyström M (2014) Coral reefs as novel ecosystems: embracing new futures. Curr Opin Environ Sustain 7:9–14. https://doi.org/10.1016/j.cosust.2013.11.023

Graham NAJ, Jennings S, MacNeil MA, Mouillot D, Wilson SK (2015) Predicting climate-driven regime shifts versus rebound potential in coral reefs. Nature 518:94–97. https://doi.org/10.1038/nature14140

Grandcourt EM, Cesar HS (2003) The bio-economic impact of mass coral mortality on the coastal reef fisheries of the Seychelles. Fish Res 60:539–550. https://doi.org/10.1016/S0165-7836(02)00173-X

Gratwicke B, Speight MR (2005) The relationship between fish species richness, abundance and habitat complexity in a range of shallow tropical marine habitats. J Fish Biol 66:650–667. https://doi.org/10.1111/j.0022-1112.2005.00629.x

Green AL (1998) Spatio-temporal patterns of recruitment of labroid fishes (Pisces: Labridae and Scaridae) to damselfish territories. Environ Biol Fish 51:235–244. https://doi.org/10.1023/A:1007389206099

Guest JR, Baird AH, Maynard JA, Muttaqin E, Edwards AJ, Campbell SJ et al (2012) Contrasting patterns of coral bleaching susceptibility in 2010 suggest an adaptive response to thermal stress. PLoS One 7:e33353. https://doi.org/10.1371/journal.pone.0033353

Halford AR, Caley MJ (2009) Towards an understanding of resilience in isolated coral reefs. Glob Change Biol 15:3031–3045. https://doi.org/10.1111/j.1365-2486.2009.01972

Hamilton RJ, Almany GR, Brown CJ, Pita J, Peterson NA, Choat JH (2017) Logging degrades nursery habitat for an iconic coral reef fish. Biol Conserv 210:273–280. https://doi.org/10.1016/j.biocon.2017.04.024

Hart AM, Klumpp DW, Russ GR (1996) Response of herbivorous fishes to crown-of-thorns starfish Acanthaster planci outbreaks. II. Density and biomass of selected species of herbivorous fish and fish habitat correlations. Mar Ecol Prog Ser 132:21–30. https://doi.org/10.3354/meps132021

Hempson TN, Graham NAJ, MacNeil MA, Williamson DH, Jones GP, Almany GR (2017a) Coral reef mesopredators switch prey, shortening food chains, in response to habitat degradation. Ecol Evol 7:2626–2635. https://doi.org/10.1002/ece3.2805

Hempson TN, Graham NAJ, MacNeil MA, Bodin N, Wilson SK (2017b) Regime shifts shorten food chains for mesopredators with potential sublethal effects. Funct Ecol. https://doi.org/10.1111/1365-2435.13012

Heron SF, Maynard J, van Hooidonk R, Eakin CM (2016) Warming trends and bleaching stress of the world's coral reefs 1985–2012. Sci Rep 6:38402. https://doi.org/10.1038/srep38402

Hoegh-Guldberg O (1999) Climate change, coral bleaching and the future of the world's coral reefs. Mar Freshwater Res 50:839–866. https://doi.org/10.1071/MF99078

Hoey AS, Bellwood DR (2009) Limited functional redundancy in a high diversity system: single species dominates key ecological process on coral reefs. Ecosystems 12:1316–1328. https://doi.org/10.1007/s10021-009-9291-z

Hoey AS, Bellwood DR (2011) Suppression of herbivory by macroalgal density: a critical feedback on coral reefs? Ecol Lett 14:267–273. https://doi.org/10.1111/j.1461-0248.2010.01581.x

Hoey AS, Brandl SJ, Bellwood DR (2013) Diet and cross-shelf distribution of rabbitfishes (f. Siganidae) on the northern Great Barrier Reef: implications for ecosystem function. Coral Reefs 32:973–984. https://doi.org/10.1007/s00338-013-1043-z

Hoey AS, Howells E, Johansen JL, Hobbs JPA, Messmer V, McCowan DM et al (2016) Recent advances in understanding the effects of climate change on coral reefs. Diversity 8:12. https://doi.org/10.3390/d8020012

Holbrook SJ, Schmitt RJ, Brooks AJ (2008) Resistance and resilience of a coral reef fish community to changes in coral cover. Mar Ecol Prog Ser 371:263–271. https://doi.org/10.3354/meps07690

Holbrook SJ, Schmitt RJ, Messmer V, Brooks AJ, Srinivasan M, Munday PL et al (2015) Reef fishes in biodiversity hotspots are at greatest risk from loss of coral species. PLoS One 10: e0124054. https://doi.org/10.1371/journal.pone.0124054

Houlahan JE, Currie DJ, Cottenie K, Cumming GS, Ernest SKM, Findlay CS et al (2007) Compensatory dynamics are rare in natural ecological communities. Proc Natl Acad Sci USA 104:3273–3277. https://doi.org/10.1073/pnas.0603798104

Hughes TP, Baird AH, Bellwood DR, Card M, Connolly SR, Folke C et al (2003) Climate change, human impacts, and the resilience of coral reefs. Science 301:929–933. https://doi.org/10.1126/science.1085046

Hughes TP, Rodrigues MJ, Bellwood DR, Ceccarelli DM, Hoegh-Guldberg O, McCook LJ et al (2007) Phase shifts, herbivory and the resilience of coral reefs to climate change. Curr Biol 17:360–365. https://doi.org/10.1016/j.cub.2006.12.049

Hughes TP, Graham NA, Jackson JB, Mumby PJ, Steneck RS (2010) Rising to the challenge of sustaining coral reef resilience. Trends Ecol Evol 25:633-642. doi:https://doi.org/10.1016/j.tree.2010.07.011

Hughes TP, Kerry JT, Alvarez-Noriega M, Alvarez-Romero JC, Anderson KD, Baird AH et al (2017) Global warming and recurrent mass bleaching of corals. Nature 543:373–377. https://doi.org/10.1038/nature21707

Hughes TP, Anderson KD, Connolly SR, Heron SF, Kerry JT, Lough JM et al (2018) Spatial and temporal patterns of mass bleaching of corals in the Anthropocene. Science 359:80–83. https://doi.org/10.1126/science.aan8048

Jokiel PL, Coles SL (1990) Response of Hawaiian and other Indo-Pacific reef corals to elevated temperature. Coral Reefs 8:155–162. https://doi.org/10.1007/BF00265006

Jones GP, McCormick MI, Srinivasan M, Eagle JV (2004) Coral decline threatens fish biodiversity in marine reserves. Proc Natl Acad Sci USA 101:8251–8253. https://doi.org/10.1073/pnas.0401277101

Kerry JT, Bellwood DR (2012) The effect of coral morphology on shelter selection by coral reef fishes. Coral Reefs 31:415–424. https://doi.org/10.1007/s00338-011-0859-7

Kok JE, Graham NAJ, Hoogenboom MO (2016) Climate-driven coral reorganisation influences aggressive behaviour in juvenile coral-reef fishes. Coral Reefs 35:473–483. https://doi.org/10.1007/s00338-016-1411-6

Kokita T, Nakazono A (2001) Rapid response of an obligately corallivorous filefish *Oxymonacanthus longirostris* (Monacanthidae) to a mass coral bleaching event. Coral Reefs 20:155–158. https://doi.org/10.1007/s00338100153

Komyakova V, Munday PL, Jones GP (2013) Relative importance of coral cover, habitat complexity and diversity in determining the structure of reef fish communities. PLoS One 8:e83178. https://doi.org/10.1371/journal.pone.0083178

Kramer MJ, Bellwood O, Bellwood DR (2013) The trophic importance of algal turfs for coral reef fishes: the crustacean link. Coral Reefs 32:575–583. https://doi.org/10.1007/s00338-013-1009-1

Light PR, Jones GP (1997) Habitat preference in newly settled coral trout (*Plectropomus leopardus*, Serranidae). Coral Reefs 16:117–126. https://doi.org/10.1007/s003380050065

Linares C, Pratchett MS, Coker MS (2011) Recolonisation and growth of *Acropora hyacinthus* following climate-induced coral bleaching on the Great Barrier Reef. Mar Ecol Prog Ser 438:97–104. https://doi.org/10.3354/meps09272

Lönnstedt OM, McCormick MI, Chivers DP, Ferrari MC (2014) Habitat degradation is threatening reef replenishment by making fish fearless. J Anim Ecol 83:1178–1185. https://doi.org/10.1111/1365-2656.12209

Loya Y, Sakai K, Yamazato K, Nakano Y, Sambali H, van Woesik R (2001) Coral bleaching: the winners and losers. Ecol Lett 4:122–131. https://doi.org/10.1046/j.1461-0248.2001.00203.x

Madin JS, Connolly SR (2006) Ecological consequences of major hydrodynamic disturbances on coral reefs. Nature 444:477–480. https://doi.org/10.1038/nature05328

McClanahan TR, Maina J, Pet-Soede L (2002) Effects of the 1998 coral mortality event on Kenyan coral reefs and fisheries. Ambio 31:543–550. https://doi.org/10.1579/0044-7447-31.7.543

McClanahan TR, Baird AH, Marshall PA, Toscano MA (2004) Comparing bleaching and mortality responses of hard corals between southern Kenya and the Great Barrier Reef, Australia. Mar Pollut Bull 48:327–335. https://doi.org/10.1016/j.marpolbul.2003.08.024

McCormick MI, Allan BJ (2017) Interspecific differences in how habitat degradation affects escape response. Sci Rep 7:426. https://doi.org/10.1038/s41598-017-00521-0

McCormick MI, Moore JAY, Munday PL (2010) Influence of habitat degradation on fish replenishment. Coral Reefs 29:537–546. https://doi.org/10.1007/s00338-010-0620-7

McCowan DM, Pratchett MS, Baird AH (2012) Variation in bleaching susceptibility and mortality attributable to growth forms. In: Proceedings of 12th international coral reef symposium, Cairns, vol 9A, pp 1–6

McIlwain JL, Jones GP (1997) Prey selection by an obligate coral-feeding wrasse and its response to small-scale disturbance. Mar Ecol Prog Ser 155:189–198. https://doi.org/10.3354/meps155189

Messmer V, Jones GP, Munday PL, Holbrook SJ, Schmitt RJ, Brooks AJ (2011) Habitat diversity as a determinant of fish community structure on coral reefs. Ecology 92:2285–2298. https://doi.org/10.1890/11-0037.1

Mouillot D, Villéger S, Parravicini V, Kulbicki M, Arias-González JE, Bender M et al (2014) Functional over-redundancy and high functional vulnerability in global fish faunas on tropical reefs. Proc Natl Acad Sci USA 111:13757–13762. https://doi.org/10.1073/pnas.1317625111

Mumby PJ, Steneck RS (2008) Coral reef management and conservation in light of rapidly evolving ecological paradigms. Trends Ecol Evol 23:555–563. https://doi.org/10.1016/j.tree.2008.06.011

Munday PL (2001) Fitness consequences of habitat use and competition among coral-dwelling fishes. Oecologia 128:585–593. https://doi.org/10.1007/s004420100690

Munday PL (2004) Habitat loss, resource specialization, and extinction on coral reefs. Glob Change Biol 10:1642–1647. https://doi.org/10.1111/j.1365-2486.2004.00839.x

Munday PL, Jones GP, Pratchett MS, Williams A (2008) Climate change and the future for coral reef fishes. Fish Fish 9:261–285. https://doi.org/10.1111/j.1467-2979.2008.00281

Nash KL, Graham NAJ, Jennings S, Wilson SK, Bellwood DR (2016) Herbivore cross-scale redundancy supports response diversity and promotes coral reef resilience. J Appl Ecol 53:646–655. https://doi.org/10.1111/1365-2664.12430

Noonan SHC, Jones GP, Pratchett MS (2012) Coral size, health and structural complexity: effects on the ecology of a coral reef damselfish. Mar Ecol Prog Ser 456:127–137. https://doi.org/10.3354/meps09687

Pisapia C, Pratchett MS, Cole A (2012) Butterflyfishes feeding responses to bleached corals. In: Proceedings of 12th international coral reef symposium, Cairns, vol 13C, pp 1–5

Pratchett MS (2014) Feeding preferences and dietary specialization among obligate coral-feeding butterflyfishes. In: Pratchett MS, Berumen ML, Kapoor B (eds) Biology of butterflyfishes. CRC, Boca Raton, pp 140–179

Pratchett MS, Wilson SK, Berumen ML, McCormick MI (2004) Sub-lethal effects of coral bleaching on an obligate coral feeding butterflyfish. Coral Reefs 23:352–356. https://doi.org/10.1007/s00338-004-0394-x

Pratchett MS, Wilson SK, Baird AH (2006) Declines in the abundance of Chaetodon butterflyfishes (Chaetodontidae) following extensive coral depletion. J Fish Biol 69:1269–1280. https://doi.org/10.1111/j.1095-8649.2006.01161.x

Pratchett MS, Munday PL, Wilson SK, Graham NAJ, Cinner JE, Bellwood DR et al (2008a) Effects of climate-induced coral bleaching on coral-reef fishes: ecological and economic consequences. Oceanogr Mar Biol 46:251–296. https://doi.org/10.1201/9781420065756.ch6

Pratchett MS, Berumen ML, Marnane MJ, Eagle JE, Pratchett DJ (2008b) Habitat associations of juvenile versus adult butterflyfishes. Coral Reefs 27:541–551. https://doi.org/10.1007/s00338-008-0357-8

Pratchett MS, Wilson SK, Graham NAJ, Munday PL, Jones GP, Polunin NVC (2009) Coral bleaching and consequences for motile reef organisms: past, present and uncertain future effects. In: van Oppen MJH, Lough JM (eds) Coral bleaching: patterns, processes, causes and consequences. Springer, Heidelberg, pp 139–158

Pratchett MS, Bay LK, Gehrke PC, Koehn J, Osborne K, Pressey RL et al (2011a) Contribution of climate change to degradation and loss of critical fish habitats in Australian aquatic environments. Mar Freshwater Res 62:1062–1081. https://doi.org/10.1071/MF10303

Pratchett MS, Hoey AS, Wilson SK, Messmer V, Graham NAJ (2011b) Changes in the biodiversity and functioning of reef fish assemblages following coral bleaching and coral loss. Diversity 3:424–452. https://doi.org/10.3390/d3030424

Pratchett MS, Munday PL, Graham NAJ, Kronen M, Pinica S, Friedman K et al (2011c) Vulnerability of coastal fisheries in the tropical Pacific to climate change. In: Bell JD, Johnson JE, Hobday AJ (eds) Vulnerability of tropical pacific fisheries and aquaculture to climate change. Secretariat for the Pacific Community, Noumea, New Caledonia, pp 493–576

Pratchett MS, Coker DJ, Jones GP, Munday PL (2012) Specialization in habitat use by coral reef damselfishes and their susceptibility to habitat loss. Ecol Evol 2:2168–2180. https://doi.org/10.1002/ece3.321

Pratchett MS, McCowan DM, Heron S, Maynard JA (2013) Changes in bleaching susceptibility among corals subject to ocean warming and recurrent bleaching in Moorea, French Polynesia. PLoS One 8:e70443. https://doi.org/10.1371/journal.pone.0070443

Pratchett MS, Hoey AS, Wilson SK (2014) Reef degradation and the loss of critical ecosystem goods and services provided by coral reef fishes. Curr Opin Environ Sustain 7:37–43. https://doi.org/10.1016/j.cosust.2013.11.022

Pratchett MS, Anderson KD, Hoogenboom MO, Widman E, Baird AH, Pandolfi JM et al (2015a) Spatial, temporal and taxonomic variation in coral growth: implications for the structure and function of coral reef ecosystems. Oceanogr Mar Biol 53:215–295. https://doi.org/10.1201/b18733-7

Pratchett MS, Blowes SA, Coker D, Kubacki E, Nowicki J, Hoey AS (2015b) Indirect benefits of scleractinian corals for non-corallivorous butterflyfishes. Coral Reefs 34:665–672. https://doi.org/10.1007/s00338-014-1254-y

Pratchett MS, Cameron D, Donelson J, Evans L, Frisch AJ, Hobday AJ et al (2017) Effects of climate change on coral grouper (Plectropomus spp.) and possible adaptation options. Rev Fish Biol Fish 27:297–316. https://doi.org/10.1007/s11160-016-9455-9

Rasher DB, Hoey AS, Hay ME (2013) Consumer diversity interacts with prey defenses to drive ecosystem function. Ecology 94:1347–1358. https://doi.org/10.1890/12-0389.1

Richardson LE, Graham NA, Pratchett MS, Hoey AS (2017) Structural complexity mediates functional structure of reef fish assemblages among coral habitats. Environ Biol Fish 100:193–207. https://doi.org/10.1007/s10641-016-0571-0

Rogers A, Blanchard JL, Mumby PJ (2014) Vulnerability of coral reef fisheries to a loss of structural complexity. Curr Biol 24:1000–1005. https://doi.org/10.1016/j.cub.2014.03.026

Sano M (2001) Short term responses of fishes to macroalgal overgrowth on coral rubble on a degraded reef at Iriomote Island, Japan. Bull Mar Sci 68:543–556

Sano M, Shimizu M, Nose Y (1984) Changes in structure of coral-reef fish communities by destruction of hermatypic corals — observational and experimental views. Pac Sci 38:51–79

Sano M, Shimizu M, Nose Y (1987) Long-term effects of destruction of hermatypic corals by Acanthaster planci infestation of reef fish communities at Iriomote Island, Japan. Mar Ecol Prog Ser 37:191–199. https://doi.org/10.3354/meps037191

Sheppard CRC, Spalding S, Bradshaw C, Wilson SK (2002) Erosion vs recovery of coral reefs after 1998 El Niño: Chagos reefs, Indian Ocean. Ambio 31:40–48. https://doi.org/10.1579/0044-7447-31.1.40

Smith SV, Buddemeier RW (1992) Global change and coral reef ecosystems. Annu Rev Ecol Syst 23:89–118. https://doi.org/10.1146/annurev.es.23.110192.000513

Speers AE, Besedin EY, Palardy JE, Moore C (2016) Impacts of climate change and ocean acidification on coral reef fisheries: an integrated ecological–economic model. Ecol Econ 128:33–43. https://doi.org/10.1016/j.ecolecon.2016.04.012

Stella JS, Pratchett MS, Hutchings PA, Jones GP (2011) Coral-associated invertebrates: diversity, ecological importance and vulnerability to disturbance. Oceanogr Mar Biol 49:43–104

Syms C, Jones GP (2000) Disturbance, habitat structure, and the dynamics of a coral-reef fish community. Ecology 81:2714–2729. https://doi.org/10.1890/0012-9658(2000)081

Walther G-R, Post E, Convey P, Menzels A, Parmesan C, Beebee TJC et al (2002) Ecological responses to recent climate change. Nature 416:389–395. https://doi.org/10.1038/416389a

Wen C, Pratchett MS, Almany G, Jones GP (2013) Patterns of recruitment and microhabitat associations for three predatory coral reef fishes on the southern Great Barrier Reef, Australia. Coral Reefs 32:389–398. https://doi.org/10.1007/s00338-012-0985-x

Wen CKC, Bonin MC, Harrison HB, Williamson DH, Jones GP (2016) Dietary shift in juvenile coral trout (*Plectropomus maculatus*) following coral reef degradation from a flood plume disturbance. Coral Reefs 35:451–455. https://doi.org/10.1007/s00338-016-1398-x

Wilkinson CR (2000) World-wide coral reef bleaching and mortality during 1998: a global climate change warning for the new millennium? In: Sheppard CRC (ed) Seas at the millennium: an environmental evaluation. Elsevier Science, Amsterdam, pp 43–57

Williamson DH, Ceccarelli DM, Evans RD, Jones GP, Russ GR (2014) Habitat dynamics, marine reserve status, and the decline and recovery of coral reef fish communities. Ecol Evol 4:337–354. https://doi.org/10.1002/ece3.934

Wilson SK, Graham NAJ, Pratchett MS, Jones GP, Polunin N (2006) Multiple disturbances and the global degradation of coral reefs: are reef fishes at risk or resilient? Glob Change Biol 12:1–15. https://doi.org/10.1111/j.1365-2486.2006.01252.x

Wilson SK, Burgess SC, Cheal AJ, Emslie M, Fisher R, Miller I et al (2008) Habitat utilization by coral reef fish: implications for specialists vs. generalists in a changing environment. J Anim Ecol 77:220–228. https://doi.org/10.1111/j.1365-2656.2007.01341.x

Wilson SK, Depczynski M, Fisher R, Holmes TH, O'Leary R, Tinkler P (2010) Habitat associations of juvenile fish at Ningaloo Reef, Western Australia: the importance of coral and algae. PLoS One 5:e15185. https://doi.org/10.1371/journal.pone.0015185

Wilson SK, Graham NAJ, Pratchett MS (2014) Susceptibility of butterflyfish to habitat disturbance: Do 'chaets' ever prosper? In: Pratchett MS, Berumen ML, Kapoor B (eds) Biology of butterflyfishes. CRC, Boca Raton, pp 226–245

Wilson SK, Depczynski M, Holmes TH, Noble MM, Radford BT, Tinkler P et al (2017) Climatic conditions and nursery habitat quality provide indicators of reef fish recruitment strength. Limnol Oceanogr. https://doi.org/10.1002/lno.10540

Chapter 12
Ocean Acidification and Coral Bleaching

R. Albright

12.1 Introduction

Simultaneous with the increases in global sea surface temperature, increasing atmospheric carbon dioxide (CO_2) is driving changes in the chemistry of the oceans—a process known as ocean acidification. Over the last two decades, reef-related ocean acidification research has focused primarily on the consequences of elevated CO_2 on calcification. The impacts of ocean acidification on other critical processes such as coral-algal symbioses and bleaching thresholds are less well known. In this chapter, I review the available literature on the impacts of ocean acidification on coral bleaching. I begin by providing context for ocean acidification and its impacts on coral reefs. I focus primarily on primary literature investigating the effects of CO_2 on photophysiology, coral-algal symbioses, and bleaching responses while shedding light on information needs and unresolved issues. I also briefly touch on environmental factors other than temperature and ocean acidification that have the potential to influence coral bleaching responses (e.g., nutrients).

12.1.1 Ocean Acidification

Simultaneous with the increases in global sea surface temperature (SST), increasing atmospheric carbon dioxide (CO_2, the main greenhouse gas) is driving changes in the seawater chemistry of the oceans. At present, the ocean absorbs about one-third of fossil fuel CO_2 emissions and will eventually sequester up to 90% of anthropogenic CO_2 (Archer et al. 2009; Sabine et al. 2011), causing measurable shifts in

R. Albright (✉)
California Academy of Sciences, San Francisco, CA, USA
e-mail: ralbright@calacademy.org

M. J. H. van Oppen, J. M. Lough (eds.), *Coral Bleaching*, Ecological Studies 233,
https://doi.org/10.1007/978-3-319-75393-5_12

seawater carbonate chemistry (Bates et al. 2014; Canadell et al. 2007; Le Quéré et al. 2015). On entry into the ocean, CO_2 reacts with seawater via the following net chemical reaction:

$$H_2O + (CO_2)aq + CO_3^{2-} \rightarrow 2HCO_3^- \tag{12.1}$$

As a result, concentrations of aqueous carbon dioxide, $[CO_2]aq$, and bicarbonate $\left[HCO_3^-\right]$ increase, while the concentration of carbonate $\left[CO_3^{2-}\right]$ and the pH of seawater decrease (Broecker et al. 1979; Caldeira and Wickett 2003; Sabine et al. 2004); this process is referred to as ocean acidification (OA). In the last 200 years, the global average pH of ocean surface waters has declined by about 0.1 units, from pH ~8.2 to pH 8.1 (Rhein et al. 2013), which equates to an increase in acidity (i.e., hydrogen ion concentrations) of approximately 30%. Under a business-as-usual scenario, a further decrease of 0.3–0.4 units (to pH 7.7–7.8) is expected by the end of the twenty-first century. Another important outcome of OA for calcifying organisms, such as reef-building corals, is the decrease in the saturation state of calcium carbonate (Ω), defined as $\Omega = [Ca^{2+}]\left[CO_3^{2-}\right]/K'sp$, where $K'sp$ is the solubility product for a particular mineral phase of $CaCO_3$ (e.g., aragonite, calcite). Aragonite is the dominant biogenic form of $CaCO_3$ secreted by many reef-building organisms, including corals. If $\Omega > 1$, seawater is supersaturated with respect to $CaCO_3$, and conditions are favorable for $CaCO_3$ precipitation; conversely, if $\Omega < 1$, seawater is undersaturated with respect to $CaCO_3$, and the dissolution of $CaCO_3$ is favored. The surface waters of the tropical oceans are currently supersaturated with respect to aragonite; however, the saturation state of aragonite (Ω_{arag}) of tropical Pacific surface waters is estimated to have decreased from values of about 4.5 in preindustrial times (Kleypas et al. 1999; Cao and Caldeira 2008) to about 3.8 by 1995 (Feely et al. 2009) and is expected to continue declining to approximately 3.0 by the middle of this century and 2.3 by the end of the century (Feely et al. 2009).

12.1.2 Variability in Seawater Carbonate Chemistry of Coral Reefs

OA projections are based on trends from data collected in open ocean environments (Doney et al. 2009; Feely et al. 2009; Zeebe 2012), and their implications for shallow, nearshore environments, such as coral reefs, are poorly understood. In coastal regions, OA can interact with other natural and anthropogenic environmental processes to hasten local declines in pH and carbonate mineral saturation states (Duarte et al. 2013; Feely et al. 2010). In the Great Barrier Reef, Australia, for example, inshore reefs are subjected to elevated pCO_2 levels compared to offshore reefs, and the rate of increase in inshore pCO_2 is faster than offshore and atmospheric values (Cyronak et al. 2014; Uthicke et al. 2014). Changes in pH in coastal ecosystems result from a multitude of drivers, including oceanic uptake of anthropogenic CO_2 emissions, nutrient inputs, changes in metabolism (the balance between primary production, respiration, and calcification), impacts from watershed processes, organic matter, and hydrodynamics. For example, organic carbon metabolism (photosynthesis and respiration) and inorganic carbon metabolism (calcification and dissolution) can drive strong diel and seasonal fluctuations in seawater chemistry (Ohde and van Woesik 1999; Anthony et al. 2011a; Bates et al. 2010; Kleypas et al. 2011; Shamberger et al. 2011; Gray et al. 2012; Shaw et al. 2012; Albright et al. 2013, 2015; Koweek et al. 2015). Characteristic ranges are on the order of 0.3 pH units (Duarte et al. 2013) but can be as large as 1 pH unit in some environments [e.g., Lady Elliot Island reef flat, southern Great Barrier Reef (Shaw et al. 2012)]. The extent to which reef metabolism alters the carbonate chemistry of the overlying water column is a function of numerous factors, including benthic community composition (Anthony et al. 2013; Koweek et al. 2014), biological activity (which can vary with temperature, light, and nutrient availability), physical forcing (e.g., temperature, salinity), tidal regime, water depth, and residence time (Falter et al. 2008, 2012).

Changes in the adjacent watershed can also influence alkalinity and CO_2 fluxes that, together with metabolic processes and oceanic dynamics, can yield decadal changes of up to 0.5 units in coastal pH (Duarte et al. 2013). Within many coastal ecosystems, anthropogenic nutrient inputs enhance microbial respiration, driving concurrent acidification and hypoxia (Altieri et al. 2017). These interactions between OA and local to regional drivers yield complex regulation of pH in coastal waters that have the capacity to strongly impact ecosystem health and performance. Accordingly, many reef environments experience variable ocean chemistry that often differs from open ocean conditions and, over relatively short time scales (days and months), range more widely than the difference in mean conditions between preindustrial to future OA scenarios (Hofmann et al. 2011; Shaw et al. 2012; Albright et al. 2013). Understanding the significance of OA projections in the context of this background variability is central to gauging the susceptibility of reef ecosystems to projected changes in ocean chemistry.

12.1.3 Impacts of Ocean Acidification on Coral Reefs

Major changes in ocean chemistry can have profound effects on marine ecosystems (Doney 2010) and have even been implicated in creating conditions leading to past mass extinction events (Veron 2008; Clarkson et al. 2015). pH plays a key role in many physiological processes such as ion transport, enzyme activity, and protein function, and as such, changes in CO_2 can influence a wide range of physiological processes including acid-base regulation, metabolism, energetics, and dependent processes (Pörtner 2005, 2008). While extracellular and intracellular pH is usually tightly regulated, the capacity of regulatory mechanisms can be overwhelmed. OA also has the capacity to alter physiological processes that depend on carbon species as reactants, including calcification and photosynthesis. Coral reefs are widely regarded as one of the most vulnerable marine ecosystems to OA, in part because the very architecture of the ecosystem is reliant on carbonate-secreting organisms. OA will lead to variable but predominantly adverse biological and ecological responses for key species of coral reef organisms (Kroeker et al. 2010, 2013), including slowed reef growth, altered competitive interactions, and impaired population replenishment. Slower calcification results in slower coral growth, more fragile structures (Madin et al. 2008), and potentially a shift from net accretion to net dissolution (Silverman et al. 2009; Andersson and Gledhill 2013; Perry et al. 2013), with implications for greater susceptibility to storm damage, slower recovery rates between disturbances, less habitat-forming structures, and overall reduced reef resilience (Anthony et al. 2011b). Both laboratory and field studies provide evidence that coral reefs have already lost significant calcification capacity due to OA (Dove et al. 2013; Albright et al. 2016).

Model results suggest that if CO_2 emissions continue to follow a business-as-usual path, tropical coral reefs are likely to shift toward conditions that are marginal for reef growth (i.e., net dissolution) this century (Hoegh-Guldberg et al. 2007; Silverman et al. 2009). In addition to impacting corals, OA lowers the abundance of crustose coralline algae (Kuffner et al. 2007), a key player in solidifying reef framework and providing settlement cues for numerous reef invertebrate larvae (Harrington et al. 2004). Other impacts include shifts in competitive interactions between corals and macroalgae (Diaz-Pulido et al. 2011), shifts in the competitive hierarchy of corals (Horwitz et al. 2017), synergistic effects of temperature and OA on coral mortality (Prada et al. 2017), and impairment of behavioral responses critical for fish recruitment (Munday et al. 2010, 2014). Coral communities around natural CO_2 seeps show shifts in community composition from highly diverse and structurally complex systems to those characterized by much lower diversity and structural complexity (Fabricius et al. 2011, 2014), leading to loss of ecological function and associated services.

A wide range of impacts are likely on coastal human communities including reduced food, income, and well-being, as well as longer-term impacts such as increasing vulnerability as coral reefs become less able to protect coastal areas from storms and waves (Pendleton 1995; Cooley et al. 2009; Pascal et al. 2016). While local adaptation over evolutionary time scales (involving genetic isolation and strong selective pressure) renders certain communities more resistant to low-Ω seawater (e.g., Golbuu et al. 2016; Barkley et al. 2017), there is little evidence that coral reef organisms can adapt or acclimate to future OA scenarios (but see Putnam and Gates 2015). The possibility of OA acting to lower coral bleaching thresholds has been suggested (e.g., Anthony et al. 2008); however acidification effects on coral bleaching are highly uncertain. Whether or not OA influences the bleaching response of corals remains a topic of debate and is the primary focus of this chapter.

12.2 Ocean Acidification and Coral Bleaching

In reef-building corals, *Symbiodinium* density and physiology are regulated by host- and symbiont-driven mechanisms that vary in response to environmental conditions such as light intensity, temperature, and nutrients (Lesser 2004, 2011). Abrupt changes in environmental conditions can disrupt coral-algal symbioses and cause bleaching (i.e., loss of *Symbiodinium* and/or photosynthetic pigments) (Lesser 2011). While the effects of elevated temperature on coral bleaching are well known (Fitt et al. 2001; Jokiel 2004; Lesser 2011), the effects of elevated pCO_2 on bleaching responses are less clear. Compared to calcification studies, less attention has been placed on other aspects of holobiont physiology, including symbiont photophysiology and primary productivity. Where these effects have been investigated, outcomes are equivocal (Table 12.1). To inform our understanding of potential mechanisms by which changing seawater chemistry could influence coral-algal symbioses, it is helpful to briefly review proposed bleaching mechanisms and relationships to dissolved inorganic carbon (DIC).

12.2.1 Bleaching Mechanisms and Dissolved Inorganic Carbon

The mechanism behind warmwater bleaching is generally accepted to involve accumulation of oxidative stress at photosystem II (PSII) in the symbiont, a process known as photoinhibition (Lesser 1996). Photoinhibition occurs when the rate of photodamage to PSII exceeds the rate of repair and can result in reduced photosynthetic rates and elevated reactive oxygen species (ROS), further damaging the symbiont and coral. This stress can change the energetic/metabolic demands of the symbiont, reducing the amount of photosynthate translocated to the host. Physiological changes in the host, such as reduced tissue thickness and apoptosis of gastrodermal cells, can precede changes in symbionts when corals are exposed to stress. Various photoprotective mechanisms exist to provide alternate electron pathways to divert excess excitation energy that may otherwise lead to the formation of ROS and photooxidative damage of proteins, lipids, and pigments. These include pathways such as the water-water cycle and photorespiration.

Zooxanthellae use the Calvin-Benson cycle (dark reaction) to fix CO_2. *Symbiodinium* contain a form II Rubisco (the enzyme ribulose bisphosphate carboxylase/oxygenase), which has a poor ability to discriminate between CO_2 and O_2 (Rowan et al. 1996). Consequently, maintaining an elevated ratio of CO_2:O_2 at the site of photosynthesis benefits overall productivity. In theory, physiological processes that depend on carbon species as reactants (e.g., photosynthesis, calcification) may be directly influenced by acidification-induced changes in the inorganic carbon supply. For organisms that are CO_2 limited, increasing dissolved CO_2 and HCO_3^- associated with OA could potentially alter photosynthetic kinetics and "fertilize"

Table 12.1 Studies examining effects of OA and/or OA and temperature on coral bleaching or photophysiology

Reference	Species	T (°C)	pH, pCO_2	Response variable	Observed effect(s)
Anlauf et al. (2011)	*Porites panamensis*	29, 30	7.8, 8.03 (546, 900–1000 ppm)	Growth, algal density, biomass, survival, larval settlement, bleaching	No effect of low pH on larval settlement, survival, or bleaching of primary polyps. Temp reduced algal density. Temp × CO_2 increased algal density. Temp exacerbated CO_2 effects on growth. Temp × CO_2 reduced biomass of primary polyps.
Anthony et al. (2008)	*Acropora intermedia, Porites lobata*	25–26, 28–29	8.0–8.4, 7.85–7.95, 7.6–7.7	Productivity, calcification	CO_2-induced bleaching observed in two species. *Acropora*: P_{net} unaffected by intermediate CO_2, but P_{net} increased at CO_2 × temp. Large reductions in P_{net} at high CO_2. Calcification decreased by 60% at high-CO_2 × high-temp *Porites*: Calcification and net productivity decreased with decreasing pH. Intermediate CO_2 and high-temp increased calcification.
Baghdasarian et al. (2017)	*Seriatopora caliendrum*	27.6, 30.4	466 ppm, ~895 ppm	Algal density, chl $cell^{-1}$, algal division	Temperature-induced bleaching (decreased algal density), reduced algal division, and altered chl $cell^{-1}$, but no CO_2 effect. Intraspecific variation observed in bleaching responses, unrelated to algal mitotic index or pigment content.
Comeau et al. (2016)	*Porites rus, Acropora pulchra, Pocillopora damicornis, Pavona cactus,* massive *Porites* sp., *Psammocora profundacell, Porites irregularis, Pocillopora verrucosa*	27	280, 400, 550, 700, 1000, 2000 µatm	Photosynthesis (P_{net}, P_{gross}), respiration (R_{dark}, LEDR)	Overall, respiration and photosynthesis of eight coral spp. and seven calcified algal spp. was relatively insensitive to increasing pCO_2. No effect of OA on P_{net} in seven of eight coral spp. and seven of seven calcified algal spp., P_{gross} in six of eight coral spp. and six of seven calcified algal spp., R_{dark} in six of eight coral spp. and six of seven calcified algal spp., and LEDR in eight of eight coral spp. and seven of seven calcified algal spp. Minor effects detected for a few traits in some species.

Comeau et al. (2017)	Mixed community	24, 27	~400, ~1300 μatm	P_{net}, G_{net}	No effect of pCO_2 on P_{net}-PAR relationship. pCO_2 altered G_{net}-PAR relationship by suppressing G_{net}. OA may alter balance between net community calcification (ncc) and net community production (ncp) by depressing ncc without affecting ncp.
Crawley et al. (2010)	*Acropora formosa*	23	7.55–7.65, 7.80–7.90, 8.00–8.20 (380, 560, 1100 ppm)	Photorespiration (PGPase activity), symbiont productivity, photosynthetic capacity, photoprotection	Chl *a* $cell^{-1}$ increased, while photosynthetic capacity per chlorophyll decreased with increasing CO_2. No change in algal density. Dark respiration remained constant, whereas LEDR increased with increasing CO_2. Under high CO_2, energy dissipation pathways were over-activated (e.g., nonphotochemical quenching through xanthophyll de-epoxidation) or partially closed due to a reduction in enzyme synthesis. Decreased expression of PGPase (phosphoglycolate phosphatase).
Dove et al. (2013)	Mixed community	25, 26, 28, 30	7.7, 7.9, 8.1, 8.2, (301, 405, 611, 1009 μatm)	G_{net}, P_{net}, R_{net}	OA decreased daytime calcification and increased nighttime dissolution (overall reduction in G_{net}). No effect on P_{net} or R_{net}. Bleaching and mortality observed in all scenarios (including controls and preindustrial treatment).
Fine and Tchernov (2007)	*Oculina patagonica, Madracis pharencis*	17–30	7.3–7.6, 8.0–8.3	Tissue biomass, algal density, gametogenesis	Low pH induced partial loss of symbionts (bleaching) in some, but not all *O. patagonia* corals. No effect on gametogenesis. Enhanced protein content and polyp biomass for some corals.
Hoadley et al. (2015)	*Acropora millepora, Pocillopora damicornis, Montipora monasteriata, Turbinaria reniformis*	26.5, 31.5	382, 607, 741 μatm	Algal cellular volume, protein, lipid, carbohydrate content, maximal photochemical efficiency, intracellular carbonic anhydrase	Complex host- and symbiont-specific responses. Overall, temp had greater influence on photophysiology (algal cellular volume, protein, lipid, carbohydrate content, maximal photochemical efficiency, intracellular carbonic anhydrase) than CO_2.
Iguchi et al. (2012)	*Porites australiensis*	27	7.4, 7.6, 8.0	Calcification, algal density, chl *a* $cell^{-1}$, fluorescence yield (*Fv/Fm*)	pCO_2 decreased calcification and fluorescence yield, but no effect on algal density or chl *a* $cell^{-1}$. Intraspecific differences in response.

(continued)

Table 12.1 (continued)

Reference	Species	T (°C)	pH, pCO_2	Response variable	Observed effect(s)
Kaniewska et al. (2012)	*Acropora millepora*	25	7.6–7.7, 7.8–7.9, 8.0–8.2 (260–440, 600–790, 1010–1350 µatm)	Algal density, P_{net}, R_{net}, gene expression	High CO_2 decreased algal densities (>50%), photosynthesis, and respiration. Changes in gene expression were consistent with metabolic suppression, increase in oxidative stress, apoptosis, and symbiont loss. Other expression patterns demonstrate upregulation of membrane transporters, as well as regulation of genes involved in membrane cytoskeletal interactions and cytoskeletal remodeling.
Kavousi et al. (2015)	*Acropora digitifera, Montipora digitata, Porites cylindrical*	28, 31	400, 1000 µatm	Calcification, protein content, maximum photosynthetic efficiency, algal density, chl content, *Symbiodinium* type, *Fv/Fm*	Significant intra- and interspecific differences in responses. Different responses to pCO_2, temp, and pCO_2 × temp between hosts with different *Symbiodinium* type and between similar *Symbiodinium* types in colonies of different species. No effect of pCO_2 or pCO_2 × temp on protein content; pCO_2 decreased algal density in some, but not all combinations. pCO_2 × temp reduced *Fv/Fm* in some, but not all combinations. Mixed effects on chl *a* $cell^{-1}$.
Krueger et al. (2017)	*Stylophora pistillata*	+1–2	7.8	Pigmentation, net O_2 production, primary production	No visual signs of bleaching. No change in algal density or protein or carbohydrate content in symbiont or host tissue. Low pH had additive positive effect on net O_2 production at elevated temp. Low pH significantly enhanced the efficiency of gross O_2 generation per unit chl, independent of temp. pH × temp increased total chl concentration, predominantly driven by temp (+45%), compared to pH (−9%). Synergistic effects of pH x temp on electron flow in PSII. Temp increased catalase activity in host tissue. No effect on symbiont antioxidant enzyme activity.
Krief et al. (2010)	*Porites* sp., *Stylophora pistillata*	25	8.09, 7.49, 7.19 (pCO_2 > 2000 ppm)	Algal density, skeletal growth, coral tissue biomass (protein concentration) chl *a*, isotopic skeletal composition	pCO_2 decreased algal density and skeletal growth but increased coral tissue biomass (protein concentration) and chl *a*. Altered isotopic skeletal composition.

Langdon et al. (2003)	Mixed community	26.5	404, 658 μatm	Calcification, net community production, dark respiration, light respiration	pCO_2 negatively impacted calcification. No change in net community production or dark respiration. pCO_2 increased light respiration.
Noonan et al. (2013)	*Acropora millepora, Pocillopora damicornis, Seriatopora hystrix, Porites cylindrical,* massive *Porites* sp., *Galaxea fascicularis*	NA	7.8–7.9, 8.0–8.05 (~390, 500–900, >1000 ppm)	*Symbiodinium* type	Strong differences in *Symbiodinium* type by coral species, but no differences along CO_2 gradients at volcanic CO_2 seeps.
Noonan and Fabricius (2016)	*Acropora millepora, Seriatopora hystrix*	28, 31	7.79 (~780 μatm), 7.95	Bleaching, maximum PSII quantum yields, light-limited electron transport rates, gross photosynthesis, pigment concentrations	No effect of pCO_2 on bleaching during transient bleaching event along CO_2 gradient near volcanic CO_2 seeps. In tank experiments, temp, but not CO_2, affected bleaching. pCO_2 increased maximum photosystem II quantum yields and light-limited electron transport rates. pCO_2 increased gross photosynthesis and pigment concentrations in *S. hystrix*. Overall, stronger temp effects than CO_2 and species-specific responses.
Reynaud et al. (2003)	*Stylophora pistillata*	25, 28	460, 760 μatm	Chl c_2, chl *a*, protein, photosynthesis, respiration, calcification, algal density	No effects on chl *c2* or protein. Chl *a* increased under temp × pCO_2. Algal density increased at high pCO_2. P_{net} affected by both temp (increased) and pCO_2 (decreased), but no impacts on respiration. Temp × pCO_2 decreased calcification (no change with pCO_2 at normal temp).
Rodolfo-Metalpa et al. (2010)	*Cladocora caespitosa*	~15–20, ~20–25	400, 700 μatm	Photosynthesis, respiration, photosynthetic efficiency (*Fv/Fm*), zooxanthellae density	No effect of pCO_2 or pCO_2 × temp on photosynthesis, photosynthetic efficiency, or calcification.
Schneider and Erez (2006)	*Acropora eurystoma*	24	7.9–8.5	Calcification, photosynthesis, respiration	Calcification positively correlated to CO_3^{2-} and Ω_{arag}. Reduction in CO_3^{2-} by 30% reduced calcification by 50%. Photosynthesis and respiration did not show any significant response to changes in seawater CO_2.

(continued)

Table 12.1 (continued)

Reference	Species	T (°C)	pH, pCO_2	Response variable	Observed effect(s)
Schoepf et al. (2013)	*Acropora millepora, Montipora monasteriata, Pocillopora damicornis, Turbinaria reniformis*	26.5, 29	382, 607, 741 µatm	Coral energy reserves (lipid, protein, carbohydrate), chl *a*, algal density	Coral energy reserves (lipid, protein, carbohydrate) showed species-specific responses to elevated pCO_2 and temp. Nonlinear, species-specific responses of chl *a* and algal density to pCO_2. Temp modulated pCO_2 response, sometimes mitigating and worsening pCO_2 effects.
Strahl et al. (2016)	massive *Porites* sp., *Acropora millepora*	29.5	7.8, 8.1 (323, 803 µatm)	Tissue biomass, lipid, protein, tissue energy content, fatty acid content, pigment content, oxidative stress parameters	No effect of pCO_2 on tissue biomass, lipid, protein and tissue energy content, fatty acid content, pigment content, and oxidative stress parameters. Differences between species and location much greater than effect of pCO_2 In *Porites* sp., only one of the biochemical parameters investigated (ratio of photoprotective to light-harvesting pigments) responded to pCO_2.
Takahashi and Kurihara (2013)	*Acropora digitifera*	29	7.56, 7.97 (744, 2142 µatm)	Calcification, respiration, photosynthesis, algal density, photosynthetic efficiency (*Fv/Fm*)	No effect of CO_2 on calcification, bleaching, or productivity (photosynthesis, respiration, photosynthetic efficiency, algal density).
Wall et al. (2013)	*Seriatopora caliendrum*	27.7, 30.5	445, 840 ppm	Bleaching, maximum photochemical efficiency (*Fv/Fm*), effective photochemical efficiency ($\Delta F/Fm'$) of PSII, Pnet, photosynthetic efficiency (α), chl *a*, symbiont density	Temp-induced bleaching and negatively impacted photosynthetic performance of symbionts as measured by maximum photochemical efficiency (*Fv/Fm*) and effective photochemical efficiency ($\Delta F/Fm'$) of PSII, net photosynthesis (P_{net}) and photosynthetic efficiency (α), chl *a* and symbiont density. High pCO_2 had no direct or synergistic effect on *Symbiodinium* photophysiology or productivity and did not cause bleaching.

photosynthesis, with some benefits generally expected for plants and algae. Seagrasses, for example, which appear carbon limited, respond positively to increasing seawater CO_2 (Beer and Koch 1996; Zimmerman et al. 1997; Jiang et al. 2010; Russell et al. 2013; Ow et al. 2015). For most investigated species, however, photosynthetic responses to OA are relatively small and are highly variable among taxa (Mackey et al. 2015). For zooxanthellae, the source and reliability of CO_2 is complicated by their intracellular location. While molecular CO_2 can freely diffuse across cell membranes and lipid bilayers, at typical seawater pH (~8.1), molecular CO_2 represents only a small fraction (<1%) of the available DIC in seawater, with the majority in the form of HCO_3^- (Zeebe and Wolf-Gladrow 2001), which is largely inhibited from diffusing into the host cells due to its ionic charge. To enhance the delivery of CO_2 to the endosymbiont, many coral species implement a range of CO_2-concentrating mechanisms (CCMs), which facilitate the dehydration of HCO_3^- into CO_2 in the presence of carbonic anhydrase and H^+-ATPase (Furla et al. 2000; Al-Horani et al. 2003). At low irradiance, respiratory CO_2 is sufficient to meet photosynthetic demand (Muscatine et al. 1989). At high solar irradiance, however, zooxanthellae rely on the host to supplement the DIC supply by converting HCO_3^- from bulk seawater (Goiran et al. 1996; Marubini et al. 2008). Consequently, CO_2 delivery to the symbiont is heavily regulated by host CCM activity. In turn, CCM activity relies on energy (ATP) that is, in part, derived from photosynthate transferred from the zooxanthellae (Al-Horani et al. 2003) (though note that many species utilize heterotrophic carbon to mitigate intracellular CO_2 limitation during periods of autotrophic stress, enhancing bleaching resistance). Feedbacks between host-regulated DIC delivery to the symbiont and symbiont-regulated energy to the host complicate our ability to predict effects of increased CO_2 on photophysiology. Present information suggests that *Symbiodinium* in hospite are DIC limited (e.g., Goiran et al. 1996). As $CO_2(aq)$ increases with OA, symbiont productivity could increase due to a release from carbon limitation; however, the effect of elevated CO_2 on photosynthesis is likely to be minimal for endosymbiotic *Symbiodinium* owing to host regulation of DIC supply and the presence of CCMs (Mackey et al. 2015).

An alternate bleaching mechanism was proposed by Wooldridge (2009a) whereby the expulsion of zooxanthellae is triggered by CO_2 limitation around Rubisco resulting from a functional breakdown in CCM function such as carbonic anhydrase and/or Ca^{2+}-ATPase activity (i.e., CO_2 demand exceeds available supply). In this scenario, inadequate CCM activity (due to either direct damage to the CCM or insufficient energy to fuel CCM activity) limits the ability of the coral host to maintain a sufficient supply of CO_2 for the endosymbiont, particularly during periods of high photosynthetic demand (e.g., high solar irradiance or thermal stress). This leads to photoinhibition, oxidative damage, and eventual expulsion of zooxanthellae. Whether this model explains the breakdown of coral-algae symbiosis in response to other known bleaching triggers, such as low temperature, low salinity, high sedimentation, aerial exposure, etc. is not yet known.

12.2.2 Evidence of Ocean Acidification-Induced Bleaching

To date, there have been no documented cases of acidification-induced bleaching in the natural environment; most of what we know has been determined experimentally and not verified ecologically. Only a few experimental studies have reported coral bleaching in response to acidification stress. For example, a causal link between CO_2 exposure and bleaching response was reported in a symbiotic sea anemone (Pecheux 2002), although these results were not published in the peer-reviewed literature. Anthony et al. (2008) compared bleaching, productivity, and calcification responses to acidification (pH 7.6–8.4) and warming (25–29 °C) at ecologically relevant irradiances (~1000 μmol photons m^{-2} s^{-1}) and showed that elevated CO_2 induced bleaching (loss of pigmentation) in two key groups of reef-building organisms—crustose coralline algae (CCA) and branching (*Acropora*) and massive (*Porites*) coral species. These results are often scrutinized because bleaching was evaluated using a luminance colorimetric scale rather than quantifying algal (*Symbiodinium*) densities or pigment content. The mechanism underlying the observed bleaching response was not explicitly investigated; however the authors hypothesize that changes in seawater chemistry influence bleaching thresholds by altering the functioning of the carbon-concentrating mechanism, photoprotective mechanisms (e.g., photorespiration), and/or direct impacts of acidosis. Kaniewska et al. (2012) investigated the phenotypic and transcriptional responses of *Acropora millepora* colonies exposed to OA and showed that exposure to high CO_2 drives major changes in gene expression, respiration, photosynthesis, and symbiosis. Elevated pCO_2 (>1000 μatm) resulted in a loss (>50%) of *Symbiodinium* cells, an associated decrease in photosynthesis and respiration, and an increase in transcripts of genes involved in/responsible for alleviating oxidative stress, suggesting that the photosynthetic apparatus of the zooxanthellae was compromised. Changes in gene expression were consistent with metabolic suppression, an increase in oxidative stress, apoptosis, and symbiont loss. Based on the transcriptomics results, the authors suggest that similar cellular events occur during acidosis-induced bleaching as those reported for thermally induced bleaching (e.g., Weis 2008). These include increased ROS (and/or reactive nitrogen species, RNS) production, which in turn disrupts calcium homeostasis, a condition that has been linked to coral bleaching (DeSalvo et al. 2008). Additional cellular impacts of acidosis included changes to acid-base regulation and mitochondrial ATPase activity.

In contrast, mounting experimental evidence from laboratory, field, and modeling studies suggests that the influence of elevated CO_2 on coral bleaching may be trivial. For example, thermal bleaching (loss of *Symbiodinium* and/or chlorophyll content) in *Seriatopora caliendrum* was unaffected by high pCO_2 (>800 ppm) in two studies (Wall et al. 2013; Baghdasarian et al. 2017). Hoadley et al. (2015) report various physiological impacts in four coral species (*Acropora millepora*, *Pocillopora damicornis*, *Montipora monasteriata*, *Turbinaria reniformis*) exposed to increased temperature (31.5 °C) and pCO_2 (~740 ppm), including changes in maximal photochemical efficiency and biochemical composition of the symbionts (e.g., algal cellular volume, protein, and lipid content). However, elevated temperature played a greater role in altering physiological responses than pCO_2. Interestingly, the photophysiological response and biochemical composition of the symbionts differed among clades and influenced holobiont responses, drawing attention to the need to understand symbiont, host, and symbiont × host (holobiont) responses (Hoadley et al. 2015). In the Gulf of Aqaba, *Stylophora pistillata* colonies showed no signs of bleaching despite spending 1.5 months at 1–2 °C above long-term summer maximum SST and a seawater pH of 7.8 (Krueger et al. 2017). Symbiotic dinoflagellates did, however, show improved photochemistry with higher pigmentation and a doubling in net oxygen production, leading to a 51% increase in primary productivity. In a *Symbiodinium* energetics study, Bedwell-Ivers et al. (2016) found no significant differences in zooxanthellae density or Chl *a* content in two Caribbean coral species (*Acropora cervicornis* and *Porites divaricata*) exposed to elevated pCO_2 (~1000 μatm). They did, however, report reductions in the maximum rate of net photosynthesis (P_{max}) and dark respiration (R_d), a response the authors attribute to metabolic suppression (an adaptive response to conserve energy) as opposed to bleaching. At natural volcanic CO_2 seeps in Papua New Guinea, where corals are chronically exposed to elevated CO_2 up to 800 μatm, the majority of variation in important biochemical measures such as tissue biomass, energy storage, pigmentation, cell protection, and cell damage was attributed to species (massive *Porites* vs. *Acropora millepora*) and location, with little effect of pCO_2 (Strahl et al. 2016). At these same CO_2 seeps, Noonan and Fabricius (2016) surveyed four common coral families (Acroporidae, Faviidae, Pocilloporidae, or Poritidae) and a thermally sensitive species *Seriatopora hystrix* along CO_2 gradients during a minor regional bleaching event and found little indication that elevated pCO_2 influenced bleaching susceptibility of the broader coral community. Concurrent tank experiments also showed no effect of elevated pCO_2 (780 μatm) on bleaching sensitivity in *Acropora millepora* or *Seriatopora hystrix* (Noonan and Fabricius 2016). Finally, a modeling study assessed the sensitivity of coral bleaching projections under different Ω_{arag} sensitivities and found that Ω_{arag} exhibits limited influence on bleaching sensitivity under RCP 2.6 and 4.5 scenarios. By the year 2050, RCP 4.5 results in >95% of global reefs experiencing annual bleaching regardless of Ω_{arag} sensitivity. Even under the high mitigation scenario RCP 2.6, >90% of global reefs are projected to experience annual bleaching by the mid-twenty-first century regardless of Ω_{arag} sensitivity (Kwiatkowski et al. 2015).

12.2.3 Photoacclimation and Photoprotection

Photoacclimation refers to the physiological acclimation by an organism to a certain light environment. Similar to plants, corals demonstrate photoacclimatory responses such as changes in symbiont density and/or chlorophyll content per cell. To date, the effects of CO_2 enrichment on photoacclimation are equivocal. It is well-established that the coral-algal symbiosis is a dynamic reciprocal relationship, in which the symbiotic interaction can change depending on environmental conditions that differentially benefit either partner (Wooldridge 2017). It has been suggested that there is an optimum zooxanthellae density that optimizes autotrophic capacity (P:R) by maximizing light harvesting and minimizing intraspecific competition for resources such as intracellular CO_2 (Wooldridge 2017). Environmentally triggered increases in algal density may cause resource limitation and influence photosynthetic capacity while increasing respiratory and maintenance costs. Nutrient enrichment, for example, can inhibit the ability of the coral host to maintain demographic control of its algal symbionts, resulting in increased algal densities that act as net carbon sinks and limit energetic resources of the coral (Wooldridge 2013, 2016). While high *Symbiodinium* densities have been suggested to buffer corals from thermal stress (Stimson et al. 2002), Cunning and Baker (2012) showed that increases in symbiont density actually lowered coral bleaching thresholds. This may be because more *Symbiodinium* produce more reactive oxygen species under stressful conditions (Lesser 1996). Elevated $p\mathrm{CO}_2$ promotes enlarged zooxanthellae populations in some (e.g., Reynaud et al. 2003; Crawley et al. 2010; Anlauf et al. 2011), but not all (Rodolfo-Metalpa et al. 2010; Bedwell-Ivers et al. 2016) cases (Table 12.1). While increases in CO_2 supply may initially release symbionts from DIC limitation and facilitate growth, an enlarged endosymbiont population may increase the risk of CO_2 limitation during periods of high irradiance, theoretically making corals more susceptible to bleaching.

Photorespiration is one of several photoprotective mechanisms that provides alternate electron pathways to divert excess excitation energy that could otherwise lead to ROS formation and photooxidative damage of proteins, lipids, and pigments. Compared to photosynthesis, photorespiration is generally viewed as energetically wasteful because of its higher consumption of NADPH and ATP per unit of sugar produced, but it can be an important physiological pathway for mitigating oxidative stress during periods of excess excitation energy. During photorespiration, Rubisco binds with O_2 (as opposed to CO_2), resulting in the production of phosphoglycolate (PG). Excess PG can inhibit the Calvin cycle, so PGPase breaks down PG to glycolate, allowing the Calvin cycle to continue. Glycolate is either excreted or enzymatically broken down, adding to the fixed carbon supply for photosynthesis. Crawley et al. (2010) investigated the effect of increasing CO_2 on photosynthetic capacity, photoacclimation, and photoprotection in *Acropora formosa* and found that CO_2 enrichment increased chlorophyll *a* per cell but did not affect symbiont cell density. PGPase expression was reduced by 45% at high CO_2 (1160–1500 ppm). The authors suggest that OA has the capacity to influence ROS formation and subsequent oxidative stress by compromising enzymatic activity of key photoprotective pathways. Given that many intracellular enzymes are pH-sensitive, more studies are needed on the effects of CO_2 enrichment on enzymatic pathways that underpin coral-algal symbioses.

12.2.4 Phylotype-Specific Responses and Symbiont Shuffling

Thermal stress is known to differentially affect phylotypes of *Symbiodinium*, resulting in host-specific responses. Compared to temperature, little work has been done to assess phylotype-specific responses to CO_2. Brading et al. (2011) investigated the effect of pCO_2 on the photosynthesis and growth of four coral-associated phylotypes of *Symbiodinium* (cultured cells) and found the response to be phylotype-specific. Whereas certain phylotypes (A1 and B1) were largely unaffected by a doubling of pCO_2, the growth rate (A13) and photosynthetic capacity (A2) of other phylotypes are doubled. This variability may be linked to differences in carbon acquisition as well as preference for dissolved inorganic carbon species (CO_2 vs. HCO_3^-) and may partially explain species-specific responses observed in other studies (e.g., Bedwell-Ivers et al. 2016). Symbiont shuffling—i.e., increased abundance of heat-tolerant symbionts following thermal bleaching (Buddemeier et al. 2004)—has been shown to reduce coral susceptibility to recurrent warming (e.g., Cunning et al. 2015). Whether coral-algal associations respond to changes in pCO_2 has not been thoroughly explored; however, Noonan et al. (2013) found that coral-symbiont associations remained stable regardless of proximity to volcanic CO_2 seeps in Papua New Guinea, indicating that acclimatization through symbiont shuffling may not be an option to cope with ocean acidification.

12.2.5 Photosynthesis-Respiration

Experiments investigating the effect of elevated CO_2 on coral photosynthesis and/or carbon production show complex and species-specific responses with variable results (Table 12.1). The vast majority of studies on coral reef organisms (e.g., corals and calcified algae) and communities suggest that photosynthetic rate is relatively unaffected by elevated CO_2 (Leclercq et al. 2002; Langdon et al. 2003; Schneider and Erez 2006; Rodolfo-Metalpa et al. 2010; Dove et al. 2013; Takahashi and Kurihara 2013; Comeau et al. 2016). A meta-analysis of 11 studies found that the mean effect size of CO_2 on coral photosynthesis was not statistically discernible from zero (Kroeker et al. 2013). Comeau et al. (2016) used a large range of pCO_2 values (280–2000 μatm) and 15 species of common reef calcifiers (eight coral species and seven calcifying algae) on the shallow reefs of Moorea, to show that net photosynthesis, dark respiration, light-enhanced dark respiration (LEDR), and gross photosynthesis of corals and calcified algae are largely insensitive to pCO_2 during short-term incubations. The general lack of a "CO_2 fertilization" effect on photosynthesis may be, in part, because zooxanthellae primarily use external HCO_3^- (Goiran et al. 1996; Gattuso et al. 1999; Schneider and Erez 2006). In contrast, Kaniewska et al. (2012) report net decreases in both photosynthesis and respiration of *Acropora millepora* colonies exposed to elevated CO_2, and Anthony et al. (2008) report that high CO_2 levels (1000–1300 μatm) induced productivity loss and bleaching of *Acropora intermedia*. Iguchi et al. (2012) report reduced photosynthetic efficiency of the massive coral *Porites australiensis* at high CO_2 (1175–1439 and 1801–2193 μatm), although zooxanthella density was not affected. Meanwhile, greater net productivity with elevated CO_2 was reported for symbiotic sea anemones in laboratory experiments and near natural CO_2 seeps (Suggett et al. 2012; Towanda and Thuesen 2012). Langdon and Atkinson (2005) found a 20–50% increase in carbon production, but not oxygen production, of coral assemblages composed of *Porites compressa* and *Montipora capitata*. It is important to note that CO_2 enrichment does not automatically result in increased productivity, as other factors such as nitrogen, phosphorus, and iron may limit photosynthesis. Consequently, it is valuable to understand the interplay between the influence of OA on primary productivity under different nutrient regimes.

It has been suggested that aerobic respiration in corals will increase under OA to compensate for increased energetic demands associated with maintaining calcification rates in a thermodynamically challenging environment (e.g., McCulloch et al. 2012). However, empirical evidence shows that the effects of elevated pCO_2 on aerobic respiration are ambiguous, with, for example, no effects of high pCO_2 reported on dark respiration of *Stylophora pistillata* (Reynaud et al. 2003), *Acropora eurystoma* (Schneider and Erez 2006), and *A. formosa* (Crawley et al. 2010), while a decrease in respiration has been reported for massive *Porites* spp. (Edmunds 2012), *A. millepora* (Kaniewska et al. 2012), and larvae of *P. astreoides* (Albright and Langdon 2011).

12.3 Ocean Acidification and Coral Reef Resilience

While the link between OA and coral bleaching is tenuous, it is increasingly clear that OA has the capacity to influence post-bleaching recovery by acting on a variety of processes that underpin coral reef resilience, namely, population replenishment and growth.

12.3.1 Reproduction and Recruitment

Ocean acidification has been shown to negatively impact multiple, sequential early life history stages which may severely compromise sexual recruitment and the ability of coral reefs to recover from disturbance. For example, laboratory experiments have found negative impacts of OA on three sequential life history phases necessary for successful coral recruitment: (1) larval availability, by compromising fertilization (Albright et al. 2010; Albright and Mason 2013) but see Chua et al. (2013); (2) settlement ecology, by altering the availability of known settlement cues such as crustose coralline algae (Albright 2011; Albright and Langdon 2011; Doropoulos et al. 2012; Doropoulos and Diaz-Pulido 2013); and (3) post-settlement ecology, by impeding post-settlement growth and survival (Albright et al. 2008, 2010; de Putron et al. 2010; Albright 2011; Albright and Langdon 2011; Anlauf et al. 2011; Moya et al. 2012; Foster et al. 2015, 2016). Field observations from volcanic CO_2 seeps in Papua New Guinea validate laboratory findings, indicating altered settlement substrata and reduced coral recruitment at high CO_2 (Fabricius et al. 2017).

12.3.2 Growth and Calcification

Over the last two decades, OA research has focused primarily on the consequences of shifting ocean chemistry on coral calcification (Kroeker et al. 2010, 2013; Riebesell and Gattuso 2014; Andersson et al. 2015). While some species appear insensitive over the range of conditions investigated (Comeau et al. 2013; Takahashi and Kurihara 2013), the majority of field and laboratory studies show declines in coral calcification with increasing CO_2 (e.g., Gattuso et al. 1998; Langdon et al. 2000, 2003; Marubini et al. 2001, 2003; Reynaud et al. 2003; Langdon and Atkinson 2005; Silverman et al. 2009; Kroeker et al. 2010, 2013; Anthony et al. 2011b; Pandolfi et al. 2011; Dove et al. 2013; Albright et al. 2016). According to a meta-analysis of 25 studies, the mean response of coral calcification to a unit change in Ω_{arag} is approximately 15% (Chan and Connolly 2013). In addition to direct impacts on reef builders, OA and warming have been shown to accelerate decalcification of coral communities, with microbial communities (Dove et al. 2013), endolithic algae (Reyes-Nivia et al. 2013), and excavating sponges (Fang et al. 2013) being the primary agents of erosion.

Both laboratory and field studies provide evidence that coral reefs have already lost significant calcification capacity due to OA (Dove et al. 2013; Albright et al. 2016). Model results suggest that if CO_2 emissions continue to follow a business-as-usual path, tropical coral reefs are likely to shift toward conditions that are marginal for reef growth (i.e., net dissolution) this century (Hoegh-Guldberg et al. 2007; Silverman et al. 2009). As the reef framework and carbonate balance are compromised, a wide range of impacts are likely on coastal human communities (Fabricius 2005; Fabricius et al. 2013; Kroeker et al. 2013; Hoegh-Guldberg 2014; Wong et al. 2014; Albright et al. 2016; Edmunds et al. 2016). These include reduced food, income, and well-being, as well as longer-term impacts such as increasing vulnerability as coral reefs become less able to protect coastal areas from storms and waves (Pendleton 1995; Hoegh-Guldberg et al. 2007; Cooley et al. 2009; Pascal et al. 2016).

12.4 Other Environmental Factors (Nutrients)

While a combination of thermal stress and high irradiance is the primary trigger for modern mass-bleaching events (e.g., Hoegh-Guldberg 1999), at a local to regional scale, other environmental stressors can cause bleaching independently and act synergistically by effectively lowering the threshold temperature at which coral bleaching occurs (Lesser 2004, 2011). These factors include changes in salinity, seawater chemistry, disease, sedimentation, cyanide fishing, pollution, unusually low temperatures, excess ultraviolet (UV) radiation, aerial exposure, bacterial pathogens and pollutants, nutrients, and solar radiation (reviewed in Brown 1997; Lesser 2011). Most of these have been determined experimentally in the laboratory and not verified ecologically. In contrast to the recent global bleaching events associated with global climate change (Hughes et al. 2017), nonthermal bleaching tends to occur on smaller spatial scales in response to localized and/or pulsed stress. For this reason, the majority of these factors are not dealt with here. However, due to the vulnerability of coastal ecosystems to terrestrial inputs, nutrients have the potential to operate on chronic and regional scales, thereby influencing bleaching thresholds in nearshore waters.

The mechanism by which excess nutrients influence bleaching thresholds is not dissimilar to that proposed for CO_2—release from N or C limitation fuel algal densities and lead to excess ROS under stressful conditions. *Symbiodinium* are typically nitrogen-limited at high irradiance (Fabricius 2005). *Symbiodinium* densities typically increase in response to elevated DIN, which is preferentially used for zooxanthellae growth, as opposed to heterotrophically derived nutrients which increase both algal and host tissue growth (Hoegh-Guldberg and Smith 1989; Muscatine et al. 1989; Marubini and Davies 1996; Fabricius 2005). Increased algal populations produce more ROS under stressful conditions (Lesser 1996), thereby making corals more susceptible to bleaching when sea surface temperatures rise (Wooldridge 2009b; Cunning and Baker 2012; Wiedenmann et al. 2012). Thus, the temperature threshold for bleaching has the potential to fluctuate as a function of nutrient levels and their influence on symbiont densities and/or growth rates (Wooldridge 2009a; Cunning and Baker 2012; Wiedenmann et al. 2012).

Despite these proposed links between nutrient availability and bleaching thresholds, there is little empirical evidence that nutrients increase bleaching prevalence in the field. Bleaching severity in inshore environments can be exacerbated relative to offshore (e.g., Wooldridge (2009b), a phenomenon that is often attributed to environmental stress associated with nutrient loading (e.g., Wooldridge 2009b; Wagner et al. 2010). Using a large-scale dataset from the Great Barrier Reef, Wooldridge and Done (2009) investigated geographic patterns of coral bleaching in 1998 and 2002 and show a synergism between heat stress and nutrient flux as a causative mechanism for observed bleaching patterns. Wiedenmann et al. (2012) showed that increased DIN, in combination with limited phosphate concentrations, increases the susceptibility of corals to temperature- and light-induced bleaching. Using a manipulative field experiment in the Florida Keys, Vega Thurber et al. (2014)

showed that coastal nutrient loading—at levels commonly found on many degraded reefs worldwide—increases both bleaching severity and disease prevalence. Encouragingly, one year after termination of nutrient enrichment, there were no differences in bleaching or disease prevalence, suggesting that improvements to water quality may be an effective lever to mitigate some coral bleaching and disease. It is certain that local processes such as nitrogen pollution and eutrophication exacerbate the effects of OA, both chemically and physiologically. Changing sediment loads from terrestrial sources and using controls on nutrient inputs as a policy lever for mitigating coastal water acidification can also modify the carbonate chemistry of surface waters by altering the balance between autotrophy and heterotrophy, thereby helping to alleviate coastal acidification (Bille et al. 2013).

12.5 Conclusions

Overall, our understanding of the impacts of OA on coral-algal symbioses, and associated bleaching dynamics, is incomplete. While studies yield mixed results, mounting experimental evidence suggests that bleaching will not be accentuated at ecologically meaningful levels by the expected increase in pCO_2 over the next century. In theory, changes in CO_2 have the capacity to influence a variety of physiological processes that are integral to host-algal dynamics and photophysiology with the most obvious avenues being direct impacts on algal population dynamics (e.g., algal density and chlorophyll concentrations), downstream effects on energy allocation, and/or impacts on enzymatic pathways that underpin photosynthesis, oxidative stress, and/or photoprotection. However, empirical evidence is dominated by mixed responses, suggesting that the influence of OA on bleaching thresholds is equivocal. Generally, temperature seems to have a greater influence on productivity and photophysiology than CO_2, and intra- and interspecific variation in both host and symbiont responses outweighs CO_2 effects. The lack of a clear signal may, in part, be due to differences in experimental design (e.g., outdoor/indoor, closed versus flow-through systems, duration which affects acclimation potential), levels of other abiotic factors such as light and nutrients, phylotype-specific responses in *Symbiodinium,* and species-specific responses. Certainly, comparisons among species and determination of functional relationships between pCO_2 and photophysiology are complicated by the wide range of experimental conditions that dominate the literature. Studies that couple observations at the phenotypic level with underlying molecular mechanisms (e.g., Crawley et al. 2010) show promise to elucidate relationships between CO_2 enrichment and bleaching physiology. Specifically, studies evaluating connections between CO_2 and oxidative stress (ROS production), apoptosis, photoinhibition (and associated repair pathways such as D1 proteins), enzymatic activity (e.g., antioxidant enzymes, Rubisco), and associated processes such as PSII function are needed, as are investigations into phylotype-specific responses, host-symbiont interactions, and the influence on holobiont responses.

While thermal stress remains the primary concern regarding acute impacts to coral reefs, it is clear that both temperature and OA act synergistically to erode coral reef health and performance. As discussed, direct links between OA and bleaching responses are, at present, tenuous; however, it is certain that OA impedes post-disturbance recovery by slowing growth and reproduction. Given that warming and OA share a root cause—increasing atmospheric CO_2—it is increasingly clear that deep and rapid emissions reductions are critical to secure the future of coral reefs.

References

Albright R (2011) Reviewing the effects of ocean acidification on sexual reproduction and early life history stages of reef-building corals. J Mar Biol 2011:1–14. https://doi.org/10.1155/2011/473615

Albright R, Langdon C (2011) Ocean acidification impacts multiple early life history processes of the Caribbean coral *Porites astreoides*. Global Change Biol 17:2478–2487. https://doi.org/10.1111/j.1365-2486.2011.02404.x

Albright R, Mason B (2013) Projected near-future levels of temperature and pCO_2 reduce coral fertilization success. PLoS One 8. https://doi.org/10.1371/journal.pone.0056468

Albright R, Mason B, Langdon C (2008) Effect of aragonite saturation state on settlement and post-settlement growth of *Porites astreoides* larvae. Coral Reefs 27:485–490. https://doi.org/10.1007/s00338-008-0392-5

Albright R, Mason B, Miller M, Langdon C (2010) Ocean acidification compromises recruitment success of the threatened Caribbean coral *Acropora palmata*. Proc Natl Acad Sci USA 107:20400–20404. https://doi.org/10.1073/pnas.1007273107

Albright R, Langdon C, Anthony KRN (2013) Dynamics of seawater carbonate chemistry, production, and calcification of a coral reef flat, central Great Barrier Reef. Biogeosciences 10:6747–6758. https://doi.org/10.5194/bg-10-6747-2013

Albright R, Benthuysen J, Cantin N, Caldeira K, Anthony K (2015) Coral reef metabolism and carbon chemistry dynamics of a coral reef flat. Geophys Res Lett 42. https://doi.org/10.1002/2015GL063488

Albright R, Caldeira L, Hosfelt J, Kwiatkowski L, Maclaren JK, Mason BM et al (2016) Reversal of ocean acidification enhances net coral reef calcification. Nature 531:362–365. https://doi.org/10.1038/nature17155

Al-Horani FA, Al-Moghrabi SM, de Beer D (2003) Microsensor study of photosynthesis and calcification in the scleractinian coral, *Galaxea fascicularis*: active internal carbon cycle. J Exp Mar Biol Ecol 288:1–15. https://doi.org/10.1016/s0022-0981(02)00578-6

Altieri AH, Harrison SB, Seemann J, Collin R, Diaz RJ, Knowlton N (2017) Tropical dead zones and mass mortalities on coral reefs. Proc Natl Acad Sci USA 114:3660–3665. https://doi.org/10.1073/pnas.1621517114

Andersson AJ, Gledhill D (2013) Ocean acidification and coral reefs: effects on breakdown, dissolution, and net ecosystem calcification. Annu Rev Mar Sci 5:321–348. https://doi.org/10.1146/annurev-marine-121211-172241

Andersson AJ, Kline DI, Edmunds PJ, Archer SD, Bednaršek N, Carpenter RC et al (2015) Understanding ocean acidification impacts on organismal to ecological scales. Oceanography 25:16–27. https://doi.org/10.5670/oceanog.2015.27

Anlauf H, D'Croz L, O'Dea A (2011) A corrosive concoction: the combined effects of ocean warming and acidification on the early growth of a stony coral are multiplicative. J Exp Mar Biol Ecol 397:13–20. https://doi.org/10.1016/j.jembe.2010.11.009

Anthony KRN, Kline DI, Diaz-Pulido G, Dove S, Hoegh-Guldberg O (2008) Ocean acidification causes bleaching and productivity loss in coral reef builders. Proc Natl Acad Sci USA 105:17442–17446. https://doi.org/10.1073/pnas.0804478105

Anthony KRN, Kleypas A, JA GJ-P (2011a) Coral reefs modify their seawater carbon chemistry - implications for impacts of ocean acidification. Global Change Biol 17:3655–3666. https://doi.org/10.1111/j.1365-2486.2011.02510.x

Anthony KRN, Maynard JA, Diaz-Pulido G, Mumby PJ, Marshall PA, Cao L et al (2011b) Ocean acidification and warming will lower coral reef resilience. Global Change Biol 17:1798–1808. https://doi.org/10.1111/j.1365-2486.2010.02364.x

Anthony KRN, Diaz-Pulido G, Verlinden N, Tilbrook B, Andersson AJ (2013) Benthic buffers and boosters of ocean acidification on coral reefs. Biogeosciences 10:4897–4909. https://doi.org/10.5194/bg-10-4897-2013

Archer D, Eby M, Brovkin V, Ridgwell A, Cao L, Mikolajewicz U et al (2009) Atmospheric lifetime of fossil fuel carbon dioxide. Annu Rev Earth Planet Sci 37:117–134. https://doi.org/10.1146/annurev.earth.031208.100206

Baghdasarian G, Osberg A, Mihora D, Putnam HM, Gates RD, Edmunds PJ (2017) Effects of temperature and pCO_2 on population regulation of *Symbiodinium* spp. in a tropical reef coral. Biol Bull 232:123–139. https://doi.org/10.1086/692718

Barkley HC, Cohen AL, McCorkle DC, Golbuu Y (2017) Mechanisms and thresholds for pH tolerance in Palau corals. J Exp Mar Biol Ecol 489:7–14. https://doi.org/10.1016/j.jembe.2017.01.003

Bates NR, Amat A, Andersson AJ (2010) Feedbacks and responses of coral calcification on the Bermuda reef system to seasonal changes in biological processes and ocean acidification. Biogeosciences 7:2509–2530. https://doi.org/10.5194/bg-7-2509-2010

Bates NR, Astor YM, Church MJ, Currie K, Dore JE, González-Dávila M et al (2014) A time-series view of changing ocean chemistry due to ocean uptake of anthropogenic CO_2 and ocean acidification. Oceanography 27:126–141. https://doi.org/10.5670/oceanog.2014.16

Bedwell-Ivers HE, Koch MS, Peach KE, Joles L, Dutra E, Manfrino C (2016) The role of *in hospite* zooxanthellae photophysiology and reef chemistry on elevated pCO_2 effects in two branching Caribbean corals: *Acropora cervicornis* and *Porites divaricata*. ICES J Mar Sci 74:1103–1112. https://doi.org/10.1093/icesjms/fsw026

Beer S, Koch E (1996) Photosynthesis of marine macroalgae and seagrasses in globally changing CO_2 environments. Mar Ecol Prog Ser 141:199–204. https://doi.org/10.3354/meps141199

Bille R, Kelly R, Biastoch A, Harrould-Kolieb E, Herr D, Joos F et al (2013) Taking action against ocean acidification: a review of management and policy options. Environ Manage 52:761–779. https://doi.org/10.1007/s00267-013-0132-7

Brading P, Warner ME, Davey P, Smith DJ, Achterberg EP, Suggett DJ (2011) Differential effects of ocean acidification on growth and photosynthesis among phylotypes of *Symbiodinium* (Dinophyceae). Limnol Oceanogr 56:927–938. https://doi.org/10.4319/lo.2011.56.3.0927

Broecker WS, Takahashi T, Simpson HJ, Peng TH (1979) Fate of fossil fuel carbon dioxide and the global carbon budget. Science 206:409–418. https://doi.org/10.1126/science.206.4417.409

Brown BE (1997) Coral bleaching: causes and consequences. Coral Reefs 16:S129–S138. https://doi.org/10.1007/s003380050249

Buddemeier RW, Baker AC, Fautin DG, Jacobs JR (2004) The adaptive hypothesis of bleaching. In: Rosenberg E, Loya Y (eds) Coral health and disease. Springer, Berlin, pp 427–444

Caldeira K, Wickett ME (2003) Anthropogenic carbon and ocean pH. Nature 425:365. https://doi.org/10.1038/425365a

Canadell JG, Le Quéré C, Raupach MR, Field CB, Buitenhuis ET, Ciais P et al (2007) Contributions to accelerating atmospheric CO_2 growth from economic activity, carbon intensity, and efficiency of natural sinks. Proc Natl Acad Sci USA 104:18866–18870. https://doi.org/10.1073/pnas.0702737104

Cao L, Caldeira K (2008) Atmospheric CO_2 stabilization and ocean acidification. Geophys Res Lett 35. https://doi.org/10.1029/2008gl035072

Chan NC, Connolly SR (2013) Sensitivity of coral calcification to ocean acidification: a meta-analysis. Global Change Biol 19:282–290. https://doi.org/10.1111/gcb.12011

Chua CM, Leggat W, Moya A, Baird AH (2013) Temperature affects the early life history stages of corals more than near future ocean acidification. Mar Ecol Prog Ser 475:85–92. https://doi.org/10.3354/meps10077

Clarkson MO, Kasemann SA, Wood RA, Lenton TM, Daines SJ, Richoz S et al (2015) Ocean acidification and the Permo-Triassic mass extinction. Science 348:229–232. https://doi.org/10.1126/science.aaa0193

Comeau S, Carpenter RC, Edmunds PJ (2013) Coral reef calcifiers buffer their response to ocean acidification using both bicarbonate and carbonate. Proc R Soc B 280. https://doi.org/10.1098/rspb.2012.2374

Comeau S, Carpenter RC, Edmunds PJ (2016) Effects of pCO_2 on photosynthesis and respiration of tropical scleractinian corals and calcified algae. ICES J Mar Sci 74:1092–1102. https://doi.org/10.1093/icesjms/fsv267

Comeau S, Edmunds PJ, Lantz CA, Carpenter RC (2017) Daily variation in net primary production and net calcification in coral reef communities exposed to elevated pCO_2. Biogeosciences 14:3549–3560. https://doi.org/10.5194/bg-14-3549-2017

Cooley SR, Kite-Powell HL, Doney SC (2009) Ocean acidification's potential to alter global marine ecosystem services. Oceanography 22:172–181. https://doi.org/10.5670/oceanog.2009.106

Crawley A, Kline DI, Dunn S, Anthony KRN, Dove S (2010) The effect of ocean acidification on symbiont photorespiration and productivity in *Acropora formosa*. Global Change Biol 16:851–863. https://doi.org/10.1111/j.1365-2486.2009.01943.x

Cunning R, Baker AC (2012) Excess algal symbionts increase the susceptibility of reef corals to bleaching. Nat Clim Change 3:259–262. https://doi.org/10.1038/nclimate1711

Cunning R, Silverstein RN, Baker AC (2015) Investigating the causes and consequences of symbiont shuffling in a multi-partner reef coral symbiosis under environmental change. Proc R Soc B 282. https://doi.org/10.1098/rspb.2014.1725

Cyronak T, Schulz KG, Santos IR, Eyre BD (2014) Enhanced acidification of global coral reefs driven by regional biogeochemical feedbacks. Geophys Res Lett 41:5538–5545. https://doi.org/10.1002/2014GL060849

de Putron SJ, McCorkle DC, Cohen AL, Dillon AB (2010) The impact of seawater saturation state and bicarbonate ion concentration on calcification by new recruits of two Atlantic corals. Coral Reefs 30:321–328. https://doi.org/10.1007/s00338-010-0697-z

DeSalvo MK, Voolstra CR, Sunagawa S, Schwarz JA, Stillman JH, Coffroth MA et al (2008) Differential gene expression during thermal stress and bleaching in the Caribbean coral *Montastraea faveolata*. Mol Ecol 17:3952–3971. https://doi.org/10.1111/j.1365-294X.2008.03879.x

Diaz-Pulido G, Gouezo M, Tilbrook B, Dove S, Anthony KRN (2011) High CO_2 enhances the competitive strength of seaweeds over corals. Ecol Lett 14:156–162. https://doi.org/10.1111/j.1461-0248.2010.01565.x

Doney SC (2010) The growing human footprint on coastal and open-ocean biogeochemistry. Science 328:1512–1516. https://doi.org/10.1126/science.1185198

Doney SC, Fabry VJ, Feely RA, Kleypas JA (2009) Ocean acidification: the other CO_2 problem. Annu Rev Mar Sci 1:169–192. https://doi.org/10.1146/annurev.marine.010908.163834

Doropoulos C, Diaz-Pulido G (2013) High CO_2 reduces the settlement of a spawning coral on three common species of crustose coralline algae. Mar Ecol Prog Ser 475:93–99. https://doi.org/10.3354/meps10096

Doropoulos C, Ward S, Diaz-Pulido G, Hoegh-Guldberg O, Mumby PJ (2012) Ocean acidification reduces coral recruitment by disrupting intimate larval-algal settlement interactions. Ecol Lett 15:338–346. https://doi.org/10.1111/j.1461-0248.2012.01743.x

Dove SG, Kline DI, Pantos O, Angly FE, Tyson GW, Hoegh-Guldberg O (2013) Future reef decalcification under a business-as-usual CO_2 emission scenario. Proc Natl Acad Sci 110:15342–15347. https://doi.org/10.1073/pnas.1302701110

Duarte CM, Hendriks IE, Moore TS, Olsen YS, Steckbauer A, Ramajo L et al (2013) Is ocean acidification an open-ocean syndrome? Understanding anthropogenic impacts on seawater pH. Estuaries Coast 36:221–236. https://doi.org/10.1007/s12237-013-9594-3

Edmunds PJ (2012) Effect of pCO_2 on the growth, respiration, and photophysiology of massive *Porites* spp. in Moorea, French Polynesia. Mar Biol 159:2149–2160. https://doi.org/10.1007/s00227-012-2001-y

Edmunds PJ, Comeau S, Lantz C, Andersson A, Biggs C, Cohen A et al (2016) Integrating the effects of ocean acidification across functional scales on tropical coral reefs. BioScience 66:350–362. https://doi.org/10.1093/bi-osci/biw023

Fabricius KE (2005) Effects of terrestrial runoff on the ecology of corals and coral reefs: review and synthesis. Mar Pollut Bull 50:125–146. https://doi.org/10.1016/j.marpolbul.2004.11.028

Fabricius KE, Langdon C, Uthicke S, Humphrey C, Noonan S, De'ath G et al (2011) Losers and winners in coral reefs acclimatized to elevated carbon dioxide concentrations. Nat Clim Change 1:165–169. https://doi.org/10.1038/nclimate1122

Fabricius KE, Cséke HC, De'ath G (2013) Does trophic status enhance or reduce the thermal tolerance of scleractinian corals? A review, experiment and conceptual framework. PLoS One 8. https://doi.org/10.1371/journal.pone.0054399

Fabricius KE, De'ath G, Noonan S, Uthicke S (2014) Ecological effects of ocean acidification and habitat complexity on reef-associated macroinvertebrate communities. Proc R Soc B 281. https://doi.org/10.1098/rspb.2013.2479

Fabricius KE, Nonnan SHC, Abrego D, Harrington L, De'ath G (2017) Low recruitment due to altered settlement substrata as primary constraint for coral communities under ocean acidification. Proc R Soc B 284. https://doi.org/10.1098/prsb.2017.1536

Falter JL, Lowe RJ, Atkinson MJ, Monismith SG, Schar DW (2008) Continuous measurements of net production over a shallow reef community using a modified Eulerian approach. J Geophys Res 113. https://doi.org/10.1029/2007jc004663

Falter JL, Lowe RJ, Atkinson MJ, Cuet P (2012) Seasonal coupling and de-coupling of net calcification rates from coral reef metabolism and carbonate chemistry at Ningaloo Reef, Western Australia. J Geophys Res 117. https://doi.org/10.1029/2011jc007268

Fang JKH, Mello-Athayde MA, Schönberg CHL, Kline DI, Hoegh-Guldberg O, Dove S (2013) Sponge biomass and bioerosion rates increase under ocean warming and acidification. Global Change Biol 19:3581–3591. https://doi.org/10.1111/gcb.12334

Feely RA, Doney SC, Cooley SR (2009) Ocean acidification: present conditions and future changes in a high-CO_2 world. Oceanography 22:36–47. https://doi.org/10.5670/oceanog.2009.95

Feely RA, Alin SR, Newton J, Sabine CL, Warner M, Devol A et al (2010) The combined effects of ocean acidification, mixing, and respiration on pH and carbonate saturation in an urbanized estuary. Estuaries Coast Shelf Sci 88:442–449. https://doi.org/10.1016/j.ecss.2010.05.004

Fine M, Tchernov D (2007) Scleractinian coral species survive and recover from decalcification. Science 315:1881. https://doi.org/10.1126/science.1137094

Fitt WK, Brown BE, Warner ME, Dunne RP (2001) Coral bleaching: interpretation of thermal tolerance limits and thermal thresholds in tropical corals. Coral Reefs 20:51–65. https://doi.org/10.1007/s003380100146

Foster T, Gilmour JP, Chua CM, Falter JL, McCulloch MT (2015) Effect of ocean warming and acidification on the early life stages of subtropical *Acropora spicifera*. Coral Reefs 34:1217–1226. https://doi.org/10.1007/s00338-015-1342-7

Foster T, Falter JL, McCulloch MT, Clode PL (2016) Ocean acidification causes structural deformities in juvenile coral skeletons. Sci Adv 2. https://doi.org/10.1126/sciadv.1501130

Furla P, Galgani I, Durand I, Allemand D (2000) Sources and mechanisms of inorganic carbon transport for coral calcification and photosynthesis. J Exp Biol 203:3445–3457

Gattuso J-P, Frankignoulle M, Bourge I, Romaine S, Buddemeier RW (1998) Effect of calcium carbonate saturation of seawater on coral calcification. Global Planet Change 18:37–46. https://doi.org/10.1016/S0921-8181(98)00035-6

Gattuso J-P, Allemand D, Frankignoulle M (1999) Photosynthesis and calcification at cellular, organismal and community levels in coral reefs: a review on interactions and control by carbonate chemistry. Am Zool 39:160–183. https://doi.org/10.1093/icb/39.1.160

Goiran C, Al-Moghrabi S, Allemand D, Jaubert J (1996) Inorganic carbon uptake for photosynthesis by the symbiotic coral/dinoflagellate association I. Photosynthetic performances of symbionts and dependence on sea water bicarbonate. J Exp Mar Biol Ecol 199:207–225. https://doi.org/10.1016/0022-0981(95)00201-4

Golbuu Y, Gouezo M, Kurihara H, Rehm L, Wolanski E (2016) Long-term isolation and local adaptation in Palau's Nikko Bay help corals thrive in acidic waters. Coral Reefs 35:909–918. https://doi.org/10.1007/s00338-016-1457-5

Gray SEC, DeGrandpre MD, Langdon C, Corredor JE (2012) Short-term and seasonal pH, pCO_2 and saturation state variability in a coral-reef ecosystem. Global Biogeochem Cycles 26. https://doi.org/10.1029/2011gb004114

Harrington L, Fabricius KE, De'ath G, Negri AP (2004) Recognition and selection of settlement substrata determine post-settlement survival in corals. Ecology 85:3428–3437. https://doi.org/10.1890/04-0298

Hoadley KD, Pettay DT, Grottoli AG, Cai WC, Melman TF, Schoepf V et al (2015) Physiological response to elevated temperature and pCO_2 varies across four Pacific coral species: understanding the unique host+symbiont response. Sci Rep 5. https://doi.org/10.1038/srep18371

Hoegh-Guldberg O (1999) Climate change, coral bleaching and the future of the world's coral reefs. Mar Freshwater Res 50:839–866. https://doi.org/10.1071/MF99078

Hoegh-Guldberg O (2014) Coral reef sustainability through adaptation: glimmer of hope or persistent mirage? Curr Opin Environ Sustain 7:127–133. https://doi.org/10.1016/j.cosust.2014.01.005

Hoegh-Guldberg O, Smith GJ (1989) Influence of the population density of zooxanthellae and supply of ammonium on the biomass and metabolic characteristics of the reef corals *Seriatopora hystrix* and *Stylophora pistillata*. Mar Ecol Prog Ser 57:173–186. https://doi.org/10.3354/meps057173

Hoegh-Guldberg O, Mumby PJ, Hooten AJ, Steneck RS, Greenfield P, Gomez E et al (2007) Coral reefs under rapid climate change and ocean acidification. Science 318:1737–1742. https://doi.org/10.1126/science.1152509

Hofmann GE, Smith JE, Johnson KS, Send U, Levin LA, Micheli F et al (2011) High-frequency dynamics of ocean pH: a multi-ecosystem comparison. PLoS One 6. https://doi.org/10.1371/journal.pone.0028983

Horwitz R, Hoogenboom MO, Fine M (2017) Spatial competition dynamics between reef corals under ocean acidification. Sci Rep 7. https://doi.org/10.1038/srep40288

Hughes TP, Kerry JT, Álvarez-Noriega M, Álvarez-Romero JG, Anderson KD, Baird AH et al (2017) Global warming and recurrent mass bleaching of corals. Nature 543:373–377. https://doi.org/10.1038/nature21707

Iguchi A, Ozaki S, Nakamura T, Inoue M, Tanaka Y, Suzuki A et al (2012) Effects of acidified seawater on coral calcification and symbiotic algae on the massive coral *Porites australiensis*. Mar Environ Res 73:32–36. https://doi.org/10.1016/j.marenvres.2011.10.008

Jiang ZJ, Huang XP, Zhang JP (2010) Effects of CO(2) enrichment on photosynthesis, growth, and biochemical composition of seagrass *Thalassia hemprichii* (Ehrenb.) Aschers. J Integr Plant Biol 52:904–913. https://doi.org/10.1111/j.1744-7909.2010.00991.x

Jokiel PL (2004) Temperature stress and coral bleaching. In: Rosenberg E, Loya Y (eds) Coral health and disease. Springer, Berlin, pp 401–425

Kaniewska P, Campbell PR, Kline DI, Rodriguez-Lanetty M, Miller DJ, Dove S et al (2012) Major cellular and physiological impacts of ocean acidification on a reef building coral. PLoS One 7. https://doi.org/10.1371/journal.pone.0034659

Kavousi J, Reimer JD, Tanaka Y, Nakamura T (2015) Colony-specific investigations reveal highly variable responses among individual corals to ocean acidification and warming. Mar Environ Res 109:9–20. https://doi.org/10.1016/j.marenvres.2015.05.004

Kleypas JA, Buddemeier RW, Archer D, Gattuso J-P, Langdon C, Opdyke BN (1999) Geochemical consequences of increased atmospheric carbon dioxide on coral reefs. Science 284:118–120. https://doi.org/10.1126/science.284.5411.118

Kleypas JA, Anthony KRN, Gattuso J-P (2011) Coral reefs modify their seawater carbon chemistry - case study from a barrier reef (Moorea, French Polynesia). Global Change Biol 17:3667–3678. https://doi.org/10.1111/j.1365-2486.2011.02530.x

Koweek D, Dunbar RB, Rogers JS, Williams GJ, Price N, Mucciarone D et al (2014) Environmental and ecological controls of coral community metabolism on Palmyra Atoll. Coral Reefs 34: 3339-351. doi:https://doi.org/10.1007/s00338-014-1217-3

Koweek DA, Dunbar RB, Monismith SG, Mucciarone DA, Woodson CB, Samuel L (2015) High-resolution physical and biogeochemical variability from a shallow back reef on Ofu, American Samoa: an end-member perspective. Coral Reefs 34:979–991. https://doi.org/10.1007/s00338-015-1308-9

Krief S, Hendy EJ, Fine M, Yam R, Meibom A, Foster GL et al (2010) Physiological and isotopic responses of scleractinian corals to ocean acidification. Geochim Cosmochim Acta 74:4988–5001. https://doi.org/10.1016/j.gca.2010.05.023

Kroeker KJ, Kordas RL, Crim RN, Singh GG (2010) Meta-analysis reveals negative yet variable effects of ocean acidification on marine organisms. Ecol Lett 13:1419–1434. https://doi.org/10.1111/j.1461-0248.2010.01518.x

Kroeker KJ, Kordas RL, Crim R, Hendriks IE, Ramajo L, Singh GS et al (2013) Impacts of ocean acidification on marine organisms: quantifying sensitivities and interaction with warming. Global Change Biol 19:1884–1896. https://doi.org/10.1111/gcb.12179

Krueger T, Horwitz N, Bodin J, Giovani ME, Escrig S, Meibom A et al (2017) Common reef-building coral in the Northern Red Sea resistant to elevated temperature and acidification. R Soc Open Sci 4. https://doi.org/10.1098/rsos.170038

Kuffner IB, Andersson AJ, Jokiel PL, Rodgers KS, Mackenzie FT (2007) Decreased abundance of crustose coralline algae due to ocean acidification. Nat Geosci 1:114–117. https://doi.org/10.1038/ngeo100

Kwiatkowski L, Cox P, Halloran PR, Mumby PJ, Wiltshire AJ (2015) Coral bleaching under unconventional scenarios of climate warming and ocean acidification. Nat Clim Change 5:777–781. https://doi.org/10.1038/nclimate2655

Langdon C, Atkinson MJ (2005) Effect of elevated pCO_2 on photosynthesis and calcification of corals and interactions with seasonal change in temperature/irradiance and nutrient enrichment. J Geophys Res 110. https://doi.org/10.1029/2004jc002576

Langdon C, Takahashi T, Sweeney C, Chipman D, Goddard J, Marubini et al (2000) Effect of calcium carbonate saturation state on the calcification rate of an experimental coral reef. Global Biogeochem Cycles 14:639–654. https://doi.org/10.1029/1999gb001195

Langdon C, Broecker WS, Hammond DE, Glenn E, Fitzsimmons K, Nelson SG et al (2003) Effect of elevated CO_2 on the community metabolism of an experimental coral reef. Global Biogeochem Cycles 17. https://doi.org/10.1029/2002gb001941

Le Quéré C, Moriarty R, Andrew RM, Peters GP, Friedlingstein P, Jones SD et al (2015) Global carbon budget 2014. Earth Syst Sci Data 7:47–85. https://doi.org/10.5194/essd-7-47-2015

Leclercq N, Gattuso J-P, Jaubert J (2002) Primary production, respiration, and calcification of a coral reef mesocosm under increased CO_2 partial pressure. Limnol Oceanogr 47:558–564. https://doi.org/10.4319/lo.2002.47.2.0558

Lesser MP (1996) Elevated temperatures and ultraviolet radiation cause oxidative stress and inhibit photosynthesis in symbiotic dinoflagellates. Limnol Oceanogr 41:271–283. https://doi.org/10.4319/lo.1996.41.2.0271

Lesser MP (2004) Experimental biology of coral reef ecosystems. J Exp Mar Biol Ecol 300:217–252. https://doi.org/10.1016/j.jembe.2003.12.027

Lesser MP (2011) Coral bleaching: causes and mechanisms. In: Dubinsky Z, Stambler N (eds) Coral reefs: an ecosystem in transition. Springer, Berlin, pp 405–419

Mackey KRM, Morris JJ, Morel FMM, Kranz SA (2015) Response of photosynthesis to ocean acidification. Oceanography 25:74–91. https://doi.org/10.5670/oceanog.2015.33

Madin JS, O'Donnell MJ, Connolly SR (2008) Climate-mediated mechanical changes to post-disturbance coral assemblages. Biol Lett 4:490–493. https://doi.org/10.1098/rsbl.2008.0249

Marubini F, Davies PS (1996) Nitrate increases zooxanthellae population density and reduces skeletogenesis in corals. Mar Biol 127:319–328. https://doi.org/10.1007/BF00942117

Marubini F, Barnett H, Langdon C, Atkinson MJ (2001) Dependence of calcification on light and carbonate ion concentration for the hermatypic coral *Porites compressa*. Mar Ecol Prog Ser 220:153–162. https://doi.org/10.3354/meps220153

Marubini F, Ferrier-Pagès C, Cuif J-P (2003) Suppression of skeletal growth in scleractinian corals by decreasing ambient carbonate-ion concentration: a cross-family comparison. Proc R Soc B 270:179–184. https://doi.org/10.1098/rspb.2002.2212

Marubini F, Ferrier-Pagès C, Furla P, Allemand D (2008) Coral calcification responds to seawater acidification: a working hypothesis towards a physiological mechanism. Coral Reefs 27:491–499. https://doi.org/10.1007/s00338-008-0375-6

McCulloch M, Falter J, Trotter J, Montagna P (2012) Coral resilience to ocean acidification and global warming through pH up-regulation. Nat Clim Change 2:623–627. https://doi.org/10.1038/nclimate1473

Moya A, Huisman L, Ball EE, Hayward DC, Grasso LC, Chua CM et al (2012) Whole transcriptome analysis of the coral *Acropora millepora* reveals complex responses to CO_2-driven acidification during the initiation of calcification. Mol Ecol 21:2440–2454. https://doi.org/10.1111/j.1365-294X.2012.05554.x

Munday PL, Dixson DL, McCormick MI, Meekan M, Ferrari MCO, Chivers DP (2010) Replenishment of fish populations is threatened by ocean acidification. Proc Natl Acad Sci USA 107:12930–12934. https://doi.org/10.1073/pnas.1004519107

Munday PL, Cheal AJ, Dixson DL, Rummer JL, Fabricius KE (2014) Behavioural impairment in reef fishes caused by ocean acidification at CO_2 seeps. Nat Clim Change 4:487–492. https://doi.org/10.1038/nclimate2195

Muscatine L, Falkowski PG, Dubinsky Z, Cook PA, McCloskey LR (1989) The effect of external nutrient resources on the population dynamics of zooxanthellae in a reef coral. Proc R Soc B 236. https://doi.org/10.1098/prsb.1989.0025

Noonan SHC, Fabricius KE (2016) Ocean acidification affects productivity but not the severity of thermal bleaching in some tropical corals. ICES J Mar Sci 73:715–726. https://doi.org/10.1093/icesjms/fsv127

Noonan SHC, Fabricius KE, Humphrey C (2013) *Symbiodinium* community composition in scleractinian corals is not affected by life-long exposure to elevated carbon dioxide. PLoS One 8. https://doi.org/10.1371/journal.pone.0063985

Ohde S, van Woesik R (1999) Carbon dioxide flux and metabolic processes of a coral reef, Okinawa. Bull Mar Sci 65:559–576

Ow YX, Collier CJ, Uthicke S (2015) Responses of three tropical seagrass species to CO_2 enrichment. Mar Biol 162:1005–1017. https://doi.org/10.1007/s00227-015-2644-6

Pandolfi JM, Connolly SR, Marshall DJ, Cohen AL (2011) Projecting coral reef futures under global warming and ocean acidification. Science 333:418–422. https://doi.org/10.1126/science.1204794

Pascal N, Allenbach M, Brathwaite A, Burke L, Le Port G, Clua E (2016) Economic valuation of coral reef ecosystem service of coastal protection: a pragmatic approach. Ecosyst Services 21:72–80. https://doi.org/10.1016/j.ecoser.2016.07.005

Pecheux M (2002) CO_2 increase, a direct cause of coral reef mass bleaching? Mar Life 12:63–68

Pendleton LH (1995) Valuing coral reef protection. Ocean Coast Manage 26:119–131. https://doi.org/10.1016/0964-5691(95)00007-O

Perry CT, Murphy GN, Kench PS, Smithers SG, Edinger EN, Steneck RS et al (2013) Caribbean-wide decline in carbonate production threatens coral reef growth. Nat Commun 4. https://doi.org/10.1038/ncomms2409

Pörtner HO (2005) Synergistic effects of temperature extremes, hypoxia, and increases in CO_2 on marine animals: from Earth history to global change. J Geophys Res 110. https://doi.org/10.1029/2004jc002561

Pörtner H (2008) Ecosystem effects of ocean acidification in times of ocean warming: a physiologist's view. Mar Ecol Prog Ser 373:203–217. https://doi.org/10.3354/meps07768

Prada F, Caroselli E, Mengoli S, Brizi L, Fantazzini P, Capaccioni B et al (2017) Ocean warming and acidification synergistically increase coral mortality. Sci Rep 7. https://doi.org/10.1038/srep40842

Putnam HM, Gates RD (2015) Preconditioning in the reef-building coral *Pocillopora damicornis* and the potential for trans-generational acclimatization in coral larvae under future climate change conditions. J Exp Biol 218:2365–2372. https://doi.org/10.1242/jeb.123018

Reyes-Nivia C, Diaz-Pulido G, Kline D, Guldberg OH, Dove S (2013) Ocean acidification and warming scenarios increase microbioerosion of coral skeletons. Global Change Biol 19:1919–1929. https://doi.org/10.1111/gcb.12158

Reynaud S, Leclercq N, Romaine-Lioud S, Ferrier-Pages C, Jaubert J, Gattuso J-P (2003) Interacting effects of CO_2 partial pressure and temperature on photosynthesis and calcification in a scleractinian coral. Global Change Biol 9:1660–1668. https://doi.org/10.1046/j.1529-8817.2003.00678.x

Rhein M, Rintoul SR, Aoki S, Campos E, Chambers D, Feely RA et al (2013) Observations: ocean. In: Stocker TF, Qin D, Plattner G-K, Tignor M, Allen SK, Boschung J et al (eds) Climate change 2013: the physical science basis. Contribution of working group I to the fifth assessment report of the intergovernmental panel on climate change. Cambridge University Press, Cambridge, UK

Riebesell U, Gattuso J-P (2014) Lessons learned from ocean acidification research. Nat Clim Change 5:12–14. https://doi.org/10.1038/nclimate2456

Rodolfo-Metalpa R, Martin S, Ferrier-Pages C, Gattuso J-P (2010) Response of the temperate coral *Cladocora caespitosa* to mid- and long-term exposure to pCO_2 and temperature levels projected for the year 2100 AD. Biogeosciences 7:289–300. https://doi.org/10.5194/bg-7-289-2010

Rowan R, Whitney SM, Fowler A, Yellowlees D (1996) Rubisco in marine symbiotic dinoflagellates: form II enzymes in eukaryotic oxygenic phototrophs encoded by a nuclear multigene family. Plant Cell 8:539–553. https://doi.org/10.2307/3870331

Russell BD, Connell SD, Uthicke S, Muehllehner N, Fabricius KE, Hall-Spencer JM (2013) Future seagrass beds: can increased productivity lead to increased carbon storage? Mar Pollut Bull 73:463–469. https://doi.org/10.1016/j.marpolbul.2013.01.031

Sabine CL, Feely RA, Gruber N, Key RM, Lee K, Bullister JL et al (2004) The oceanic sink for anthropogenic CO_2. Science 305:367–371. https://doi.org/10.1126/science.1097403

Sabine CL, Feely RA, Wanninkhof R, Takahashi T, Khatiwala S, Park G-H (2011) The global ocean carbon cycle. State of the Climate 2010. Bull Am Meteorol Soc 92:S100–S108. https://doi.org/10.1175/1520-0477-92.6.S1

Schneider K, Erez J (2006) The effect of carbonate chemistry on calcification and photosynthesis in the hermatypic coral *Acropora eurystoma*. Limnol Oceanogr 51:1284–1293. https://doi.org/10.4319/lo.2006.51.3.1284

Schoepf V, Grottoli AG, Warner ME, Cai WJ, Melman TF, Hoadley KD et al (2013) Coral energy reserves and calcification in a high-CO_2 world at two temperatures. PLoS One 8. https://doi.org/10.1371/journal.pone.0075049

Shamberger KEF, Feely RA, Sabine CL, Atkinson MJ, DeCarlo EH, Mackenzie FT et al (2011) Calcification and organic production on a Hawaiian coral reef. Mar Chem 127:64–75. https://doi.org/10.1016/j.marchem.2011.08.003

Shaw EC, McNeil BI, Tilbrook B (2012) Impacts of ocean acidification in naturally variable coral reef flat ecosystems. J Geophys Res. 117. https://doi.org/10.1029/2011jc007655

Silverman J, Lazar B, Cao L, Caldeira K, Erez J (2009) Coral reefs may start dissolving when atmospheric CO_2 doubles. Geophys Res Lett 36. https://doi.org/10.1029/2008gl036282

Stimson J, Sakai K, Sembali H (2002) Interspecific comparison of the symbiotic relationship in corals with high and low rates of bleaching-induced mortality. Coral Reefs 21:409–421. https://doi.org/10.1007/s00338-002-0264-3

Strahl J, Francis DS, Doyle J, Humphrey C, Fabricius KE (2016) Biochemical responses to ocean acidification contrast between tropical corals with high and low abundances at volcanic carbon dioxide seeps. ICES J Mar Sci 73:897–909. https://doi.org/10.1093/icesjms/fsv194

Suggett DJ, Hall-Spencer JM, Rodolfo-Metalpa R, Boatman TG, Payton R, Pettay DT et al (2012) Sea anemones may thrive in a high CO_2 world. Global Change Biol 18:3015–3025. https://doi.org/10.1111/j.1365-2486.2012.02767.x

Takahashi A, Kurihara H (2013) Ocean acidification does not affect the physiology of the tropical coral *Acropora digitifera* during a 5-week experiment. Coral Reefs 32:305–314. https://doi.org/10.1007/s00338-012-0979-8

Towanda T, Thuesen EV (2012) Prolonged exposure to elevated CO_2 promotes growth of the algal symbiont *Symbiodinium muscatinei* in the intertidal sea anemone *Anthopleura elegantissima*. Biol Open 1:615–621. https://doi.org/10.1242/bio.2012521

Uthicke S, Furnas M, Lønborg C (2014) Coral reefs on the edge? Carbon chemistry on inshore reefs of the Great Barrier Reef. PLoS One 9. https://doi.org/10.1371/journal.pone.0109092

Vega Thurber RL, Burkepile DE, Fuchs C, Shantz AA, McMinds R, Zaneveld JR (2014) Chronic nutrient enrichment increases prevalence and severity of coral disease and bleaching. Global Change Biol 20:544–554. https://doi.org/10.1111/gcb.12450

Veron JEN (2008) Mass extinctions and ocean acidification: biological constraints on geological dilemmas. Coral Reefs 27:459–472. https://doi.org/10.1007/s00338-008-0381-8

Wagner DE, Kramer P, van Woesik R (2010) Species composition, habitat, and water quality influence coral bleaching in southern Florida. Mar Ecol Prog Ser 408:65–78. https://doi.org/10.3354/meps08584

Wall CB, Fan TY, Edmunds PJ (2013) Ocean acidification has no effect on thermal bleaching in the coral *Seriatopora caliendrum*. Coral Reefs 33:119–130. https://doi.org/10.1007/s00338-013-1085-2

Weis VM (2008) Cellular mechanisms of Cnidarian bleaching: stress causes the collapse of symbiosis. J Exp Biol 211:3059–3066. https://doi.org/10.1242/jeb.009597

Wiedenmann J, D'Angelo C, Smith EG, Hunt AN, Legiret F-E, Postle AD et al (2012) Nutrient enrichment can increase the susceptibility of reef corals to bleaching. Nat Clim Change 3:160–164. https://doi.org/10.1038/nclimate1661

Wong PP, Losada IJ, Gattuso J-P, Hinkel J, Khattabi A, McInnes KL et al (2014) Coastal systems and low-lying areas. In: Field CB, Barros VR, Dokken DJ, Mach KJ, Mastrandrea MD, Bilir TE et al (eds) Climate change 2014: impacts, adaptation, and vulnerability. Part A: Global and sectoral aspects. Contribution of working group II to the fifth assessment report of the intergovernmental panel on climate change. Cambridge University Press, Cambridge, UK, pp 361–409

Wooldridge SA (2009a) A new conceptual model for the warm-water breakdown of the coral–algae endosymbiosis. Mar Freshwater Res 60:483–496. https://doi.org/10.1071/mf08251

Wooldridge SA (2009b) Water quality and coral bleaching thresholds: formalising the linkage for the inshore reefs of the Great Barrier Reef, Australia. Mar Pollut Bull 58:745–751. https://doi.org/10.1016/j.marpolbul.2008.12.013

Wooldridge SA (2013) Breakdown of the coral-algae symbiosis: towards formalizing a linkage between warm-water bleaching thresholds and the growth rate of intracellular zooxanthellae. Biogeosciences 10:1647–1658. https://doi.org/10.5194/bg-10-1647-2013

Wooldridge SA (2016) Excess seawater nutrients, enlarged algal symbiont densities and bleaching sensitive reef locations: 1. Identifying thresholds of concern for the Great Barrier Reef, Australia. Mar Pollut Bull. https://doi.org/10.1016/j.marpolbul.2016.04.054

Wooldridge SA (2017) Instability and breakdown of the coral–algae symbiosis upon exceedence of the interglacial pCO_2 threshold (>260 ppmv): the "missing" Earth-system feedback mechanism. Coral Reefs 36:1025–1037. https://doi.org/10.1007/s00338-017-1594-5

Wooldridge SA, Done TJ (2009) Improved water quality can ameliorate effects of climate change on corals. Ecol Appl 19:1492–1499. https://doi.org/10.1890/08-0963.1

Zeebe RE (2012) History of seawater carbonate chemistry, atmospheric CO_2, and ocean acidification. Annu Rev Earth Planet Sci 40:141–165. https://doi.org/10.1146/annurev-earth-042711-105521

Zeebe RE, Wolf-Gladrow D (2001) CO_2 in seawater: equilibrium, kinetics, isotopes. Elsevier oceanography series, Vol 65. Elsevier, New York

Zimmerman RC, Kohrs DG, Steller DL, Alberte RS (1997) Impacts of CO_2 enrichment on productivity and light requirements of eelgrass. Plant Physiol 115:599–607. https://doi.org/10.1104/pp.115.2.599

Chapter 13
Future Scenarios: A Review of Modelling Efforts to Predict the Future of Coral Reefs in an Era of Climate Change

S. D. Donner, S. F. Heron, and W. J. Skirving

13.1 Introduction

Observations of mass coral bleaching and mortality over the past 4 decades are linked to periods of anomalously warm sea surface temperatures, which have become more frequent as the climate has warmed (Glynn 1991; Hoegh-Guldberg 1999; Donner et al. 2017; Hughes et al. 2017). Global climate models predict that the planet's climate could warm as much as 5 °C by the end of the century, without substantial effort to reduce greenhouse gas emissions far below current levels (IPCC 2014). This continued climate warming poses a serious threat to the long-term health of coral reef ecosystems (Hughes et al. 2003). At the same time, roughly a third of the carbon dioxide emitted by human activity is being absorbed by the oceans causing ocean acidification, which is expected to reduce the rates of coral calcification and reef accretion (Kleypas et al. 1999; Guinotte et al. 2003).

This chapter reviews efforts to predict the effect of future climate change on coral reefs, focussing on coral bleaching. The chapter includes an introduction to climate modelling, a review of the application of climate models to coral bleaching, a case study on the Great Barrier Reef, and a discussion of future research needs.

S. D. Donner (✉)
Department of Geography, University of British Columbia, Vancouver, BC, Canada
e-mail: simon.donner@ubc.ca

S. F. Heron
U.S. National Oceanic and Atmospheric Administration, Coral Reef Watch, College Park, MD, USA

W. J. Skirving
U.S. National Oceanic and Atmospheric Administration, Coral Reef Watch, College Park, MD, USA

ReefSense Pty Ltd., Townsville, QLD, Australia

M. J. H. van Oppen, J. M. Lough (eds.), *Coral Bleaching*, Ecological Studies 233,
https://doi.org/10.1007/978-3-319-75393-5_13

13.2 Modelling Future Climates

Global climate models, which can simulate the response of the Earth to the emissions of greenhouse gases, provide the basis for predictions of the response of coral reefs to climate change. The earliest climate models were simple representations of radiative properties of the atmosphere and the earth's surface, based on the energy from the sun and the composition of the atmosphere. These simple one-dimensional models evolved into today's complex general circulation models (GCMs) that use physical principles to describe the transfer of heat, moisture, and momentum in a three-dimensional grid representing the global climate system.

The current generation of GCMs and "earth systems models" used in the Intergovernmental Panel on Climate Change (IPCC) assessments link the major components of the climate system—atmosphere, ocean, land surface, cryosphere, and biosphere—to best capture the range of physical and biological feedbacks associated with climate variability and change. The models are continually updated to improve the representation of important physical and chemical processes and are rigorously tested against observed data. One generic metric for contrasting GCMs is their equilibrium "climate sensitivity", the predicted equilibrium change in global average temperature caused by a doubling of atmospheric carbon dioxide (CO_2) concentrations, compared to the pre-industrial level. The GCMs used in the IPCC Fifth Assessment Report (AR5) have equilibrium climate sensitivities ranging from 1.5 to 4.5 °C, with a mean of roughly 3 °C (IPCC 2014).

Projections of sea surface temperatures (SSTs) from GCMs are the most reliable information available for predicting the thermal environment that will influence coral reefs in the future. However, there are several key limitations to the modelled representation of future climates that are particularly relevant to the study of coral reefs. These include (1) the coarse spatial resolution of GCMs, (2) the representation of natural modes of climate variability, and (3) the uncertainty over future greenhouse gas emissions.

First, the coarse spatial resolution of climate models limits their ability to provide forecasts for coral reefs. For example, many current models have horizontal resolution of around 1° (~100 km) in the atmosphere and ocean and vertical resolution on the order of 10–100 m in the surface ocean (i.e. depth of each ocean grid cell). Without a representation of the complex bathymetry and hydrodynamics of individual coral reefs, neither of these sets of GCMs can capture processes such as the local upwelling of cooler deep waters or heating of shallow waters on the reef flat (Skirving and Guinotte 2001; Wooldridge and Done 2004). The direct GCM output is better suited to represent the mean temperature of an area of the ocean containing coral reefs rather than the temperature surrounding an individual coral reef or an individual coral.

Higher-resolution regional information can be obtained by downscaling GCM output using dynamical or statistical methods. For example, predictions for an individual reef could be made by forcing a high-resolution hydrodynamic model with the coarser output from a GCM (see Sect. 13.5). Alternatively, statistical relationships between the average temperature for a region and the temperature at specific reef locations could be used to translate GCM output to a higher resolution (Donner et al. 2005; van Hooidonk et al. 2015, 2016). Kwiatkowski et al. (2014), however, suggest that GCM skill at simulating warm season SST and historically forced SST trends is currently limited at less than 16° latitude–longitude resolution. This raises questions about the fidelity of high-resolution bleaching projections produced using statistical downscaling, rather than dynamic downscaling via hydrodynamic models (e.g. van Hooidonk et al. 2015).

Second, the ability to project the variability of future ocean temperatures for many coral reefs depends on model representation of natural modes of climate variability. Mass coral bleaching events have been linked to large-scale oscillations in the atmosphere–ocean system, most notably the El Niño–Southern Oscillation (ENSO; Hughes et al. 2017), but also the Atlantic Multi-decadal Oscillation (Donner et al. 2007) and other modes of variability. Therefore, the reliability of future projections for coral reefs will depend on model ability to represent the periodicity, spatial patterns, and teleconnections of such natural modes of climate variability in the existing climate and the response of these natural modes of variability to human-induced climate warming. For example, overestimation of ENSO variability, common in current models, could lead to overprediction of bleaching frequency in parts of the Pacific (e.g. see bias correction in Logan et al. 2014).

Third, future climate projections depend on assumptions about future changes in climate "forcings", including rates of emission of primary greenhouse gases. In order to represent a range of possible climate futures, GCM simulations are conducted with different emission "scenarios" that are developed based on different sets of assumptions about demographic, economic, and technologic change. The IPCC AR5 employed a set of scenarios called Representative Concentration Pathways (RCPs) which describe four different pathways for future atmospheric greenhouse gas concentrations, expressed in units of radiative forcing (IPCC 2014). These include a fossil fuel-intensive emission scenario (RCP8.5), two intermediate scenarios (RCP6.0 and RCP 4.5), and a strong mitigation scenario (RCP2.6). RCP4.5 roughly represents the future emission trajectory if the parties to the Paris Climate Agreement achieve their national emission reduction targets; RCP2.6 represents a likely ($>$66%) change of avoiding 2 °C of global average surface warming above the pre-industrial level.

The RCPs provide the GCMs with different trajectories of radiative forcing (caused by changes in atmospheric greenhouse gas and aerosol concentrations) rather than the emissions themselves (e.g. RCP8.5 depicts 8.5 W m^{-2} of radiative forcing in 2100 relative to 1750). The concentration of greenhouse gases in the atmosphere and the implied radiative forcing depend on the emissions but also the response of the planet's ecosystems (e.g. CO_2 uptake) and feedback mechanisms (e.g. melting permafrost leading to methane release). Since the current rate of CO_2 emissions exceeds the rate of CO_2 uptake by the oceans and terrestrial vegetation, freezing emissions at today's rate will cause a continued increase in atmospheric CO_2 concentrations. Due to the long residence time of CO_2 and other greenhouse gases in the atmosphere and the capital lock-in of existing fossil fuel infrastructure, a certain level of climate warming is expected to occur because of past emissions and energy decisions. For example, GCMs indicate a likely "committed" warming of 0.3–0.7 °C for the period 2016–2035 relative to 1985–2005 (IPCC 2014).

13.3 Predicting Coral Bleaching from Climate Models

The heterogeneity of coral response to temperature stress poses a central challenge in predicting the impact of climate warming on coral reefs. Hughes et al. (2003) summarised the possible theoretical models describing temperature thresholds for coral reefs under climate change. The simplest model (Hoegh-Guldberg 1999) assumes that mass coral bleaching or mortality will occur when the temperature exceeds some threshold (Fig. 13.1a). An alternative model recognises that different corals (by taxa, growth form, symbiont community composition, etc.) can have different thermal tolerances (Fig. 13.1b). An adaptive model envisions thresholds that might increase over time due to acclimatisation, adaptation, or community changes (Fig. 13.1c).

Most studies to date have used a single temperature-based threshold of 1–2 °C above the recent average summer maximum to predict whether mass coral bleaching or coral mortality will occur under future climate scenarios (Hoegh-Guldberg 1999; Sheppard 2003; Donner et al. 2005; van Hooidonk et al. 2013). More recent studies also investigated the effect of possible temperature adaptation on the projected frequency of bleaching events, a variant of the approach depicted in Fig. 13.1c (Donner 2009; Logan et al. 2014).

The first group of climate change and coral bleaching studies determined temperature thresholds for individual sites from historical observations of coral bleaching or mortality (Hoegh-Guldberg 1999; Sheppard 2003; Jones 2004). Hoegh-Guldberg (1999) did the first major review of the effect of climate change on the occurrence of mass coral bleaching. In that study, the projected future SSTs from three different GCMs under a business-as-usual scenario (IS92a) were used to predict the occurrence of coral bleaching at sites in French Polynesia, Jamaica, Rarotonga, and Thailand and at three sites on the Great Barrier Reef, Australia (GBR). For each site, in situ records of monthly SST and observations of mass coral bleaching were used to determine a monthly averaged temperature threshold. That study assumed that if the projected future SST[1] during a given year exceeded the temperature threshold for the site, mass coral bleaching would occur that year. The results suggested that coral bleaching at the level observed during the 1997–1998

[1]To eliminate systemic differences between modelled and observed temperatures, the future temperatures are estimated as the sum of model "anomalies" (i.e. modelled January, 2050 SST minus modelled mean January SST for today's climate) and observations (i.e. January in today's climate).

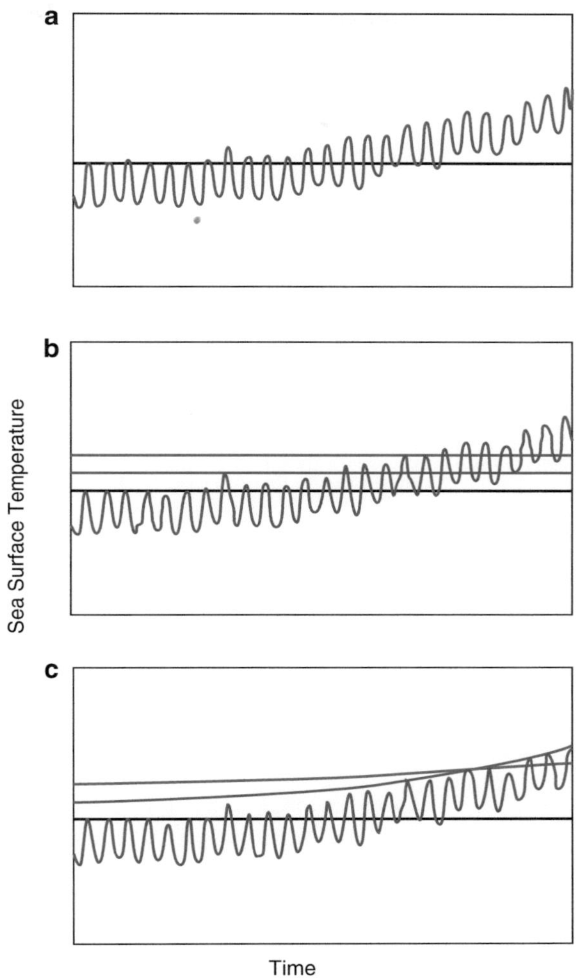

Fig. 13.1 Three models for describing temperature thresholds (blue and green lines) for predicting coral bleaching: (**a**) a single constant threshold across all species; (**b**) multiple constant thresholds, reflecting either differences in bleaching susceptibility (e.g. between species or growth forms) or severity (bleaching vs. mortality); and (**c**) multiple thresholds that increase in time, due to acclimation and/or evolution by corals and their symbionts (adapted from Hughes et al. 2003). The red line represents sea surface temperature as it increases over time

mass bleaching event at the seven sites would occur biannually within 20–40 years under business-as-usual conditions.

Sheppard (2003) used observations from the 1998 bleaching event to define an "extinction date" for 33 Indian Ocean reefs, the year in which the probability of SST for the warmest month or 3 months surpassing that of 1998 exceeded 20%. This was based on evidence that Indian Ocean reefs affected in 1998 had required a minimum of 5 years to recover. Sheppard (2003) controlled for GCM bias in mean SST and seasonal SST amplitude in each grid cell by adding GCM anomalies (i.e. future projection for January 2050 minus the present-day January climatological mean, from the HadISST $1^\circ \times 1^\circ$ resolution global dataset) to observed climatology and by fitting the seasonal amplitude to that in the observed data.

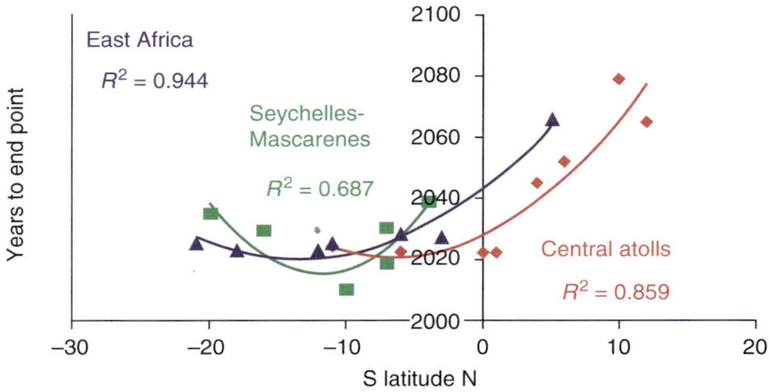

Fig. 13.2 "Extinction dates" for coral reefs across the Indian Ocean. The data points represent the year that the probability of exceeding the SST threshold (the warmest 3 months of 1998) reaches 20%, or once every 5 years, for the individual site. The curves are significant fits for the three main coral reef regions in the eastern Indian Ocean (from Sheppard 2003)

The study found the extinction date should occur between 2010 and 2030 for most southern Indian Ocean coral reefs, but not until the latter half of the century for some coral reefs north of the equator (Fig. 13.2). The date for high-latitude reefs in the Arabian Sea may be delayed until the end of the century due to cold-water upwelling. Notably, the projected increase in SSTs suggested that adaptation or acclimatisation by corals and their symbionts by 2 °C could delay the extinction date beyond the end of this century.

A second generation of studies modelling future coral bleaching used average maximum temperatures in observed climate data to determine bleaching thresholds (Hoegh-Guldberg 2001, 2005; Donner et al. 2005; Donner 2009). The NOAA Coral Reef Watch satellite-based coral bleaching prediction programme predicts the onset of bleaching using the degree heating week (DHW), a measure of the accumulated temperatures in excess of the usual summer maximum (Liu et al. 2006; Skirving et al. 2006; Chap. 4). Donner et al. (2005) took advantage of these satellite-derived data products to estimate the frequency of coral bleaching and required rates of temperature adaptation for thousands of coral reef locations worldwide for upper and lower business-as-usual climate scenarios (SRES A2, SRES B2). The historical satellite SST and DHW data provided a reliable way to develop algorithms for predicting the occurrence of thermal stress that can cause bleaching at sites worldwide from GCMs.

Algorithms were developed by transforming the observed 36 × 36 km resolution twice-weekly satellite SST data for 1985–2002 into monthly SSTs at the same satellite resolution and into monthly SSTs at the resolution of two different GCMs (HadCM3 from the UK and parallel climate model (PCM) from the USA). From these, a degree heating month (DHM) index was calculated. Historical data analysis showed that DHM > 1 °C·month and DHM > 2 °C·month were the best proxies for the lower and upper bleaching thresholds, DHW > 4 °C·week and

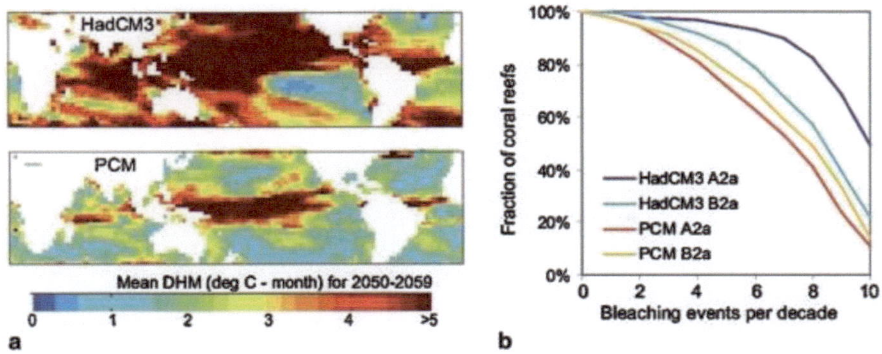

Fig. 13.3 Projected thermal stress measured as degree heating months (DHM) for 2050–2059 (from Donner et al. 2005): (**a**) annual mean DHM for 2050–2059 according to HadCM3 and PCM in the SRES A2 scenario, (**b**) number of times per decade that thermal stress exceeds the upper bleaching threshold (DHM > 2 °C·month), expressed as a fraction of world's coral reefs. Significant bleaching is expected at DHM > 1 °C·month; mass bleaching and significant mortality are expected at DHM > 2 °C·month

DHW > 8 °C·week respectively, used by the NOAA Coral Reef Watch programme (Liu et al. 2006). A statistical downscaling relationship between the maximum annual SST at the satellite resolution and at the GCM resolution was also developed for each 36 × 36 km grid cell containing a coral reef.

The results of Donner et al. (2005) showed the variation in thermal stress and the required temperature adaptation for extended survival of corals under different climate models and emission scenarios. The annual mean DHMs exceed the upper bleaching thresholds across much of the tropics by the 2050s in both models and under either scenario (Fig. 13.3a). Warming is projected to be greatest in the central equatorial Pacific, as both GCMs predict that ENSO-like conditions prevail in a warmer climate. The lower (DHM > 1 °C·month) and upper (DHM > 2 °C·month) bleaching thresholds are surpassed at the majority of the world's reefs every 2 years by the 2050s (Fig. 13.3b).

Following a method similar to Sheppard (2003), Donner et al. (2005) estimated the rate of temperature adaptation or acclimatisation required to avoid surpassing the coral bleaching thresholds in future decades. The results indicated the majority of the world's coral reefs would require adaptation of at least 0.2–0.3 °C per decade to ensure that low-intensity bleaching events (DHM > 1 °C·month) do not occur more than once or twice a decade by the 2030s to 2050s (Table 13.1). The required rates of adaptation vary widely across the tropics, with values of up to 0.5–1.0 °C per decade in parts of the central Pacific and Polynesia, even in the GCM with low climate sensitivity.

A different modelling approach is to examine SST or thermal stress indices averaged over a large region as representative of the extent of bleaching in the region. In an early example, McWilliams et al. (2005) contrasted historical SSTs in the Caribbean from the MOHSST6 historical dataset with historical data on the

Table 13.1 Percentage of reefs requiring thermal adaptation by 2030–2039. Shown is the percent of coral reef grid cells, in the SRES A2 and B2 emission scenarios that require a 0.5 °C or 1.0 °C increase in the temperature threshold at which degree heating months begin to accumulate in order to avoid mass bleaching more than once every 5 years

	Coral reefs (%)			
	HadCM3		PCM	
Ocean region	+0.5 °C	+1.0 °C	+0.5 °C	+1.0 °C
Indian Ocean	83–92	46–55	38–57	10–19
SE Asia	58–62	16–17	39–40	6–9
Micronesia	58–79	13–54	58–79	6–7
GBR/Coral Sea	29–67	7–40	17–53	4–6
Polynesia	69–81	24–39	58–82	19–31
Caribbean	75–78	22–30	13–40	0–11

spatial extent of bleaching determined from the reports to ReefBase (http://www.reefbase.org). Regressions estimated that a 0.1 °C increase in Caribbean-average SST would cause a 35% increase in the number of 1° × 1° coral reef cells reporting some bleaching and a 42% increase in the fraction of coral colonies bleached. Although this type of regional bleaching prediction is less meaningful for individual coral reefs, it is well-suited to the application of coarse GCM predictions for the future.

In a study of the role of climate change in the 2005 Caribbean bleaching event, Donner et al. (2007) also examined the mean thermal stress over the broad affected region rather than the SSTs at individual sites or grid cells. By using a large region, the study was able to use historical datasets and GCMs (CM2.0, CM2.1, from the US Geophysical Fluid Dynamics Laboratory) to examine the probability of the 2005 bleaching event occurring with and without the effect of past greenhouse gas emissions on the climate. The analysis showed that anthropogenic forcing increased the chance of a coral bleaching event, such as that observed in the Caribbean in 2005, by at least an order of magnitude.

The Donner et al. (2007) study and a similar analysis at the global scale (Donner 2009) also provided further insight into the effect of different emission scenarios and possible adaptation or acclimatisation on the frequency of mass bleaching events in the future. The GCMs predict that the DHM would exceed 2 °C at least biannually by the 2020s or 2030s in all regions under both a business-as-usual scenario (SRES A1b) and a lower emission scenario (SRES B1) in which atmospheric CO_2 concentrations stabilise at double the pre-industrial levels in the year 2100. However, the results change if corals and their symbionts are able to adapt or acclimatise by 1.0–1.5 °C. In the business-as-usual scenario, such adjustment would postpone mass coral bleaching from occurring once every 5 years until the latter half of the century. The study implied that, in the absence of any adaptation or acclimatisation, dangerously frequent mass bleaching events are likely to occur within decades due to committed warming in the climate system (Donner 2009).

A number of further modelling studies have been undertaken since the first edition of this book. The global-scale projections in recent studies that employed

GCMs used in IPCC-AR5 (Frieler et al. 2013; van Hooidonk et al. 2013, 2014, 2016) resemble those of earlier studies, due in large part to the similarity in global mean SST projections between the GCMs used in the IPCC AR4 and AR5 assessments. Among the key advances in the recent studies were the use of more GCM simulations, the treatment of model data, and the inclusion of the possible synergistic or opposing effects of ocean acidification. Frieler et al. (2013) employed the relationship between regional and global mean surface temperature change in GCM output to estimate the effect of different proposed global temperature thresholds on future bleaching frequencies. The study concluded that, in the absence of adaptation and given potential synergistic effects of ocean acidification, global mean surface temperature warming would need to be limited to less than 1.5 °C above pre-industrial levels to avoid degradation of >90% of coral reefs. In another analysis using a suite of IPCC AR5 models, Van Hooidonk et al. (2014) noted that severe annual mass bleaching is projected to occur a decade or more later at high-latitude reefs, yet this benefit may be countered by changes in coral calcification due to the more rapid projected decline in aragonite saturation state in cooler, high-latitude waters.

Another key development in recent studies is the use of bleaching prediction methods that consider other variables such as the historical climate experience and mechanisms by which corals may adjust to climate warming. For example, Donner (2011), Teneva et al. (2011), and Logan et al. (2012) used bleaching observations from ReefBase to test prediction methods in which the threshold or alert level (i.e. DHW > 4 °C·week) was determined by historical SST variability. Although such methods may not significantly alter global-scale bleaching projections (Donner 2009), there is evidence that recent thermal history is an important variable for predicting future bleaching in regions such as the western Pacific with low seasonal variability in SST (Teneva et al. 2011; Kleypas et al. 2015, 2016).

Logan et al. (2014) tested the effect of algorithms representing different possible adaptive responses by corals and their symbionts—genetic adaptation to recent thermal history, symbiont shuffling, and transient community shifts—on bleaching projections. Adaptation to recent thermal history via genetic directional selection had a larger effect on future bleaching frequency than temporary changes in bleaching thresholds due to symbiont shuffling. The thermal history model results also implied that corals and their symbionts have likely already adapted or acclimated to some climate warming since the pre-industrial period. Further development of such mechanistic models may help identify how different coral reef communities will change as ocean temperatures rise.

13.4 The Great Barrier Reef: A Case Study

The Great Barrier Reef (GBR) is the world's largest coral reef system, stretching over 2000 km along the northeast coast of Australia at the edge of the Indo-Pacific biodiversity hotspot. It comprises over 3000 individual reefs found inshore, mid-shelf, and up to 200 km offshore. The reefs are subject to a variety of disturbances, including agricultural runoff, riverine flood plumes, crown-of-thorns starfish outbreaks, severe tropical storms, fishing, tourism, and marine shipping. The GBR is also one of the world's most protected reef regions. The Representative Areas Program and the Reef Water Quality Protection Plan were implemented as part of a strategy to maintain biodiversity and support the ecosystem's resilience to aid survival of the reef through climate change (www.gbrmpa.gov.au). In 2015, national and state governments released the Reef 2050 Plan to provide a framework for protecting and managing the GBR through until 2050 (www.environment.gov.au/marine/gbr/long-term-sustainability-plan) although the plan is focussed on water quality issues and is relatively silent about climate change impacts.

Prior to 1979, no mass bleaching events were reported for the GBR. Since that time, several mass bleaching events have been recorded: 1980, 1982, 1987, 1992, 1994, 1998, 2002, 2006, and most recently the back-to-back events of 2016 and 2017 (Chap. 3; Berkelmans and Oliver 1999; Hoegh-Guldberg and Hoegh-Guldberg 2004; Lough et al. 2006; Hughes et al. 2017; Hughes and Kerry 2017). Average SST decreases to the south (poleward) along the GBR, and, due to localised adaptation, the threshold temperature at which corals bleach also decreases to the south (Berkelmans 2002). The bleaching events have generally increased in their intensity and extent over time. The 1998 event (40% of reefs bleached) and subsequently the 2002 event (50% of reefs bleached) were each described as the most severe bleaching events then recorded on the GBR (Berkelmans and Oliver 1999; Berkelmans et al. 2004). These impacts were superseded in 2016, during which more than 90% of the surveyed shallow-water corals were bleached and an estimated 30% died (Hughes and Kerry 2017). The greatest heat stress and impacts were observed in the northernmost one-third of the GBR where effects were seen across all hard coral species, including fast-growing branching species and slow-growing massive species. Intense heat stress returned in 2017, with the highest levels in the central sector resulting in an estimated additional coral mortality of 19% (Hughes and Kerry 2017).

Several studies have specifically examined projected increase in SST on the GBR. Lough et al. (2006), using historical data and model projections, suggested that the waters of the GBR will be 1–3 °C warmer than at present by the end of this century (Fig. 13.4). The Hoegh-Guldberg (1999) study concluded that temperatures would regularly exceed those observed in 1998 at sites in the northern, central, and southern GBR by the year 2020 (Fig. 13.5)—consistent with the levels of heat stress and bleaching in 2016 and 2017 (Hughes et al. 2017; Hughes and Kerry 2017). Hoegh-Guldberg (1999) and more recent studies concluded such temperature increases may be sufficient to induce annual bleaching events across the GBR by 2050. For example, the recent van Hooidonk et al. (2016) model study projected that annual Bleaching Alert Level II conditions (DHW $>$ 8 °C·week) would occur, on average, across GBR grid cells by the year 2051 in RCP8.5 and the year 2064 in RCP4.5. Even the most optimistic climate scenario predicts that catastrophic thermal events are possible at mid- and outer-shelf reefs by 2050 (Done et al. 2003).

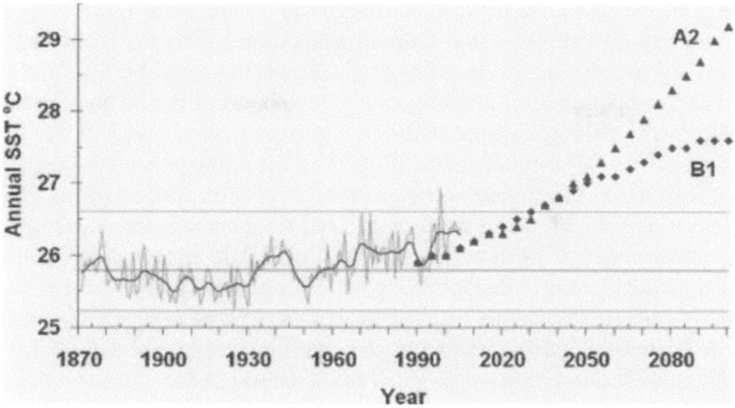

Fig. 13.4 Observed and projected annual mean SST for the Great Barrier Reef. The thin line is the annual mean instrumental SST record, 1871–2005 (HadISST and NOAA OI.v2 SST); the thick black line is the 10-year Gaussian filter. The horizontal lines denote the 1871–1989 mean SST (25.8 °C) and the observed range (25.2–26.6 °C). Projected SSTs are given for GBR 1990–2100 (ReefClim, Roger Jones, CSIRO) for the B1 (diamonds) and A2 (triangles) middle of the road climate scenarios. Both scenarios suggest that, by 2035, average GBR SSTs will be outside the range observed in the instrumental record prior to 1990 (adapted from Lough et al. 2006)

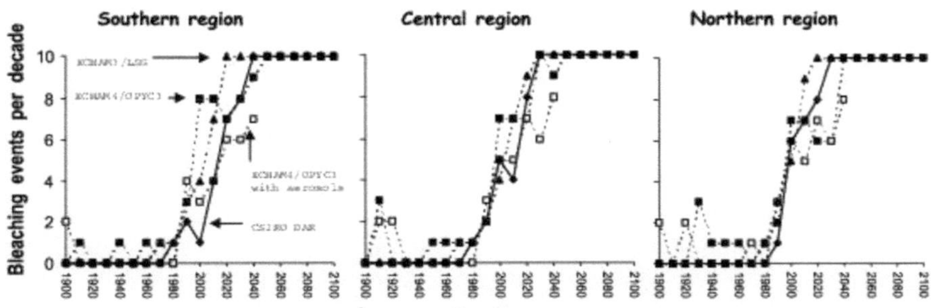

Fig. 13.5 The number of times (per decade) that predicted SSTs will exceed coral bleaching thresholds for: (**a**) southern (23.5°S, 149.5°E), (**b**) central (18.0°S, 147.5°E), and (**c**) northern (11.0°S, 143.0°E) sites on the Great Barrier Reef. The models are ECHAM4/OPYC3 (black squares), ECHAM4/OPYC3 with aerosol effect added (white squares), ECHAM3/LSG (black triangles), and CSIRO DAR GCM (black dots; from Hoegh-Guldberg 1999)

Several studies have examined the impact of temperature-induced bleaching on coral community structure. For example, using a Bayesian network model linking key system variables—coral habitat, community type, local SST, climatological SST, and potential for cooling by upwelling of deep water—Wooldridge and Done (2004) correctly predicted the coral mortality category (low, medium, high) for 71% of field observations following the 2002 bleaching event on the GBR.

A subsequent modelling study that considered two hard coral types, varying constraints on algal growth rates, and thermal adaptation by corals, found that natural (i.e. herbivory) or managed constraint of algal growth would be essential to allow recovery of coral populations after bleaching episodes, but would not halt long-term coral reef decline (Wooldridge et al. 2005).

Donner et al. (2005) estimated that 17–67% of the coral reefs across the GBR will require at least a 0.5 °C increase in their thermal tolerance, while 4–40% will require an increase of 1.0 °C, by the year 2030 to avoid frequent harmful bleaching events. Multiple occurrences of bleaching at sites on the GBR are cited as evidence that corals at these sites are not developing greater thermal tolerance (Hoegh-Guldberg 1999). However, field and laboratory evidence suggest that the common GBR species *Acropora millepora* can increase its thermal tolerance level by 1.0–1.5 °C by shuffling the dominant symbiont in its tissue (Berkelmans and van Oppen 2006; Chap. 9). Such an increase in thermal tolerance might help GBR corals avoid predicted bleaching events in the next several decades, but it is insufficient to meet the larger temperature increase predicted for the latter half of the century (Hoegh-Guldberg 1999; Done et al. 2003; Berkelmans and van Oppen 2006; Lough et al. 2006). Furthermore, recent analysis has indicated that within-summer temperature trajectories have conferred short-term tolerance to GBR corals and that with projected warming their protective mechanism will be lost (Ainsworth et al. 2016). An increased understanding of adaptation and acclimatisation would aid in determining management strategies for the GBR (Hoegh-Guldberg and Hoegh-Guldberg 2004).

13.5 Future Improvements in Physical Modelling

The general projected increase in thermal stress on coral reefs under future emission scenarios is so rapid and global in scale that it is unlikely to change with future improvement in GCMs (Donner et al. 2005). This has been evidenced by the recent fulfilment of the projections by Hoegh-Guldberg (1999) and the global extent of the 2014–2017 bleaching event (NOAA Coral Reef Watch 2017). Future models operating at high horizontal resolution (<1 km) may be instrumental in making specific predictions for individual coral reef complexes, which could inform management strategies to support the capacity of reefs to resist and recover from disturbance events. While regional climate models are available for a few specific areas (e.g. van Hooidonk et al. 2015), statistical downscaling techniques have been applied to the IPCC AR5 GCM output for reefs worldwide to support management efforts (van Hooidonk et al. 2016). Advances in the application of fluid dynamics theory and computing power are already increasing the resolution of GCMs and hydrodynamic models. For example, the Hybrid Coordinate Ocean Model (HYCOM) can presently operate globally at a horizontal resolution as fine as 2/25° (Metzger et al. 2006) and regionally at horizontal resolution of 1/25° (Kourafalou and Balotro 2006; Prasad and Hogan 2007).

Modelling efforts are being conducted at very high spatial resolutions (<1 km) in and around coral reefs. Skirving et al. (2004) describe a hydrodynamic model for Palau with ~250 m resolution. Mapped ocean currents have been used to determine the vertical mixing of water across the Palau lagoon during periods of low wind speeds that are characteristic of bleaching events (Skirving and Guinotte 2001). Skirving et al. (2006) describe the cooling of SST due to vertical mixing and link this to the capacity of a water column to absorb solar radiation. Well-mixed regions distribute heat throughout the water column, incurring a small temperature increase throughout the column. In stratified regions, heat is contained near the surface, causing a significant rise in SST. This suggests different temperature climates in which corals exist and, therefore, a different level of acclimatisation to thermal events. The design of marine protected areas can include the modelled thermal capacitance to provide protection for corals during climate-induced bleaching events (Skirving et al. 2006).

Bode et al. (1997) describe a parameterisation scheme for sub-resolution features, such as those seen in and around coral reefs. This scheme was applied to a tidal model of the southern GBR with a resolution of ~8 km that successfully simulated the tidal amplitude and phase throughout the region. Development of such modelling techniques at various horizontal resolutions will improve the accuracy of forecasts of the effects of climate change and assist design of marine protected areas for coral reef ecosystems.

13.6 Conclusions

Various model studies present an overall picture of the effect of climate change on the frequency and severity of mass coral bleaching and bleaching-induced mortality. Several studies confirm the original conclusion of Hoegh-Guldberg (1999) that mass coral bleaching could become a biannual event by the 2020s or 2030s at many coral reefs without any thermal adaptations by corals and their symbionts, although important local and regional exceptions have been noted (Sheppard 2003; Donner et al. 2005; van Hooidonk et al. 2013). Human-induced warming has already increased the likelihood of mass coral bleaching events in some regions (Donner et al. 2007).

This result is generally robust across different emission scenarios. Due to time lags in the climate system (between emissions and climate impact) and in the economic system (between a decision to reduce emissions and actual emission reduction), the simulated climate in different RCPs does not diverge until the latter half of the century. There is a greater range in future projections between different GCMs, with different climate sensitivities, than between different emission scenarios. Nevertheless, even the lowest estimate of business-as-usual projected ocean warming from available studies indicates that mass coral bleaching could occur biannually on the majority of coral reefs worldwide by 2050.

Corals and their symbionts will have to adapt to rising temperatures to avoid bleaching events that are too frequent to allow the reef communities to recover. The rate and magnitude at which corals and their symbionts will have to adapt to increasing temperatures appears to vary across the globe (Donner et al. 2005; van Hooidonk et al. 2014). Climate projections for this century indicate that temperature adaptation might postpone the occurrence of frequent harmful bleaching events and allow corals to survive, provided that policies and technologies alter the path of greenhouse gas emissions and magnitude of future warming. However, in that case, long-term "committed warming" beyond 2100 could still represent a serious threat to coral reefs (Donner 2009).

An important area for additional research is the effect of climate-induced coral bleaching and other disturbances on coral reef community structure. More frequent coral bleaching events, especially when combined with local disturbances such as fishing, pollution, or sedimentation, are expected to keep coral and fish species richness low (Wilson et al. 2006; Chap. 11). Ecological models will be instrumental in describing the effect of higher bleaching frequencies and other disturbances on coral and macroalgal cover (e.g. Baskett et al. 2009, 2010; Ortiz et al. 2014). As more field data have become available, these models have begun to include multiple coral taxa with different temperature tolerances, growth rates, and reproduction rates. Combined physical and ecological models will be critical in estimating the ability of reefs to adapt or acclimatise to warmer ocean temperatures, as well as other local (e.g. fishing pressure) and global (e.g. rising pCO_2) stressors.

In recent years, projected impacts of climate change on coral reefs have been incorporated into local-to-regional management strategies and have begun to influence the policy sphere (e.g. Heron et al. 2017). In an era when the projected impacts articulated two decades ago are already being seen, response actions that include the best available information on future impacts will be essential to support the ongoing maintenance of coral reefs and the myriad goods and services they provide in the face of predicted increases to both the intensity and frequency of disturbance events.

Acknowledgements The manuscript contents are solely the opinions of the author(s) and do not constitute a statement of policy, decision, or position on behalf of NOAA or the US Government.

References

Ainsworth TM, Heron SF, Ortiz JC, Grech A, Ogawa D, Eakin CM et al (2016) Climate change disables coral bleaching protection on the Great Barrier Reef. Science 352:338–342. https://doi.org/10.1126/science.aac7125

Baskett ML, Gaines SD, Nisbet RM (2009) Symbiont diversity may help coral reefs survive moderate climate change. Ecol Appl 19:3–17. https://doi.org/10.1890/08-0139.1

Baskett ML, Nisbet RM, Kappel CV, Mumby PJ, Gaines SD (2010) Conservation management approaches to protecting the capacity for corals to respond to climate change: a theoretical comparison. Glob Chang Biol 16:1229–1246. https://doi.org/10.1111/j.1365-2486.2009.02062.x

Berkelmans R (2002) Time-integrated thermal bleaching thresholds of reefs and their variation on the Great Barrier Reef. Mar Ecol Prog Ser 229:73–82. https://doi.org/10.3354/meps229073
Berkelmans R, Oliver JK (1999) Large-scale bleaching of corals on the Great Barrier Reef. Coral Reefs 18:55–60. https://doi.org/10.1007/s003380050154
Berkelmans R, van Oppen MJH (2006) The role of Zooxanthellae in the thermal tolerance of corals: a "nugget of hope" for coral reefs in an era of climate change. Proc R Soc B 273:2305–2312. https://doi.org/10.1098/rspb.2006.3567
Berkelmans R, De'ath G, Kininmonth S, Skirving W (2004) A comparison of the 1998 and 2002 coral bleaching events on the Great Barrier Reef: spatial correlation, patterns, and predictions. Coral Reefs 23:74–83. https://doi.org/10.1007/s00338-003-0353-y
Bode L, Mason LB, Middleton JH (1997) Reef parameterisation schemes with applications to tidal modeling. Prog Ocean 40:285–324. https://doi.org/10.1016/S0079-6611(98)00006-8
Done TP, Whetton P, Jones R, Berkelmans R, Lough J, Skirving W et al (2003) Global climate change and coral bleaching on the Great Barrier Reef. Final report to the State of Queensland Greenhouse Task Force, Queensland Department of Natural Resources and Mining, Queensland
Donner SD (2009) Coping with commitment: projected thermal stress on coral reefs under different future scenarios. PLoS One 4:e5712. https://doi.org/10.1371/journal.pone.0005712
Donner SD (2011) An evaluation of the effect of recent temperature variability on the prediction of coral bleaching events. Ecol Appl 21:1718–1730. https://doi.org/10.1890/10-0107.1
Donner SD, Skirving WJ, Little CM, Oppenheimer M, Hoegh-Guldberg O (2005) Global assessment of coral bleaching and required rates of adaptation under climate change. Glob Chang Biol 11:2251–2265. https://doi.org/10.1111/j.1365-2486.2005.01073.x
Donner SD, Knutson TR, Oppenheimer M (2007) Model-based assessment of the role of human-induced climate change in the 2005 Caribbean coral bleaching event. Proc Natl Acad Sci U S A 104:5483–5488. https://doi.org/10.1073/pnas.0610122104
Donner SD, Rickbeil GJM, Heron SF (2017) A new, high-resolution global mass coral bleaching database. PLoS One 12:e0175490. https://doi.org/10.1371/journal.pone.0175490
Frieler K, Meinshausen M, Golly A, Mengel M, Lebek K, Donner SD et al (2013) Limiting global warming to 2°C is unlikely to save most coral reefs. Nat Clim Chang 3:165–170. https://doi.org/10.1038/nclimate1674
Glynn PW (1991) Coral-reef bleaching in the 1980s and possible connections with global warming. Trends Ecol Evol 6:175–179. https://doi.org/10.1016/0169-5347(91)90208-F
Guinotte JM, Buddemeier RW, Kleypas JA (2003) Future coral reef habitat marginality: temporal and spatial effects of climate change in the Pacific basin. Coral Reefs 22:551–558. https://doi.org/10.1007/s00338-003-03331-4
Heron SF, Eakin CM, Douvere F, Anderson K, Day JC, Geiger E et al (2017) Impacts of climate change on world heritage coral reefs: a first global scientific assessment. UNESCO World Heritage Centre, Paris
Hoegh-Guldberg O (1999) Climate change, coral bleaching and the future of the world's coral reefs. Mar Freshw Res 50:839–866. https://doi.org/10.1071/MF99078
Hoegh-Guldberg O (2001) The future of coral reefs: integrating climate model projections and the recent behaviour of corals and their dinoflagellates. In: Proceedings of 9th international coral reef symposium, Bali 2, pp 1105–1110
Hoegh-Guldberg O (2005) Low coral cover in a high-CO_2 world. J Geophys Res 110:C09S06. https://doi.org/10.1029/2004JC002528
Hoegh-Guldberg H, Hoegh-Guldberg O (2004) The implications of climate change for Australia's Great Barrier Reef: people and industries at risk. World Wildlife Fund/Queensland Tourism Industry Council, Sydney
Hughes TP, Kerry JT (2017) Back-to-back bleaching has now hit two-thirds of the Great Barrier Reef, The Conversation, published online April 12 2017. http://theconversation.com/back-to-back-bleaching-has-now-hit-two-thirds-of-the-great-barrier-reef-76092

Hughes TP, Baird AH, Bellwood DR, Card M, Connolly SR, Folke C et al (2003) Climate change, human impacts, and the resilience of coral reefs. Science 301:929–933. https://doi.org/10.1126/science1085046

Hughes TP, Kerry JT, Álvarez-Noriega M, Álvarez-Romero JG, Anderson KD, Baird AH et al (2017) Global warming and recurrent mass bleaching of corals. Nature 543:373–377. https://doi.org/10.1038/nature21707

IPCC (2014) Climate Change 2014: Synthesis Report. In: Core Writing Team, Pachauri RK, Meyer LA (eds) Contribution of Working Groups I, II and III to the fifth assessment report of the Intergovernmental Panel on Climate Change. IPCC, Geneva, Switzerland, 151 p

Jones R (2004) Managing climate risk. In: Corfee-Morlot J, Agrawala S (eds) The benefits of climate change policies. OECD, Paris, pp 249–297

Kleypas JA, Buddemeier RW, Archer D, Gattuso J-P, Langdon C, Opdyke BN (1999) Geochemical consequences of increased atmospheric CO_2 on coral reefs. Science 284:118–120. https://doi.org/10.1126/science.284.5411.118

Kleypas JA, Castruccio FS, Curchitser EN, McLeod E (2015) The impact of ENSO on coral heat stress in the western equatorial Pacific. Glob Chang Biol 21:2525–2539. https://doi.org/10.1111/gcb.12881

Kleypas JA, Thompson DM, Castruccio FS, Curchitser EN, Pinksy M, Watson WR (2016) Larval connectivity across temperature gradients and its potential effect on heat tolerance in coral populations. Glob Chang Biol 22:3539–3549. https://doi.org/10.1111/gcb.13347

Kourafalou VH, Balotro RS (2006) Connecting the US Florida Keys coral reef ecosystem to the hydrodynamics. In: Proceedings of 10th international coral reef symposium, Japan, vol 10, pp 890–895

Kwiatkowski L, Halloran PR, Mumby PJ, Stephenson DB (2014) What spatial scales are believable for climate model projections of sea surface temperature? Clim Dyn 43:1483–1496. https://doi.org/10.1007/s00382-013-1967-6

Liu G, Strong AE, Skirving W, Arzayus LF (2006) Overview of NOAA coral reef watch program's near-real-time satellite global coral bleaching monitoring activities. In: Proceedings of 10th international coral reef symposium, Japan, vol 10, pp 1783–1793

Logan CA, Dunne JP, Eakin CM, Donner SD (2012) A framework for comparing coral bleaching thresholds. In: Proceedings of 12th international coral reef symposium, Cairns, 10B

Logan CA, Dunne JP, Eakin CM, Donner SD (2014) Incorporating adaptive responses into future projections of coral bleaching. Glob Chang Biol 20:125–139. https://doi.org/10.1111/gcb.12390

Lough J, Berkelmans R, van Oppen M, Wooldridge S, Steinberg C (2006) The Great Barrier Reef and climate change. Bull Aust Meteorol Ocean Soc 19:53–58

McWilliams JP, Côté IM, Gill JA, Sutherland WJ, Watkinson AR (2005) Accelerating impacts of temperature-induced coral bleaching in the Caribbean. Ecology 86:2055–2060. https://doi.org/10.1890/04-1657

Metzger E, Wallcraft AJ, Hurlburt HE, Chassignet EP, Schmitz WJ (2006) 1/12° global HYCOM: initial development and evaluation. EOS 87[Suppl]:OS441-02

NOAA Coral Reef Watch (2017). Global coral bleaching 2014–2017: status and an appeal for observations. Available at: https://coralreefwatch.noaa.gov/satellite/index.php. Accessed 13 Aug 2017

Ortiz JC, Bozec Y-M, Wolff NH, Doropoulos C, Mumby PJ (2014) Global disparity in the ecological benefits of reducing carbon emissions for coral reefs. Nat Clim Chang 4:1090–1094. https://doi.org/10.1038/nclimate2439

Prasad TG, Hogan PJ (2007) Upper-ocean response to Hurricane Ivan in a 1/25° nested Gulf of Mexico HYCOM. J Geophys Res 112:C04013. https://doi.org/10.1029/2006JC003695

Sheppard CRC (2003) Predicted recurrences of mass coral mortality in the Indian Ocean. Nature 425:294–297. https://doi.org/10.1038/nature01987

Skirving W, Guinotte J (2001) The sea surface temperature story on the Great Barrier Reef during the coral bleaching event of 1998. In: Wolanski E (ed) Oceanographic processes of coral reefs: physical and biological links in the Great Barrier Reef. CRC, Boca Raton, pp 301–313

Skirving W, Heron S, Steinberg C, Strong AE, McLean C, Heron M et al (2004) Palau modeling final report. National Oceanic and Atmospheric Administration/Australian Institute of Marine Science, Townsville, 46 pp

Skirving W, Heron M, Heron S (2006) The hydrodynamics of a bleaching event: implications for management and monitoring. In: Phinney JT, Hoegh-Guldberg O, Kleypas J, Skirving W, Strong A (eds) Coral reefs and climate change: science and management. American Geophysical Union, Washington, DC, pp 145–161

Teneva L, Karnauskas M, Logan C, Bianucci L, Currie J, Kleypas J (2011) Predicting coral bleaching hotspots: the role of regional variability in thermal stress and potential adaptation rates. Coral Reefs 31:1–12. https://doi.org/10.1007/s00338-011-0812-9

van Hooidonk R, Maynard JA, Planes S (2013) Temporary refugia for coral reefs in a warming world. Nat Clim Chang 3:508–511. https://doi.org/10.1038/nclimate1829

van Hooidonk RJ, Maynard JA, Manzello D, Planes S (2014) Opposite latitudinal gradients in projected ocean acidification and bleaching impacts on coral reefs. Glob Chang Biol 20:103–112. https://doi.org/10.1111/gcb.12394

van Hooidonk R, Maynard JA, Liu Y, Lee S-K (2015) Downscaled projections of Caribbean coral bleaching that can inform conservation planning. Glob Chang Biol 21:3389–3401. https://doi.org/10.1111/gcb.12901

van Hooidonk R, Maynard J, Tamelander J, Gove J, Ahmadia G, Raymunod L et al (2016) Local-scale projections of coral reef futures and implications of the Paris Agreement. Sci Rep 6:39666. https://doi.org/10.1038/srep39666

Wilson SK, Nichols GAJ, Pratchett MS, Jones GP, Polunin NVC (2006) Multiple disturbances and the global degradation of coral reefs: are reef fishes at risk or resilient? Glob Chang Biol 12:2220–2234. https://doi.org/10.1111/j.1365-2486.2006.01252.x

Wooldridge S, Done T (2004) Learning to predict large-scale coral bleaching from past events: a Bayesian approach using remotely sensed data, in-situ data, and environmental proxies. Coral Reefs 23:96–108. https://doi.org/10.1007/s00338-003-0361-y

Wooldridge S, Done T, Berkelmans R, Jones R, Marshall P (2005) Precursors for resilience in coral communities in a warming climate: a belief network approach. Mar Ecol Prog Ser 295:157–169. https://doi.org/10.3354/meps295157

Chapter 14
Synthesis: Coral Bleaching: Patterns, Processes, Causes and Consequences

M. J. H. van Oppen and J. M. Lough

There is no doubt that ocean temperatures have increased as a result of human activities that began with the industrial revolution (Chap. 4). These rapidly rising temperatures are subjecting the world's biota to unprecedented levels of stress, with many species experiencing negative impacts. Coral reef organisms are no exception, and thermal stress is causing a breakdown of the close association between two groups of organisms of key importance to coral reefs, scleractinian corals and microalgae in the genus *Symbiodinium*. This breakdown of symbiosis results in the expulsion of *Symbiodinium* from, and a paling of the coral tissues, a phenomenon called coral bleaching. Reef-building corals are responsible for the three-dimensional framework that provides habitat to many other reef dwellers and are also the main primary producers of these beautiful biomes; both functions are only possible through the symbiosis between coral and *Symbiodinium*. Disruption of the symbiosis thus has consequences for ongoing maintenance of these structurally complex, biologically diverse, charismatic and both economically and socially important ecosystems. This volume brings together various aspects of coral bleaching, ranging from the genes involved in the bleaching response to global patterns of bleaching and the evolution of symbiosis over geological time frames. Topics covered are the evolution of photosymbiosis in corals (Chap. 2), a review of spatial and temporal patterns of past coral bleaching events (Chap. 3), an explanation of how coral bleaching is monitored (Chap. 5), an analysis of the environmental

M. J. H. van Oppen (✉)
Australian Institute of Marine Science, Townsville MC, QLD, Australia

School of BioSciences, The University of Melbourne, Parkville, Victoria, Australia
e-mail: madeleine.van@unimelb.edu.au

J. M. Lough
Australian Institute of Marine Science, Townsville MC, QLD, Australia

ARC Centre of Excellence for Coral Reef Studies, James Cook University, Townsville, QLD, Australia

M. J. H. van Oppen, J. M. Lough (eds.), *Coral Bleaching*, Ecological Studies 233,
https://doi.org/10.1007/978-3-319-75393-5_14

factors involved in bleaching (Chap. 4, temperature; Chap. 12, ocean acidification and eutrophication) and the cellular causes underpinning bleaching (Chap. 8), the roles of microbial symbionts in coral bleaching and acclimatisation to climate change and the consequences of thermal stress on microbial communities (Chaps. 6 and 7), the consequences of bleaching for hard and soft corals (Chap. 10) and fish assemblages (Chap. 11), how well bleaching can be predicted from accumulated thermal stress (Chap. 9) and projecting the incidence of coral bleaching over this century using climate models (Chap. 13).

Climate models based on unmitigated greenhouse gas emissions predict that most coral reefs will experience thermal anomalies that cause mass coral bleaching annually by the middle of this century. But even under the most stringent emission scenarios, oceans will continue to warm and corals will decline further unless they can acclimatise or adapt to the rapidly changing environmental conditions. While a small number of instances of increased bleaching tolerance over repeated bleaching events have been reported (Chap. 9; Maynard et al. 2008; Guest et al. 2012; Penin et al. 2013), the generally rapid decline of coral cover and diversity across the globe indicates natural rates of acclimatisation and adaptation are too slow for corals to keep up with the fast pace of global climate warming. The challenge for managers, scientists and policy makers, therefore, is to find innovative solutions that will ensure the persistence of coral reefs into the future (Hughes et al. 2017).

14.1 Climate Change Refuges and Assisted Evolution

As extensively discussed in this volume, mass bleaching events have already caused devastation on many coral reefs across the world. Patterns of thermal stress and coral bleaching are almost always spatially patchy, suggesting "climate change refuges" may exist (Chap. 10). Such refuges are likely to be located either in areas with stable cool water or in those that show historically large fluctuations in sea surface temperatures (and pCO_2) and have comparatively high temperature maxima (McClanahan et al. 2007; Carilli et al. 2012; Barshis et al. 2013; Kenkel and Matz 2016; Camp et al. 2017). It is important to understand which corals and coral reefs are most resistant to bleaching (Camp et al. 2016; Jin et al. 2016; Camp et al. 2017) and which reefs have a relatively lower probability of experiencing thermal anomalies (https://50reefs.org/). Degraded reefs may be reseeded from such "climate change refuges", and it is therefore also critical to understand the potential pathways of dispersal of corals and other coral reef organisms over ecological time scales (van Oppen and Gates 2006; Hellberg 2007).

Additionally, coral stock with enhanced climate resilience may be developed using a variety of bioengineering strategies (van Oppen et al. 2015, 2017; Anthony et al. 2017). The use of such stock for restoration of degraded reefs may increase the chances that corals and coral reefs will survive this century while actions are taken to halt ocean warming. Assisted evolution (AE), the enhancement of certain traits by accelerating naturally occurring evolutionary processes, is one such intervention (van Oppen et al. 2015, 2017). AE includes a set of approaches targeted at both the coral host animal and its microbial symbionts, i.e. selective breeding and assisted gene flow (AGF), conditioning or epigenetic programming and the (genetic) manipulation and laboratory evolution of the coral microbiome.

Selective breeding in the context of AE is defined sensu lato and includes both the crossing of conspecific colonies selected based on their genotypes or phenotypes, and the formation of interspecific hybrids which have increased genetic diversity and new gene combinations compared to their parents, which serve as new substratum for selection. Coral hybrids can be as fit as or fitter than their parents in certain environments (Willis et al. 2006; Fogarty 2012) and may have gene combinations that perform better under predicted future ocean conditions. AGF is the managed movement of individuals with favourable traits (alleles/genotypes) into populations (unidirectional) to reduce local maladaptation to climate or other environmental change (either current or future change) (Aitken and Whitlock 2013). To increase climate resilience, corals would be moved from comparatively warm environments—to which they have adapted—to cooler reefs where they are expected to propagate and interbreed with the native corals. In this manner, thermal tolerance could be bolstered in the regional hybrid offspring compared to the native purebred corals. As a variation on AGF sensu stricto, warm-adapted corals may be crossed with cold-adapted conspecifics ex situ, followed by deployment of the regional hybrid offspring at the cooler location, thus combining the principles of AGF and selective breeding. The motivation for this alternative approach is that transplanted adult colonies may be maladapted to the transplant environment (Kenkel et al. 2015), resulting in high rates of mortality post-transplantation (Howells et al. 2013). Regional hybrids may outperform the native genotypes under conditions of environmental change and stress (van Oppen et al. 2014; Dixon et al. 2015). Further, deployment at the larval or early recruit stage provides an opportunity for developmental acclimatisation to increase survival rates at the target reef.

Conditioning of corals and reef fish using sublethal stress levels may induce epigenetic changes that increase environmental stress tolerance through developmental and/or transgenerational acclimatisation (Donelson et al. 2012; Putnam and Gates 2015). This field of research is very much in its infancy, particularly for corals, and considerable research investment is required to assess whether conditioning can increase climate resilience across multiple generations (Torda et al. 2017).

The coral microbiome is highly diverse and comprises members from multiple eukaryotic and prokaryotic phyla, as well as viruses (Chaps. 6 and 7). Analogous to probiotic methods such as faecal transplantation to treat irritable bowel syndrome or the inoculation of crops to increase tolerance to temperature and drought, corals can be inoculated with microbiomes isolated from hardier species (Damjanovic et al. 2017) or with artificial cocktails of culturable microbes (dos Santos et al. 2015), with the aim to make them more resistant to environmental stress. Such microbes can be enhanced prior to delivery to the coral via genetic engineering (Levin et al. 2017) or experimental evolution (Chakravarti et al. 2017).

Innovative approaches for coral reef restoration, such as those discussed above, were not yet considered when the first edition of this book was published in 2009.

However, they are now being actively explored due to the urgency of developing options for coral reef management in an era of rapid environmental change.

14.2 Conclusion

Coral bleaching is a biological response to changes in the physical environment of present-day coral reefs, and in some instance in response to pathogen infection. Several mass coral bleaching events in recent years, most notably 1997–1998, 2010 and 2014–2017, have prompted enhanced research efforts into many aspects of this phenomenon. We now know much more about the causes and consequences of coral bleaching than we did only two decades ago. However, there is still much more to learn, and, unfortunately, many of the experiments are happening in real time in the real world. Changes in the physical environment, in particular warming of the tropical oceans, are, without drastic greenhouse gas mitigation strategies, set to continue into the foreseeable future. Continued warming of the tropical oceans, along with other climate change and ocean acidification impacts on coral reefs, is an added human-induced burden on already seriously compromised ecosystems due to direct local and regional stresses. In combination, this panoply of human-induced pressures does not bode well for the maintenance of the world's coral reefs into the future. New interventions, such as shading or cooling of the reef and assisted evolution, are being explored as possible solutions to mitigate climate change impacts. Such interventions will buy time, but the ultimate solution for curtailing ocean warming, acidification and coral bleaching over the next 50–100 years lies in the reduction of greenhouse gas concentrations in the atmosphere.

References

Aitken SN, Whitlock MC (2013) Assisted gene flow to facilitate local adaptation to climate change. Annu Rev Ecol Evol Syst 44:367–388. https://doi.org/10.1146/annurev-ecolsys-110512-135747

Anthony K, Bay LK, Costanza R, Firn F, Gunn J, Harrison P et al (2017) New interventions are needed to save coral reefs. Nat Ecol Evol 1:1420–1422. https://doi.org/10.1038/s41559-017-0313-5

Barshis DJ, Ladner JT, Oliver TA, Seneca FO, Traylor-Knowles N, Palumbi SR (2013) Genomic basis for coral resilience to climate change. Proc Natl Acad Sci U S A 110:1387–1392. https://doi.org/10.1073/pnas.1210224110

Camp EF, Suggett DJ, Gendron G, Jompa J, Manfrino C, Smith DJ (2016) Mangrove and seagrass beds provide different biogeochemical services for corals threatened by climate change. Front Mar Sci 3:52. https://doi.org/10.3389/fmars.2016.00052

Camp EF, Nitschke MR, Rodolfo-Metalpa R, Houlbreque F, Gardner SG, Smith DJ et al (2017) Reef-building corals thrive within hot-acidified and deoxygenated waters. Sci Rep 7:2434. https://doi.org/10.1038/s41598-017-02383-y

Carilli J, Donner SD, Hartmann AC (2012) Historical temperature variability affects coral response to heat stress. PLoS One 7:e34418. https://doi.org/10.1371/journal.pone.0034418

Chakravarti LJ, Beltran VH, van Oppen MJH (2017) Rapid thermal adaptation in photosymbionts of reef-building corals. Glob Chang Biol 23:4675–4688. https://doi.org/10.1111/gcb.13702

Damjanovic K, Blackall LL, Webster NS, van Oppen MJH (2017) The contribution of microbial biotechnology to mitigating coral reef degradation. Microb Biotechnol 10:1236–1243. https://doi.org/10.1111/1751-7915.12769

Dixon GB, Davies SW, Aglyamova GA, Meyer E, Bay LK, Matz MV (2015) Genomic determinants of coral heat tolerance across latitudes. Science 348:1460–1462. https://doi.org/10.1126/science.1261224

Donelson JM, Munday PL, Mccormick MI, Pitcher CR (2012) Rapid transgenerational acclimation of a tropical reef fish to climate change. Nat Clim Chang 2:30–32. https://doi.org/10.1038/nclimate1323

dos Santos HF, Duarte GAS, da Costa Rachid CT, Chaloub RM, Calderon EN, de Barros Marangoni LF et al (2015) Impact of oil spills on coral reefs can be reduced by bioremediation using probiotic microbiota. Sci Rep 5:18268. https://doi.org/10.1038/srep18268

Fogarty ND (2012) Caribbean acroporid coral hybrids are viable across life history stages. Mar Ecol Prog Ser 446:145–159. https://doi.org/10.3354/meps09469

Guest JR, Baird AH, Maynard JA, Muttaqin E, Edwards AJ, Campbell SJ et al (2012) Contrasting patterns of coral bleaching susceptibility in 2010 suggest an adaptive response to thermal stress. PLoS One 7:e33353. https://doi.org/10.1371/journal.pone.0033353

Hellberg ME (2007) Footprints on water: the genetic wake of dispersal among reefs. Coral Reefs 26:463–473. https://doi.org/10.1007/s00338-007-0205-2

Howells EJ, Willis BL, Berkelmans R, van Oppen MJH, Bay LK (2013) Limits to thermal acclimatisation in corals. Ecology 94:1078–1088. https://doi.org/10.1890/12-1257.1

Hughes TP, Barnes ML, Bellwood DR, Cinner JE, Cumming GS, Jackson JBC et al (2017) Coral reefs in the Anthropocene. Nature 546:82–90. https://doi.org/10.1038/nature22901

Jin YK, Lundgren P, Lutz A, Raina J-P, Howells EJ, Paley AS et al (2016) Genetic markers for antioxidant capacity in a reef-building coral. Sci Adv 2:e1500842. https://doi.org/10.1126/sciadv.1500842

Kenkel CD, Matz MV (2016) Gene expression plasticity as a mechanism of coral adaptation to a variable environment. Nat Ecol Evol 1:0014. https://doi.org/10.1038/s41559-016-0014

Kenkel CD, Almanza AT, Matz MV (2015) Fine-scale environmental specialization of reef-building corals might be limiting reef recovery in the Florida Keys. Ecology 96:3197–3212. https://doi.org/10.1890/14-2297.1

Levin RA, Voolstra C, Agrawal S, Steinberg PD, Suggett D, van Oppen MJH (2017) Engineering strategies to decode and enhance the genomes of coral symbionts. Front Microbiol 8:1220. https://doi.org/10.3389/fmicb.2017.01220

Maynard JA, Anthony KRN, Marshall PA, Masiri I (2008) Major bleaching events can lead to increased thermal tolerance in corals. Mar Biol 155:173–182. https://doi.org/10.1007/s00227-008-1015-y

McClanahan TR, Ateweberhan M, Muhando C, Maina J, Mohammed SM (2007) Effects of climate and temperature variation on coral bleaching and mortality. Ecol Monogr 77:503–525. https://doi.org/10.1890/06-1182.1

Penin L, Vidal-Dupiol J, Adjeroud M (2013) Response of coral assemblages to thermal stress: are bleaching intensity and spatial patterns consistent between events? Environ Monit Assess 185:5031–5042. https://doi.org/10.1007/s10661-012-2923-3

Putnam HM, Gates RD (2015) Preconditioning in the reef-building coral *Pocillopora damicornis* and the potential for trans-generational acclimatization in coral larvae under future climate change conditions. J Exp Biol 218:2365–2372. https://doi.org/10.1242/jeb.123018

Torda G, Donelson JM, Aranda M, Barshis DJ, Bay L, Berumen ML et al (2017) Rapid adaptive responses to climate change in corals. Nat Clim Chang 7:627–636. https://doi.org/10.1038/nclimate3374

van Oppen MJH, Gates RD (2006) Conservation genetics and the resilience of reef-building corals. Mol Ecol 15:3863–3883. https://doi.org/10.1111/j.1365-294X.2006.03026.x

van Oppen MJH, Puill-Stephan E, Lundgren P, De'ath G, Bay LK (2014) First generation fitness consequences of interpopulational hybridisation in a Great Barrier Reef coral and its implications for assisted migration management. Coral Reefs 33:607–611. https://doi.org/10.1007/s00338-014-1145-2

van Oppen MJH, Oliver JK, Putnam HM, Gates RD (2015) Building coral reef resilience through assisted evolution. Proc Natl Acad Sci U S A 112:2307–2313. https://doi.org/10.1073/pnas.1422301112

van Oppen MJ, Gates RD, Blackall LL, Cantin N, Chakravarti LJ, Chan WY et al (2017) Shifting paradigms in restoration of the world's coral reefs. Glob Chang Biol 23:3437–3448. https://doi.org/10.1111/gcb.13647

Willis BL, van Oppen MJH, Miller DJ, Vollmer SV, Ayre DJ (2006) The role of hybridization in the evolution of reef corals. Annu Rev Ecol Evol Syst 37:489–517. https://doi.org/10.1146/annurev.ecolsys.37.091305.110136

Index

M. J. H. van Oppen, J. M. Lough (eds.), *Coral Bleaching*, Ecological Studies 233,
https://doi.org/10.1007/978-3-319-75393-5

MIX
Papier aus verantwortungsvollen Quellen
Paper from responsible sources
FSC® C105338

If you have any concerns about our products,
you can contact us on
ProductSafety@springernature.com

In case Publisher is established outside the EU,
the EU authorized representative is:
Springer Nature Customer Service Center GmbH
Europaplatz 3, 69115 Heidelberg, Germany

Printed by Libri Plureos GmbH
in Hamburg, Germany